GONDWANA AND TETHYS

GEOLOGICAL SOCIETY SPECIAL PUBLICATIONS

Series Editor: Kenneth Coe

This volume is published by Oxford University Press
Other volumes in this series are published by Blackwell Scientific Publications

31 P. P. E. Weaver and J. Thompson (eds): Geology and geochemistry of abyssal plains
32 A. C. Scott (ed): Coal and coal-bearing strata
33 T. C. Pharaoh *et al.* (eds): Geochemistry and mineralization of proterozoic volcanic suites
34 J. C. Goff and B. P. J. Williams (eds): Fluid flow in sedimentary basins and aquifers
35 L. E. Frostick and I. Reid (ed): Desert sediments; ancient and modern
36 J. D. Marshall (ed): Diagenesis of sedimentary sequences
37 M. G. Audley-Charles and A. Hallam (eds): Gondwana and Tethys

Gondwana and Tethys

Edited by

M. G. AUDLEY-CHARLES
Department of Geological Sciences
University College London

and

A. HALLAM
Department of Geological Sciences
University of Birmingham

Published for
THE GEOLOGICAL SOCIETY
by
OXFORD UNIVERSITY PRESS
1988

Oxford University Press, Walton Street, Oxford OX2 6DP

Oxford New York Toronto
Delhi Bombay Calcutta Madras Karachi
Petaling Jaya Singapore Hong Kong Tokyo
Nairobi Dar es Salaam Cape Town
Melbourne Auckland

and associated companies in
Berlin Ibadan

Oxford is a trade mark of Oxford University Press

Published in the United States
by Oxford University Press, New York

British Library Cataloguing in Publication Data
Audley-Charles, M. G.
Gondwana and Tethys.
1. Tectonic features. Geomorphology
I. Title II. Hallam, A. (Anthony)
III. Geological Society of London
IV. Series 551.1'36
ISBN 0–19–854448–0

Library of Congress Cataloging in Publication Data
Gondwana and Tethys / edited by M. G. Audley-Charles and A. Hallam.
p. cm.
1. Gondwana (Geology)—Congresses. 2. Tethys (Paleogeography)—Congresses. I. Audley-Charles, Michael Geoffrey. II. Hallam, A. (Anthony), 1933– . III. Geological Society of London.
QE511.5.G657 1988 551.1'36—dc19 88–1558 CIP
ISBN 0–19–854448–0

Typeset by Wyvern Typesetting Ltd, Bristol
Printed in Great Britain
at the University Printing House, Oxford
by David Stanford
Printer to the University

Preface

Understanding the crustal mechanisms and processes involved in the break-up of the supercontinent Gondwana and the creation of the succession of evolving Tethys ocean basins is one of the challenging themes of the geological sciences today. South-east Asia may now be viewed as a collage of continental fragments that rifted from Gondwana and were carried northwards by spreading oceans to collide with the accreting Asian continent. These new Tethys oceans spread across tropical and equatorial latitudes. The associated rifted continental margins provided the habitat for much of the Tethyan marine life that has been fossilized. These fossil remains and associated sediments preserve evidence of the response of evolving marine organisms to changing sea levels and environments as the Gondwana continental margin habitats were carried across a wide range of latitudes from Late Palaeozoic times until the mid-Cainozoic. During the later part of the Palaeozoic era, the great Gondwana continent with its southern cold interior stretched northwards into latitudes of sub-tropical seas. This warm Tethys ocean seems likely to have produced climatic conditions in maritime Gondwana that may have favoured a land flora different from the *Glossopteris* flora able to tolerate the winters of the chilly hinterland. With these thoughts in mind, the editors of this volume arranged a Joint Meeting of the Geological Society of London and the Palaeontological Association. This book is the result of the first annual Lyell Meeting of these two societies.

London M. G. A.-C.
Birmingham A. H.
December 1987

GEOLOGICAL SOCIETY SPECIAL PUBLICATION NO. 37

Contents

Introduction

M. G. Audley-Charles and A. Hallam

In the last few years it has become apparent that the concept of a single end-Palaeozoic supercontinent, Pangaea, breaking up progressively through the Mesozoic and Tertiary into the present continental array, is seriously oversimplified. While the Atlantic Ocean was opening up, causing disintegration of western Pangaea, a series of substantial continental fragments was being swept across Panthalassa, to dock on the western margins of North America and the eastern and south-eastern margins of Asia. Some of the more southerly continental blocks that now form part of eastern 'Laurasia' had their origin on the other side of Tethys, in the Gondwanan supercontinent; the process of accretion probably began in Palaeozoic times. There is no problem in geotectonics more challenging than obtaining a fuller understanding of this process. It requires collaborative research by tectonicians, palaeomagnetists, stratigraphers, and palaeobiogeographers, and with this in mind it was decided to hold a joint meeting, the first Lyell Meeting, between the Geological Society and the Palaeontological Association, on 7 and 8 May 1986. This volume represents the published proceedings, with the chapters corresponding substantially with the sequence of lectures presented.

The first part is concerned with two aspects of Gondwana and Tethys evolution: (1) the tectonic (structural and metamorphic) processes associated with the splitting and dispersal of continents; and (2) evaluation of the palaeomagnetic and stratigraphical records associated with the splitting of Gondwana and evolution of the Tethys oceans.

Price *et al.* find the popular mechanisms for plate movement, namely basal traction, slab pull, and ridge push inadequate to explain known plate movements. They argue that the driving force for plate motion is the body weight of the lithosphere, generating gravity glide on the lithosphere–asthenosphere boundary. They postulate that the continental splitting and subduction initiation result from this mechanism, as well as effects requiring high horizontal stress. Dewey, on the other hand, argues that Pangaea-type continental configurations generate deviatoric tension, and hence are predisposed to splitting along weak zones and where tension is locally enhanced by thermal uplifts. He recognizes episodic continental assembly, disruption, and reassembly associated with tectonic and magmatic events and sea-level changes.

Hall puts forward a case for considering that high-temperature/lower-pressure metamorphism in parts of the eastern Mediterranean was the result of thermal effects on the deep crust during lithospheric extension, and not the consequence of subduction-related compression. Furthermore, he shows how the generation of flysch deposits could result from the peripheral uplift associated with lithospheric stretching. The preservation of high-T and low-P metamorphic rocks associated with crustal extension will occur where subsequent collision thrusts the metamorphic basement over its former cover. He identifies an example from eastern Tethys, where the Gondwana margin (Australia) has collided with a volcanic arc in Timor.

Hall's paper provides the link between the extension processes of rifted Gondwana and collision activities associated with subduction of Tethys. Mann and Vita-Finzi are concerned with analysing the forces involved at such collisional plate boundaries. Using the geometry of the frontal folds, they have determined the horizontal force equivalent to 0.134–0.790 kbar (value is model-dependent) to have been responsible for these folds where the mountain front has advanced at 17 mm/yr.

Tarling reviews the palaeomagnetic data for the Mesozoic configuration of Gondwana and its breakup. He emphasizes the importance of matching ocean-floor magnetic anomalies that formed within the same transform segment. On this basis, he proposes modifications to the configuration and breakup of Mesozoic Gondwana put forward by Norton and Sclater (1979). He points out the need for many more palaeomagnetic data, especially from the collided continental fragments rifted from Gondwana.

The rifted continental margin of northern Australia has been well documented by the petroleum exploration companies, so that, together with the magnetic spreading lineations in the floor of the north-east Indian Ocean, there

From AUDLEY-CHARLES, M. G. & HALLAM, A. (eds) *Gondwana and Tethys* Geological Society Special Publication No. 37, pp. 1–4.

is a body of evidence indicating that a major continental block was rifted from northern Australia in the Jurassic. This rifted margin has been greatly modified during the Cainozoic by collision with the arc–trench systems. Nevertheless, this Jurassic rifting episode is traced by Audley-Charles in the stratigraphic record of the fold and thrust belt collision zone of the Banda Arc, Sulawesi, and New Guinea. The importance of this uplifted fold and thrust belt is that it exposes the stratigraphic sequences that record the rifting episode of Mesozoic Gondwana breakup, and it also presents an example of a collision suture at a rifted continental margin. This may be valuable in guiding our interpretation of the palaeogeographical significance of tectonic sutures. How valuable it is depends in part on whether or not this Banda–New Guinea suturing is typical of continental collision zones.

Metcalfe interprets the pre-Tertiary core of south-east Asia in terms of at least four continental blocks rifted from Gondwana. He sets out the stratigraphic, palaeontological, and palaeomagnetic evidence for his reconstruction that these blocks are now separated by tectonic collision sutures. The presence of glacial-marine deposits and associated cold-water faunas provides diagnostic evidence for these Asian continental blocks having been part of eastern Gondwana in the late Carboniferous and early Permian. Palaeomagnetic data are not adequate in quality or range to provide much guidance. The fossil flora, designated as warm Cathaysian (*Gigantopteris*) or cold Gondwanan (*Glossopteris*), are generally regarded as characteristic of distinct palaeogeographical provinces. However, the degree to which the warmer climate around parts of maritime Gondwana could have been associated with Cathaysian floras is unknown.

The evaluation of the palaeogeographical significance of structural sutures is one of the outstanding problems of Asian geology. Adjacent regions of contrasting facies and faunas require explanation, but what kind of suture should we expect to find where two continental blocks have collided? How does the obliquity of collision and the relative role of strike-slip movement, some of which may post-date collision, influence the suture? In the last chapter of this first part of the volume, Şengör *et al.* present their latest review and reappraisal of the palaeogeographic evolution of the whole Gondwana–Tethys belt. Based largely on regional geology derived from reviewing an extensive literature, they start from the basic tenet that major structural sutures containing ophiolite fragments represent tectonic sutures between continental blocks where oceanic crust has been subducted. They test this against available magmatic and structural evidence for subduction together with faunal, facies, or palaeomagnetic indications of large-scale movements of the blocks involved. Inevitably, in such a broad regional approach to this huge and complex region there will remain disagreements over local and not so local interpretations. Nevertheless, they offer the beginning of an explanation for the remarkable continuity of facies and faunas that can be observed, for example in the late Triassic, late Cretaceous, and Eocene: from Crete in the eastern Mediterranean to Timor and Seram in eastern Indonesia. A feature to emerge is the variety of continental blocks to have been transferred from Gondwana to Laurasia. Another conclusion that may be drawn from this review is that an enormous amount of ocean crust and its sedimentary cover must have been subducted, and an impressively large volume of continental margin deposits (slope and rise), as well as forearc deposits (perhaps with basement), seem to have disappeared in the process of reassembling these Gondwana continental fragments at the Laurasian margin.

The second part of this volume from Chapter 10 onward, is devoted principally to a presentation and evaluation of palaeontological evidence relevant to the Gondwana–Tethys theme, and is dealt with in stratigraphical order as far as is feasible.

Taking account of the distribution of Ordovician and Silurian marine invertebrates, Cocks and Fortey present a new base map for ancient Gondwana. Whereas the core comprising South America, Africa, peninsular India, Madagascar, and Antarctica has remained uncontroversial as a coherent supercontinent through much of Phanerozoic time, there has been dispute about peripheral regions along the Tethyan margin. In the Cocks and Fortey reconstruction, a belt extending from southern Europe and the Middle East to southern China and south-east Asia is incorporated into the early Palaeozoic Gondwana supercontinent.

Chaloner and Creber review the current status of the well-known late Palaeozoic *Glossopteris*, Cathaysia, and Angara floral provinces, urging caution in the interpretation of fragmentary leaf remains, misidentification of which has led to confusion in the past. Confining their attention to the more reliable data, they use it to test an early Permian plate reconstruction based on palaeomagnetic results. Note is taken by Waterhouse of the distribution of Permian marine faunas and sediments, and the climatic zones inferred from them, to throw into question some recent palaeogeographical reconstructions that demand a vast wedge-shaped Tethys in Asia, and movement of allochthonous terranes over substantial distances.

Turning to the Mesozoic, Kristan-Tollmann adds to her earlier published examples of Triassic fossils that appear strikingly similar, even at species level, throughout the length of the Old World Tethys. What is especially interesting is that the same species occur also in western North America. This has been interpreted as the consequence of free larval migration westwards across Panthalassa, by means of an equatorial palaeocurrent (Tollmann and Kristan-Tollmann 1985) but account should also be taken of the fact that the western margin of North America consists of a series of allochthonous terranes. Indeed, much of Panthalassa might have contained numerous large islands in Triassic times (Tozer 1982) which would have greatly facilitated marine invertebrate dispersal. However, crustacean coprolites similar to those described by Kristan-Tollmann have also been recorded by Senowbari-Daryan and Stanley (1986) from Peru, which is unlikely to have been composed of far-travelled terranes (Hallam 1986).

Jurassic ammonite distributional data are dealt with in two short articles by Thierry and Westermann, dealing respectively with the western and eastern Tethys. Thierry attempts to disentangle historical from ecological factors in accounting for ammonite distributions, in order to use these to test a palaeogeographical reconstruction. This, however, involves having to make assumptions about the life habits of particular groups. A conflict emerges concerning the interpretation of the Caucaso-Dobrodgean furrow. Another contributor to this book, Şengör, considers this to have been closed in mid-Jurassic times, whereas Thierry requires it to have remained open at that time, in order to account for the confinement of certain ammonites to the north of the Pontides. This conflict points out the need for a better understanding both of Jurassic palaeogeography and the factors controlling ammonite distributions. In contrast, Westermann finds that Jurassic ammonite distributional data support recent plate-tectonic reconstructions indicating a separation of the Lhasa and Quamdo blocks of Tibet from the Tethyan Himalaya.

Mesozoic brachiopod distributions suggest to Ager that Mesozoic Turkey should be placed on the northern side of Tethys rather than the southern side, as in some reconstructions, because the Turkish faunas are European rather than African or Middle Eastern in their affinities. Skelton's article raises once more the question of trans-Pacific migration, this time utilizing data from mid- to late-Cretaceous benthic organisms. He infers an island-studded ocean, related to extensive mid-plate volcanicity, which allowed Caribbean benthos to spread westward by larval drift into the eastern Old World Tethys, utilizing the islands as a series of staging posts.

The only article on terrestrial vertebrate distributions, by Rage, presents a comprehensive review of Mesozoic and Cainozoic data. Free dispersal was possible in Triassic times across the whole of Pangaea, and land communications between Gondwana and Laurasia persisted into the Jurassic. Subsequently, relations between the Pangaea components became more complicated, and after the Eocene, exchanges between the now-separated Gondwana continents disappeared, whereas intermittent exchanges continued between some Gondwana continents and the northern continents. Rage presents an interesting analysis of distributions changing with continental breakup. The outstanding problem remains the lack of proven late Cretaceous endemic fauna in India, although all plate-tectonic reconstructions agree that the subcontinent was isolated by ocean at this time.

Rosen and Smith present a new method of analysis of late Cretaceous to Recent reef coral and sea-urchin distributions, and argue the case for using cladistic biogeography as the most rigorous method available. By using parsimony analysis of taxa shared between sample localities, one should be able to determine the relative recency of faunal contact between different regions. The method provides a means of testing independently derived palaeogeographical reconstructions and could prove of great value for older periods, for which the inferred palaeogeographies are considerably more conjectural. Whitmore's concluding chapter deals with angiosperm distributions in the Malay Archipelago, which was created in mid-Miocene times by eastern Gondwana–Laurasia collision. Some plant taxa support the geological inference that Gondwana fragments moved north before the main collision, to provide island 'stepping-stones'.

As is characteristic of a symposium volume presenting a wide variety of techniques and viewpoints, it is perhaps most useful in providing a series of 'state of the art' reviews and identifying areas of potential or actual conflict that need to be resolved by further research, rather than in solving major problems. Clearly, one such area is the timing of separation and collision of particular continental blocks, particularly adjacent to the eastern Tethys. What does seem to be beyond serious dispute now, is that a succession of continental fragments has been peeled off from the northern rim of Gondwana, transported across Tethys, and plastered against the southern margin of Laurasia. (Owen's (1983) alternative, involving the creation of Tethys by Earth expansion since the Triassic, can be decisively ruled out (Weijermars 1986).)

Some sort of large-scale geotectonic explanation is required. Bearing in mind the gravity-gliding interpretation of plate tectonics favoured by Price *et al.* in this volume, the most useful way of proceeding might be to pursue further the stimulating idea of Anderson (1981), who notes the remarkable correspondence of the Wegenerian reconstruction of continents and the present location of geoid anomalies and hot-spots. A supercontinent such as Pangaea could have acted to insulate a large part of the underlying mantle for periods of well over 10^8 years. The excess heat trapped beneath the supercontinent would have caused uplift by thermal expansion and partial melting; a large geoid high would have been generated by this expansion. Eventually, the supercontinent would disintegrate, and fragments disperse from the geoid anomaly. The main implication in the present context is that the anomaly would have been centred under Gondwana.

References

Anderson, D. L. (1981). Hotspots, polar wander, Mesozoic convection and the geoid. *Nature* **297**, 391–3.

Hallam, A. (1986). Evidence of displaced terranes from Permian to Jurassic faunas around the Pacific margins. *J. Geol. Soc. London* **143**, 209–16.

Norton, I. O. and Sclater, J. G. (1979). A model for the evolution of the Indian Ocean and the breakup of Gondwanaland. *J. Geophys. Res.* **84**, 6803–30.

Owen, H. G. (1983). *Atlas of continental displacement, 200 million years to the present.* Cambridge University Press, Cambridge. (159 pp.)

Senowbari-Daryan, B., and Stanley, G. D. (1986). Thalassinid anomuran microcoprolites from Upper Triassic carbonate rocks of central Peru. *Lethaia* **19**, 343–54.

Tollmann, A. and Kristan-Tollmann, E. (1985). Paleogeography of the European Tethys from Paleozoic to Mesozoic and the Triassic relations of the eastern part of Tethys and Panthalassa. In *The Tethys: her paleogeography and paleobiogeography from Paleozoic to Mesozoic* (ed. K. Nakazawa and J. M. Dickins), pp. 3–22, Tokai University Press, Tokyo.

Tozer, E. T. (1982). Marine Triassic faunas of North America: their significance in assessing plate and terrane movements. *Geol. Rundsch.* **78**, 1077–1104.

Weijermars, R. (1986). Slow but not fast global expansion may explain the surface dichotomy of Earth. *Phys. Earth & Planet. Int.* **43**, 67–89.

M. G. Audley-Charles and A. Hallam, Department of Geological Sciences, University College London, Gower Street, London WC1E 6BT, and Department of Geological Sciences, University of Birmingham, Birmingham BI5 2TT, UK.

Gravity glide and plate tectonics

Neville J. Price, Geoffrey D. Price, and Sarah L. Price

ABSTRACT: We briefly discuss the three groups of mechanisms which are generally held to be responsible for plate motions, namely basal traction, slab pull, and ridge push. We conclude that individually, or in combination, these mechanisms are insufficient to account for known plate behaviour. Instead, we reintroduce and expand upon an idea proposed by Hales (1969) that the driving force for plate motion is the body weight of the lithosphere, which gives rise to gravity glide on the inclined interface between the lithosphere and asthenosphere. It is shown that this mechanism is capable of producing an average horizontal stress of as much as 6 kbar and is able to explain (1) continental splitting, (2) the initiation of subduction, and (3) observable plate motions, as well as a variety of effects requiring high lateral stresses. This is the first model to be proposed that successfully provides a quantitatively and physically viable mechanism to explain plate-tectonic processes and the behaviour of the Earth's lithosphere.

Introduction

It is generally accepted that the Earth consists of an outer, rigid lithosphere, subdivided into several plates, overlying a dynamic, convecting asthenosphere. However, the theory of plate tectonics lacks a generally accepted, detailed, or quantitative model to describe the dynamics and kinematics of plate motion and mantle convection. We must identify and quantify the forces which determine plate motion, if we are to progress beyond completely qualitative models for major events such as the breakup of Gondwana.

In this paper we briefly discuss some current models of the various forces which are thought, by some, to determine the motion of the lithospheric plates, and hence the evolution of continental crust. We subsequently present a simple analysis which indicates that gravity glide, resulting from the body weight of the lithosphere itself, can generate large lateral stresses, acting normal to, and directed away from, the oceanic ridges and continental rifts. These stresses are of such a magnitude that they will play a vital role in determining the initiation of continental rifting, the development of subduction zones and the kinetics of plate motion.

Forces acting on lithospheric plates

Studies directed at determining the origin of the forces acting on plates, tend to be performed either from a fluid dynamical perspective or from a plate-tectonic point of view (Loper 1985). In the former approach, the plates are considered merely to behave passively as an upper, cold, thermal boundary layer of a convecting mantle, while in the latter approach, the forces and moments which act upon a plate are analysed in an effort to identify the dominant forces and hence the nature of the driving force for plate motion. Such plate-oriented studies invariably conclude that it is highly improbable for plates to be driven by basal, mantle-derived drag (e.g. Forsyth and Uyeda 1975; Richardson *et al.* 1979; Bott 1982), and that plates must, to some extent, be decoupled from the underlying convective mantle, as proposed by such workers as Parsons and Richter (1981). However, the behaviour of lithospheric plates is not expected to be totally unrelated to mantle convection, since subduction of rigid lithosphere will undoubtedly have an effect upon the dynamics of the mantle through local viscous coupling. Nevertheless, current thinking appears to favour the contention that the details of plate motion, particularly their translational velocities and rates of rotation, are determined by the forces acting at the margins of the plates, rather than by the flow of the underlying mantle.

The forces that act at the edges of plates have been discussed in a number of works (e.g. Forsyth and Uyeda 1975; Bott 1982; Dewey 1988), and can be divided into forces which might move plates and forces which would resist plate motion. The two main forces which act at the edge of plates to produce motion are a pushing force at the ridges, i.e. constructive plate boundaries, and a pulling force at the trenches. The latter force is thought to result from the

From AUDLEY-CHARLES, M. G. & HALLAM, A. (eds) *Gondwana and Tethys* Geological Society Special Publication No. 37, pp. 5–21.

negative buoyancy associated with the sinking of cold lithosphere at subduction zones. There are many forces which resist plate motion. The most significant are probably those associated with slab subduction, continental collision and transform-fault friction (Bott 1982).

The hypothesis that plates are driven by edge forces acting at trenches and ridges is acceptable on thermodynamic grounds (e.g. Runcorn 1980; Bott 1982). There have been several detailed, quantitative studies of the feasibility of driving plates by edge forces (e.g. Forsyth and Uyeda 1975; Chapple and Tullis 1977). From an analysis of current plate geometry and motion, these workers concluded that the forces acting on the downgoing slab are an order of magnitude larger than any other force, and so control the velocity of the oceanic plates. Forsyth and Uyeda (1975) suggested that oceanic plates which are attached to substantial amounts of subducting lithosphere, approach a terminal velocity. At this velocity the slab-pull force that results from the negative buoyancy of the plate is balanced by the viscous and frictional drag forces which resist plate subduction, irrespective of all other driving forces acting on the plate. This model, however, breaks down when extrapolated back in time. Firstly, there is the problem of initiating subduction, and secondly, their analysis of current oceanic plate velocities in terms of terminal velocities appears to be incompatible with the high velocity (>150 mm/yr) believed to be associated with the Indian plate before 55 Ma BP (Peel, 1982).

An analysis of intra-plate regional stresses led Richardson *et al.* (1979) to conclude that the forces within the mantle, which resisted slab subduction, approximately balance the large gravitational potential of the slab, and hence significant ridge-push forces must exist to explain the ubiquitously observed compressive intra-plate stresses. In addition, Price and Audley-Charles (1987) have argued that detachment of the downgoing slab has occurred below Timor and that, if it is now sinking under its own weight, it is doing so very slowly (<20 mm/yr). Thus it can be inferred that the large slab-pull forces are generally absorbed by the underlying mantle, by means of viscous drag and coupling, rather than being transmitted to the rest of the trailing plate, and hence the motion of surface plates must be largely determined by ridge-push forces.

The nature of the 'ridge-push' forces is as contentious as their role in plate tectonics. An early suggestion for the origin of the 'ridge-push' force (Hales 1969), was that it arose from the gravity sliding of young oceanic lithosphere down an inclined and 'lubricated' lithosphere–asthenosphere interface. This mechanism is not strictly a 'ridge-push' force, but is the result of the distributed body weight of the lithosphere. McKenzie (1972), Turcotte and Schubert (1982), and others concentrated on 'gravity glide' resulting from the topographic relief of ocean ridges. Bott (1982, p. 357) dismissed the suggestion that gravity glide can play any role as a plate driving force, and stated instead that the ridge-push force is caused by the progressive upwelling of low-density asthenosphere beneath ridge crests. His arguments against gravity gliding are aimed only at the attempts to explain ridge-push forces as resulting from ridge topography, and are not applicable to Hales' original suggestion that gravity glide is the result of sliding down the inclined surface at the base of the lithosphere. We later reintroduce and expand upon the idea briefly outlined by Hales, and deal with certain criticisms of his mechanism.

The various concepts so far put forward in the literature are piecemeal approaches to the problem of the evolution of a plate. Ideally, the whole history of the development of the plate, its motion and associated features should be explicable in terms of one or a combination of mechanisms. The main steps involved in such an evolution are represented in Fig. 1.

Initially, (Fig. 1a) we consider continental or oceanic lithosphere to extend around the Earth, without perturbations such as ridges or subduction zones. This uniform state is disturbed (Fig. 1b) by the rising of the thermal plume which causes an increase in temperature and results in an upward deflection of the geotherms. We take the junction between the lithosphere and the asthenosphere to be thermally defined, and to occur at such a temperature that there is a marked change in the rheological behaviour of the mantle material. Consequently, the plume causes an upward migration of the lithosphere–asthenosphere boundary. It is above the high point of the plume, we suggest, that a ridge will develop. The establishment of the ridge with a well-developed, low-velocity zone beneath is shown in Fig. 1c. However, it is not until subduction is initiated that a plate is defined and set free to move (Fig. 1d). At this stage the ridge becomes a 'spreading centre'.

It is our contention that the whole of this cycle of events can be explained in terms of gravity glide and associated gravitational effects. However, as we shall see, the horizontal stresses in the lithosphere, in the initial, or 'at rest', condition are extremely important to our thesis, so we shall discuss the factors which control the

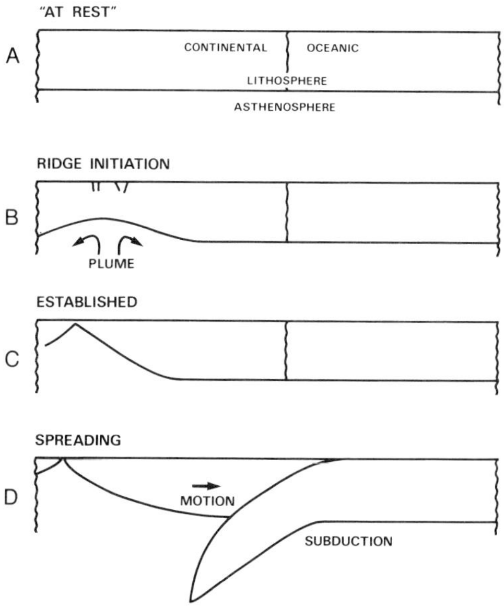

FIG. 1. (a) The 'at rest' configuration of lithosphere and asthenosphere prior to a perturbation caused by a plume. (b) The initiation of a ridge above a plume. (c) Ridge established, but significant plate motion inhibited. (d) Initiation of subduction and significant plate motion.

magnitude of these lateral stresses before we proceed to consider the gravity-glide mechanism.

Magnitude of lateral lithospheric stresses

In the initial state represented in Fig. 1a, the lithosphere will be in a state of zero lateral strain and the lateral stresses will be induced only as the result of gravitational loading (i.e. the total vertical stress S_z) where:

$$S_z = \varrho \cdot g \cdot z \tag{1}$$

and where ϱ is the density, z is the depth and g is the gravitational acceleration. If the lithosphere does not fail, the lateral stresses will be the result of elastic deformation, so that the horizontal stress (S_x) is given by:

$$S_x = S_z/(m-1) \tag{2}$$

where m is Poisson's number (Price 1966). This ranges from a minimum of 2.0 [for a liquid], to infinity for a completely rigid solid. The value we shall use for strong lithospheric material is $m = 4.4$. If failure occurs, it may be the result of brittle deformation (this mode is likely in the upper levels of the lithosphere) or, at high temperatures (in the lower continental crust or in the lower levels of the lithosphere), failure may occur in a ductile manner. For brittle shear failure (i.e. the development of normal faults), the ratio between the greatest and least principal stresses (S_1/S_3) is given by the relationship:

$$S_1/S_3 = (1 + \sin\phi)/(1 - \sin\phi) \tag{3}$$

where ϕ is the angle of sliding friction (Price 1966). In the environment under consideration, $S_1 = S_z$, and $S_3 = S_x$, so that S_3 can be readily estimated from the value of ϕ. If ductile failure occurs however, then in order to estimate the lateral stress, it is necessary to use the equations of state of the various rock types that are likely to deform by diffusion mechanisms (Fyfe *et al.* 1978).

The lateral stresses (S_x) for the elastic or brittle failure conditions can readily be assessed. The value of m for olivine is approximately 4.4, hence the increase in lateral stress with depth is as indicated by line B in Fig. 2a. The vertical stress (S_z) is represented by line A, so it will be seen that the lateral stress is 70 per cent smaller than the vertical stress at any level z.

The value of ϕ for many strong rocks is about 45°, so that if failure occurs by brittle shear, the ratio of S_1/S_3 is 5.8. The lateral stress for these rocks is represented by line C in Fig. 2a. It can be inferred from this figure that the lateral stresses, induced elastically, are too high to permit failure to occur in rocks with $\phi = 45°$. (In order for shear to occur ϕ must be less than 30°, for the lateral stress conditions represented by line B.)

The problem of assessing the differential stresses that are consistent with ductile deformation is much more difficult, for it requires quantifying temperature and strain rates, and it depends upon whether the rocks are wet or dry. In addition, the stresses depend upon lithology and grain size. Hence, the lateral stress/depth relationship in oceanic and continental lithosphere will be different.

The probable differential stress that wet and dry olivine can withstand at a strain rate of 10^{-15} $\sec^{-1}$ (derived from Dewey 1988) for a given depth, is indicated in Fig. 2b. The combined stress curves of Fig. 2a and b are represented in Fig. 2c, where the 'probable' lateral stress/depth relationship is shown as the bold solid line. The bold dashed line represents the probable lateral stress/depth relationship for continental crust. The shaded area indicates the amount by which the lateral stress falls short of the lithostatic pressure (for which condition the horizontal and vertical stresses are equal and given by line A in

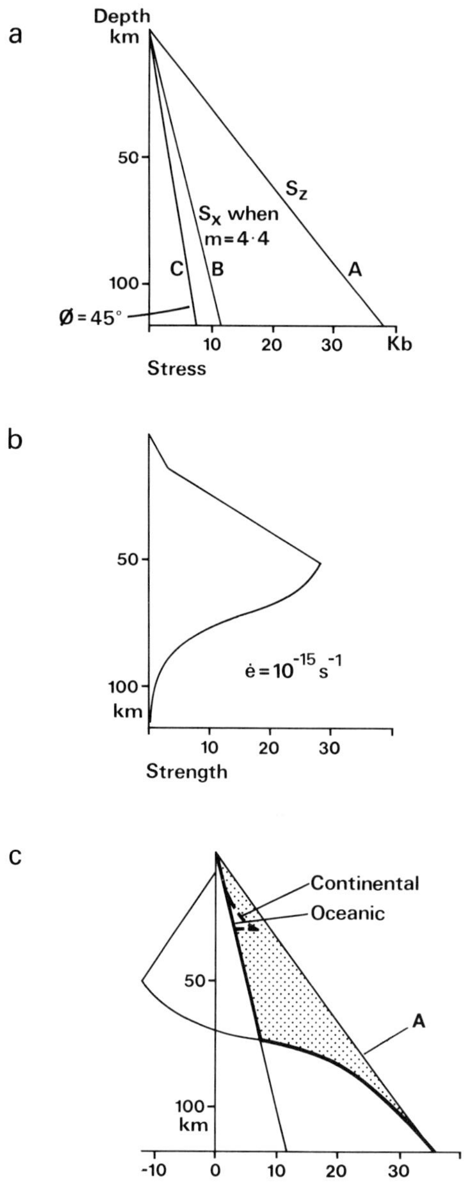

FIG. 2. (a) Stress/depth relationships. A = geostatic stress, B = lateral stress in an elastic body ($m = 4.4$), and C = lateral stress after brittle shear failure. (b) Depth/differential stress conditions for failure (after Dewey 1987). (c) Combination of curve A from Fig. 2(a) and 2(b), with probable depth/lateral stress conditions shown by bold line.

Fig. 2c). It can be inferred from this figure that the average horizontal stress, for the 'at rest' condition which we are considering, is between 30 and 35 per cent lower than the lithostatic pressure. The lithostatic pressure at a depth of 116.5 km (i.e. at the bottom of the lithosphere) is 38.4 kbar, so that the average lithostatic pressure for the lithosphere is 19.2 kbar. Hence, the average horizontal stress in the 'at rest' condition is 30–35 per cent of this latter figure, i.e. approximately 6 kbar. Let us now proceed to discuss the importance of these conclusions.

Gravity-glide mechanism

It was during a visit to Iceland, some years ago, that the senior author reached the tentative conclusion that gravity glide may be an important element in determining plate movements. An extremely schematic profile through this island is indicated in Fig. 3a. The essential elements of Iceland are that it is cut by a main (and some minor) volcanic zone(s), and that the rest of the island is mainly composed of lava flows. It is reasonable to assume that lava flows mainly downhill from its source in the volcanic zone. However, a feature of Iceland is that—except for those most recently deposited—the flows dip inward towards the volcanic zone, and the more ancient the flow, the steeper the dip (Jacoby *et al.* 1980).

Iceland lies astride the Mid-Atlantic ridge and it is known that the low-velocity zone is close to the surface beneath—and dips away to either side of—the volcanic zone. If it is assumed that the low-velocity zone is also a 'low-viscosity zone' (LVZ), then the junction between the low-velocity zone and the plate will mark the top of a zone of weakness (e.g. Parsons and Richter 1981). From the configuration of the lava flows and the top of the LVZ (Fig. 3b), the similarity of these features to 'roll-over' structures associated with gravity gliding in deltas is obvious. The east and west parts of Iceland would, accordingly, form parts of large-scale, roll-over anticlines.

The Hales (1969) model, in which he shows a portion of the lithosphere and the shear stresses that are induced by body weight at an inclined junction with a viscous asthenosphere, is shown in Fig. 4a. However, doubts have been cast regarding the validity of such a model, by Peel (1982) and Jacoby (1980). Peel suggests, correctly, that a small iceberg with the geometry of a 'plate and slab' (Fig. 4b), in completely calm conditions of air and water, would show no

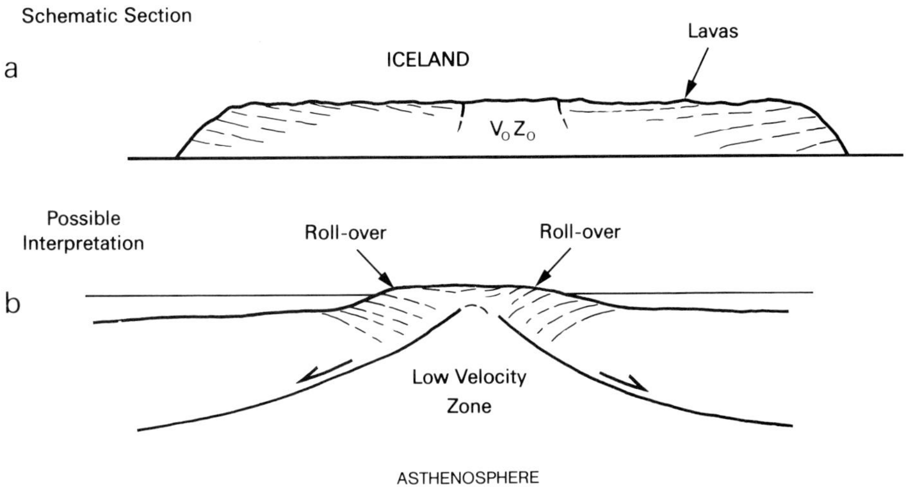

FIG. 3. (a) Schematic E–W section through Iceland, showing central volcanic zone (V_0Z_0) and inward dipping lava flows. (b) Possible interpretation of features shown in (a), as part of roll-over structures.

tendency to forward motion, because of its shape. Hence, he argues that the Hales model would not give rise to forward plate motion. Jacoby proposes another model, represented in Fig. 4c, in which two rigid blocks (cut by a plane of weakness, i.e. one possessed of low viscosity) are surrounded by a fluid with the same density as that of the blocks. The blocks would have boundary stresses as shown, and consequently, there would be no tendency for gravity glide of the upper block along the plane of weakness.

We consider that neither of the models proposed by Peel and Jacoby are germane to the point originally raised by Hales. Thus, if the density of the 'iceberg' or the rigid block is assumed to be greater than the surrounding fluid, then forward motion of the iceberg would occur as it sank, or the upper block in Fig. 4c would experience gravity glide. Indeed, this is a point made by Jacoby, who contends that in such situations, the stress driving the gravity glide would be related to the difference in density between the block and the surrounding liquid.

Although the model proposed by Peel and the points made by Jacoby may appear to be persuasive, we argue that they are inappropriate and misleading when considering lithospheric motion. As already discussed, we do not believe that the lithosphere is in a state of hydrostatic equilibrium, as assumed in Jacoby's model in Fig. 4c. Rather, we suggest that a modified version of the Jacoby model (Fig. 4d) represents a better approach to simulating lithospheric processes. In this model, two of the blocks shown in Fig. 4c are placed back to back and the lower blocks are a firm base. It is clear that in this case the two upper prisms, A and B will exert a pressure P against each other, and will behave in a manner comparable with that of a gigantic 'nutcracker'. Moreover, this stress will be related to the total density of the prisms and will be unaffected by the density of the underlying blocks. The stresses generated by the prisms could be transmitted over considerable distances if, as in Fig. 4e, the triangular end prisms are separated by a rectangular block which is unrestricted along its base. If we apply periodic boundary conditions, we suggest that such a configuration could be described as a 'flat-Earth' model. In the Jacoby model (Fig. 4c) the material underlying the plane of weakness is taken to be rigid. We contend that in models represented by Figs 4d and e, the material beneath the weakness plane need not be completely rigid, provided that it is a stiff, contained fluid, and that the plane or zone of weakness has a coefficient of viscosity which is significantly (i.e. an order of magnitude) smaller than the fluid beneath it.

Let us now translate this simple conceptual model into one which is thought to represent real conditions. Our model considers a strong, mainly elastic, oceanic lithosphere with a density of 3300 kg m^{-3}, which is created at a mid-oceanic ridge, and thickens with time as a result of cooling. Unfortunately, there is no agreed definition of the exact base of the lithosphere, so for the sake of simplicity, we shall assume that

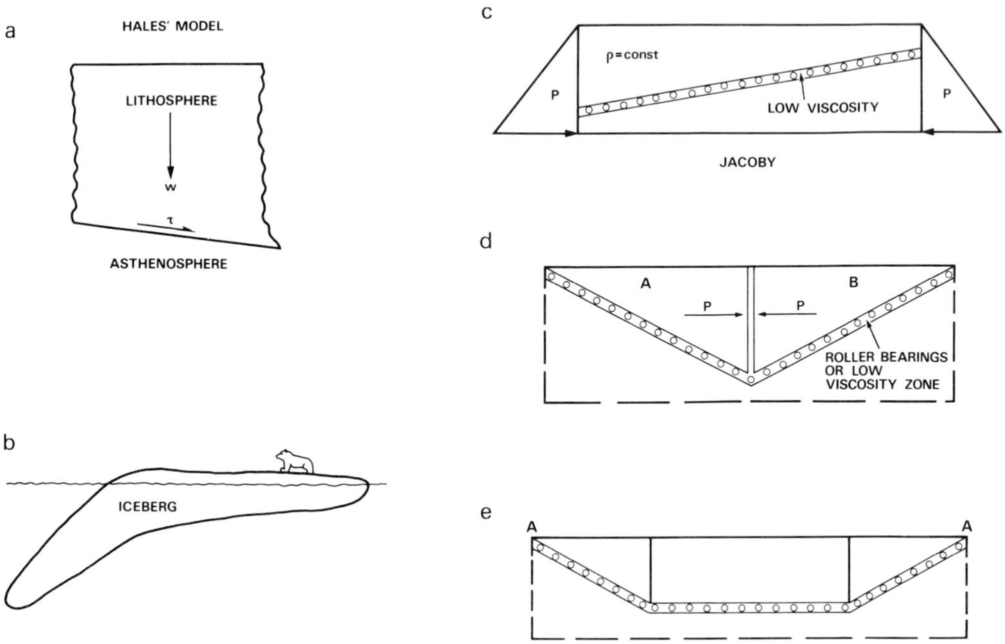

FIG. 4. (a) Hales' model. (b) Peel's model. (c) Jacoby's model. (d) Rigid block model. (e) Rigid block and plate model with periodic boundary conditions.

oceanic lithosphere younger than 80 Ma, has a thickness ($|z|$) given by the empirical relationship:

$$|z| = 2.32(kt)^{1/2} \tag{4}$$

where k (taken as $1 \times 10^{-6}\,\mathrm{m^2\,s^{-1}}$) is the thermal conductivity of the lithosphere and t is the time measured in Ma (Turcotte and Schubert 1982). We assume that the base of oceanic lithosphere older than 80 Ma is parallel to the Earth's surface.

The exact nature of the asthenosphere beneath lithospheric material is also poorly defined. Cathles (1975) suggests that a 70–75-km thick zone of mantle material with a viscosity (η) of 4×10^{19} Pa separates the lithosphere from the rest of the more viscous mantle ($\eta = 10^{21}$ Pa). Although there is undoubted evidence that this LVZ exists under the lithosphere near spreading centres (Bott 1982), it is unclear how far it extends under older oceanic lithosphere or under continental lithosphere. However, consider the situation shown in Fig. 5, in which, initially, only the upper part of the approximately triangular prism is underlain by a low-viscosity zone, which tapers to zero thickness at point A. The portion of the prism upslope from point A will be free to glide and will, therefore, exert an extremely high shear stress at point A, which will result in shear of the interface. Strain softening, caused by the local shear-induced generation of heat or the reduction of grain size in the shear zone, will result in a downslope extension of the LVZ. We propose that this self-propagating mechanism will almost certainly result in the LVZ spreading laterally, until the lithosphere–LVZ–asthenosphere interface closely resembles the situation modelled in Fig. 4e.

Let us now proceed to calculate the lateral stresses that can ensue from such a gravity-glide mechanism, and to determine what effects may result from such stresses.

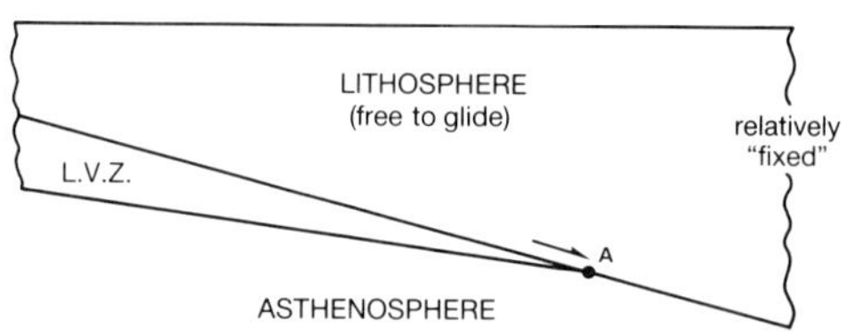

FIG. 5. High shear stresses will form at point A, as the result of the lithosphere's freedom to glide over the LVZ, relative to the fixed or slower gliding asthenosphere.

Calculation of potential gravity-glide force and associated stresses

Considering the model as set out in Fig. 6, then rearranging Eqn (4) it follows that:

$$t = z^2/a^2k \tag{4a}$$

(where $a = 2.32$), so that the volume (V) of the element shown in Fig. 6 is

$$V = -z\mathrm{d}y\mathrm{d}t \tag{5}$$

$$V = a(kt)^{1/2}\mathrm{d}y\mathrm{d}t \tag{6}$$

Therefore the vertical force (F) due to gravity at point t is

$$F = \varrho ga(kt)^{1/2}\mathrm{d}y\mathrm{d}t \tag{7}$$

where ϱ is the density and g is the gravitational acceleration constant. The gradient of the lower surface of the lithosphere at point t is given by:

$$\mathrm{d}z/\mathrm{d}t = a^2k/2z = -ak/2(kt)^{1/2} \tag{8}$$

and hence

$$\tan\theta = -\mathrm{d}z/\mathrm{d}t = ak/2(kt)^{1/2} \tag{9}$$

The vertically acting force induced by body weight does not of itself have a resolvable horizontal component. However, when that body impinges on an inclined surface (dip: θ) and has a potential to slide, then it can be restrained from sliding by a horizontal force (F_{H}) which is given by:

$$F_{\mathrm{H}} = F\sin\theta\cos\theta \tag{10}$$

If it is assumed that θ is small, so that $\theta \sim \sin\theta \sim \tan\theta$ and $\cos\theta \sim 1$, then

$$F_{\mathrm{H}} = \varrho ga^2 k\mathrm{d}y\mathrm{d}t/2 \tag{11}$$

For most of the lithosphere the value of θ is indeed small, so that, assuming $\mathrm{d}y$ = unit extent,

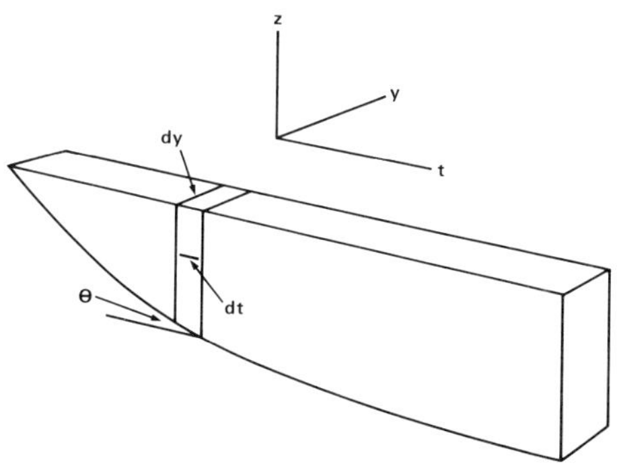

FIG. 6. Axes and quantities used in calculus analysis.

the total horizontal force (F_{TH}) exerted by oceanic lithosphere of age t (<80 Ma) is given by

$$F_{\mathrm{TH}} = \varrho ga^2kt/2 \tag{12}$$

Assuming stress channelling and that the viscous restraint to gravity gliding is zero, then the potential horizontal stress (P) that can be generated at the end of the block is

$$P = \varrho ga^2kt/2|z| \tag{13}$$

Alternatively, Eqn (13) can be expressed as

$$P = \varrho g|z|/2 \tag{13a}$$

where $|z|$ is the thickness of the lithosphere of age t<80 Ma.

From Eqn (4), $|z| = 116.5$ km at $t = 80$ Ma, thus from Eqn (13a), it follows that the horizontal pressure generated at the 'footwall' of 80 Ma lithosphere has a potential value of 19.2 kbar, for a density of 3300 kg m^{-3}, which for the assumed age is independent of the length of the plate (L) or its velocity (v).

It must be emphasized that this potential value of 19.2 kbar would only be attained if the existing 'at rest' lateral stress is zero, i.e. the lithosphere is assumed to be perfectly rigid. If the 'at rest' lateral stresses are already equal to the lithostatic, none of the gravity-glide potential is realized (this is equivalent to the hydrostatic stress state discussed by Jacoby (1980)). However, as we have seen, the average horizontal stress available to produce plate motion is only about 6 kbar.

However, before we can deal with the plate motion, it is first necessary to show how the splitting of a continent and the initiation of a subduction zone can take place and thereby provide space for plate motion.

Splitting of continents by gravity glide

It is apposite at this point to quote two statements made by Bott (1982) regarding the mechanism of continental splitting. In the first of these he poses the question: 'What mechanism causes the persistent splitting of continents which has probably occurred episodically over much of geological time? At each stage of continental splitting, a new divergent plate boundary forms within a continental interior margin and develops into an active spreading centre as the new ocean along the split starts to widen. Thus, the early Mesozoic supercontinent Pangaea has broken up by stages over the last 280 Ma the Atlantic and Indian Oceans have opened up between separating fragments'. He continues to summarize the requirements for splitting and the mechanisms that have been proposed by stating: 'a new continental split appears to require a substantial tensile stress system and a sufficient source of basaltic magma in the underlying upper mantle to initiate the sea-floor spreading process. The following four sources of tension have been suggested: (i) drag on the base of the lithosphere exerted by upwelling and diverging convection currents; (ii) membrane stresses developing in the interior of a plate migrating towards the equator; (iii) stresses associated with surface loading and associated isostatic upthrust in uplifted continental regions such as East Africa; and (iv) the trench suction force acting at convergent plate boundaries. It is suggested . . . that the tension which predominantly causes continental splitting is caused by the trench suction force'.

We would completely disagree with this final statement and suggest that continental splitting can also be attributed to gravity glide, though item (iii) may also contribute to graben development and rifting. We would, however, agree with Bott that the generation of rifting and the eventual formation of ridges is related to a thermal anomaly caused by a 'plume' or 'hot-spot'. Here it is envisaged that such a plume causes an upward migration of a critical isotherm which defines the base of the lithosphere (i.e. a thinning of the lithosphere), at which there is a marked change in rheological properties of the mantle material. Below this critical temperature (i.e. physically above the isotherm) the lithosphere exists as an essentially strong, solid unit, while at higher temperatures (physically below the isotherm) a low-viscosity zone (perhaps in this case caused by partial melting) separates the lithosphere from the asthenosphere. As a result of the limited extent of the plume, we postulate that the isotherm surface is inclined away from some central point or line; and it is along this inclined plane that gravity glide may result (Fig. 7a).

In the following paragraphs, we consider simple models to demonstrate how the stresses induced by gravity glide will initiate a ridge. The first, illustrative example is a simple quantitative assessment of a two-dimensional model. The second example is a qualitative treatment of a three-dimensional model, indicating how a thermal plane, with circular plan, may develop to form a ridge and cause splitting of a plate. Here, of course, we refer only to very large energy plumes that are capable of bringing about regional upwelling of the lithosphere–asthenosphere interface (for example the hot-spot under Iceland rather than the one under Hawaii).

In the previous discussion it was assumed that the LVZ approached close to the surface at the axis of spreading. However, if we now assume that a thermal ridge exists as indicated in Fig. 7a (see also Fig. 2.23 in Bott 1982), then the base of the lithosphere will, initially, only be locally

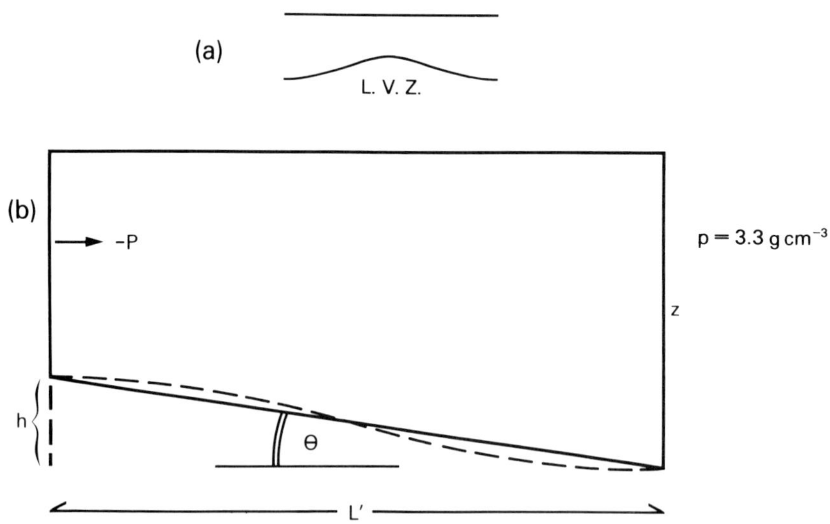

FIG. 7. (a) Diagrammatic representation of an embryonic ridge. (b) Model used to estimate reduction in lateral pressure (–P).

$$-P = \frac{\varrho g(2z-h)\, L \sin\theta \cos\theta}{2(z-h)} = \frac{\varrho g(2z-h)\, h \cos^2\theta}{2(z-h)}$$

deflected upwards by a few kilometres. We know of no assessment of the form and aspect ratio of the base of the lithosphere which a hot-spot or plume may engender at this early stage. Consequently, we assume that this unknown curve can be adequately represented by the uniform slope shown in Fig. 7b. (A treatment by Turcotte and Schubert (1982) indicates that a 'plume' rising through a porous medium does, in fact, exhibit only a small rounded crest and that the limbs are approximately linear.) For a two-dimensional situation (as would be appropriate for a thermal ridge), the value $(-P)$ by which the horizontal stress in Fig. 7b is reduced is given by

$$-P = \varrho g(2|z| - h)L \sin\theta \cos\theta / 2(|z| - h) \qquad (14)$$

where the various symbols are defined in Fig. 7b.

Values of $-P$ for a range of values of h and θ are listed in Table 1, where the values of L (in kilometres) for the various specific situations are given in brackets. It will be seen that, for the range of values of θ, $-P$ is constant for any given value of h. The values of $-P$ are large, for quite small values of h. These stresses would, of course be superimposed upon existing lateral stresses (which in the elastic zone would be in the range 3–7 kbar), so that the *in situ* tensile stresses will be smaller than those listed in Table 1.

If the plume is circular in plan then, by the time the two-dimensional model represented in Fig. 7 becomes operative, it is likely that lithospheric failure would have already been initiated. This follows because the plume with such radial symmetry would give rise to a boundary deflection with a radial slope (Fig. 8a). It can be inferred that the mass of a segment of lithosphere will, in part, be restrained from free gliding by the strength of the lithosphere along the vertical line OO′. The area of a line is zero, so that the glide force of any segment of plate (Fig. 8a) would give rise to an infinite stress at the line OO′, or at least a very high stress in its immediate vicinity. Consequently, no matter how large the tensile strength of the rock may be, failure would be initiated for a very small value of vertical migration (h) of the LVZ to form a series of radiating fractures about OO′. The 'least energy' requirement is met if a triple point develops. The zones of weakness of such a triple point would permit what was originally a circular section plume to become a linear feature. The fracture, or fractures, which would become most pronounced are those which are most nearly normal to the direction of easiest plate motion (Fig. 8b). 'Easy' plate motion can

TABLE 1. *Potential reduction in horizontal stress for average dip of slope of the upper surface of the LVZ and height of this zone at the axis; the length of the block (L, in km) is given in brackets.*

	Average dip of slope, θ, (degrees)			
	1	2	4	8
height (h in km)				
10	−3.4	−3.4	−3.4	−3.4
	(573)	(286)	(143)	(71)
20	−7.3	−7.3	−7.3	−7.3
	(1146)	(573)	(286)	(142)
30	−11.8	−11.8	−11.8	−11.8
	(1719)	(859)	(429)	(213)

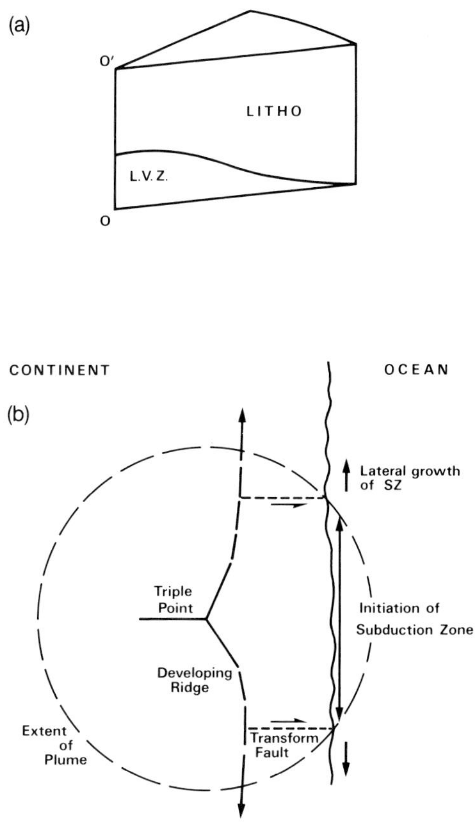

FIG. 8. (a) Diagrammatic representation of segment of lithosphere above a circular plume. Mass of rock in segment will, as the result of potential gravity glide, cause extremely high tensile stresses to develop in the vicinity of the axial line OO′. (b) Schematic diagram showing plate rupture development.

best occur when subduction has been initiated: and the place most favoured for this is at, or near, the nearest junction between the continental and oceanic plates.

Comment on lateral compressive stresses induced by gravity glide

Once a ridge has formed, the lithosphere is released to develop a small amount of glide, until the forces induced by elastic deformation attain a state of equilibrium with the potential gravity-glide forces. These strains result in stresses which are approximately 30 per cent of the average lithostatic pressure, so that the average increase in lateral stress is about 6 kbar. However, the lithosphere is not a homogeneous isotropic body, and consequently, the level of stress induced internally by the gravity-glide mechanism will not be uniformly distributed. The stress distribution will be influenced by two factors, namely, 1) the elastic modulus, and 2) the 'strength' of the different parts of the lithosphere. The Young's modulus of olivine is approximately 2×10^6 bar, while the average modulus for the continental crust rocks is about an order of magnitude smaller, ranging as it does from about 10^4 bar for weak clay-rocks, to $0.5-0.8 \times 10^6$ bar for the majority of crystalline and metamorphic rocks. However, crystalline and metamorphic rocks are dominantly in the lower crustal regions, where the differential stresses that can exist are controlled by the equations of state. From Fig. 2, it follows that this is also true of the lower 40 to 50 km of the lithosphere. Accordingly, for continental lithosphere we represent the levels of increase in stress that can be induced by gravity gliding against a completely resistant block, as shown in Fig. 9. We attribute a potential average increase in the crust and lower lithosphere of 0.5 kbar. Consequently, the 'strong' portion of the lithosphere from 30 to 70 km would experience an average increase of about 16.5 kbar. This increase in lateral stress must be added to the 'at rest' lateral stresses which ranged from 3 kbar at a depth of 30 km, to 7 kbar at a depth of 70 km. These 'at rest' lateral stresses are 7.5 kbar and 16.5 kbar, respectively, below the corresponding lithostatic pressures. Consequently, if the potential gravity-glide stresses are added to the 'at rest' lateral stress, the result is as indicated in Fig. 10 where, at a depth of 30 km, there exists a differential stress of 9.0 kbar (with S_1 horizontal), and at 70 km the stresses are close to 'hydrostatic'.

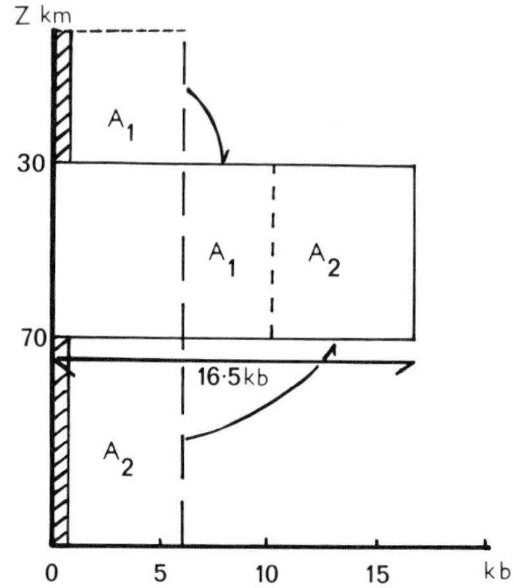

FIG. 9. Average lateral stresses induced by gravity glide. Distribution of stresses that result in a lithosphere of different elastic moduli and/or strengths are also shown.

It should be noted that a topographically derived ridge-push stress of about 0.5 kbar would also exist which would augment these differential stresses (Turcotte and Schubert 1982). However, against this factor one should set the conclusions of Turcotte and Schubert (1982), who assessed the thickness of the 'elastic lithosphere' in continental plates to be about 54 km. Such increase in thickness of the 'elastic' portion of the plate would have the effect of

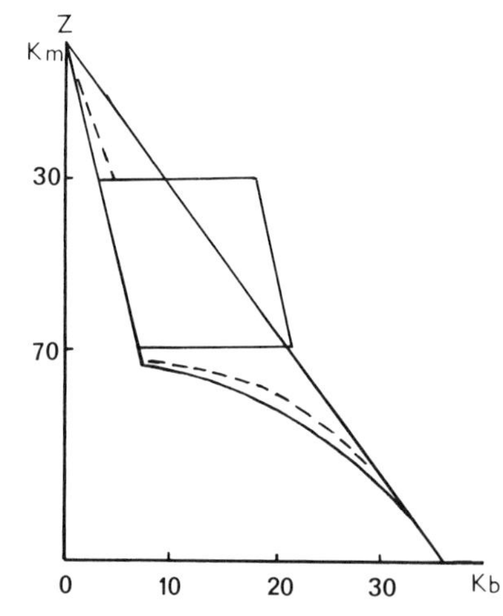

FIG. 10. Superposition of the lateral stresses from Figs 2 and 9.

reducing the magnitude of the estimated differential stress.

Estimates can be made from Fig. 2, for oceanic lithosphere, which indicate that differential stresses may be 2 to 3 kbar lower than for continental lithosphere. Again there will be an increase because of the ridge-push stresses. Moreover, Turcotte and Schubert (1982) indicate that in an oceanic plate the 'elastic lithosphere' is only about 28 km thick. Consequently, despite a slight degree of uncertainty, it is reasonable to conclude from our model that, at a depth of about 30 km, the lithosphere would, as a result of the gravity-glide potential, experience a maximum differential stress of about 10 kbar, with S_1 acting approximately horizontally.

The largest differential stress forecast by this analysis relates to the 30–70 km deep layer in 80-Ma-old lithosphere. This stress is reduced in lithosphere of greater age, and attains relatively small values as the result of flexure, in the vicinity of a convergent margin, and reaches extremely small values in the downgoing slab. It is in these last two regions that seismic activity in the lithosphere occurs most frequently. Our model predicts that the stress drop associated with such seismic activity in these areas will be relatively small, so there is no conflict between our model and the differential stresses inferred from seismic activity. Elsewhere in the plate, high differential stresses occur only in the 30–70 km deep zone. The differential stresses in the upper layers of crust will be an order of magnitude lower than in this zone, so that the stress drops associated with any crustal seismicity will be relatively small.

It should be noted that our model is based on a specific strength/depth relationship (after Dewey 1988). It is possible to change such a relationship, within quite broad limits, without invalidating our conclusions. Hence, we may now consider the problem of plate rupture at a subduction zone.

Initiation of subduction zones

It will be clear from Fig. 2a that the differential stress required to cause failure of this strong portion of the lithosphere is probably in the range 20 to 28 kbar, so that the lithosphere is in no danger of rupture (solely as a result of the lateral stresses) of about 10 kbar, induced by the potential for gravity glide. Therefore, the gravity-glide mechanism alone cannot explain the generation of subduction zones. However, these stresses, in combination with other compressive stresses induced by bending of the lithosphere, can certainly give rise to failure.

A quantitative representation of the combined mechanisms envisaged is given in Fig. 11. It is indicated that at the junction between continental and oceanic lithosphere, the deposition of sediments forming a continental slope and rise will result in isostatic downwarp. In so doing, the strong section of the lithosphere will react in an elastic fashion, so that compressive stresses will develop above the neutral surface (NS) (Fig. 12a). The magnitude of the stresses increases with distance from the neutral surface, such that the maximum stress (S_f) (termed fibre stress by engineers) is given by:

$$S_f = wE/2R \qquad (15)$$

where 'w' is the thickness of the elastic zone (which we have earlier taken to be 40 km), E is Young's modulus, and R is the radius of curvature of the 'syncline' caused by isostatic sinking.

If these fibre stresses are sufficiently large, they will combine with the stresses caused by

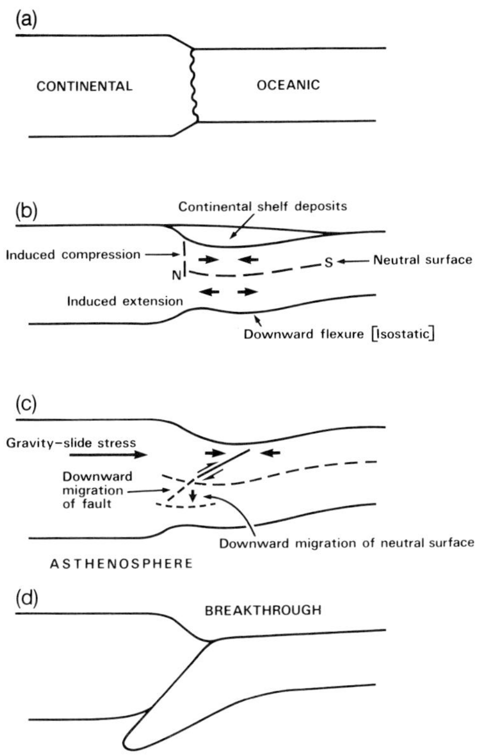

FIG. 11. Stages in the development of a subduction zone (see text).

gravity gliding (up to 10 kbar) so that together they can cause brittle failure of the strong, elastic portion of the lithosphere (at a differential stress of about 20–28 kbar). Once shear failure has occurred, near the top of the strong zone, this will reduce the effective thickness of that zone. Flexure will become easier, the fibre stresses reach higher magnitudes in lower regions of the lithosphere, so that the thrust, which is initiated in the uppermost part of the strong zone, migrates downward, and eventually results in a thrust that cuts completely through the lithosphere and so initiates a subduction zone.

It is of interest now to establish what conditions of lithospheric flexure can induce fibre stresses of 10 to 18 kbar. The situation envisaged is as indicated in Fig. 12b, in which sediments on and adjacent to a continental margin (the shaded area) cause a downwarp curve ACFEB which, for simplicity, we assume to be made up of a series of arcs of circles. The distance AB = L, and the total vertical displacement is U. (Point B is on the ocean floor and A is near sea-level, hence the tilt in the diagram). Here we are specifically concerned with the section CE (length $L/2$) and the vertical displacement DF = $U/2$. To calculate the radius of curvature (R) of this section (Fig. 12b) we use the construction in Fig. 12c, from which it follows:

$$\tan\theta = 2U/L = L/4\{2R - (U/2)\}$$

so that

$$R = (L^2 + 4U^2)/16U \tag{16}$$

But

$$S_f = wE/2R$$

so that

$$S_f = 8wEU/(L^2 + 4U^2) \tag{17}$$

If we set $L = 400$ km, $w = 40$ km, $U = 4$ km and $E = 2.10^6$ bar, then the fibre stresses generated at the uppermost level of the strong elastic zone will be approximately 16 kbar. Thus, if we add these fibre stresses to the maximum differential stress of about 10 kbar, caused by the potential gravity glide, the combined stresses will be 26 kbar, and the initiation of a subduction zone, when L is about 400 km and $U = 4$ km, is inevitable. It is emphasized that this is an illustrative model. Failure is related to curvature and also to the magnitude of E. Thus, for example if E is less than the value assumed above, the depth U will need to be commensurately greater than 4 km for the given value of L.

It must not be thought that these are special conditions, for they constitute the equivalent of a self-fulfilling prophecy. Once the continental margin deposits produce a sufficiently significant downward curvature, the perturbation thus formed will be amplified, and the radius of curvature reduced, as the result of buckling by the gravity-glide-induced lateral stresses. Thus, for example, as the Atlantic Ocean opens and the gravity-glide stresses increase in magnitude at the continental margins, the probability of lithospheric failure under either the American or European continental rise becomes greater. Does, for example the Lisbon Earthquake of 1755, mark the precursor of such an event? If it does, then the magnitude of the earthquake(s), with the attendant devastation, which will result from the initiation of one or more thrusts, at differential stresses of about 25 kbar, must be in excess of anything experienced during the recorded history of mankind.

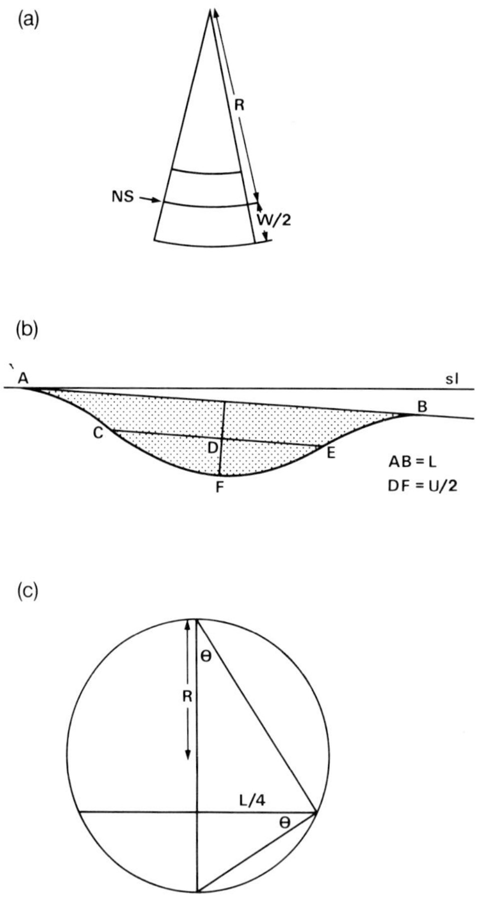

FIG. 12. (a) Diagram used to estimate fibre stresses. (b) Assumed geometry of downward curvature induced by continental rise sediments. (c) Construction used to estimate R.

Dynamics of plate motion

Let us now indicate how the gravity-glide mechanism outlined previously, with its attendant lateral stresses, can drive plates. We do this by considering three examples, the first two of which are merely illustrative, while the third represents, in a simplistic, two-dimensional model, the way in which a complete plate may move.

Example 1

In this first example we consider the portion of the lithosphere which is less than 80 Ma old and make the completely unrealistic assumption that it is unrestrained at the vertical plane of the wedge, in order that we can calculate the free-glide velocity that such a prismatic block could attain. This physically unrealistic example is included, so that an estimate can be made of the proportion of the glide potential which is needed to overcome viscous resistance to motion at normal plate velocities, and what proportion remains to generate a horizontal stress at the end of the prismatic block.

We assume that the total force acting parallel to the slope of the lower boundary given by Eqn (10) is balanced by the viscous drag, (F_R) on the base of the block, given by:

$$F_{TH} = F_R = \eta\dot{\gamma}L \tag{17}$$

where $\dot{\gamma}$ is the shear strain rate, η the viscosity of the LVZ, and L the length of the lithospheric unit.

If we assume that the viscous asthenosphere behaves as a Newtonian liquid, then (see Fig. 13a)

$$\dot{\gamma} = v/h \tag{18}$$

where v is the free-glide velocity of the block and h is the thickness of the asthenosphere which acts as a shear zone.

From Eqn (4), the lengths (L) of oceanic lithosphere younger than 80 Ma in age, for plate velocities of 30, 50, and 70 mm/yr are, respectively, 2400, 4000, and 5600 km. As we shall see, 'free-glide' velocities for lithospheric prisms are of the order of 1 m/yr. To use such velocities in association with Eqn (4) would indicate lithospheric lengths of up to 80 000 km. Hence we feel that it is unrealistic and inappropriate to use Eqn (12) in this example, as it is obtained from the empirically derived Eqn (4) and is based upon observed plate velocities. Here, instead we shall calculate the free glide velocity down slopes corresponding to lithospheric lengths of 2400, 4000, and 5600 km. Thus, from Eqns (13a), (17) and (18):

$$v = \varrho g z^2 h/2nL \tag{19}$$

If we assume that the oceanic lithosphere is underlain by a 75 km thick LVZ with a viscosity $\eta = 4.10^{19}$ Pa, then the free-glide velocities on slopes of 2400, 4000, and 5600 km are 5.5, 3.3, and 2.4 m/yr. If free gliding is restrained and slowed to 50 mm/yr, then only 1 to 2 per cent of the potential force is used to overcome viscous resistance to glide. Consequently, at least 98 per cent is available to cause plate motion of lithospheric slabs of an age greater than 80 Ma. This aspect is considered in the next example.

Example 2

In this example we estimate the thickness of shear zone which would develop beneath the plate in the situation represented in Fig. 13b, in which an 80-Ma old oceanic lithospheric plate is pushing a continental plate, which is 5000 km in extent, in the direction of movement. We will assume that this continental plate is underlain by mantle material with a viscosity of 10^{21} Pa (although as discussed above it is probably less), and we will also assume that the plate velocity is 50 mm/yr, and that the horizontal stress at the junction between the oceanic and continental plate is 6 kbar. If the force which generates this stress is balanced by retardational forces due to viscous drag at the base of the continental lithosphere, then using Eqn 19 and a subcontinental asthenosphere viscosity of 10^{21} Pa, the calculated width of the sublithospheric shear zone is 113.5 km, and could of course, if η is significantly smaller, be an order of magnitude thinner.

Example 3

Consider now the example represented in Fig. 13c in which an 80 Ma old block of oceanic lithosphere is driving not only a 5000-km long continental unit, but is also forcing a subducting slab to move at the same velocity. Here we assume that the subducting slab has neutral buoyancy, and ignore the frontal resistance of this slab to movement. We further assume that the downgoing slab is in contact with a shear zone on the upper and lower surface, and that these shear zones have the same thickness (h) as that which underlies the oceanic lithosphere. If the length of the subduction zone is taken as

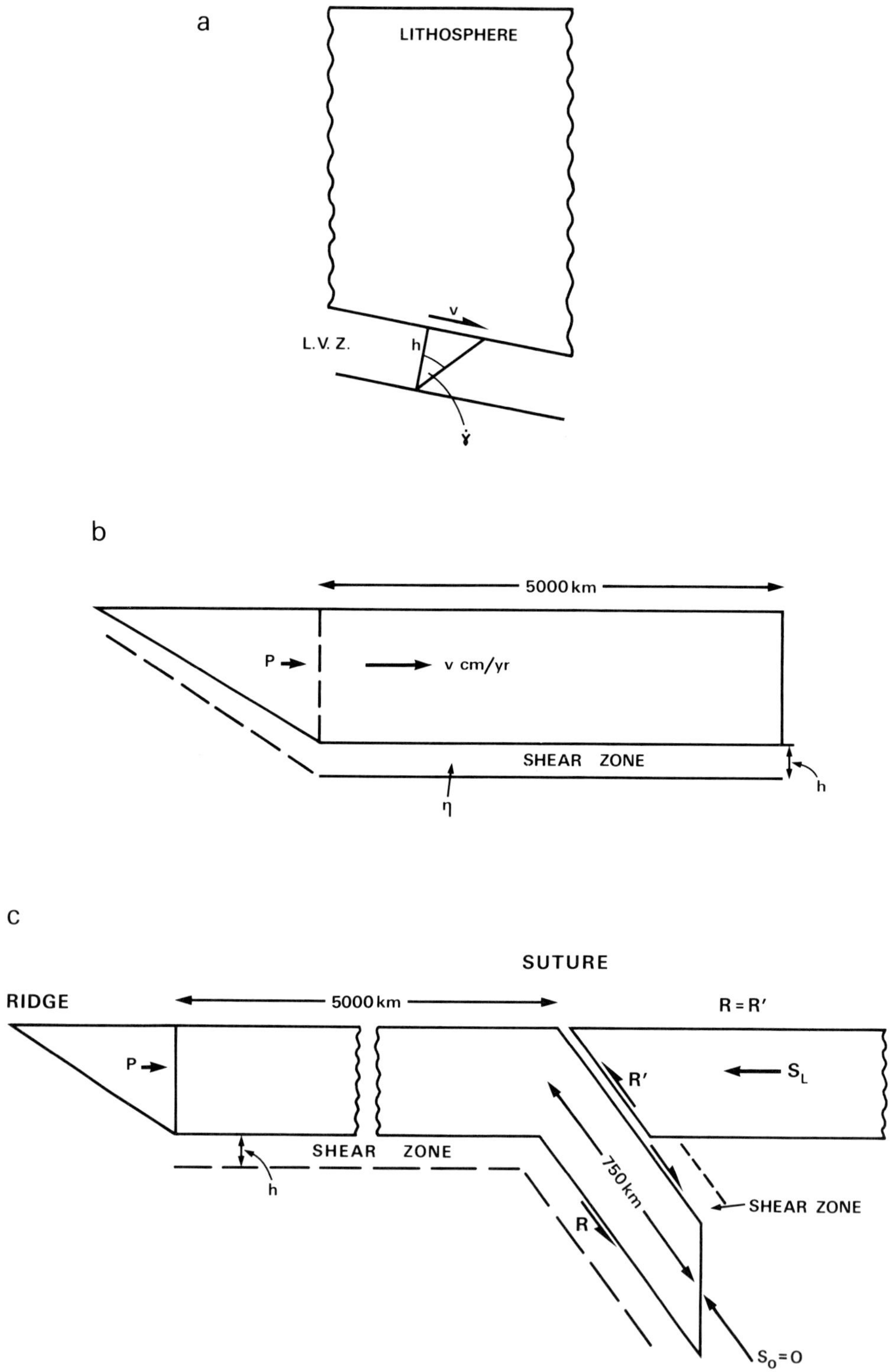

FIG. 13. (a) Model used to estimate free-glide velocity for Example 1. (b) Model for Example 2. (c) Model for Example 3.

750 km, this is equivalent to increasing the length of the continental mass in the previous example by 1500 km (from 5000 km to 6500 km). The thickness (h) of the shear zone will commensurately need to be increased from 113.5 to 147.5 km. (It is interesting to compare this modest thickness of shear zone with the wavelength of 4000 km, used by Turcotte and Schubert (1982), for a convection cell required to drive a subducting slab.)

As will be noted in Fig. 13c, we assume that the shear resistance R' of the suture, between the two lithospheric plates, is the same as that between the base of the lithosphere and the asthenosphere, R. Despite the high lateral stress in this region, it is believed that this assumption is valid, because if the downgoing slab is oceanic with clay rock or soft sediment as its top few hundred metres, then high fluid pressures will be generated in these sediments as a result of subduction, thereby significantly reducing the shear resistance. The shear stress at the base of the lithosphere is about 100 bar. This is equivalent to a differential stress of 200 bar. This is the same order of magnitude as the average differential stress that can develop in crustal rocks. Consequently, the assumption that $S = R$ is not likely to cause a significant error.

In the 'steady-state' model considered in this third example, resistance to movement is spread uniformly throughout the plate. Consequently, it follows that there will be a build-up in magnitude of the horizontal stress, from small values near the ridge to a maximum mean value of about 6 kbar, where the lithosphere attains an age of 80 Ma. Thereafter, in lithosphere older than 80 Ma, the component of lateral stress resulting from gravity glide decreases in the direction away from the ridge until it reaches zero at the tip of the downgoing slab.

However, resistance to plate motion is unlikely to be uniformly distributed throughout the plate. If at any point, or zone, in the 'old' (>80 Ma) lithosphere, resistance is so great that plate motion would cease at the stress level which would be associated with the steady-state model; this would bring about a redistribution of the stresses in the plate. The stresses resisting plate motion are strain-rate dependent, and therefore a reduction in the plate velocity will result in a higher proportion of the potential maximum horizontal stress migrating forward into the 'old' lithosphere, so that it can break down points or zones of resistance.

Some implications and applications of the gravity-glide mechanism

We believe that the high, lateral compressive stresses, predicted by our model to develop in lithosphere older than 80 Ma, are capable of causing the 'multiple pile-up' of India and Tibet, and of causing lithospheric plate buckling, to form 'compression' features such as the Timor Trough Foreland Basin. In addition, the model can explain the development of regional 'tectonic inversions' observed at continental margins.

Near sites of continental rifting, the lithosphere is in an extensional phase which is likely to give rise to a family of subparallel grabens, only one of which, the main, central structure, will eventually open up to form oceanic lithosphere. As time progresses, the subsidiary grabens will be carried away from the central ridge and so will migrate from an extensional regime at or near the rift, to a compressional regime a thousand kilometres or so away from the newly formed ridge. In such a transition, the subsidiary graben complex may undergo a compressive strain of about 2 per cent and so will exhibit a so-called 'tectonic inversion'.

During rifting, a small amount of oceanic crustal material may be emplaced between the blocks of a subsidiary graben unit. If these grabens eventually undergo tectonic inversion and are subjected to extremely high, lateral compression then, provided that the normal faults are suitably oriented, one individual fault, or set of faults, may be reactivated in the reverse mode and result in the emplacement of 'ophiolites' at a high crustal level, well before any major continental–continental collision occurs. Such ophiolitic bodies would not therefore mark the sutures between collided continents, but rather would be indicators of a failed rift system (see also Hall 1984).

Finally, we believe that the approach outlined in this paper has applications in all aspects of plate tectonics, including such events as the breakup of Gondwana and the closing of Tethys. Gondwana experienced several events, when sections of the continent rifted away from their original sites. The shortest dimension (L) of such 'breakaway' sections ranges from about three hundred to several thousand kilometres. These dimensions are in good agreement with the figures given in Table 1, where it will be seen that this distance (L), which is the length of the slope of the displaced boundary, will be determined by the value of θ. For small values of θ, L may range from several hundreds to one or more thousands of kilometres. Hence, the breakup of a continental mass is not sensitive to the distance of the centre of a plume to the edge of that continent. However, a more detailed mechanistic analysis of the breakup of this continental mass must await further palaeogeographical reconstructions.

Conclusions

We contend that the lateral stresses in the lithosphere, when it is 'at rest' (i.e. before the generation of a thermal perturbation and the onset of potential or actual gravity glide) are generated wholly by the effects of gravitational loading. Failure may occur in the crust, or the lower levels of the lithosphere, so that in these upper and lower zones, the lateral stresses, at a specific level, are only a little smaller in magnitude than the vertical stress. In the central zone (which, from rock mechanics data, we infer to be approximately 40 km thick), the lateral stresses are determined by the value of Poisson's number ($m = 4.4$), so that $S_x = 0.3\,S_z$. However, at higher and lower levels in the lithosphere the horizontal stresses are much closer to the geostatic stress, consequently, the average lateral stress, throughout the whole lithosphere, is approximately $0.7\,S_z$.

When the 'at rest' state of the lithosphere is disturbed by the development of a thermal plume, the upward migration of the lithosphere–asthenosphere interface results in a potential for gravity glide. In the text of the paper, we establish the magnitudes of the reduction of horizontal stress above the plume and the increase in magnitude of the horizontal stress away from a ridge, once it has developed. Prior to ridge development, this potential for gravity glide is shared. Part of it causes a reduction of lateral stresses above the plume and part causes an increase in lateral stresses away from the plume. For plumes with circular plan it has been indicated that there is a considerable stress magnification in the lithosphere above the axis of the plume, so that ridge initiation occurs at a relatively early stage of 'interface' uplift. This release of the whole of the gravity-glide potential by the initiation of the ridge, permits a relatively small amount of gravity glide to take place. This ceases when the potential for gravity glide is balanced by elastic and other strains in the lithosphere away from the ridge. The total potential for such movement is approximately 30 per cent of the average geostatic pressure (i.e. about 6 kbar).

However, because the strong central zone of the lithosphere has a high elastic modulus (E), while the upper and lower zones are weak and/or exhibit a low elastic modulus, the stress distribution is concentrated in the central zone of the lithosphere. As a result of this stress channelling, at a depth of about 30 km, a lateral stress will develop which may be as much as 10 kbar in excess of the geostatic stress. These high stresses are contained by elastic strains until subduction is initiated.

Subduction is set off by the superposition of these gravity-glide induced stresses, upon lateral stresses induced by downward flexing of the lithosphere beneath continental rise deposits. Once breakthrough of the lithosphere has occurred, subduction is initiated. Moreover, the development of this plane of weakness enables all, or a significant proportion of the elastically stored stresses, induced by gravity glide, to be released, so that the average lateral stress attempts to approach the 'at rest' state of $0.7\,S_z$. This decrease in the elastically stored stresses results in a freed potential for gravity glide to start plate motion and its attendant effects.

We therefore concur with earlier statements that plate tectonics is related to heat and gravitational potential. However, we emphasize the role played by gravity glide. We do not suggest that ridge push, and even an element of slab pull, have no role to play, but submit that such forces and stresses are of minor importance when compared with the gravity-glide mechanism in initiating ridges, subduction zones and driving plates. Moreover, the gravity-glide mechanism is capable of explaining the origins of the large stresses required to pile up continental plates, to give rise to regions such as Tibet, to buckle the lithosphere and to produce features such as the Timor Trough, and tectonic inversion of basins and graben systems.

Geologists who, several decades ago, believed in continental drift, were held to ridicule because they could not suggest a mechanism which was able to cause such 'drift'. It is ironic, therefore, that plate tectonics has been widely accepted (and correctly so) for the last twenty years, without a viable mechanism which was capable of generating the various features associated with plate tectonics, being proposed in the literature. We submit that the concepts put forward in this paper provide a coherent and comprehensive philosophy and a mechanism which rectifies that lamentable state of disarray which, hitherto, has existed regarding the theories relating to plate initiation and movements.

ACKNOWLEDGEMENTS: The authors would like to thank Joan Price for her patience and assistance, and to express their indebtedness to Don Weidner for his critical appraisal of the paper and to Michael Audley-Charles, without whose insistence the paper would not have been written. The senior author wishes to thank British Petroleum for support, while both the junior authors wish to thank the Royal Society for the award of 1983 University Research Fellowships.

References

BOTT, M. H. P. (1982). *The interior of the Earth.* Edward Arnold, London.

CATHLES, L. M. (1975). *The viscosity of the Earth's mantle.* Princeton University Press, Princeton, NJ.

CHAPPLE, W. M. and TULLIS, T. E. (1977). Evaluation of the forces that drive the plates. *J. Geophys. Res.* **49**, 1967–84.

DEWEY, J. F. (1988). Lithospheric stress, deformation, and tectonic cycles: the disruption of Pangaea and the closure of Tethys. This volume, p. 23.

FORSYTH, D. and UYEDA, S. (1975). On the relative importance of the driving forces of plate motion. *Geophys. J. Roy. Astron. Soc.* **43**, 163–200.

FYFE, W. S., PRICE, N. J., and THOMPSON, A. B. (1978). *Fluids in the Earth's crust.* Elsevier, Amsterdam (383 pp.).

HALES, A. L. (1969). Gravitational sliding and continental drift. *Earth Planet. Sci. Lett.* **6**, 31–4.

HALL, R. (1984). Ophiolites: figments of oceanic lithosphere? In *Ophiolites and ocean lithosphere* (ed. I. G. Gass, S. L. Leppard, and A. W. Shelton), pp. 393–403, Special Publ. Geol. Soc. Lond. 13.

JACOBY, A. R. (1980). Plate sliding and sinking in mantle convection and the driving mechanism. In *Mechanisms of continental drift and plate tectonics* (ed. P. A. Davies and S. K. Runcorn), pp. 159–72, Academic Press, New York.

JACOBY, W., BJORNSSON, A., and MULLER, D. (1980). Iceland: evolution, active tectonics and structure. *J. Geophys.* **47**, 1–277.

LOPER, D. E. (1985). A simple model of whole-mantle convection. *J. Geophys. Res.* **90**, 1809–36.

MCKENZIE, D. P. (1972). Plate tectonics. In *Nature of the solid Earth* (ed. E. C. Robertson), pp. 323–60, McGraw-Hill, New York.

PARSONS, B. and RICHTER, F. (1981). Mantle convection and the oceanic lithosphere. In *The sea.* Vol. 7 (ed. C. Emiliani), pp. 73–117, Wiley, New York.

PEEL, F. J. (1982). The driving mechanisms of plate motion and deformation. M.Sc. dissertation. University of London.

PRICE, N. J. (1966). *Fault and joint development in brittle and semi-brittle rocks.* Pergamon, Oxford.

PRICE, N. J. and AUDLEY-CHARLES, M. G. (1987). Tectonic collision processes after plate rupture. *Tectonophysics* **140**, 121–9.

RICHARDSON, R. M., SOLOMON, S. C., and SLEEP, N. H. (1979). Tectonic stresses in the plates. *Rev. Geophys. Space Phys.* **17**, 981–1019.

RUNCORN, S. K. (1980). Mechanics of plate tectonics: mantle convection currents, plumes, gravity sliding or expansion? *Tectonophys.* **63**, 297–307.

TURCOTTE, D. L. and SCHUBERT, G. (1982). *Geodynamics.* Wiley, New York (450 pp.).

NEVILLE J. PRICE, GEOFFREY D. PRICE, Department of Geological Sciences, University College London, Gower Street, London WC1E 6BT.

SARAH L. PRICE, Department of Theoretical Chemistry, University Chemical Laboratory, Lensfield Road, Cambridge CB2 1EW, UK.

Lithospheric stress, deformation, and tectonic cycles: the disruption of Pangaea and the closure of Tethys

J. F. Dewey

ABSTRACT: Intra-plate stress is governed on a global scale by plate boundary configuration and the disposition of continents. When continents are assembled into Pangaea-type masses, they are in deviatoric tension and predisposed to rifting and splitting along weak zones and where tension is locally enhanced by compensated uplifts. When continents are dispersed, they are, mainly, in deviatoric compression when local and minor thrusting and strike-slip faulting occur. Cyclic assembly, disruption, separation and reassembly of continental fragments causes a globally episodic, perhaps even periodic, repetition and synchroneity of tectonic events, alkaline and mafic magmatism and sea-level changes.

Introduction

The evolution of the Alpine orogenic system from Gibraltar to south-east Asia resulted from the closing of the Tethyan ocean(s) which, in turn, was the consequence of the disruption of Pangaea (Smith 1971; Dewey *et al.* 1973). The closing of the westward terminating Tethyan oceanic wedge or gore was not the contraction of a single ocean, but rather was accompanied by the detachment of thin and irregular strips of continental lithosphere from the Gondwana margin, which converged on Asia faster than did Gondwana, thus opening up post-Triassic southerly Neo-Tethyan oceans as more northerly Palaeo-Tethyan oceans closed (Şengör 1979; Şengör and Yılmaz 1981). The closure of the Palaeo-Tethyan oceanic tract in the late Triassic/early Jurassic led to collisions that produced the Cimmerides. The later closure of Neo-Tethyan oceans gave rise to the Tethysides (Şengör 1984, 1985). This paper is concerned with the nature and origin of the lithospheric stress conditions that led to and accompanied the disruption of Pangaea and the closing of Tethys, and the relationship between global patterns of stress and tectonic cycles, and the episodicity and periodicity of intra-plate deformation.

Lithospheric strains

Large strains and displacements, concentrated in plate boundary zones, are usually approximately plane, and arise from the horizontal shortening and stretching and, to a lesser extent, flexure of the lithosphere. Horizontal shortening involves thickening, warming and softening of the lithosphere, thrust and strike-slip faulting in the upper crust, prograde metamorphism, silicic and mafic magmatism, uplift, and denudation. These processes lead to orogenic belts such as the Alps and Himalayas. Horizontal stretching involves thinning, cooling and hardening of the lithosphere, normal and strike-slip faulting in the upper crust, retrograde metamorphism, subsidence and sedimentation, and the rapid setting of isotopic clocks. Stretching may lead to continental splitting, as along the northern margin of Gondwana in Triassic/Jurassic times but, more usually, generates aborted continental rifts often with little or no magmatism. If present, magmatism is principally mafic and/or alkaline. Extension is harder to study than shortening because it is usually accompanied in older extensional regimes by subsidence and, therefore, burial by sediments.

Intra-plate deformation, although much subtler and characterized by smaller strains and displacements than those of plate boundary zones, is widespread, particularly in the continental lithosphere (Dewey 1982) as witnessed by earthquakes, rifts, zones of compressional and wrench deformation, uplifts, and basins. North-west Europe, for example, shows considerable mild tectonism (Schmidt 1981), and much of the topography of Britain may have been generated recently by the intra-plate stresses that are relieved by small earthquakes (Burton and Neilson 1980). Thermally young, thin lithosphere is particularly susceptible to intra-plate deformation; in the oceans, seismicity is commonest and shallowest in the

From AUDLEY-CHARLES, M. G. & HALLAM, A. (eds) *Gondwana and Tethys* Geological Society Special Publication No. 37, pp. 23–40.

youngest lithosphere (Richardson and Cox 1984; Okal 1984; Newmark *et al.* 1984) and, in the continents, is rarest in the oldest shields (Dewey 1982).

Extension is the commonest form of intraplate deformation; rifting has affected the continents since the earliest stabilization of Archaean shields. The continents are affected, pervasively, by extension zones of many ages in a wide variety of tectonic settings. It is clear, however, that there have been times (e.g. Triassic to early Cretaceous) when deviatoric tension has induced rifting over enormous continental areas at other times (e.g. late Cretaceous), during which deviatoric compression has induced minor but widespread wrench and thrust faulting, and basin inversion.

Stress origins and magnitudes

The Earth's lithosphere suffers naturally-induced stress at all scales, from local to plate-wide, by a wide variety of mechanisms, some of which result from body forces and some from applied forces; some of these are active or renewable, others are residual or non-renewable (Bott and Kusznir 1984). The magnitude and orientation of natural stress fields are commonly determined or inferred, with varying reliability, from geological structures such as fault and joint patterns and fault offsets (Engelder 1982; Holst 1982), volcanic lineaments (Nakamura 1969), and stress-relief buckles (Adams 1982; Williams *et al.* 1985), earthquake source mechanisms and first-motion studies (Gephart and Forsyth 1985), seismic moment and stress drop (Hanks 1977), borehole overcoring, strain relief and breakouts (Gay 1975; Zoback *et al.* 1985), hydraulic fracturing and crack orientation (Hickman *et al.* 1985; Stock *et al.* 1985; Warren and Smith 1985), offset boreholes (Bell 1985), stress-aligned microcracks deduced from the polarization of shear waves (Crampin 1985), and theoretical modelling (Brace and Kohlstadt 1980; Kirby 1980; Leary 1985). The last presents the difficulty that continental rocks, in particular, are inhomogeneous and pervaded by surfaces and zones of weakness. Such zones, where they are favourably oriented in a particular stress field, may focus slip at lower stress levels and at different orientations to those predicted (Anderson 1942) for an isotropic unfractured substance, particularly at shallow levels in the brittle regime. At deeper levels, where more ductile rheologies occur, isotropic models may be more appropriate, although seismic reflection studies show that structural inhomogeneities persist to upper-mantle depths (MacGeary and Warner 1985).

These techniques and methods have been used to study, deduce, and integrate the state of stress throughout much of the lithosphere (e.g. Lilwall 1976; Gay 1977; McGarr and Gay 1978; Richardson *et al.* 1979; Wesnousky and Scholz, 1980; Pulli and Toksoz 1981; Yang and Aggarwal 1981; McGarr 1982; Gough *et al.* 1983; Bergman *et al.* 1984, 1985; Gough 1984; Lambeck *et al.* 1984; Nunn 1985; Assumpcao *et al.* 1985; Denham 1985; Hasegawa *et al.* 1985). Extinct geological structures such as fault and joint patterns and dyke swarms have been used to deduce extinct stress fields (Engelder and Geiser 1980; Barnett *et al.* 1984; Ernst and Halls 1984; Kretz *et al.* 1985). The sources of naturally-induced stresses, their value in MPa, and whether they are renewable (in the sense of Bott and Kusznir 1984) in the lithosphere, are summarized in this section and the most important stresses (numbered 1–8) are illustrated in Fig. 1.

McTigue and Mei (1981) have shown that strong relief induces horizontal compression, up to 50 MPa, at topographic ridge axes, and tension at valley bottoms, these effects eventually diminishing downwards to lithostatic conditions. Vertical loading generates a vertical compression and unloading generates a vertical tension (17 MPa), the latter commonly expressed as pressure-relief joint sets that roughly parallel the topographic surfaces. Stresses (60 MPa) may be generated by regional lateral changes in the thickness and density of lithospheric layers (Artyushkov 1973; Park and Westbrook 1983). Similarly, on a somewhat smaller scale, stresses are induced by lateral strength variations and shapes of rock bodies. Forceful diapiric intrusion of granite, salt, and mud causes compression in the injected country rock; magma pressure at dyke-tips induces very high localized tensile stresses that cause rapid crack propagation. More passive magma intrusion causes tension that leads to collapse of magma-chamber walls and stopping. At the marginal contact regions of terrains suffering differential uplift or subsidence, shear generates oblique stress fields. Stresses (400 MPa) are generated by the cooling or heating of the lithosphere (Turcotte 1983). The lithosphere cools and subsides as it ages, and also as it thickens, towards an ambient value in regions of lithospheric stretching or advective thinning. Bratt *et al.* (1985) have shown that, as lithospheric rocks cool through an elastic 'blocking temperature' (particularly in young lithosphere where cooling rates are high)

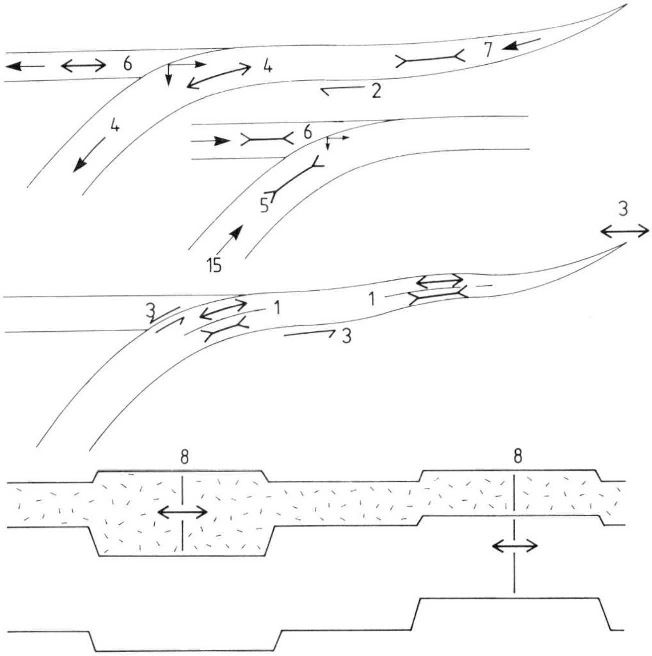

FIG. 1. Principal ways in which renewable stresses are developed in the lithosphere. Numbers refer to listed mechanisms on pp. 25–6 of text.

thermal stress is an important factor in generating earthquakes in the uppermost lithosphere older than 15 Ma.

When plates change their latitude, they are strained (membrane tectonics) because of the 1/300 ellipticity of the Earth. Movement to higher latitudes and an increased radius of curvature cause compression within the plate and tension at its edges (100 MPa). Decrease in latitude induces tension within a plate and compression at its edges (Turcotte 1974). Stress levels (10^{-3} MPa) generated by Earth tides are far too low to generate any geologically significant strains.

Structural compatibility problems at plate boundaries and triple junctions cause substantial stresses (20 MPa, renewable). Gaps and overlaps arise from geometric complexities at plate boundaries, and the growth of these induces tension and compression respectively. Examples are transpressive and transtensional segments of, and flake rotation within, continental transform zones, gaps at transform–transform–transform triple junctions, and gaps at the termination of strike-slip bounded wedges in convergent zones (Dewey *et al.* 1986). The resulting stresses cause relatively localized deformation and can be considered as local tectonic weather in a more regional tectonic climate.

Elastic or visco-elastic flexure or bending of the lithosphere (*1*) causes tension on the convex (extrados), and compression on the concave (intrados) sides, respectively, of a neutral surface of no longitudinal strain in the flexural bend; maximum stresses are between 500 and 1000 MPa. This is expressed as extensional strains on the outer walls of trenches, below thrust sheets, and also by rift valleys on arches.

If plates are driven by mantle convection (*2*) tractional stresses (50 MPa, renewable) would be developed at the base of the lithosphere. Such drag could be expressed as plate basal-shear, or by tension or compression within the plate, but basal tractional forces are likely to be unimportant (Turcotte 1983).

If plate motion is driven by body forces inherent in the plate mosaic geometry (Hager 1978; Hager and O'Connell 1979; Forsyth and Uyeda, 1975) such as ridge push and slab pull, then resistive viscous drag forces (*3*) balancing the torque on plates are in better agreement with intra-plate stress data than convective drag on the base of plates (Richardson *et al.* 1979). Resisting forces (20 MPa, renewable) arise from the strength and buoyancy of rocks and sliding friction on faults at plate boundaries. Rock deformation at plate boundaries and plate boundary zones is largely driven by stresses resulting from these forces. Stress diffusion from plate boundaries at rates of 0.1 to 10 km/yr (Bott and Dean 1973) may widen convergent plate boundary zones in continents to extensive regions of complex 'foreland' deformation (Vilotte *et al.* 1982), such as the widespread late Mesozoic–Cainozoic deformation and basin inversion of north-west Europe in the Alpine foreland (Dewey 1982).

The weight of a subducting slab (*4*) exerts a pulling force (50 MPa, renewable) upon the attached non-subducted portion, that is proportional to the length—up to about 300 km—of slab in the asthenosphere (Forsyth and Uyeda 1975). Isacks and Molnar (1969) adduced evidence from focal mechanism data that down-slab tension (*4*) characterizes those subducting slabs whose leading edge penetrates up to about 300 km into the asthenosphere, whereas down-slab compression occurs in those that have moved to 500 km or deeper. Thus, shallow slabs may exert a pull and deep slabs may exert a resisting push on subducting plates.

Molnar and Atwater (1978), Uyeda and Kanamori (1979) and Dewey (1980) argued that whether oceanic plate subduction leads to compression (Andes) or tension (30 MPa, renewable) (Marianas) in the overriding plate (*6*), depends upon its motion in the reference frame of the trench—which is controlled by the plate bend through which the plate passes before subduction. The position of the bend is, in turn, controlled by the age of the subducting lithosphere. Old, cold, thick lithosphere sinks readily and induces a roll-back of the bend, while young, warm, thin lithosphere is more buoyant. Hence, arc style is extensional, with back-arc basins where roll-back proceeds faster than the overriding slab can advance or wherever the latter retreats from the trench, and compressional where override advance is faster than roll-back. Şengör (1979) has argued that Neo-Tethyan oceanic tracts may have originated as back-arc basins above south-dipping subduction zones: the north-west suction and detachment of thin continental strips being caused by roll-back.

Stresses (30 MPa, renewable) are generated by ridge-induced plate compression (*7*) caused by the difference in vertically-integrated mass between ridge axes and the deeper, older portions of the oceanic lithosphere (Forsyth and Uyeda 1975); this is the so-called ridge-push force. Where the excess mass of an elevated region is compensated at depth by a mass-deficient root, the weight of the former and buoyancy of the latter oppose to produce a vertical compression and horizontal tension (Bott 1982*a*,*b*; Bott and Kusznir 1979). Thus, a tensional force (50 MPa renewable) (*8*) is induced throughout the continents (Bott, 1982*a*,*b*), in mountain belts (Andes; Dalmayrac and Molnar 1981), domes and arches (Bott 1981), and in plateaux produced by continental collision (Tibet; England and McKenzie 1982), basaltic underplating (McKenzie 1984), or by advective thinning of the lithosphere (Spohn and Schubert 1982; Parker *et al.* 1984). Where compensation occurs within the crust, the resultant horizontal tension is thin-skinned and confined to the crust; sub-Moho stress patterns may be quite different and related to plate-wide driving forces. In contrast, where an uplift is induced by advective thinning of the lithosphere, compensation is in the asthenosphere, and horizontal tension occurs to the base of the lithosphere and can result in whole lithosphere failure. This may be happening in the Basin and Range, possibly by the subduction of anomalous mantle (Gough 1984), at rifted margins by lateral heat flow, and by convective thinning such as at the Gulf of Suez margins (Steckler 1985), in the eastern Australian highlands (G. D. Karner and J. Weissel pers. comm.), and in orogens where rapid uplift is induced by lithospheric delamination (Bird 1978). Therefore, grabens and rift valleys will be deep-rooted in an extending lithosphere where compensation is in the mantle and associated magmatism is therefore more likely, and shallow-rooted where compensation is in the crust.

The state of stress in the lithosphere (at plate boundary, large regional, and plate-wide scales), which induces widespread strains, is a complex interaction of renewable stresses. This is because renewable stresses are long-term, persistent, and subject to stress amplification (Mithen 1982), while non-renewable stresses last only as long as the feature causing them and are subject to viscous dissipation (Bott and Kusznir 1984). However, local non-renewable stresses may be of great importance in reducing or enhancing renewable stresses, and preconditioning a particular area or zone in the lithosphere to failure.

Limits and vertical distribution of stress

The strength of the lithosphere under tension or compression is governed by the vertically-integrated breaking stress and creep yield stress of the rocks of which it is composed. In an unfractured brittle rock mass, the shear strength is the shear stress at which faults originate (Anderson 1942). In fractured brittle rocks, the shear strength is the shear stress at which favourably oriented existing planes of weakness begin to slip, and it is generally much lower than the unfractured strength. The shear strength rises with increasing confining pressure; at low temperatures, the coefficient of friction and hence the fracture and sliding strength of most rocks, is proportional to depth (Byerlee 1968). Strength is lowered in the presence of water. Shearing strengths for dry and wet rocks in compression and extension (Byerlee, 1968) are plotted in Fig. 2. At deeper, hotter levels where rocks begin to creep, strength is temperature rather than pressure dependent, lower in wet than in dry phases, and lower in quartz than in olivine—these minerals being principal phases of continental and oceanic crust respectively (Fig. 2). Where the creep curves for a mineral phase intersects the Byerlee fracture line, a strength maximum exists at a depth at which the largest earthquakes and stress drops occur (Dewey *et al.* 1986). A correlation exists between theoretically-deduced strength maxima and the depth–frequency distribution of earthquakes (Chen and Molnar 1983; Meissner and Strehlau 1982; Das and Scholz 1983). Hypocentres in the continental lithosphere are mostly at relatively shallow levels in the crust, generally less than 12 km in a crust of average thickness with a normal heat flow (Pulli and Toksoz 1981; Yang and Aggarwal 1981; Basham and Adams 1984).

In Fig. 3, ten tensional strength profiles, derived from the data of Fig. 2, are shown for various combinations of wet and dry crust (quartz) and upper mantle (olivine), crustal thickness and geothermal gradient, to illustrate the significant differences in integrated lithospheric tensional strength produced by these variations. A wet crust and topmost mantle (*A*) weakens the lithosphere by 17 per cent compared with a completely dry lithosphere (*B*), whereas, if the lower crust is dry (*D*), weakening is by only 8 per cent. Rapid stretching of a lithosphere with a wet crust and dry mantle by 100 per cent (*F*), reduces the strength by almost 70 per cent at the end of stretching, but the thermally-recovered lithosphere (*G*) is about 14 per cent stronger than the unstretched lithosphere. Thus, rapid stretching and thermal recovery (Dewey 1982) or slow stretching (England 1983) increase the strength of the lithosphere and make that zone less likely to be stretched again. Advective thinning (*I*) by 50 per cent decreases the strength even more than does stretching. Continental (*B*) and oceanic (*J*) lithosphere of the same thickness have the same strength as if they are dry, but a lithosphere with a substantially wet upper crust (*D*) is 8 per cent weaker. A 5 km increase in the thickness of the continental crust (*D–H*) causes a 7 per cent decrease in strength because dry quartz begins to creep in the lowest crust. These variations control not only integrated strength, but also rheological layering of brittle and ductile phases and, therefore, faulted layers and potential crustal *décollement* or detachment levels. For

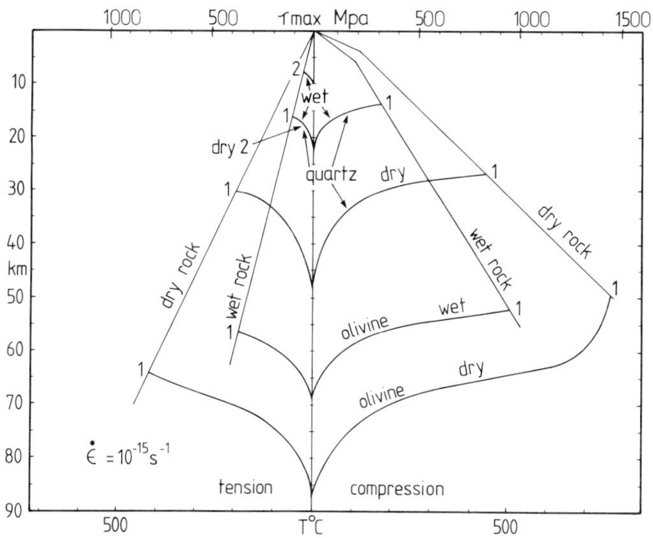

FIG. 2. Shear strength envelopes in tension and compression for brittle failure (straight lines), and for quartz and olivine (curved lines) under wet and dry conditions. Data from Kirby (1980), and Meissner and Strehlau (1982).

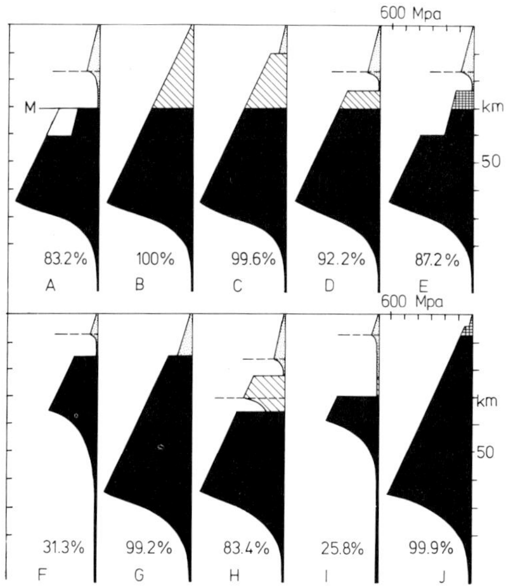

FIG. 3. Shear strength envelopes in tension for lithospheric profiles with various combinations of dry and wet quartz and olivine, crustal thicknesses and geothermal gradient. Black—olivine; lines—dry quartz; dots—wet quartz; cross-hatching—mafic rocks. Percentages refer to strengths relative to B (100 per cent). Depth in kilometres on vertical axes, shear strength in MPa on horizontal axes.

example, in *B*, *D*, *G*, and *J*, there are no crustal detachment levels, whereas in *A*, *D* and *E*, potential detachment occurs at about 16 km, and in *F* and *I* at about 8 km. In *H*, detachment levels occur at 16 and 30 km. In a region in which the lithosphere is extremely thin, either by stretching or in proximity to a ridge axis, the olivine creep curve could be brought to depths of 15 km or less, and allow the potential thrusting of thin continental lithosphere flakes (Dewey 1982) or the obduction of ophiolite nappes (Dewey and Shackleton 1984).

Theory suggests that stress levels must, and stress regimes should, change with depth. Flexural bending of the lithosphere at trenches and by loads such as ice, sediments, volcanics, and thrust sheets causes peripheral arching whose wavelength and amplitude depends upon the thickness and thermal age of the lithosphere (Karner *et al.* 1983). Small longitudinal elastic bending strains in lithospheric flexural arches induce shallow stresses as high as 1000 MPa, that diminish downwards to zero at the neutral surface. A modest amount of extensional faulting results from these strains, as dome and arch crustal faults and as normal fault systems on the outer walls of trenches and beneath thrust complexes in foreland basins. Similarly, flake rotation in shear zones (Karner and Dewey 1986) generates complex variation between shallow tension and compression. Also, shallow deviatoric tension can be developed in uplifts that are compensated in the crust, with a different stress regime in the upper mantle. For example, Gay (1975, 1977) and McGarr and Gay (1978) show that, although at shallow levels σ_h is usually greater than σ_z, in Caledonian and Alpine mountain belts $\sigma_z > \sigma_h$, emphasizing the role of compensated uplifts in generating horizontal tension.

Observation shows that stress regimes can vary vertically in the crust. In parts of the western Basin and Range, a wrench regime (σ_2 vertical) above 6 km gives way to a gravity regime (σ_1 vertical) from 9 to 20 km (Vetter and Ryall 1983) and, in central Australia—a continent predominantly in deviatoric compression—fault plane solutions (Lambeck *et al.* 1984) indicate a shallow wrench regime giving way at about 15 km to a thrust regime (σ_3 vertical).

Some ways in which stress levels and regimes may change with depth are developed in this section and illustrated in Fig. 4. If we assume a vertically uniform elastic rheology in the upper lithosphere, with a Poisson's ratio of 0.25 and a linear increase in stress due to loading (σ_z), horizontal stress (σ_x) is $\sigma_{z/3}$ and radially uniform in the horizontal plane. If we now superimpose a 1 kbar horizontal compressional (σ_{hc}) or tensional (σ_{ht}) stress, we obtain a gravity regime ($\sigma_z > \sigma_x > \sigma_{ht}$) and normal faulting throughout the lithosphere except, perhaps, at very shallow levels where σ_z and σ_x converge and where strike-slip faulting could occur; strike-slip faulting occurs above 6 km in parts of the Basin and Range (Vetter and Ryall 1983). For a superimposed 100 MPa horizontal compression (σ_{hc}), a wrench regime ($\sigma_{hc} > \sigma_z > \sigma_x$) is predicted for depths less than about 5 km and has been observed by Okal (1984) and by Newmark *et al.* (1984) in the young oceanic lithosphere.

The implications are that thrusting is only possible at very shallow depths where σ_x approaches σ_z, a stress condition demonstrated by Gay (1975, 1977) for southern Africa. Below 5 km, $\sigma_z > \sigma_{hc} > \sigma_x$, and what might be termed a cross-gravity regime (σ_{hc} is σ_2) is developed. Such a regime has been observed by Bergman *et al.* (1984, 1985) in young oceanic lithosphere in the southern Indian Ocean. Such a rheology, however, does not allow thrusting except in the uppermost few kilometres. Also, in compression, this rheology yields differential and shear-

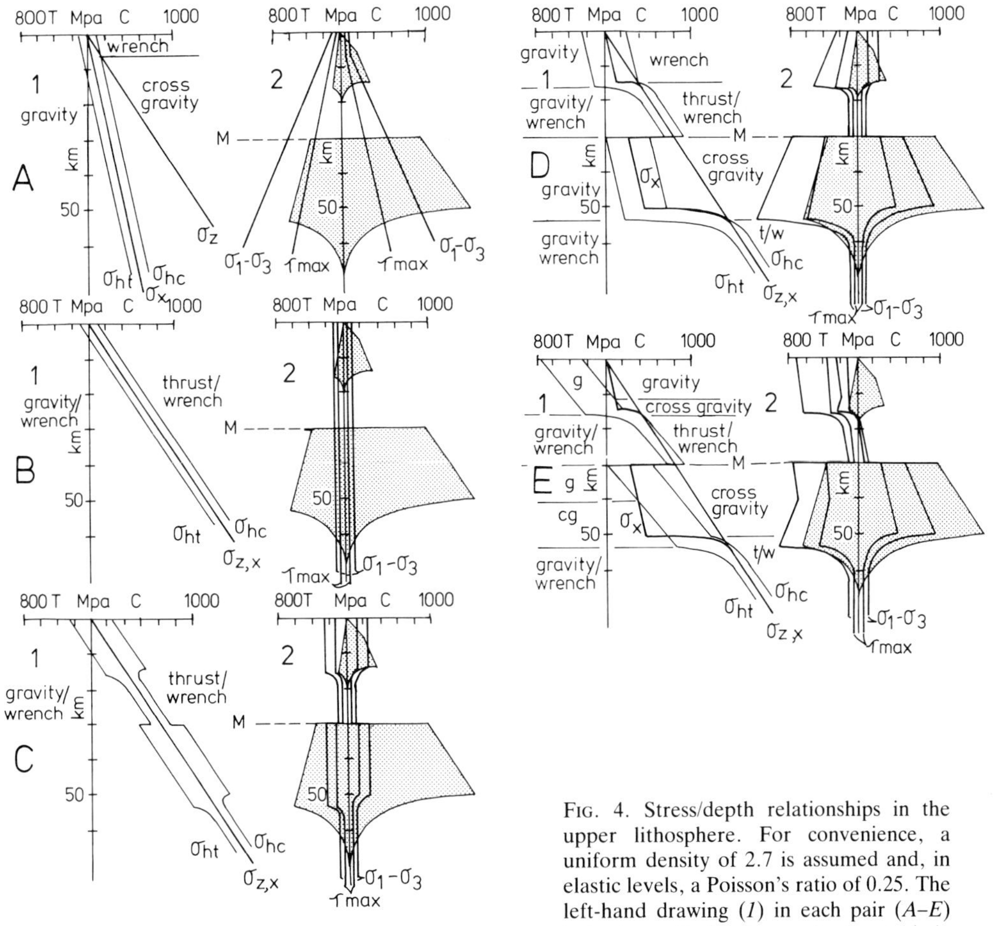

FIG. 4. Stress/depth relationships in the upper lithosphere. For convenience, a uniform density of 2.7 is assumed and, in elastic levels, a Poisson's ratio of 0.25. The left-hand drawing (*1*) in each pair (*A–E*) indicates vertical (σ_z), cross-horizontal (σ_x), and main horizontal (σ_{hc} in compression, σ_{ht} in tension) stresses. The right-hand drawing (*2*) indicates differential stress ($\sigma_1 - \sigma_3$) and maximum shearing stress ($\sigma_{max} = \sigma_1 - \sigma_3/2$). Dotted area is where shear failure is possible with a wet crust and dry mantle and a geothermal gradient corresponding with a surface heat flow of 1 μcal cm^2 s^{-1}. Abbreviations: *M*—Moho; *cg*—cross gravity; *g*—gravity; *t/w*—thrust/wrench; *T*—tension; *C*—compression.

ing stresses sufficient to allow brittle failure in only the uppermost three or four kilometres (*A2*). In tension, stresses appear ample to allow failure of the upper crust, and sufficient to cause failure of the uppermost mantle, if frictional resistance to sliding is reduced slightly from the theoretical value (*A2*).

In Fig. 4B, a hydrostatic spherical ambient stress field is assumed, that is the whole lithosphere behaves as a fluid with zero long-term strength, so that $\sigma_z = \sigma_x = \sigma_h$. If we apply an additional horizontal stress to cause either deviatoric compression (σ_{hc}) or tension (σ_{ht}), the tectonic regime is thrust/wrench (either equally likely) in compression and gravity/wrench (either equally likely) in tension (*B1*). In compression, stresses are only sufficient to cause brittle failure in the top two or three kilometres of the crust (*B2*). In tension, shear stresses are sufficient to cause normal faulting in about the uppermost 10 km, but not in the uppermost mantle. If, however, we allow viscous dissipation of deviatoric stress in the weak ductile layers of the crust and mantle, amplication of stress occurs (Kusznir and Bott 1977; Kusznir and Park 1982; Mithen 1982) in the stronger brittle layers (Fig. 4C). A 150 per cent amplification does not alter the stress regime, but increases differential and shearing stress to increase the depth in compression to which failure can occur in the shallow crust and to render easy the failure of the whole thickness of the crust in tension.

The situation, whereby stress conditions are likely to generate observable geological structures, can be improved by allowing stress amplification in long-term elastic layers in the upper crust and upper mantle (Fig. 4D). In compression, a wrench regime in the upper 15 km gives way to a thrust/wrench regime in the lower crust, a transition observed by Lambeck *et al.* (1984) in central Australia. At very shallow levels, thrusting becomes likely as σ_z and σ_x values converge. Nicholson *et al.* (1984) deduced shallow thrust solutions in the New Madrid seismic zone, and shallow strike-slip faulting is common in many present-day regimes of intra-plate deviatoric compression, for example in Brazil (Assumpcao *et al* 1985; R. Mosmann pers. comm.) with shallow, dextral north-east-striking faults and sinistral north-west-striking faults. The dihedral angle of these faults is bisected by an east–west compression generated by the push force from the Mid-Atlantic Ridge and by stress diffusion from the compressive Andean subduction plate boundary. In tension (Fig. 4D), differential and shear stresses are sufficiently large to cause extension. A gravity regime is displaced by a gravity/wrench regime in the lower crust and below about 55 km, and wrenching becomes increasingly likely at very shallow levels where σ_x and σ_z values converge. If arch bending stresses with a neutral surface at 30 km are superimposed (Fig. 4E) on the stress condition of Fig. 4D, very large tensional, differential, and shear stresses can be generated in the upper crust, but are unlikely to be significant in causing or enhancing major extension because they give way, below the neutral surface, to increasing compression. A principal conclusion is that, although vertically-integrated stress is enough to cause the lithosphere to fail in tension, the theoretical strength is an unrealistic approximation of the real strength, which is probably much lower because of inhomogeneities and zones of weakness in the crust and mantle.

Lithospheric stress conditions

The present-day intra-plate stress regime is mainly one of deviatoric compression (Richardson *et al.* 1979), and stress regimes may be coherent over very large areas while maintaining stress compatibility at stress province boundaries (McGarr 1982; Gough *et al.* 1983). The state of within-plate stress depends, to a large extent, upon plate boundary configuration and, therefore, upon the ways in which the main renewable stresses are combined (Richardson *et al.* 1979; Bott and Kusznir 1984), some of which are shown in Fig. 5.

Where a plate is trapped between subduction zones, as is the Caribbean Plate in east–west section, trench suction forces give rise to intraplate deviatoric tension with stresses up to 30 MPa (Fig. 5A). The superimposition of compensated uplifts enhances deviatoric tension and probably localizes the site of lithospheric rupture where compensation is below the lithosphere, and crustal stretching where compensation is in the crust. Such a deviatoric tensional regime would be widespread in continents assembled into Pangaea-type configurations (Bott 1982*b*). Where a continent lies between ridges (Fig. 5B), the ridge-push forces induce deviatoric compression, which can be reduced but not eliminated in compensated oceanic uplifts, and eliminated and replaced by deviatoric tension in compensated continental uplifts. Thus, the African Plate is in deviatoric compression from its surrounding ridges and from compressional plate boundary forces in the Alpine system, but the East African Rift system is forming by east–west tension induced by a major arch compensated in the mantle. The regional elevation of the African continent can be ascribed to a slight, probably elastic, crustal thickening caused by its compression by ridge-push and collisional forces. Thus, Pangaea-type continental configurations in deviatoric tension seem certain to self-destruct and split (Bott 1982*a*,*b*) and lead to dispersed continents in deviatoric compression: some of the geological consequences of which will be explored in the last section.

Where a continent is affected by ridge push from one side and trench suction from the other, there will be a change from diminishing compression to increasing tension (Fig. 5C) which may be eliminated or enhanced, respectively, by compensated uplifts. Where a continent lies between a ridge and a subduction boundary with trench override (e.g. South America), it is in deviatoric compression which may be reduced or

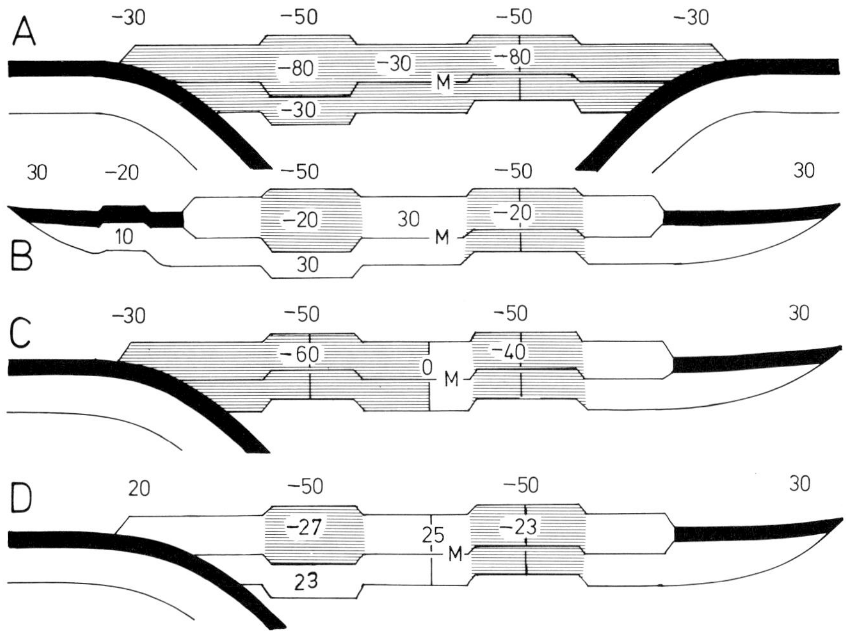

FIG. 5. Plate- and regional-scale combinations of renewable stresses that give rise to lateral and vertical stress variations. Black—oceanic crust; horizontal lines—areas in deviatoric tension; positive values—compressional stress; negative values—tensional stress.

replaced by tension in compensated uplifts (Fig. 5D).

Plateaux and uplifts, therefore, may exist in a variety of tectonic settings with different vertical stress profiles, depending upon whether compensation is in the crust or mantle and whether the plate is in background deviatoric tension or compression. An uplift is most likely to lead to rifting, lithospheric failure and continental separation where it is compensated in the mantle, and where the continent has subduction zones on two or more sides. Under such conditions, lithosphere-penetrating stresses may be as high as 80 MPa (Fig. 5A) which, in turn, is likely to lead to high extensional strain rates, and thinning and weakening of the lithosphere (England 1983). Lithospheric stretching induced by mantle-compensated uplifts in continents with a background deviatoric compression (e.g. East Africa) are less likely to lead to continental splitting after rifting, because the lower deviatoric tension gives rise to lower extensional strain rates and, perhaps therefore, buffering by lithospheric hardening (England 1983). Also, extensional regimes on mantle-compensated plateaux, such as the Basin and Range (Bott and Kusznir 1979) and East Africa, in a background regime of deviatoric compression (Fig. 5B) will be self-buffering, because extension causes lowering of the plateau, and therefore leads to the progressive elimination of the tensional stresses driving the extension before the crust is returned to normal thickness. An uplift compensated in the crust affects stress conditions only in the crust, and thus extension may be shallow-rooted. The Tibetan Plateau is maintained by north–south compression and is extending in an east–west direction. The extension is less than about 5 Ma old and accompanied rapid recent uplift of the Tibetan Plateau. In the Peruvian Andes, shallow extension (Dalmayrac and Molnar 1981) results from crustal thickening. Such extensions are not self-buffering because crustal thickness is continuously maintained by plate convergence. If lithospheric delamination (Bird 1978; Houseman *et al.* 1981) beneath an orogen occurs, shallow tension may give way to lithosphere-penetrating tension in a rapidly uplifted region and, hence, the possibility of lithospheric splitting across a thinned and weaker lithosphere. This may be one reason why continental splitting is commonly roughly along the lines of earlier orogenic belts, such as the opening of the Central and North Atlantic along the Appalachian–Caledonian orogen (Wilson 1966). Another reason for preferential rupture along an earlier orogen has been suggested by Vink *et al.* (1984), and Glazner and Bartley (1985): just as slow strain-rate stretching (or stretching followed by cooling and thermal recovery), harden the lithosphere, lithospheric shortening leads to an increase in the soft continental crustal component and therefore weakening of the lithosphere, particularly where uplift and denudation rates are low and crustal rocks are heated and softened over long periods.

Stretching of uplifts and plateaux has implications for the maximum values of extensional strains expressed by faulting and rock fabrics. Lithospheric stretching at high strain rates is buffered by lithospheric failure and separation at a stretching value (β) of about 4.5 (Dewey 1982) and hence extensional strains seem unlikely to exceed 350 per cent, except perhaps very locally in extensional shear zones. Lithospheric extensions at slow strain rates are buffered both by plateau lowering and by lithospheric hardening where extensional strains are unlikely to exceed 100 per cent. Extensions of crustally-compensated uplifts are buffered by compatibility problems between the extending crust and non-extending mantle, and are unlikely to exceed about 20 per cent.

Localization of extension

Extensional structures such as rift valleys are far commoner within plates than are compressional structures such as thrust belts and inverted rifts, for the following reasons.

1. rocks are weaker in tension than in compression (Fig. 2), and it is easier to generate differential and shearing stress profiles that will overcome lithospheric strength profiles in tension than in compression (Fig. 4).

2. uplifts generate tension and there are no obvious local or regional within-plate mechanisms that generate renewable intraplate compression, apart from that induced by the reduction of plate rigidity during loading (G. D. Karner pers. comm.).

3. in lithospheric deviatoric tension, a gravity regime persists throughout the lithosphere, whereas, in deviatoric compression, regimes may change with depth (Fig. 4).

4. in an extensional regime, hydrothermal circulation in opening cracks persists to greater depth than in compression where cracks tend to close.

5. once extension has started, the work required to continue stretching diminishes, whereas, in compression, more work is done as the crust thickens. As the lithosphere thins, tensional stress increases, and stress amplification in a thinning lithosphere further concentrates tension (Kusznir and Bott 1977).

However, an important result of stretching the lithosphere is that extension and thermal recovery strengthens it, so that stretching is likely to start again in a new place—most likely in an uplift; hence, the continents are permeated with rifts of all ages from Precambrian to present. In contrast, shortening weakens the lithosphere (Vink *et al.* 1984; Glazner and Bartley 1985) and predisposes it to subsequent rifting. Renewed stretching or compressional inversion of earlier rifts may occur where the subsequent deformation follows shortly after an earlier rapid stretching.

Both shortening and extensional structures, but especially the latter, are very common within the continents but rather rare within the oceans; the oceanic lithosphere is clearly more resistant to deformation. In the continents, plate boundaries are wider, more diffuse and complicated, and intra-plate earthquakes, uplifts, and basins are more common (Dewey 1982). The continental lithosphere is more likely to enjoy extensional strain than the oceanic, for the following reasons.

1. most continental strength profiles are weaker than oceanic, except at and near ridge axes.

2. the continents are seamed with zones of weakness and inhomogeneity, including weak orogenic zones (Glazner and Bartley 1985) and recent zones of lithospheric stretching.

3. the continents are the preferential sites of tension-inducing uplift, and plateaux produced by lithospheric shortening, advective thinning of the lithosphere, and crustal underplating (McKenzie, 1984).

4. even where hot-spot advective uplifts or volcanic edifices occur in the oceans they generate insufficient tension to offset ridge push forces, whereas those in the continents can generate deviatoric tension even against a background of deviatoric compression (Crough 1983).

A particularly good example of the relative weakness of continental and strength of oceanic lithosphere is seen in the kinematic and temporal relationship between the Red Sea, the Gulf of Suez, and the Dead Sea Transform. Initially, early Miocene stretching along the Red Sea Zone propagated north to form the Gulf of Suez Rift, which terminated as it met the continent–ocean interface along the north Egyptian continental margin. As lithospheric separation began in the Red Sea, the Gulf of Suez was isolated, and the Dead Sea left-lateral transform boundary developed in a continental weak zone parallel with the Israeli continental margin, to isolate a thin strip of continental lithosphere along the margin of the African Plate. Plate boundaries near continental margins are commonly developed parallel with the margin and just within the continent. Transform examples are the San Andreas Fault, which jumped further into the continent about 5 Ma ago, the North Anatolian Fault in northern Turkey, the Alpine Fault in New Zealand, the Dead Sea Transform, and the strips of continent between Palaeo-Tethys and Neo-Tethys (Fig. 6). Extension leading to lithospheric separation, that isolated a strip of continental lithosphere, occurred to form the Lomonosov Ridge, the Lord Howe Rise, the Seychelles, and the South China Sea fragments (such as Reed Bank), which separated from the Pearl River Basin margin of South China. Clearly, if a plate boundary is to form near a continental margin, it will choose a site just within the continental lithosphere. Thus, as Vink *et al.* (1984) have argued, a rationale is established for continental exotic terranes: strips of continental lithosphere removed by extension or strike-slip from one continental margin, and transported and attached to another (Coney *et al.* 1980). Further reasons why tension and splitting is induced near the edges of continents are that, firstly, subduction of anomalous mantle beneath ridges is likely at continental edges and causes plateau uplift (Gough 1984), and, secondly, tension is induced by mountain uplifts of Andean type at some subducting margins, where the lithosphere also becomes weaker by thermal softening.

The preferential rifting of continental lithosphere has considerable implications for regional petroleum exploration strategy. Most hydrocarbon deposits occur in rift structures, the thermal subsidence basins that may overlie rifts, and in flexurally-controlled foreland basins. These occur dominantly in continental lithosphere. Furthermore, hydrocarbon habitats are less likely to occur in deeper-water regions of strongly stretched continental lithosphere, where intense early heating, early maturation,

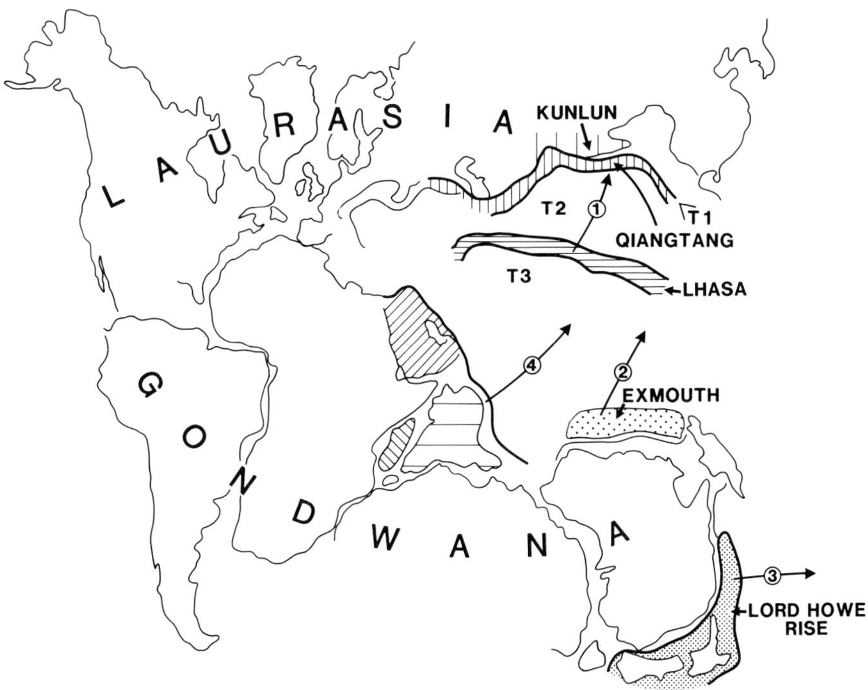

FIG. 6. Palaeotectonic reconstruction of Tethys for the early Jurassic, illustrating the various strips and blocks of continental lithosphere (ornamental) that detached from the Gondwana margin.

overcooking, and escape has occurred in and from rift basins. In deep-water regions of oceanic and severely stretched continental lithosphere, hydrocarbon deposits are likely to be developed only in giant prograding sedimentary wedges, such as the Mississippi and Niger deltas. Within continents and new continental margins, hydrocarbons are more likely to have matured and to have been trapped in rifts with β values of less than two, where volcanism and early, very rapid heating are less likely.

Calculation and buffering of extension

It is not easy to determine the amount of horizontal stretching (β) in extensional regimes. The ideal method is to compare the thickness of the stretched crust, beneath rift sedimentary infill, with the adjacent pre-stretching thickness. However, in most extensional areas, data on the geometry of the Moho are not available, the North Sea being a notable exception where the disposition of the Moho is well determined from seismic data and indicates a β factor of just less than 2.0 (Barton and Matthews 1984; Barton *et al.* 1984). If the pre-stretching thickness of the crust (C_z) and lithosphere (l_z) are known, β values may be determined from subsidence/time curves (McKenzie 1978). In the North Sea, a reasonable fit has been established between subsidence curves and seismically-determined Moho geometry (Wood and Barton 1983). High C_z/l_z ratios (e.g. in orogenic belts) give a large subsidence for a small stretch, while low C_z/l_z values (e.g. in continental shield areas) yield a small subsidence or even uplift for a large amount of extension. Usually however, crustal and lithospheric thicknesses at the time of stretching are exceedingly difficult to determine, thus prohibiting, or at least rendering difficult, the use of subsidence curves to determine β. On the Bay of Biscay continental margin of northwest France, maximum β values of about 4.5 are well-determined for the thinnest, deepest, continental crust from Moho geometry (de Charpal *et al.* 1978). Calculations of extension at this rifted margin, based upon listric fault geometry (Le Pichon *et al.* 1982) yield β values of only about 2.0. This suggests that restoration of the more obvious faults alone yields extension values that are much too low, particularly where the faults and their offsets are picked from seismic profiles.

Wernicke and Burchfiel (1982) have shown that the substantial errors which can arise in calculating extension values, when fault geometry and kinematics are improperly understood, even in well-exposed and well-mapped areas. Seismic profiles are a useful tool not a substitute for field observation of rocks. Gibbs (1984) has drawn attention to the likelihood of very gently-inclined extensional faults in the North Sea, that are extremely difficult to determine from seismic data. Gently-dipping extensional faults or detachments are now widely recognized in the Basin and Range, where they may be the shallow expression of crust-penetrating megashears, upon which very large displacements have occurred (Davis 1983; Wernicke 1981). These and the flat, rotated 'deck of cards' normal faults recognized by Proffett (1977) imply much higher crustal extensions than those predicted by the later, more obvious, relief-producing, steeper normal faults alone. Also, it is clear that sedimentary sequences, which appear virtually flat and 'undeformed' in seismic sections, may have suffered substantial small-scale strains and displacements apart from those involved in uniaxial compaction. For example, the Jurassic rocks of the English Dorset coast between Lyme Regis and Burton Bradstock, as seen in the cliffs, are gently-dipping and cut by a few large faults. Yet these rocks are also affected by small-scale syn-depositional fissuring, extensional faults and, locally more penetrative extensional strains. Therefore, because one can easily miss very low-angle structures and is bound to miss very small structures, estimates of extension from seismic sections can be taken only as a lower limit.

Wernicke *et al.* (1982) have outlined another way by which extension may be estimated. In the southern Great Basin, by adding strike-slip displacements, they estimate extension at about 65 per cent. Similarly, where a rift or extensional zone is terminated by a transform relay, the extensional displacement, from which extensional strain can be determined, should equal the displacement along the transform.

Although there is no limit to extension from a kinematic standpoint, extension is, in practice, limited or buffered in a variety of ways. First, at high strain rates, extension is terminated by lithospheric splitting at $\beta \sim 4.5$, that is about 350 per cent. This is because, at this value, large volumes of basalt rise and cause loss of cohesion in the lithosphere, and because this is the value at which an unloaded subsiding crust reaches 2.6 km below sea-level—the depth of the ridge crest (Dewey 1982). Second, slow strain-rate stretching leads to the slowing and cessation of extension at $\beta \sim 2$ (100 per cent) due to cooling and hardening of olivine in the mantle part of the lithosphere, so that stresses would need to be higher than those naturally possible to continue extension (England 1983). Sawyer (1985), however, argues that inclusion of brittle failure of the upper mantle (Fig. 3) substantially reduces the stresses needed to continue extension at slow strain rates. Many intracontinental rifts with β values of less than 2.0 are free of volcanics, suggesting a buffering mechanism at about 100 per cent extension.

A third mechanism is the lowering of compensated uplifts during stretching, which lowers the tensional stress generated by forces associated with uplift. A 67 per cent extension eliminates a 2 km uplift compensated in the crust and, therefore, the associated tensional stresses. If the lithosphere is in background deviatoric tension, that tension may be sufficient to continue extension. If the lithosphere is in background deviatoric compression of say 30 MPa, a 19 per cent extension reduces the crustal thickness to 42 km, the uplift elevation to 1.2 km, and tension to 30 MPa, thus balancing the compression at or before which stretching will cease and, perhaps, allowing rift inversion.

Fourth, extension may be terminated by plate geometry and boundary configuration changes, such as a nearby continent or arc collision. Also as continents stretch, rupture, and separate across widening oceans, background intra-plate stress changes from deviatoric tension to deviatoric compression.

Evolution of intra-plate geological regimes

Strains within plates are very much smaller than those in plate boundary zones, but are a very subtle barometer of tectonic stress conditions within the plates, and may be of great importance in making and interpreting past plate reconstructions. The global configuration of plates and the nature of their boundaries have the most profound influence on intra-plate geology. At times—notably the late Precambrian, the early Devonian, and the Permo-Triassic—Pangaea-like assembly of the continents has occurred, to be followed by rifting (Sawkins 1976) and continental splitting to dispersed configurations like the present day. The present global configuration of dispersed continents is accompanied by mostly intra-plate deviatoric compression, except in the West Pacific, and is reflected, for example, in the compressional deformation of eastern North America (Bollinger and Sibol 1985; Crone *et al.* 1985; Obermeier *et al.* 1985; Talwani and Cox 1985).

There is evidence in the geological record of broad global Wilson Cycles of continental assembly, breakup, dispersal, and then reassembly with more localized, superimposed Wilson Cycles of oceanic opening and closing. As will be argued, this lends some credence to the view of Stille (1924), that some tectonic events and phases are truly global and that there is a global plate configuration, continent distribution, and stress-based rationale for the integration of a wide variety of phenomena. These phenomena include intra-plate tectonic regimes, sea-level changes, the tectonic styles of island arcs, the timing and distribution of mountain building, unconformities, disconformities, and alkaline and mafic igneous activity.

The assembly of large continental areas, such as Pangaea, involves multiple sequential continental collisions that shorten the continents, enlarge the areas of the oceans and, therefore, cause sea-level to fall (Dewey and Windley 1981). Possibly, the mid-Jurassic regression was caused by the extensive Cimmeride collisions along the northern margin of the Tethyan ocean (Şengör 1984). Sea-level fall is enhanced by a slight continental elevation, caused by elastic thickening due to compression. Collisional assembly is followed by widespread melting of the thickened lower continental crust in thrust-shortened collisional zones, the intrusion of minimum-melting granites and the eruption of high-potash ignimbrites. The rapid, late, morphotectonic uplifts of orogens with associated magmatism in high-level extensional environments are probably related to lithospheric delamination (Bird 1978; Houseman *et al.* 1981). As during the Permo-Trias, following the assembly of Pangaea, enormous continental masses are emergent, with interior arid climates and red beds, and with high denudation rates and extensional tectonics in the mountains.

The gradual cessation of collision-induced deviatoric compression leads to a recovery of intracontinental elastic strain, and a drop in continental elevation. Trench suction forces cause a transition to deviatoric tension, which, with increasing and permanent lithospheric stretching, further reduces continental elevation. The continental enlargement that occurs as this stage develops—with the dumping of large volumes of clastic sediment derived from the denudation of an elevated continental mass with numerous collisional orogens—leads to a progressive rise in sea-level. Thus, during the late Triassic, widespread stretching and rift formation in Pangaea perhaps led to the Rhaetian/Hettangian transgression.

Deviatoric tension caused by trench suction is locally enhanced by mantle-compensated uplifts, perhaps caused by sublithospheric pluming or jetting with advective thinning of the lithosphere of a continental mass, that is stationary or moving only slowly with respect to the flow geometry of the underlying asthenosphere. These uplifts, and the weakened orogenic uplift zones, enhance tension and are the likely preferential sites of extensional tectonics, leading to lithospheric failure. Alkaline intrusions and kimberlites are encouraged by deviatoric tension and are preferentially sited in plateau uplifts and rifts (Bedard 1985). Extension of some uplifts is buffered by crustal compensation and leads to aborted rifts. Increasing deviatoric tension leads to accelerating continental extension accompanied by widespread dyke swarms and the eruption of flood basalts (Cox 1978), immediately followed by lithospheric splitting, tension being enhanced by plateaux generated by basaltic underplating (McKenzie 1984). For example, the Bathonian separation of North America and Africa followed earliest Jurassic eruption of flood basalts, sill intrusion, and dyke injection. Aptian opening of the South Atlantic was preceded by the eruption of the Neocomian Serra Geral Basalts and the intrusion of the huge Ponto Grosso dyke swarm. The separation of India from Madagascar/Seychelles immediately followed the eruption of the Deccan Traps. The opening of the North Atlantic succeeded the widespread eruption of early Tertiary basalts in Greenland and the British Isles. Continental splitting commonly occurs completely across dyke trends and follows the weakest lithospheric zones, which are, most frequently, earlier orogenic belts, such as the opening of the Central and North Atlantic along the Appalachian/Caledonian belt (Wilson 1966). Some attempts to rift and split continents have failed. For example, during the Cretaceous, prior to the separation of Greenland from North America, a substantial zone of alkaline intrusion and rifting developed from southern New England to Hudson Bay including the Monteregian Hills (Bedard 1985), representing part of a widespread region of Aptian extensional tectonics, including basalt eruption as far north-east as the Scotian Shelf (Jansa and Pe-Piper 1985).

As continental separation and oceanic growth proceeds, ridge-push forces become important and continents change from deviatoric tension to deviatoric compression, causing a slowing and cessation of rifting and, as compression increases, the inversion of some favourably-oriented rifts. This leads to the typical situation, observed at and near most mature rifted margins, of non-marine, sometimes inverted (e.g. the Benue Trough in Nigeria) rift sequences with alkaline intrusions and basalts unconformably overlain by little-deformed post-rift marine sequences (Dewey and Bird 1970), the breakup unconformity of Falvey (1974).

Commonly, the time of the breakup unconformity is marked by salt deposits and basalts, the latter indicating the accelerating extension that heralds lithospheric splitting (Cox 1978). Salt sequences, such as the Jurassic Louann of the Gulf of Mexico and the Aptian in the South Atlantic were probably deposited in stretched-continent and ocean-floored basins lying below world sea-level and developed by multiple catastrophic marine spillovers, across tectonic and volcanic barriers in narrow but widening oceans (Burke 1975). Thus, the breakup unconformity is a fundamental, continent-wide, time/stress boundary, and can be recognized in older sequences throughout the geological record (Bond *et al.* 1985).

As continents move into a phase of deviatoric compression, few rifts develop, except on compensated uplifts, and these are unlikely to lead to lithospheric splitting. Also the Jurassic to early Tertiary breakup and dispersal of Pangaea has been accompanied by a general sea-level rise, culminating in the late Cretaceous. This may be partly the result of increased rates of relative plate motion or a greater length of ridge in the oceans, but is certainly enhanced by the stretching of continental area prior to the breakup that occurred throughout the Mesozoic. Deviatoric compression causes continental elevation and intracontinental compressional deformation, such as the shallow wrenching of South America (Assumpcao *et al.* 1985), shallow thrusting and wrenching in eastern North America (Yang and Aggarwal 1981) and deformation in north-west Europe (Ahorner 1975; Burton and Neilson 1980; Turbitt *et al.* 1985). The post-Cretaceous sea-level drop may have been caused or enhanced by increasing global intra-plate compression and consequent continental uplift. Global changes in sea-level, caused by long-term changes in deviatoric tension and compression, may show superimposed shorter-term similar rises and falls caused by effects at subduction zones, such as the collision and attempted subduction of seamounts, seamount chains, oceanic plateaux, and island arcs, and change in the age and buoyancy of the subducting oceanic lithosphere. These shorter time-span effects cause minor falls and rises of continents, and could be one cause of widespread cyclicity where continents lie at or very close to world sea-level, and of more localized cyclicity at continental margins when world sea-level is low.

Some of the notions expressed and relationships explored in this paper may serve as a basis for re-examining the idea of global periodicity as well as episodicity. Although Stille (1924) undoubtedly greatly overplayed the idea of, and data for, world-wide orogenic events, global synchroneity does exist for a number of tectonic events and correlations, such as ophiolite obduction, blueschists and high sea-level during the Ordovician and the Cretaceous. Global transgressions do occur, as in the early Cambrian, early Jurassic and Aptian/Albian. Although the details may be disputed, Vail *et al.* (1977) have argued, from seismically-determined onlap and offlap stratigraphy, a detailed Phanerozoic chronology of sea-level rise and fall. The possible integration of a wide variety of geological phenomena, seen against a background of cyclically changing plate configuration and continental distribution, offers a useful tool in understanding the geological record, by putting flesh on the bones of the Wilson Cycle of ocean opening and closing, and the progressive closure of complex oceans such as Tethys.

ACKNOWLEDGEMENTS: This paper is dedicated to the memory of my dear friend Bert Quennell, whose work on the tectonics of the Levant was an inceptive and silent harbinger of much of our present understanding of rifting, wrenching, plate kinematics, and southern Tethyan evolution. I am indebted to G. D. Karner and W. C. Pitman III for many discussions and for reviewing the manuscript.

References

ADAMS, J. (1982). Stress-relief buckles in the McFarland quarry, Ottawa. *Can. J. Earth Sci.* **19**, 1883–87.

ANDERSON, E. M. (1942). *The dynamics of faulting.* Oliver & Boyd, Edinburgh.

ARTYUSHKOV, E. V. (1973). Stresses in the lithosphere caused by crustal thickness inhomogeneities. *J. Geophys. Res.* **78**, 7675–708.

ASSUMPCAO, M., SUAREZ, G., and VELOSO, J. A. (1985). Fault plate solutions of intraplate earthquakes in Brazil: some constraints on the regional stress field. *Tectonophys.* **113**, 283–93.

BARNETT, R. L., ARIMA, M., BLACKWELL, J. D., WINDER, C. G., PALMER, H. C., and HAYATSUN, A. (1984). The Picton and Varty Lake ultramafic dikes, Jurassic magnetism in the St Lawrence Platform near Belleville, Ontario. *Can. J. Earth Sci.* **21**, 1460–72.

BARTON, P. J. and MATTHEWS, D. H. (1984). Deep structure and geology of the North Sea region interpreted from a seismic refraction profile. *Ann. Geophys.* **2**, 663–8.

BARTON, P. J., MATTHEWS, D. H., HALL, J., and WARNER, M. (1984). Moho beneath the North Sea compared on normal incidence and wide angle seismic recording. *Nature* **308**, 55–6.

BASHAM, P. W. and ADAMS, J. (1984). The Miramichi, New Brunswick Earthquakes: near surface thrust faulting in the Northern Appalachians. *Geosci. Can.* **11**, 115–20.

BEDARD, J. H. (1985). The opening of the Atlantic, the Mesozoic New England Igneous Province and mechanisms of continental breakup. *Tectonophys.* **113**, 209–32.

BELL, J. S. (1985). Offset boreholes in the Rocky mountains of Alberta, Canada. *Geology* **13**, 734–7.

BERGMAN, E. A., NABELEK, J. L., and SOLOMON, S. C. (1984). An extensive region of off-ridge normal-faulting earthquakes in the southern Indian Ocean. *J. Geophys. Res.* **89**, 2425–43.

BERGMAN, E. A., NABALEK, J. L., and SOLOMON, S. C. (1985). Correction to 'An extensive region of off-ridge normal faulting earthquakes in the southern Indian Ocean'. *J. Geophys. Res.* **90**, 2066–67.

BIRD, P. (1978). Initiation of intracontinental subduction in the Himalayas. *J. Geophys. Res.* **83**, 4975–87.

BOLLINGER, G. A. and SIBOL, M. S. (1985). Seismicity, seismic reflection studies, gravity and geology of the central Virginia seismic zone: Part I, Seismicity. *Geol. Soc. Amer. Bull.* **96**, 49–57.

BOND, G. C., CHRISTIE-BLICK, N., KOMINZ, M. A., and DEVLIN, W. J. (1985). An early Cambrian rift to post-rift transition in the Cordillera of Western North America. *Nature* **315**, 742–6.

BOTT, M. H. P. (1981). Crustal doming and the mechanism of continental rifting. *Tectonophys.* **73**, 1–8.

BOTT, M. H. P. (1982*a*). The mechanism of continental splitting. *Tectonophys.* **81**, 301–9.

BOTT, M. H. P. (1982*b*). Origin of lithospheric tension causing basin formation. *Phil. Trans. Roy. Soc. Lond.* **305**, 319–24.

BOTT, M. H. P. and DEAN, D. S. (1973). Stress diffusion from plate boundaries. *Nature* **243**, 339–41.

BOTT, M. H. P. and KUSZNIR, N. J. (1979). Stress distributions associated with compensation plateau uplift structures with application to the continental splitting mechanism. *Geophys. J. Roy. Astron. Soc.* **56**, 451–9.

BOTT, M. H. P. and KUSZNIR, N. J. (1984). The origin of tectonic stress in the lithosphere. *Tectonophys.* **105**, 1–13.

BRACE, W. F. and KOHLSTADT, D. L. (1980). Limits on the lithospheric stress imposed by laboratory experiments. *J. Geophys. Res.* **85**, 6248–52.

BRATT, S. R., BERGMAN, E. A., and SOLOMON, S. C. (1985). Thermo-elastic stress: how important as a cause of earthquakes in young oceanic lithosphere? *J. Geophys. Res.* **90**, 10 249–10 260.

BURKE, K. (1975). Atlantic evaporites found by evaporation of water spilled from Pacific, Tethyan and Southern Oceans. *Geology* **3**, 613–16.

BURTON, P. W. and NEILSON, G. (1980). Annual catalogues of British Earthquakes recorded on LOWNET (1967–1978). *British Geol. Surv. Seismol. Bull.* **7**, 41 pp.

BYERLEE, J. D. (1968). Brittle–ductile transition in rocks. *J. Geophys. Res.* **73**, 4741–50.

CHEN, W. P. and MOLNAR, P. (1983). Focal depths of intracontinental and intraplate earthquakes and their implications for the thermal and mechanical properties of the lithosphere. *J. Geophys. Res.* **88**, 4183–4214.

CONEY, P. J., JONES, D. L., and MANGER, J. W. N. (1980). Cordilleran suspect terranes. *Nature* **288**, 329–33.

COX, K. G. (1978). Flood basalts, subduction and the breakup of Gondwanaland. *Nature* **274**, 47–9.

CRAMPIN, S. (1985). Evidence for aligned cracks in the Earth's crust. *First Break* **3**, 12–15.

CRONE, A. J., MCKEOWN, F. A., HARDING, S. T., HAMILTON, R. M., RUSS, D. P., and ZOBACK, M. D. (1985). Structure of the New Madrid seismic source zone in southeastern Missouri and northeastern Kansas. *Geology* **13**, 547–50.

CROUGH, S. T. (1983). Hotspot swells. *Ann. Rev. Earth Planet. Sci.* **11**, 165–93.

DALMAYRAC, B. and MOLNAR, P. (1981). Parallel thrust and normal faulting in Peru and constraints on the state of stress. *Earth Planet. Sci. Lett.* **55**, 473–81.

DAS, S. and SCHOLZ, C. H. (1983). Why large earthquakes do not nucleate at shallow depths. *Nature* **305**, 621–3.

DAVIS, G. H. (1983). Shear zone model for the origin of metamorphic core complexes. *Geology* **11**, 342–7.

DE CHARPAL, O., GUENNOC, P., MONTADERT, L., and ROBERTS, D. G. (1978). Rifting, crustal attenuation and subsidence in the Bay of Biscay. *Nature* **275**, 706–11.

DENHAM, D. (1985). The Tasman Sea Earthquake of 25 November 1983 and stress in the Australian Plate. *Tectonophys.* **111**, 329–38.

DEWEY, J. F. (1980). Episodicity, sequence and style at convergent plate boundaries. In *The continental crust and its mineral deposits* (ed. D. W. Strangway), pp. 553–73, Geol. Assoc. Can. Spec. Pap. 20, 804 pp.

DEWEY, J. F. (1982). Plate tectonics and the evolution of the British Isles. *J. Geol. Soc. Lond.* **139**, 371–412.

DEWEY, J. F. and BIRD, J. M. (1970). Mountain belts and the new global tectonics. *J. Geophys. Res.* **81**, 2625–47.

DEWEY, J. F. and SHACKLETON, R. M. (1984). A model for the evolution of the Grampian tract in the early Caledonides and Appalachians. *Nature* **312**, 115–21.

DEWEY, J. F. and WINDLEY, B. F. (1981). Growth and differentiation of the continental crust. *Phil. Trans. Roy. Soc. London* **301**, 189–206.

DEWEY, J. F., HEMPTON, M. R., KIDD, W. S. F., SAROGLU, F., and ŞENGÖR, A. M. C. (1986). Shortening of continental lithosphere: the tectonics of eastern Anatolia—a young collisional zone. In *Collision tectonics* (ed. M. P. Coward and A. C. Rise), Geol. Soc. Lond. Spec. Publ. 19.

DEWEY, J. F. and PITMAN, W. C. III, RYAN, W. B. F., and BONNIN, J. (1973). Plate tectonics and the evolution of the Alpine System. *Geol. Soc. Amer. Bull.* **84**, 3137–80.

ENGELDER, T. (1982). Is there a genetic relationship between selected regional joints and contemporary stress within the lithosphere of North America. *Tectonics* **1**, 161–77.

ENGELDER, T. and GEISER, P. (1980). On the use of regional joint sets as trajectories of palaeostress fields during the development of the Appalachian Plateau, New York. *J. Geophys. Res.* **85**, 6319–41.

ENGLAND, P. C. (1983). Constraints on extension of continental lithosphere. *J. Geophys. Res.* **88**, 1145–52.

ERNST, R. E. and HALLS, H. C. (1984). Palaeomagnetism of the Hearst dike swarm and implications for the tectonic history of the Kapuskasing structural zone, northern Ontario. *Can. J. Earth Sci.* **21**, 1499–1506.

FALVEY, D. A. (1974). The development of continental margins in plate tectonic theory. *Austral. J. Petrol. Expl.* **14**, 95–106.

FORSYTH, D. W. and UYEDA, S. (1975). On the relative importance of driving forces of plate motion. *Geophys. J. Roy. Astron. Soc.* **43**, 163–200.

GAY, N. C. (1975). In-situ stress measurements in Southern Africa. *Tectonophys.* **29**, 447–59.

GAY, N. C. (1977). Principal horizontal stresses in Southern Africa. *Pure & Appl. Geophys.* **115**, 3–10.

GEPHART, J. W. and FORSYTH, D. D. (1985). On the state of stress in New England as determined from earthquake focal mechanisms. *Geology* **13**, 70–2.

GIBBS, A. D. (1984). Clyde Field growth faults secondary detachment above basement faults in North Sea. *Am. Assoc. Petrol. Geol. Bull.* **68**, 1029–39.

GLAZNER, A. F. and BARTLEY, J. M. (1985). Evolution of lithospheric strength after thrusting. *Geology* **13**, 42–5.

GOUGH, D. I. (1984). Mantle upflow under North America and plate dynamics. *Nature* **311**, 428–32.

GOUGH, D. I., FORDJOR, C. K., and BELL, J. S. (1983). A stress province boundary and tractions on the North American Plate. *Nature* **305**, 619–21.

HAGER, B. H. (1978). Oceanic plate motions driven by

lithospheric thickening and subducted slabs. *Nature* **276**, 156–9.

Hager, B. H. and O'Connell, R. J. (1979). Kinematic models of large-scale flow in the earth's mantle. *J. Geophys. Res.* **84**, 1031–48.

Hanks, T. C. (1977). Earthquake stress drops, ambient tectonic stresses and stresses that drive plate motions. *Pure & Appl. Geophys.* **115**, 441–58.

Hasegawa, H. S., Adams, J., and Yamazaki, K. (1985). Upper crustal stresses and vertical stress migration in eastern Canada. *J. Geophys. Res.* **90**, 3637–48.

Hickman, S. H., Healy, J. H., and Zoback, M. D. (1985). In situ stress, natural fracture distribution, and borehole elongation in the Auburn geothermal well, Auburn, New York. *J. Geophys. Res.* **90**, 5497–512.

Holst, T. B. (1982). Regional jointing in the northern Michigan Basin. *Geology* **10**, 273–7.

Houseman, G. A., McKenzie, D. P., and Molnar, P. (1981). Convective instability of a thickened boundary layer and its relevance for the thermal evolution of continental convergent belts. *J. Geophys. Res.* **86**, 6115–35.

Isacks, B. and Molnar, P. (1969). Mantle earthquake mechanisms and the sinking of the lithosphere. *Nature* **223**, 1211–24.

Jansa, L. F. and Pe-Piper, G. (1985). Early Cretaceous volcanism on the northeastern American margin and implications for plate tectonics. *Geol. Soc. Am. Bull.* **96**, 83–91.

Karner, G. D. and Dewey, J. F. (1986). Rifting: lithospheric versus crustal extension as applied to the Ridge Basin of Southern California. *Am. Mem. Assoc. Petrol. Geol.* **40**, 317–38.

Karner, G. D., Steckler, M. S., and Thorne, J. A. (1983). Long-term thermomechanical properties of the continental lithosphere. *Nature* **304**, 250–3.

Kirby, S.H. (1980). Tectonic stresses in the lithosphere—constraints provided by the experimental deformation of rocks. *J. Geophys. Res.* **85**, 6353–63.

Kretz, R., Hartry, R., Garrett, D., and Cermignani, C. (1985). Petrology of the Grenville swarm of gabbro dikes, Canadian Precambrian Shield: *Can. J. Earth Sci.* **22**, 53–71.

Kusznir, N. J. and Bott, M. H. P. (1977). Stress concentration in the upper lithosphere caused by underlying visco-elastic creep. *Tectonophys.* **43**, 247–56.

Kusznir, N. J. and Park, R. G. (1982). Intraplate lithosphere strength and heatflow. *Nature* **29**, 540–2.

Lambeck, K., McQueen, H. W. S., Stephenson, R. A., and Denham, D. (1984). The state of stress within the Australian continent. *Ann. Geophys.* **2**, 723–42.

Leary, P. C. (1985). Near-surface stress and displacement in a layered elastic crust. *J. Geophys. Res.* **90**, 1901–10.

Le Pichon, X., Angelier, J., and Sibuet, J. C. (1982). Plate boundaries and extensional tectonics. *Tectonophys.* **81**, 239–56.

Lilwall, R. C. (1976). Seismicity and seismic hazard in Britain. *Inst. Geol. Sci. Seism. Bull.* **4**, 9 pp.

McGarr, A. (1982). Analysis of states of stress between provinces of constant stress. *J. Geophys. Res.* **87**, 9279–88.

McGarr, A. and Gay, N. C. (1978). State of stress in the Earth's crust: *Ann. Rev. Earth Planet. Sci.* **6**, 405–36.

MacGeary, S. and Warner, M. R. (1985). Seismic profiling of the continental lithosphere. *Nature* **317**, 795–7.

McKenzie, D. P. (1978). Some remarks on the origin and development of sedimentary basins. *Earth Planet. Sci. Lett.* **40**, 25–32.

McKenzie, D. P. (1984). A possible mechanism for epeirogenic uplift. *Nature* **307**, 616–18.

McTigue, D. F. and Mei, C. C. (1981). Gravity induced stresses near topography of small slope. *J. Geophys. Res.* **86**, 9268–78.

Meissner, R. and Strehlau, J. (1982). Limits of stresses in continental crust and their relationship to the depth of shallow earthquakes. *Tectonics* **1**, 73–9.

Mithen, D. P. (1982). Stress amplification in the upper crust and the development of normal faulting. *Tectonophys.* **83**, 259–73.

Molnar, P. and Atwater, T. (1978). Interarc spreading and Cordilleran tectonics as alternates related to the age of subducted oceanic lithosphere. *Earth Planet. Sci. Lett.* **41**, 330–40.

Nakamura, K. (1969). Arrangement of parasitic cones as a possible key to regional stress field. *Bull. Volcanol. Soc. Jap.* **14**, 8–20.

Newmark, R. L., Zoback, M. D., and Anderson, R. N. (1984). Orientation of *in situ* stresses in the oceanic crust. *Nature* **311**, 424–8.

Nicholson, C., Simpson, D. W., Singh, S., and Zollweg, J. E. (1984). Crustal structure, velocity inversions, and fault tectonics: results from a microearthquake survey in the New Madrid Seismic zone. *J. Geophys. Res.* **89**, 4545–58.

Nunn, J. A. (1985). State of stress in northern Gulf Coast. *Geology* **13**, 429–32.

Obermeier, S. F., John, G. S., Weems, R. E., Gelinas, R. L., and Rubin, M. (1985). Geologic evidence for recurrent moderate to large earthquakes new Charleston, South Carolina. *Science* **227**, 408–11.

Okal, E. (1984). Intraplate seismicity of the southern part of the Pacific Plate. *J. Geophys. Res.* **8**, 10 053–10 071.

Park, M. J. M. and Westbrook, G. K. (1983). Viscoelastic modelling of lithospheric stresses arising from density contrasts: *Geophys. J. Roy. Astron. Soc.* **74**, 905–14.

Parker, E. C., Davis, P. M., Evans, J. R., Lyer, H. M., and Olsen, K. W. (1984). Upwarp of anomalous asthenosphere beneath the Rio Grande Rift. *Nature* **312**, 354–6.

PROFETT, J. M. (1977). Cenozoic geology of the Yerrington district, Nevada and implications for the nature and origin of Basin and Range faulting. *Geol. Soc. Amer. Bull.* **88**, 247–66.

PULLI, J. J. and TOKSOZ, M. N. (1981). Fault plane solutions for north-eastern United States earthquakes. *Bull. Seism. Soc. Am.* **71**, 1875–82.

RICHARDSON, R. M. and COX, B. L. (1984). Evolution of oceanic lithosphere: a driving force study of the Nazca Plate. *J. Geophys. Res.* **89**, 10 043–10 052.

RICHARDSON, R. M., SOLOMON, S. C., and SLEEP, N. H. (1979). Tectonic stress in the plates. *Rev. Geophys. Space Phys.* **17**, 981–1019.

SAWKINS, F. J. (1976). Widespread continental rifting: some considerations of timing and mechanism. *Geology* **4**, 427–30.

SAWYER, D. S. (1985). Brittle failure in the upper mantle during extension of continental lithosphere. *J. Geophys. Res.* **90**, 3021–25.

SCHMIDT, T. J. (1981). The west European stress field: new data and interpretation. *J. Struct. Geol.* **3**, 309–15.

ŞENGÖR, A. M. C. (1979). Mid-Mesozoic closure of Permo-Triassic Tethys and its implications. *Nature* **279**, 590–93.

ŞENGÖR, A. M. C. (1984). The Cimmeride Orogenic System and the tectonics of Eurasia. *Geol. Soc. Amer. Spec. Pap.* **195**, 82 pp.

ŞENGÖR, A. M. C. (1985). The story of Tethys: how many wives did Okeanos have? *Episodes* **8**, 3–12.

ŞENGÖR, A. M. C. and YILMAZ, Y. (1981). Tethyan evolution of Turkey: a plate tectonic approach. *Tectonophys.* **75**, 181–241.

SMITH, A. G. (1971). Alpine deformation and the oceanic areas of the Tethys, Mediterranean and Atlantic. *Geol. Soc. Am. Bull.* **82**, 2039–70.

SPOHN, T. and SCHUBERT, G. (1982). Convective thinning of the lithosphere: a mechanism for the initiation of continental rifting. *J. Geophys. Res.* **87**, 4669–81.

STECKLER, M. S. (1985). Uplift and extension at the Gulf of Suez: indication of induced mantle convection. *Nature* **317**, 135–9.

STILLE, H. (1924). *Grundfragen der Vergleighenden Tektonik*. Gebrüder Borntraeger, Berlin, 443 pp.

STOCK, J. M., HEALY, J. H., HICKMAN, S. H., and ZOBACK, M. D. (1985). Hydraulic fracturing stress measurements at Yucca Mountain, Nevada, and relationship to the regional stress field. *J. Geophys. Res.* **90**, 8691–706.

TALWANI, P. and COX, J. (1985). Palaeoseismic evidence for recurrence of earthquakes near Charleston, South Carolina. *Science* **229**, 379–81.

TURBITT, T., *et al.* (1985). The North Wales earthquake of 19 July 1984. *J. Geol. Soc. London* **142**, 567–71.

TURCOTTE, D. L. (1974). Membrane tectonics. *Geophys. J. Roy. Astron. Soc.* **36**, 33–42.

TURCOTTE, D. L. (1983). Mechanisms of crustal deformation. *J. Geol. Soc. London* **140**, 701–24.

UYEDA, S. and KANAMORI, H. (1979). Back-arc opening and the mode of subduction. *J. Geophys. Res.* **84**, 1049–61.

VAIL, P. R., MITCHUM, R. M., and THOMPSON, S. (1977). Seismic stratigraphy and global changes of sea level. Part 4: Global cycles of relative changes of sea-level. *Am. Assoc. Petrol. Geol. Mem.* **26**, 83–97.

VETTER, U. R. and RYALL, A. S. (1982). Systematic change of focal mechanism with depth in the western Great Basin. *J. Geophys. Res.* **88**, 8237–50.

VILOTTE, J. P., DIAGNIERES, M., and MADARIAGA, R. (1982). Numerical models of intraplate deformation: simple mechanical models of continental collision. *J. Geophys. Res.* **87**, 10 709–10 728.

VINK, G. E., MORGAN, W. J., and ZHOO, W. L. (1984). Preferential rifting of continents: a source of displaced terranes. *J. Geophys. Res.* **89**, 10 072–10 076.

WARREN, W. E. and SMITH, C. W. (1985). In situ stress estimates from hydraulic fracturing and direct observation of crack orientation. *J. Geophys. Res.* **90**, 6829–39.

WERNICKE, B. (1981). Low angle normal faults in the Basin and Range Province: nappe tectonics in an extending orogen. *Nature* **291**, 645–8.

WERNICKE, B. and BURCHFIEL, B. C. (1982). Modes of extensional tectonics. *J. Struct. Geol.* **4**, 105–15.

WERNICKE, B., SPENCER, J. E., BURCHFIEL, B. C., and GUTH, P. L. (1982). Magnitude of crustal extension in the southern Great Basin. *Geology* **10**, 499–502.

WESNOUSKY, S. G. and SCHOLZ, C. H. (1980). The craton: its effect on the distribution of seismicity and stress in North America: *Earth Planet. Sci. Lett.* **48**, 348–55.

WILLIAMS, H. R., CORKERY, D., and LOREK, E. G. (1985). A study of joints and stress relief buckles in Palaeozoic rocks of the Niagara Peninsula, southern Ontario. *Can. J. Earth Sci.* **22**, 296–300.

WILSON, J. T. (1966). Did the Atlantic close and then reopen? *Nature* **211**, 676–81.

WOOD, R. and BARTON, P. (1983). Crustal thinning and subsidence in the North Sea. *Nature* **302**, 134–7.

YANG, J. P. and AGGARWAL, H. P. (1981). Seismotectonics of Northeastern United States and adjacent Canada. *J. Geophys. Res.* **86**, 4981–98.

ZOBACK, M. D., MOOS, D., MASTIN, L., and ANDERSON, R. N. (1985). Well bore breakouts and *in situ* stress. *J. Geophys. Res.* **90**, 5523–30.

J. F. DEWEY, Department of Earth Sciences, University of Oxford, Parks Road, Oxford OX1 3PR, UK

Basement and cover rock history in western Tethys: HT–LP metamorphism associated with extensional rifting of Gondwana

Robert Hall

ABSTRACT: High-temperature–low-pressure metamorphism of the deep crust is probable during continental lithospheric extension. The 75–65 Ma cooling ages of metamorphic and magmatic rocks preserved in overthrust crystalline slices in the southern Aegean are most plausibly explained by late Cretaceous extension of the Apulian continental margin, rather than by subduction-related magmatism. Metamorphism and magmatism at depth can be correlated with those stratigraphic features of the cover sequences that indicate extension. In particular, the puzzling 'premier flysch' of the region is interpreted as one consequence of peripheral uplift associated with stretching.

Introduction

Many recent models for the evolution of passive continental margins imply the possibility of a metamorphic event recorded by the middle and lower crust, associated with subsidence and probably with lithospheric extension (Sleep 1971; Falvey 1974; McKenzie 1978; Royden *et al.* 1980; Sclater *et al.* 1980). The precise character of the metamorphic rocks formed during such an extensional event will depend on the mechanism(s) of crustal extension and the thermal history of the stretched region (Middleton 1980), for example whether there is a heating event associated with extension, and also on the rate of sedimentation after extension and subsidence (Fig. 1). However, all stretching models imply certain metamorphic features. The rocks produced will be of a low-pressure facies series (Thompson and England 1984; Wickham and Oxburgh 1985) and the metamorphic ages recorded by the rocks will be cooling ages which post-date the stretching event (Oxburgh 1982). The resulting metamorphic fabric could be a regional extensional foliation, a localized foliation associated with deep crustal shear zones, or a combination of the two. Associated folding may occur in zones of relative displacement, such as shear zones.

The metamorphic rocks formed during extension of continental margins are beyond the reach of direct sampling; changes in the physical properties or fabric of the rocks may enable them to be recognized by geophysical methods (Falvey and Middleton 1981). However, they may be identified in former continental margins which have been deformed in mountain belts. Not only can such basement rocks be directly observed in overthrust terrains, but evidence of the extensional history of the margin may also be discernible from interpretation of the sedimentary sequences deposited within the former continental margin. If the basement and cover of a former continental margin can be reconstructed, then an extensional event may be recognizable which is recorded by metamorphic rocks of low-pressure facies, with cooling ages close to, but post-dating, a 'significant' event recorded stratigraphically by cover sediments. Stratigraphic indicators will vary with position in the margin, and will depend on global sea-level, but could include erosional unconformities, marked changes in sedimentation rates, and evidence of faulting where there is no evidence for compressional deformation. Basic igneous activity may occur if stretching proceeds to the point where the crust breaks. If more than one phase of extension occurs during the evolution of the continental margin the last extensional event will probably be the one recorded by the continental basement; if the effects of stretching are uniformly distributed, then early events are likely to be overprinted by later events; although if the effects are heterogeneous then several stretching events could presumably be recorded and preserved by the basement rocks.

Late Cretaceous metamorphism in the southern Aegean

The southern Aegean represents part of the former Apulian continental margin (Bonneau 1982), which was extended in several stages during the Mesozoic. The compressional events, which formed the present orogenic belt linking

From AUDLEY-CHARLES, M. G. & HALLAM, A. (eds) *Gondwana and Tethys* Geological Society Special Publication No. 37, pp. 41–50.

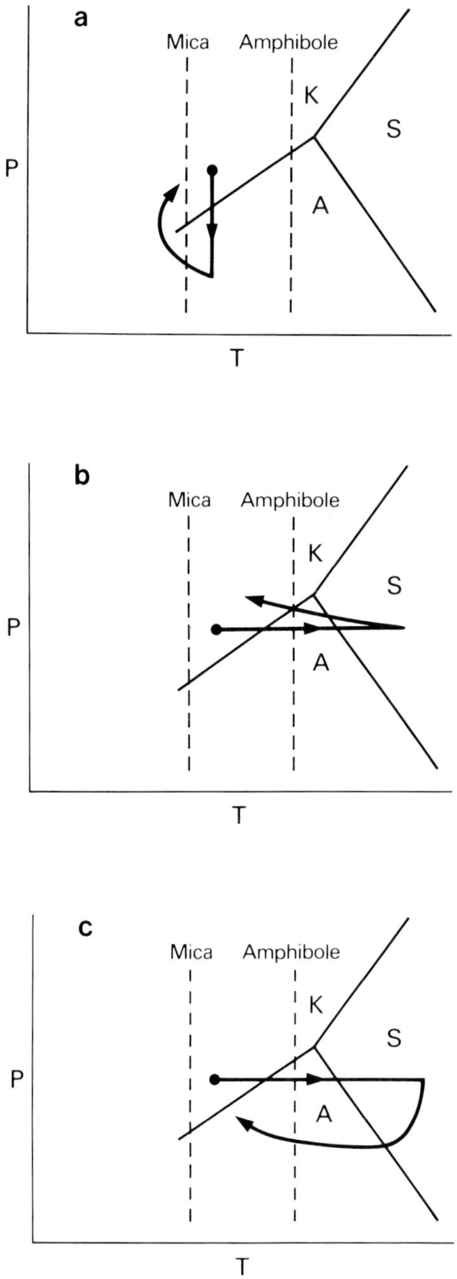

FIG. 1. Schematic pressure–temperature–time paths for rocks metamorphosed during rifting: (a) extension and cooling according to the model of McKenzie (1978), (b) extension associated with magmatism, according to the model of Royden *et al.* (1980), (c) heating followed by erosional thinning of the crust, according to the model of Sleep (1971). The stability fields of the aluminium silicate polymorphs (*A*: andalusite, *K*: kyanite, *S*: sillimanite) are shown for reference. The lines labelled mica and amphibole indicate approximate blocking temperatures for these minerals.

Greece and Turkey, began in the early Tertiary (Aubouin *et al.* 1976; Jacobshagen *et al.* 1978); siliciclastic flysch sediments were deposited during the Palaeogene and orogenic deformation culminated in the Oligocene and early Miocene (Creutzburg and Seidel 1975; Bonneau 1982; Bonneau 1984). Most of the nappes found on the islands of the southern Aegean are composed of carbonate platform, slope, and basin lithologies representing the cover sequences of the extended continental margin, which were detached from their basement and stacked, in two principal deformation events in the late Paleogene (Hall *et al.* 1984). The structurally highest unit in the southern Aegean (Fig. 2) is sporadically preserved on the islands of Crete, Anafi, Nikouria, Donoussa and possibly Syros (Bonneau 1984) and has been described as a *mélange*. It includes slices of metamorphic rocks and granitoids. The metamorphic rocks are of high-temperature–low-pressure facies, and include garnetiferous schists containing cordierite, andalusite and sillimanite; clinopyroxene amphibolites, meta-ultrabasics, calc-silicates and marbles (Bonneau 1972; Seidel *et al.* 1976, 1981; Dürr *et al.* 1978; Reinecke *et al.* 1982; Koepke and Seidel 1984). Metamorphic assemblages indicate maximum pressures of 4–5 kbar at 650–700°C. The metamorphic rocks are closely associated with I- and S-type granitoids and their contact hornfelses. Amphibole ages range from 79 to 66 Ma, whereas mica ages range 75–64 Ma. The intimate association of magmatic and contact metamorphic rocks in the same crystalline complex, and the systematically greater ages of amphiboles compared with micas, indicate that these are cooling ages. Using blocking temperatures of 500°C for amphiboles and 300°C for micas, and a linear cooling rate, indicates that the metamorphic maximum occurred before 78 Ma (Crete) and 72 Ma (Anafi). The beginning of this thermal event can only be estimated, but if the assumption is made that the heating rate to the metamorphic maximum was equal to the cooling rate, then the thermal event recorded began before at least 85 Ma (Crete) or 80 Ma (Anafi) (early Campanian or older).

Previous authors (Bonneau 1982; Reinecke *et al.* 1982) have argued that these late Cretaceous

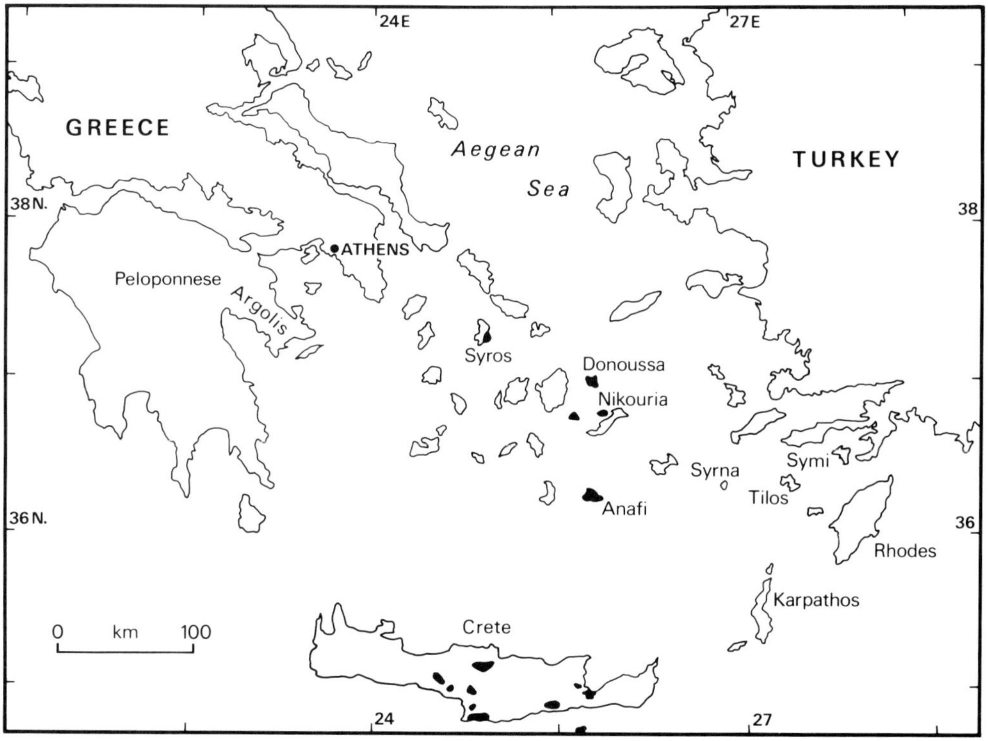

FIG. 2. The southern Aegean region. Principal localities from which high-temperature–low-pressure metamorphic rocks of late Cretaceous age have been reported are shown in black.

metamorphic and magmatic rocks are remnants of a high-temperature belt, generated by subduction processes beneath part of the Pelagonian realm. If they do record such processes, and the subsequent rapid uplift and cooling of an arc terrain as suggested, subduction must have begun before 85–80 Ma in order to subduct sufficient oceanic lithosphere to initiate melting. Reconstructions of the region vary considerably, and the continuation of the Pelagonian zone south-eastwards is still problematical, but all reconstructions now agree that any ocean basins that did exist in the Mesozoic were small (see Robertson and Dixon 1984 for review). In this case the effects of subduction of one of these ocean basins ought to be recorded over a wide area, and it is worth examining the late Cretaceous stratigraphic record of the region for some indication of this event. Associated features expected might include evidence of compressional deformation and calc-alkaline volcanism and deposition of large volumes of volcaniclastic debris eroded from the uplifted arc terrain.

Late Cretaceous stratigraphy of the southern Aegean

The most notable feature of the deep-water basinal sequences in the region is a brief siliciclastic interval of early-late Cretaceous age which interrupted otherwise pelagic carbonate–chert sedimentation. This siliciclastic interval is well known in the Hellenides, where it is known as 'premier flysch' and is considered to be characteristic of the Pindos zone; the presence of greenstone detritus led Aubouin (1965) to link it with the ophiolite complexes of the internal Hellenides, which were emplaced at the beginning of the Cretaceous. For continental Greece and the Peloponnese, Fleury (1970) suggested that an east–west facies variation indicated transport of sediment axially along the Pindos Basin because he saw no evidence to support either an internal or external origin, but he was unable to determine whether transport was from the south or the north. Although the term premier flysch is often not used, the same siliciclastic interval can be recognized throughout the southern Aegean, much further from the presumed source region, on Crete, Syrna, Tilos, Symi, and in south-west Turkey.

Throughout the Aegean, this siliciclastic interval displays a number of puzzling features. In continental Greece and the Peloponnese, the premier flysch is typically Cenomanian–Turonian in age, but may be locally as old as late Albian and as young as Senonian, with no obvious pattern to the beginning and end of siliciclastic sedimentation (Fleury 1980; Thiebault, 1982). In the southern Aegean, the age of this interval is similarly variable (Fig. 3); in west Crete it is dated as Cenomanian (Seidel 1971), in central Crete as Turonian (Bonneau and Fleury 1971, Hussin 1983), and further east as probably Turonian (Bernouilli *et al.* 1974) but locally possibly as old as Cenomanian (Roussos 1978). Considering its supposed axial origin, it shows surprisingly rapid thickness variations in the Peloponnese, is absent in some sections, and most notably is not restricted to the deepest parts of the Pindos Basin; sections recorded by Thiebault (1982) show that it is present at the basin margins but not in the axial parts of the basin, where uninterrupted chert deposition continued throughout the early Cretaceous. The facies variations observed by Fleury (1970) have been confirmed elsewhere in the Peloponnese (Maillot 1973); a change from coarse calcareous sandstones in the west to shales and marls in the east is not easily explained by an axial transport model. Piper and Pe-Piper (1980) interpreted the west to east passage in the Peloponnese from coarse channelled to fine-grained lithologies as a proximal to distal transition, and palaeocurrent evidence indicating eastward transport led them to advocate an external (western) source region for the siliciclastic debris. They suggested that lack of siliciclastic sediments in the Tripolitza carbonate platform sequences of the same age to the west, the principal reason for Fleury's (1970)

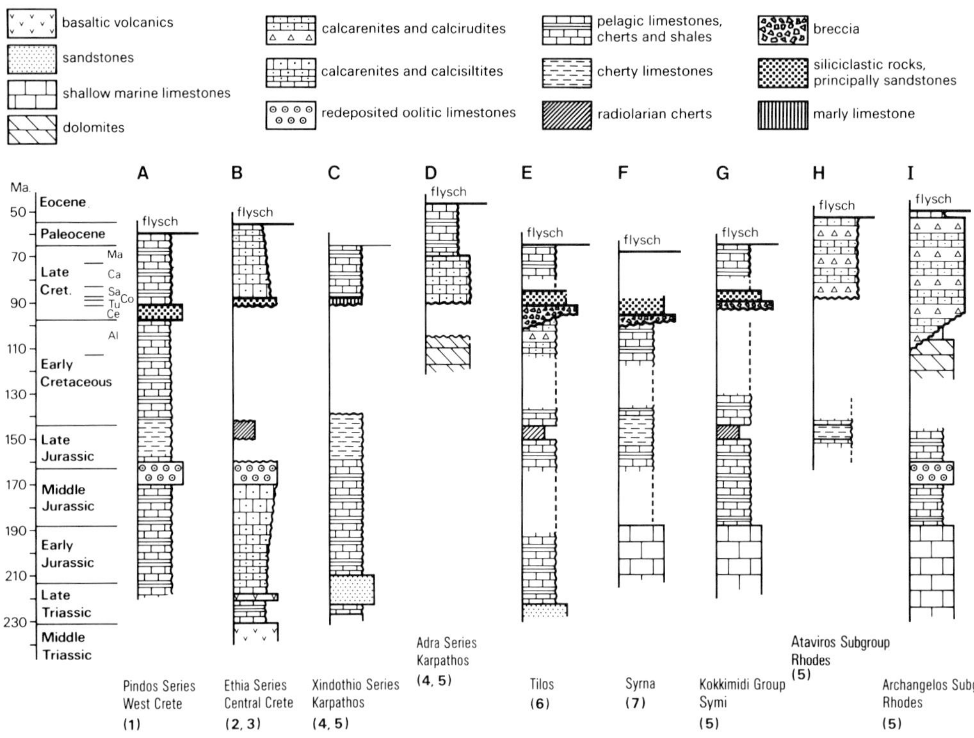

FIG. 3. Schematic stratigraphic sections through platform, slope, and basinal sequences in the southern Aegean. The early late Cretaceous extensional event is marked in different ways in platform sequences (*D*, *I*), slope sequences (*B*, *F*, *G*, *H*) and basinal sequences (*A*, *C*, *E*), principally by a brief siliciclastic interval and/or an unconformity. Based on information from: *1*: Seidel (1971), *2*: Bonneau and Fleury (1971), *3*: Hussin (1983), *4*: Davidson (1974), *5*: Harbury (1986), *6*: Roussos (1978), *7*: Marnelis (1978).

Abbreviations: *Al* Albian, *Ce* Cenomanian, *Tu* Turonian, *Co* Coniacian, *Sa* Santonian, *Ca* Campanian, *Ma* Maastrichtian. Time scale is that of Harland *et al.* 1982.

rejection of a western source region, could be explained by transport of siliciclastic sediments during times of lowered sea-level. They pointed to important disconformities in the Tripolitza limestone sequences in the early late Cretaceous in support of their hypothesis.

Many of the anomalous features of the premier flysch are also evident in the basinal sequences further east in the southern Aegean and Turkey (Fig. 3). In south-west Turkey, Syrna and Tilos, and other small islands of the southern Aegean, a coarse, angular, poorly-sorted breccia up to 10 m thick containing Jurassic and Cretaceous limestones, chert, and locally clasts of diabase, quartz, and mica (Bernouilli *et al.* 1974) marks the base of the siliciclastic interval. On Symi, breccias are common in the lower part of the siliciclastic interval (Harbury 1986). These breccias have a very proximal aspect, which is odd if they are derived either axially from the northern Hellenides or externally from west of the Tripolitza platform; they are more plausibly interpreted as submarine fault-scarp breccias. The overlying flysch sequence consists of graded conglomerates, sandstones, and shales (Bernouilli *et al.* 1974; Harbury 1986), and a common feature is the presence of shallow-water carbonate debris and clasts of continental metamorphics, serpentinites and basaltic rocks; exactly the same type of debris as found to the west on Crete and in the Peloponnese. Volcanic clasts are typically basaltic, and calc-alkaline debris has not been found anywhere. Throughout the southern Aegean the sandstones are relatively coarse grained and not typically distal deposits. As in the Peloponnese, in most sequences there is a return to carbonate deposition in the Senonian, typically pelagic with redeposited beds, and although on Symi and in Turkey the siliciclastic interval is overlain directly by wildflysch of probable Palaeocene age, the presence of pelagic limestones of late Cretaceous age in the wildflysch suggests the same return to carbonate deposition in the Senonian (Thorbecke 1976; Harbury 1986). Also as in the Peloponnese the siliciclastic interval is not identifiable in all basinal sequences; it is missing in the Profitas Ilias Subgroup of Rhodes, although it may be correlatable with Turonian marly limestones of the Xindothio Series on Karpathos (Davidson 1974). Although the siliciclastic interval may be absent, the deepest basinal sequences record renewed extension, and commonly include breccias composed entirely of basinal debris suggesting intrabasinal fault scarp deposits.

On Crete, blocks of pillow lavas (Bonneau 1973) are found in the same *mélange* unit as the metamorphic rocks. The lavas are MORB-like tholeiites (Robert and Bonneau 1982), indicating significant partial melting of the underlying upper mantle and local breaking of the continental margin crust. They are overlain by Senonian pelagic marls. On Crete, Karpathos, and Rhodes, tholeiitic dolerites and gabbros of early-late Cretaceous age (Mutti *et al.* 1970; Baranyi *et al.* 1975; Koepke *et al.* 1985) are exposed, which locally intrude recrystallized carbonates.

In shallower-water regions, carbonate deposition continued throughout the late Cretaceous with no indication of arc volcanism. Areas that remained topographically high throughout the Mesozoic are represented by sequences of platform carbonates; at a time of rising global sea-level, parts of these platforms were uplifted slightly on Rhodes (Harbury 1986), Karpathos (Davidson 1974), and in the Peloponnese (Tsaila-Monopolis 1977) eroding up to a few hundred metres of the early Cretaceous carbonates. Elsewhere the platform sequences record continuous sedimentation in shallow water on Crete (Leppig 1978), or subsidence to subtidal depths on Karpathos (Harbury 1986) on faults indicated by redeposited fault-scarp breccias, channelled calcirudites and calciturbidites. Carbonate slope regions which separated the platforms from basins, and had subsided earlier in the Mesozoic, record an increased input of redeposited shelf carbonate material into basinal areas, indicating local uplift at the basin margins. The siliciclastic interval is missing in the carbonate slope sequences of the region, but on Karpathos and Rhodes there is an abrupt change in sedimentation rate in the early late Cretaceous, and coarse redeposited carbonates of Senonian age rest directly on very early Cretaceous limestones on Rhodes (Mutti *et al.* 1970; Harbury 1986), suggesting erosion and channelling. In some of the redeposited limestones, basaltic igneous and continental metamorphic lithic clasts, similar to those in the deeper basinal sequences, can be found (Hussin 1983).

Compression or extension?

The stratigraphy of the platform, slope, and basinal sequences of the southern Aegean and surrounding regions record evidence of a significant event in the early late Cretaceous. The basinal sequences are typically interrupted by a siliciclastic interval, which is unusual in their otherwise continuous record of carbonate and chert deposition. An unconformity can be recognized in many of the shallower-water carbonate sequences. This is marked by erosion on parts of the platforms, subsidence of other parts of the platforms, and increased input of shallow-water debris on to the slopes flanking the basins. The siliciclastic interval has previously been interpreted as debris eroded and transported axially, from ophiolite nappes emplaced during the Eo-Hellenic orogenic phase in northern Greece. However this hypothesis fails to explain the variable thickness and timing of this interval, local facies variations such as those in the Peloponnese, and the very proximal character of the deposits in areas increasingly far from the supposed original source. It is odd that the Eo-Hellenic mountains, formed at the beginning of the early Cretaceous, should have revived sufficiently to provide debris to the southern Aegean and Turkey, when intervening regions such as the Argolis Peninsula (Bachmann and Risch 1978) had long since received the products of their erosion, and carbonate deposition had resumed. A very complex system of sediment erosion, transport, and distribution is required for this explanation to be satisfactory. Piper and Pe-Piper (1980), who drew attention to some of the anomalous features of the premier flysch interval in the Peloponnese, suggested a western source for siliciclastic sediment, which explains local facies variations but is less attractive when such sediment has to be transported as far east as Turkey. Their proposal that intervals of lowered sea-level allowed channelling of siliciclastic debris across a carbonate platform, is also inconsistent with evidence of a global rise in sea-level in the late Cretaceous after a possible mid-Cretaceous lowering of sea-level (Vail *et al.* 1977; Hallam 1984; Watts and Thorne 1984). In none of the sequences of the region is there evidence for a late Cretaceous episode of compression.

A simpler and more satisfactory explanation is the proposal that the stratigraphy of the region indicates an important episode of extension in the early late Cretaceous. Eastwards of the southern Aegean, the abundant ophiolites are of early-late Cretaceous age or younger. Geochemical arguments have been advanced to suggest that these ophiolites are subduction-related, although there is little stratigraphic or sedimentological support for such an interpretation (Hall 1984; Woodcock and Robertson 1984). I suggest that it was late Cretaceous extension which led to important and widespread ophiolite formation east of Greece, and stratigraphic evidence of this extensional episode is to be expected in adjacent parts of the southern Tethyan margins. Intermittent faulting, over a several million year period early in the late Cretaceous, could account for the variation in age of the siliciclastic interval. Active extensional faults shed debris, which accounts for the locally very proximal character of breccia deposits and also their rapid changes in thickness. Uplift at the basin margins caused erosion of platform and upper slope sequences, in some cases eroding through the carbonate cover to the underlying metamorphic basement. Channelling of this debris from adjacent platform regions explains local facies variations which indicate external source regions. Piper and Pe-Piper (1980) note the similarity of the heavy-mineral assemblages in the premier flysch of the Peloponnese to material from the Phyllite Series, and such a similarity is probable if, as suggested by several authors, the Phyllite Series of the Peloponnese and Crete is the likely basement of the Tripolitza and other platform carbonates of the region. Local uplift of platform regions at basin margins could be a consequence of extension of the platform carbonates on listric faults, allowing subsidence and local uplift of rotated parts of these blocks. An alternative explanation could be local peripheral uplift of thermal origin at basin margins, as predicted by several extensional models (Sclater *et al.* 1980; Royden and Keen 1980; Beaumont *et al.* 1982). Faults associated with extension provided access for basic extrusives, and account for the intrusion of basic rocks into carbonates seen on Crete, Karpathos, and Rhodes. Erosion of the basic igneous rocks which reached the surface, and the basement underlying the platform carbonates in channels at the basin margins, provided siliciclastic debris for the flysch sediments. In the deeper parts of some basins, faulting formed intraformational breccias in the basinal sequences, and locally, stretching proceeded to the point where the continental crust broke and true oceanic crust may have formed (e.g. the Arvi ocean of Bonneau (1984) on Crete). Thermal subsidence and rising sea-level restored carbonate deposition to all areas by the Campanian (?80 Ma).

Metamorphism associated with extension

There is little evidence for subduction and calc-alkaline volcanism in adjacent regions in the early late Cretaceous. No calc-alkaline volcanism is recorded, no calc-alkaline volcanic debris appears in the sedimentary sequences, and there is no evidence of compressional deformation in the region. Radiometric dating of high-P–low-T metamorphic rocks in the region indicates Jurassic, early Cretaceous and Tertiary ages, but no late Cretaceous ages (Dürr *et al.* 1978; Maluski *et al.* 1981). The absence of evidence for subduction and volcanism could be explained by supposing that all of the ocean basins were so small that they were subducted with little or no trace of the normal subduction processes. However, this requires special pleading for practically every postulated ocean in the region, and also an unusually prolonged period of very slow subduction of very small oceans. Instead of a subduction-related origin for the late Cretaceous igneous and metamorphic rocks, I suggest the alternative interpretation that the metamorphic and magmatic rocks record a thermal event associated with lithospheric extension within the Apulian continental margin. The mineral ages represent, not *uplift* and cooling, but *subsidence* and cooling after heating and extension of the margin early in the late Cretaceous (?95–90 Ma; Cenomanian–Turonian). The mineral ages record the point in the late Cretaceous where the extended and heated basement cooled to appropriate blocking temperatures.

The southern Aegean is unusual, in that continental-margin sedimentary cover sequences are still well preserved, while overthrust basement rocks have not yet been entirely eroded off the uplifted orogenic belt. The basement rocks were thrust over their cover in the late Paleogene, to a structurally high position, at an early stage in the compression of the continental margin, and thus escaped the effects of the regional metamorphism associated with collision. It is therefore possible to compare the basement and cover histories, which is not possible in undeformed passive margins or uplifted continental rift zones, where either cover-only or basement-only is seen. The deeper basement rocks of the region are generally not exposed, or where exposed are overprinted by Alpine metamorphism, and therefore the regional extent of this metamorphic event is uncertain. However, discovery of rudists in marbles forming the cover to the Menderes Massif of western Turkey indicate that it formed the basement of a Mesozoic platform sequence, and although overprinted by Alpine metamorphism, one 66 ± 4 Ma muscovite age is reported (Dürr *et al.* 1978). Dürr *et al.* regard this age as somewhat high for Alpine metamorphism, but the position of the Menderes Massif on the edge of the southern Aegean, and its reported high-T–low-P andalusite-bearing rocks (Evirgen and Ataman 1981) give rise to the speculation that the late Cretaceous metamorphic event was more widespread than its preserved relics now suggest.

Conclusions

There appears to be little stratigraphic support for a subduction-related and compressional origin of the metamorphic event recorded by continental basement metamorphic and magmatic rocks. In contrast, the interpretation of these rocks as a product of lithospheric extension is consistent with the regional stratigraphy, with the recent suggestion that continental rifts are a probable location of high-temperature–low-pressure metamorphism (Wickham and Oxburgh 1985), and with suggestions that many of the deep basins within the southern Tethyan margin were narrow (Jenkyns and Winterer 1982) and underlain largely by continental crust (D'Argenio and Alvarez 1980; Hall 1982, 1984). Not until the late Cretaceous extensional phase did many of these deep basins within the continental margin east of the south-east Aegean acquire an oceanic foundation as the Tethyan margins were stretched for the last time. Extension may have begun as early as the Albian, and may have continued until the early Senonian; in the south-east Aegean it seems to have been most important in the Turonian. Small ocean basins developed (probably largely within the continental margin of the southern Tethys) during this final extensional phase, and closed later in the Cretaceous, finally emplacing the oceanic crust as ophiolites. In mainland Greece where small ocean basins had opened and closed much earlier—by the earliest Cretaceous, the extension further east was recorded by the premier flysch. The premier flysch, which is widespread in the Hellenides and recognizable throughout the southern Aegean and in south-west Turkey, and local unconformities in the platform and slope sequences, may be the only record of regional extension in areas where Alpine metamorphism has overprinted earlier events in the basement.

High-T–low-P metamorphic rocks associated with crustal extension will be observable in orogenic belts where collision processes have thrust basement rocks over their former cover sequences. However, to escape regional metamorphism associated with orogenesis, they must be emplaced at high structural levels in the mountain belt, and therefore such rocks will be exposed only for a short period after collision, before they are eroded away. In the Aegean, post-orogenic extension and subsidence during the Neogene has resulted in the preservation of remnants of the higher thrust slices, but elsewhere only the youngest orogenic belts will expose such rocks. One such example is in

Timor, where collision between the Australian continental margin and south-east Asia began only in the last 3.5 Ma, resulting in overthrusting of basement rocks of the former rifted margin on to the Australian margin sedimentary cover rocks (Audley-Charles 1981). High-T–low-P metamorphic rocks with decompressional P–T paths have been described from Timor by Brown and Earle (1983), and like the Aegean rocks have been attributed to metamorphism associated with crustal rifting. Such extremely young orogenic belts may offer the best chance to relate the stratigraphic features of extension to the thermal processes occurring during crustal extension.

ACKNOWLEDGEMENTS: Fieldwork in the Aegean was funded by NERC and the Central Research Fund of London University. I thank Neil Harbury for help and discussion.

References

AUBOUIN, J. (1965). *Geosynclines*. Elsevier, Amsterdam. 335 pp.

AUBOUIN, J., BONNEAU, M., DAVIDSON, J., LEBOULENGER, P., MATESCO, S., and ZAMBETAKIS, A. (1976). Esquisse structurale de l'arc égéen externe: des Dinarides aux Taurides. *Bull. Soc. Géol. Fr.* (7), **18**, 327–36.

AUDLEY-CHARLES, M. G. (1981). Geometrical problems and implications of large scale overthrusting in the Banda Arc–Australian margin collision zone. In *Thrust and nappe tectonics* (ed. K. McClay and N. J. Price). Spec. Publ. Geol. Soc. London, 9, 407–16.

BACHMANN, G. H. and RISCH, H. (1978). Late Mesozoic and Paleogene development of the Argolis peninsula (Peloponnesos). In *Alps, Apennines, Hellenides* (ed. H. Closs, D. Roeder, and K. Schmidt). E. Schweizerbart'sche Verlag, Stuttgart, 424–7.

BARANYI, I., LIPPOLT, H. J., and TODT. (1975). Kalium–Argon-Datierungen an zwei Magmatiten von Kalo Chorio, Nord-Ost-Kreta. *Neues Jb. Geol. Palaeont. Mh.* **5**, 257–62.

BEAUMONT, C., KEEN, C. E., and BOUTILIER, R. (1982). A comparison of foreland and rift margin sedimentary basins. *Phil. Trans. Roy. Soc. London* **A305**, 295–317.

BERNOULLI, D., DE GRACIANSKY, P. C., and MONOD, O. (1974). The extension of the Lycian nappes (SW Turkey) into the Southeastern Aegean islands. *Eclogae Geol. Helv.* **67**, 4–90.

BONNEAU, M. (1972). Existence d'un lambeau de cristallin chevauchant sur la série du Pinde en Crète moyenne (Grèce). *C. R. Acad. Sci., Paris* **D274**, 2133–36.

BONNEAU, M. (1973). Les différentes 'séries ophiolitifères' de la Crète; une mise au point. *C. R. Acad. Sci., Paris* **D276**, 1249–52.

BONNEAU, M. (1982). Evolution géodynamique de l'arc égéen depuis le Jurassique supérieur jusqu'au Miocène. *Bull. Soc. Géol. Fr.* (7), **24**, 229–42.

BONNEAU, M. (1984). Correlation of the Hellenide nappes in the south-east Aegean and their tectonic reconstruction. In *The geological evolution of the eastern Mediterranean* (ed. J. E. Dixon and A. H. F. Robertson), pp. 499–516, Spec. Publ. Geol. Soc. London, 17.

BONNEAU, M. and FLEURY, J.-J. (1971). Précisions sur la série d'Ethia (Crète, Grèce): éxistence d'un premier flysch mésocrétacé. *C. R. Acad. Sci., Paris* **D272**, 1840–42.

BROWN, M. and EARLE, M. M. (1983). Cordierite-bearing schists and gneisses from Timor, eastern Indonesia: P–T conditions of metamorphism and tectonic implications. *J. Metam. Geol.* **1**, 183–203.

CREUTZBURG, N. and SEIDEL, E. (1975). Zum Stand der Geologie des Präneogens auf Kreta. *Neues Jb. Geol. Palaeont., Abhl.* **149**, 363–83.

DAVIDSON, J. (1974). *Contribution à l'étude géologique de l'arc égéen: l'île de Karpathos (Dodecanese méridional, Grèce)*. Thèse 3eme cycle, Université de Pierre et Marie Curie, Paris.

D'ARGENIO, B. and ALVAREZ, W. (1980). Stratigraphic evidence for crustal thickness changes on the southern Tethyan margin during the Alpine cycle. *Geol. Soc. Am. Bull.* **91**, 681–9.

DÜRR, S., ALTHERR, R., KELLER, J., OKRUSCH, M., and SEIDEL, E. (1978). The Median Aegean crystalline belt: stratigraphy, structure, metamorphism, magmatism. In: *Alps, Apennines, Hellenides* (ed. H. Closs, D. Roeder, and K. Schmidt) pp. 455–77, E. Schweizerbart'sche Verlag, Stuttgart.

EVIRGEN, M. and ATAMAN, G. (1981). Study of metamorphism of the central Menderes Massif: isograds, pressure and temperature. *Yebilimleri (Hacatepe Univ.)* **7**, 15–26.

FALVEY, D. G. (1974). The development of continental margins in plate tectonic theory. *Austral. J. Petrol. Expl.* **14**, 95–106.

FALVEY, D. G. and MIDDLETON, M. F. (1981). Passive continental margins: evidence for prebreakup deep crustal metamorphic subsidence mechanism. *Oceanologica Acta* 103–14.

FLEURY, J.-J. (1970). Sur les modalités d'installation du flysch du Pinde au passage Crétacé–Eocène (Grèce continental et Péloponnèse septentrional). *Bull. Soc. Géol. Fr.* (7), **12**, 1110–17.

FLEURY, J.-J. (1980). Les zones de Gavrovo–Tripolitza et du Pindos–Olonos (Grèce continentale et Péloponnèse du Nord). Evolution d'une plat-forme et d'un bassin dans leur cadre structurale. *Soc. Géol. Nord, Spec. Publ.* **4**, 651 pp.

HALL, R. (1982). Ophiolites and passive continental margins. *Ofioliti* **7**, 279–98.

HALL, R. (1984). Ophiolites—figments of oceanic lithosphere? In *Ophiolites and oceanic lithosphere* (ed. I. G. Gass, S. J. Lippard, and A. W. Shelton), pp. 393–403, Spec. Publ. Geol. Soc. London 13.

HALL, R., AUDLEY-CHARLES, M. G., and CARTER, D. J. (1984). The significance of Crete for the evolution of the Eastern Mediterranean. In *The geological evolution of the eastern Mediterranean* (ed. J. E. Dixon and A. H. F. Robertson), pp. 499–516, Spec. Publ. Geol. Soc. London 17.

HALLAM, A. (1984). Pre-Quaternary sea-level changes. *Ann. Rev. Earth Planet. Sci.* **12**, 205–43.

HARBURY, N. A. (1986). The stratigraphy and structural evolution of the SE Aegean. Ph.D. Thesis, University of London.

HARLAND, W. B., COX, A. V., LLEWELLYN, P. G., PICKTON, C. A. G., SMITH, A. G., and WALTERS, R. (1982). *A geologic time scale*. Cambridge University Press, 131 pp.

HUSSIN, A. H. (1983). Stratigraphy and sedimentology of the Pindos Nappe in the Asteroussia Ori, Crete. Ph.D. Thesis, University of London.

JACOBSHAGEN, V., DÜRR, ST., KOCKEL, F., KOPP, K.-O., KOWALCZYK, G., BERKHEMER, H., and BUTTNER, D. (1978). Structure and geodynamic evolution of the Aegean region. In *Alps, Apennines, Hellenides* (ed. H. Closs, D. Roeder, and K. Schmidt), pp. 537–64. E. Schweizerbart'sche Verlag., Stuttgart.

JENKYNS, H. C. and WINTERER, E. L. (1982). Palaeoceanography of Mesozoic ribbon radiolarites. *Earth Planet. Sci. Lett.* **60**, 351–75.

KOEPKE, J. and SEIDEL, E. (1984). Oberkretazisches Kristallin an der Basis von Ophiolithen der Sudägais: charakterisierung der Metamorphoze-Fazies. *Tscher. Mineral. Petrogr. Mitt.* **33**, 263–86.

KOEPKE, J., KREUZER, H., and SEIDEL, E. (1985). Ophiolites in the southern Aegean arc (Crete, Karpathos, Rhodes)—linking the ophiolite belts of the Hellenides and Taurides. *Ofioliti* **10**, 343–54.

LEPPIG, U. (1978). Microfauna in the Tripolitza limestone in northern middle Crete. In *Alps, Apennines, Hellenides* (ed. H. Closs, D. Roeder, and K. Schmidt), pp. 452–4, E. Schweizerbart'sche Verlag., Stuttgart.

MAILLOT, H. (1973). *Etude faciologiques, sedimentologiques, ébauches d'études stratonomiques et séquentielles de la Formation d'Andritsena (Péloponnèse occidental, Grece)*. Thèse 3eme cycle, Université de Lille.

MALUSKI, H., VERGELY, P., BAVAY, D., BAVAY, P., and KATSIKATSOS, G. (1981). $^{39}Ar/^{40}Ar$ dating of glaucophanes and phengites in southern Euboa (Greece) geodynamic implications. *Bull. Soc. Géol. Fr.* (7), **23**, 469–76.

MARNELIS, P. (1978). *Contribution à l'étude géologique de l'arc égéen: Les îles d'Astypalea et de Syrna (Dodecanese, Grèce)*. Thèse 3me cycle, Université de Pierre et Marie Curie, Paris.

MCKENZIE, D. P. (1978). Some remarks on the development of sedimentary basins. *Earth Planet. Sci. Lett.* **40**, 25–32.

MIDDLETON, M. F. (1980). A model of intra-cratonic basin formation, entailing deep crustal metamorphism. *Geophys. J. Roy. Astron. Soc.* **62**, 1–14.

MUTTI, E., OROMBELLI, G., and POZZI, R. (1970). Geological studies on the Dodecanese Islands. (Aegean Sea). Geological map of the Island of Rhodes (Greece), explanatory notes. *Ann. Géol. Pays Helleniques* **22**, 77–226.

OXBURGH, E. R. (1982). Heterogeneous lithospheric stretching in the early history of orogenic belts. In *Mountain building processes* (ed. K. J. Hsü), pp. 85–93, Academic Press, New York.

PIPER, D. J. W. and PE-PIPER, G. (1980). Was there a Western (external) source of terrigenous sediment for the Pindos zone of the Peloponnese (Greece)? *Neues Jb. Geol. Palaeont. Mh.* **2**, 107–15.

REINECKE, T., *et al.* (1982). Remnants of a Late Cretaceous high temperature belt on the island of Anafi (Cyclades, Greece). *Neues Jb. Geol. Palaeont., Abhl.* **145**, 157–82.

ROBERT, U. and BONNEAU, M. (1982). Les basaltes des nappes du Pinde et d'Arvi (Crete) et leur signification dans l'évolution géodynamique de la Mediterranée orientale. *Ann. Géol. Pays Helleniques* **31**, 373–408.

ROBERTSON, A. H. F. and DIXON, J. E. (1984). Introduction: aspects of the geological evolution of the Eastern Mediterranean. In *The geological evolution of the eastern Mediterranean* (ed. J. E. Dixon and A. H. F. Robertson), pp. 1–74, Spec. Publ. Geol. Soc. London 17.

ROUSSOS, N. D. (1978). *Contribution à l'étude géologique de l'arc égéen: l'île de Tilos (Dodecanèse, Grèce)*. Thèse 3eme cycle, Université de Pierre et Marie Curie, Paris.

ROYDEN, L. and KEEN, C. E. (1980). Rifting processes and thermal evolution of the continental margin of eastern Canada determined from subsidence curves. *Earth Planet. Sci. Lett.* **51**, 343–61.

ROYDEN, L., SCLATER, J. G., and VON HERZEN, R. P. (1980). Continental margin subsidence and heat flow: important parameters in formation of petroleum hydrocarbons. *Bull. Am. Assoc. Petrol. Geol.* **64**, 173–87.

SCLATER, J. G., ROYDEN, L., HORVATH, F., BURCHFIEL, B. C., SEMKIN, S. and STEGENA, L. (1980). Subsidence and thermal evolution of the intra-Carpathian basins. *Earth Planet. Sci. Lett.* **51**, 139–62.

SEIDEL, E. (1971). Die Pindos-Serie in West-Kreta, auf der Insel Gavdos und im Kedros-Gebiet (Mittel-Kreta). *Neues Jb. Geol. Palaeont., Abhl.* **137**, 443–60.

SEIDEL, E., OKRUSCH, M., KREUZER, H., RASCHKA, H., and HARRE, W. (1976). Eo-alpine metamorphism in the uppermost unit of the Cretan nappe system—petrology and geochronology. Part 1. The Lendas area (Asterousia Mountains). *Contrib. Mineral. Petrol.* **57**, 259–75.

SEIDEL, E., OKRUSCH, M., KREUZER, H., RASCHKA, H., and HARRE, W. (1981). Eo-alpine metamorphism in the uppermost unit of the Cretan nappe system—petrology and geochronology. Part 2. Synopsis of high-temperature metamorphics and associated ophiolites. *Contrib. Mineral. Petrol.* **76**, 351–61.

SLEEP, N. H. (1971). Thermal effect of the formation of Atlantic continental margins by continental break-up. *Geophys. J. Roy. Astron. Soc.* **24**, 325–39.

TSAILA-MONOPOLIS, S. (1977). Micropalaeontological study of the Tripolitza (Gavrovo) zone in the Peloponnesus. *Geol. Geophys. Res., Inst. Geol. Subsurface Res., Athens* **20**, 1–99.

THIEBAULT, F. (1982). Evolution géodynamique des Hellenides externes en Péleponnèse méridional (Grèce). *Soc. Géol. Nord, Spec. Publ.* **6** (574 pp.).

THOMPSON, A. B. and ENGLAND, P. C. (1984). Pressure–temperature–time paths of regional metamorphism II. Their inference and interpretation using mineral assemblages in metamorphic rocks. *J. Petrol.* **25**, 929–55.

THORBECKE, G. (1976). Zur Geologie der Insel Simi/Griechenland. *Neues Jb. Geol. Palaeont., Mh.* **1**, 31–40.

VAIL, R. P., *et al.* (1977). Seismic stratigraphy and global changes of sea level. *Mem. Am. Assoc. Petrol. Geol.* **26**, 49–50.

WATTS, A. B. and THORNE, J. (1984). Tectonics, global changes in sea-level and their relationship to stratigraphical sequences at the US Atlantic continental margin. *Mar. Petrol. Geol.* **1**, 319–39.

WICKHAM, S. M. and OXBURGH, E. R. (1985). Continental rifts as a setting for regional metamorphism. *Nature* **318**, 330–3.

WOODCOCK, N. H. and ROBERTSON, A. H. F. (1984). The structural variety in Tethyan ophiolite terrains. In *Ophiolites and oceanic lithosphere* (ed. I. G. Gass, S. J. Lippard, and A. W. Shelton), pp. 321–30, Spec. Publ. Geol. Soc. London 13.

ROBERT HALL, Department of Geological Sciences, University College London, Gower Street, London WC1E 6BT, UK.

Holocene serial folding in the Zagros

C. D. Mann and C. Vita-Finzi

ABSTRACT: In the south–east Zagros Mountains of Iran, deformation during the Quaternary has occurred by serial folding, with the mountain front advancing westwards at 17 mm/yr. Coastal uplift during the Holocene indicates a horizontal slip of about 30 mm/yr. The geometry of the locked and nascent folds indicates compression by a horizontal force equivalent to between 0.134 and 0.790 kbar according to which model is adopted.

Although there is general agreement about the broad pattern of movement displayed by the major lithospheric plates during the last 10^8 years, the mechanisms responsible for the movements remain uncertain. One reason is the lack of measurements of deformation over the 10^3–10^4 year range, which, by linking the kinematic narrative derived from palaeomagnetic data with the instantaneous results of seismology, would permit the forces at work to be estimated.

An area which is beginning to yield useful data is the Zagros Range, which borders the Persian Gulf and the lower Mesopotamian Plains over a distance of about 2000 km (Fig. 1). Although its formation dates mainly from Mio–Pliocene times, movement is thought to have begun in Jurassic–Lower Cretaceous times along the north–east margin of the belt and to have migrated progressively towards the south–west. Bird (1978) has used surface deformation, earthquake locations and fault plane solutions to determine the forces responsible for the Zagros orogeny. The present study places the emphasis on Holocene patterns of deformation.

Early attempts to apply the concepts of plate tectonics to the area ascribed the Persian Gulf to a downwarp in the Arabian Plate (Le Pichon *et al.* 1973). Arabia was thought to have undergone subduction beneath the stable Persian Plate until late Cretaceous times, with later northward movement being accompanied by thrusting and folding of the marginal deposits (Haynes and McQuillan 1974). Subsequently, a detailed microearthquake survey of part of the Zagros detected no shocks deeper than 20 km, and thus yielded no evidence of oceanic lithosphere under the fold belt (Niazi *et al.* 1978; Jackson and McKenzie 1984).

The stability of Arabia is often taken for granted. Hancock *et al.* (1984) have recently shown that Neogene collision on the north–east margin gave rise to north–east-striking extensional structures within the platform, but the effect appears to be largely confined to joints. Granted that an Arabian plate extends as far as the Main Zagros Reverse Fault, to what extent are the overlying Mesozoic and Cainozoic units continuous across the Gulf? Limestones correlative with the Musandam Limestone are reported at a depth of 4 km on Qeshm Island (Falcon 1973). There are numerous parallels in the stratigraphy of the two shores of the Gulf. The Aptian Dariyan formation of Iran, for example, is regarded as an extension of the Shu'aiba formation of Iraq, Kuwait, and Saudi Arabia (James and Wynd 1965).

On the other hand, surface structure in so far as it is reflected in topography and bathymetry indicates an 'Iranian' province separate from an 'Arabian' one. The margin of the 'simply folded' portion of the Zagros is commonly drawn parallel to the Iranian coast, and at most 50 km off it, so as to include Qeshm and the other major islands (see for example fig. 1 in Falcon 1974). The major offset in the alignment of the Gulf bathymetric axis, which is thereby displaced north–east relative to the median line, has even been interpreted as a large Zagros anticline (Kassler 1973). An east–west seismic line in the Strait of Hormuz reveals a basement ridge whose form and gravity signature are consistent with its being the northward extension of the Musandam Peninsula of Oman (White and Ross 1979), but it has so far not been traced into Iran. The N–S structural trends of Qatar and Bahrain are regarded as distinctively Arabian (Kassler 1973).

The contrast between the two shores of the Gulf is heightened by the pattern of seismicity, which for the period of record shows practically no activity in Arabia and a good

From AUDLEY-CHARLES, M. G. & HALLAM, A. (eds) *Gondwana and Tethys* Geological Society Special Publication No. 37, pp. 51–9.

deal of it within the Simple Folded zone of Falcon (1974) (see Fig. 1). According to Jackson and McKenzie (1984) most of the larger earthquakes nucleate at depths of 10–20 km, that is below the sedimentary layers but within the crust. Microearthquakes apparently occur within the sedimentary layers.

The shortening embodied in the Zagros folds is thought by Falcon (1974) to be about 50 km since the Miocene, 'about the same as that postulated during the separation of Arabia and Africa during the formation of the axial trough of the Red Sea'. Modelling of the structure by Bird *et al.* (1975) gave a value of about 75 km in the last 1.1 Ma. According to Falcon (1974), north-easterly movement of Arabia in conjunction with basement faulting satisfactorily explained the evolution of the Zagros belt. Lees (1952) likewise expressed the view that each of the simple anticlines of the Zagros owed its existence to a separate basement flexure. Jackson (1980) subsequently suggested that the continental margin represented by the Zagros underwent stretching and subsidence in early Mesozoic times in response to normal faulting and that reversed motion on the faults could explain the depth and nature of present-day seismicity in the region. Folding might then have proceeded as in Algeria during the El Asnam earthquake (King and Vita-Finzi 1981).

Palaeomagnetic evidence from the floor of the Red Sea and Gulf of Aden has revealed two major phases of rifting. The first lasted from about 30 Ma to 15 Ma ago. After a quiet period, spreading resumed 5 Ma ago and is still in progress (Girdler and Styles 1978). Recent work suggests that the emplacement of oceanic crust begins at regularly spaced hotspots and has propagated northwards (Bonatti 1985). The current spreading axis trends N110°E. The convergence across the Zagros, averaged over the last 1.5 Ma and allowing for the motion of Africa relative to North America is put at about 47 mm/yr about a pole near 25°N 37°W (Bird *et al.* 1975).

Rates of deformation

The rate of shortening on the Zagros coast during the Holocene has been derived by

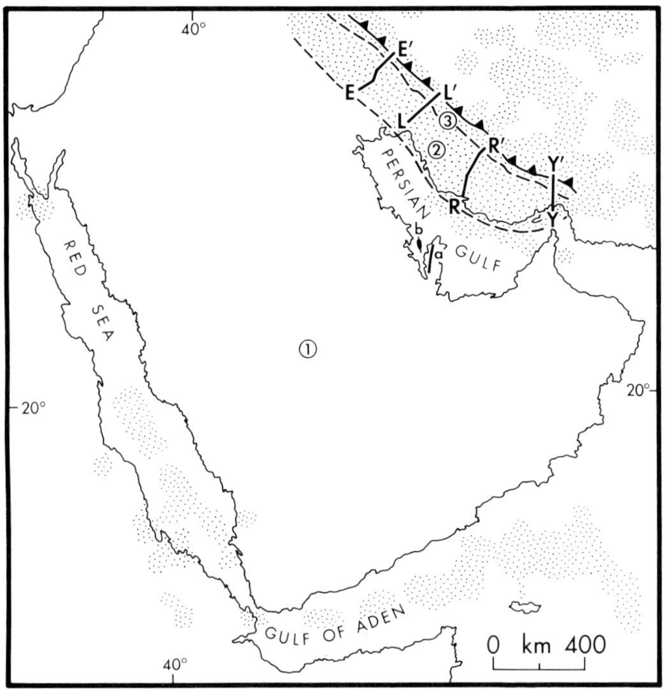

FIG. 1. Location of major regions discussed in article. The labelled lines show position of sections in Fig. 2. Toothed line represents the Zagros Main Thrust Line. (*1*) = Arabia, (*2*) = Simple Folded Zone of Falcon (1974), (*3*) = imbricated belt. Stipple = areas where earthquakes were recorded 1900–1981 (after Tübingen Atlas des Vordenen Orients, (1984), Reichert, Wiesbaden, Sheet A II 3).

geometric methods from the uplift history of three anticlines bordering the Strait of Hormuz (Vita-Finzi 1987; Mann and Vita-Finzi 1982). The resultant slip rate is 29.1 mm/yr on a bearing of N33°E after empirical sea-level correction. A reverse fault inland with a displacement of 3.5 m indicates additional shortening by 0.5 mm/yr during the Holocene on an azimuth of N35°E (Vita-Finzi and Ghorashi 1978). A total of 30 mm/yr is used in the present study.

Folding is taken to be serial (Price 1975; Shearman 1976) and the coastal anticlines are thought to take up most of the shortening. Although it is difficult to show, in the absence of a suitable reference level, that the folds inland are not deforming, structural cross-sections at various points along the belt show an inner, tightly folded belt and an outer belt in which folds are slight or absent. Measurement by Mr B. Walker of fold wavelength and amplitude depicted on the BP (1956) sections show a remarkable consistency in these dimensions (Table 1). Once the folds have locked, deformation may proceed by faulting, as in the case of the reverse fault already mentioned and, of course, the complex of reverse faults which make up the Zagros Main Thrust. The frontal fold can be seen in Section E, Fig. 2. The other three sections are truncated where they reach the coast, and are included to show that the folds are broadly consistent in their geometry at depth.

TABLE 1. *Dimensions of Zagros folds (km)*

Section*	λ	h_L	Section	λ	h_L
A	10	3.0	M	7.5	2.1
B	8	1.6	N	6	1.7
C	9	1.8	O	11.5	2.5
D	7.5	1.7	P	10.5	2.5
E	9.5	2.0	Q	13	2.5
F	10.5	3.5	R	13.5	3.0
G	13	3.0	S	12	2.3
H	18	2.7	T	16.5	4.0
I	15	3.9	U	11	3.6
J	7.5	2.4	V	13	2.8
K	7.5	2.3	W	15	5.4
L	8.5	2.0			
$\bar{x}$	10.6	2.6			

*The values for each section (some of the sections are shown in Fig. 1) represent averages for all the folds on that section. After B. Walker (*pers. comm.*) based on BP (1956).

The fold belt was initiated in the early Pliocene, some 12 Ma ago, and the folded belt is about 200 km wide (Falcon 1974). If deformation has indeed been serial, these facts indicate a south-westward advance of the folding front at about 17 mm/yr. With an average anticlinal wavelength of 10.6 km, the 'life cycle' of a single fold is 6×10^5 yr. Disregarding inevitable variations and the effects of erosion, the uplift rate at the crest required to attain the average anticlinal amplitude (h_L) of 2.7 km is ~4 mm/yr, consistent with the flank uplift rates of 1.8–6.6 mm/yr (Vita-Finzi 1982) observed on the coast. A rounded value of 5 km will be used for h_0 ($=2\,h_L$ in Fig. 3b).

Sustaining forces

In view of the incompleteness and unevenness of the seismic record and the inhomogeneity of the rocks involved, we have based our estimate of the driving force not on seismic moment but on the force required to deform and sustain the frontal fold. The remainder of the Zagros at any time t is deemed to be locked in position and, though subject to seismicity and the reactivation of existing faults, not undergoing significant new deformation (Vita-Finzi 1986).

Three simple associated models have been devised which relate the rate of uplift at the Zagros to a horizontal drive across the Arabian Plate. The basic principle is that the rate of working by the horizontal force is equated to the rate of change of potential energy where only gravity is involved and *not* viscous or elastic forces. The horizontal force is thus a lower estimate for a possible true horizontal driving force.

Basic uniform advancing front

The primitive concept is that of a plate of height t_0 moving with velocity u_0 towards a front of height $(h_0 + t_0)$, which is moving in the opposite direction with velocity v_a (Fig. 3a). Material of density ρ is first hoisted up to the plane at A, and then again at the front B, where in due course it is locked in place.

Total potential at time t is:

$$\phi(t) = \frac{1}{2}\int_0^{x_L} (t_0 + h_0)^2 \rho g \, dx + \frac{1}{2}\int_{x_L}^{x_r} t_0^2 \rho g \, dx.$$

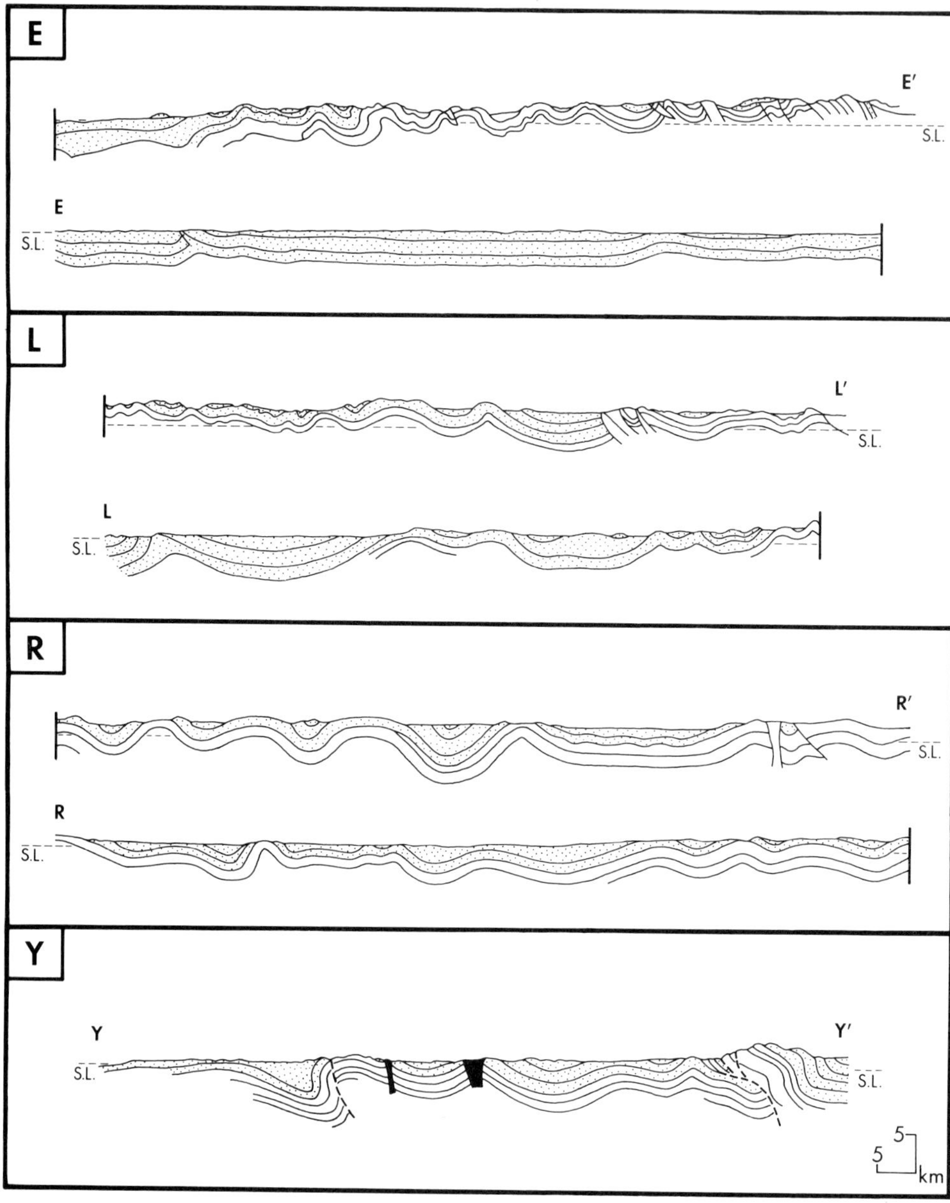

FIG. 2. Simplified versions of sections E, L, and R given in BP (1956), and cross-section (Y) from Sheet G-40 C and D of Iran 1/250 000 Series (BP 1963). Stratigraphic data have been omitted for the sake of clarity. Stipple = Tertiary, black = salt plugs, $S.L.$ = sea level.

Thus:

$$\phi(t+\Delta t) = \frac{1}{2}\int_0^{x_L+v_a\cdot\Delta t}(t_0+h_0)^2\rho g\,\mathrm{d}x + \frac{1}{2}\int_{x_L+v_a\Delta t}^{x_r} t_0^2\rho g\,\mathrm{d}x.$$

Whence:

$$\Delta\phi = \tfrac{1}{2}(t_0+h_0)^2\rho g v_a\,\Delta t - \tfrac{1}{2}t_0^2\rho g v_a\,\Delta t.$$

Thus:

$$\frac{\mathrm{d}\phi}{\mathrm{d}t} = \tfrac{1}{2}\rho g h_0(2t_0+h_0)v_a.$$

Rate of change of potential energy is equivalent to the rate of working of an external horizontal force X, i.e.

$$Xu_0 = \tfrac{1}{2}\rho g h_0(2t_0+h_0)v_a.$$

However, the rate of working ($\frac{1}{2}\rho g t_0{}^2 y_0$) required to *produce* the plate at A is intimately related to the formation of the plate at the Red Sea, and is to be subtracted from Xu_0 above.
Thus

$$F_T \cdot u_0 = \tfrac{1}{2}\rho g h_0(2t_0 + h_0)v_a - \tfrac{1}{2}\rho g t_0{}^2 u_0 \quad (1)$$

where F_T is the force on the plate per unit width producing the locked front.
Conservation of mass yields:

$$\rho u_0 t_0 = \rho v_a h_0 \quad (2)$$

Thus, from (1) and (2),

$$F_T = \tfrac{1}{2}\rho g t_0(t_0 + h_0),$$

where

$$t_0 = (v_a/u_0)h_0.$$

If R is the length of the mountain front, then the total driving force is:

$$\tfrac{1}{2}\rho g t_0(t_0 + h_0)R.$$

Substituting $u_0 = 47$ mm/yr, $v_a = 17$ mm/yr, $h_0 = 5$ km, $\rho = 2500\ \text{kg/m}^3$, $g = 9.78\ \text{m/s}^2$, $t_0 =$ 1.8 km, $2F_T/\rho g = 12.2 \times 10^6\ \text{m}^2$, and the total horizontal driving force for a mountain front 1500 km in length is estimated to be $2.24 \times 10^{17}\ \text{kg m}^{-1}\text{s}^{-2}$.

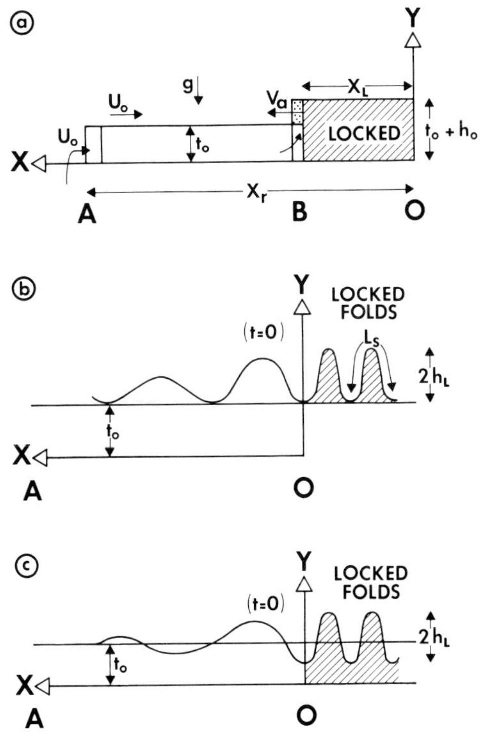

FIG. 3. (a) Schematic section across Zagros, showing symbols used in calculations in text and a set of locked folds (right). (b) Schematic section across Zagros for 'folding carpet' model. (c) Schematic section for 'pudding skin' model. For details see text.

Attenuated (implicit) sinusoidally advancing front

We now consider two related front types, the first associated with a *décollement* surface and the second without any such obstacle to vertical fold development.

Décollement (folding carpet) model

With reference to Fig. 3b,

$$\begin{cases} y = t_0 + h_L e^{-\alpha s + \alpha V_t \cdot t}[1 - \cos 2\pi s/L_s]; & s \geq (V_t \cdot t) \\ y = t_0 + h_L[1 - \cos 2\pi s/L_s]; & s < (V_t \cdot t) \end{cases}$$

'Pudding skin' model

With reference to Fig. 3c,

$$\begin{cases} y = t_0 - h_L e^{-\alpha s + \alpha V_t \cdot t}[\cos 2\pi s/L_s]; & s \geq (V_t \cdot t) \\ y = t_0 - h_L[\cos 2\pi s/L_s]; & s < (V_t \cdot t) \end{cases}$$

In both models h_L and L_s are constants estimated directly from field data. α, V_t and t_0 are constants derived from conservation equations and field data. s is the distance along a curve where $s=0$ at O. Note that the 'implicit cosine $\cos(2\pi s/L_s)$, has a different shape to the ordinary cosine, $\cos(2\pi x/L)$, and more nearly represents a physically observed fold.

As time (t) elapses, the curve moves to the right, and in due course locks in place. The time that elapses for a complete fold to lock is given by:

$$T_L = L_s/V_t.$$

The equations for the folds are empirical, being the simplest curves which can display continuous growth and folding.

Conservation principles

Conservation of length

The length x_r (see Fig. 4), being the width of the Arabian Plate plus the locked folds, is a constant, i.e.

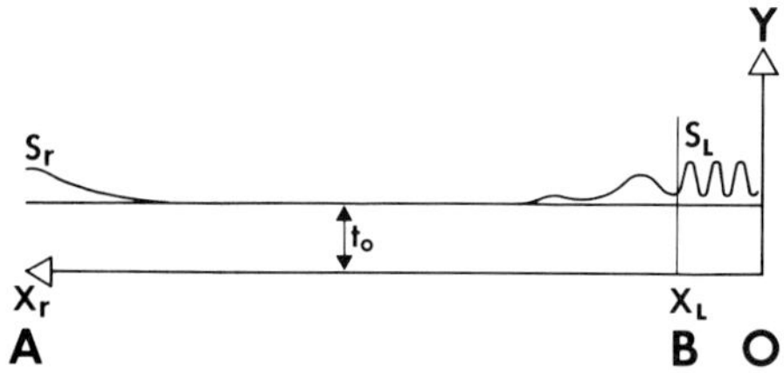

FIG. 4. Schematic section across Arabian Plate. For details see text.

$$x_r = \int_0^{x_r} dx = \int_0^{s_L} \left(\frac{dx}{ds}\right) ds + \int_{s_L}^{s_r} \left(\frac{dx}{ds}\right) ds$$

Where s_L is the location of the locking point on the surface of the fold currently undergoing folding, and s_r is the total surface distance to a distant point (at the Red Sea, see Fig. 4).

Let $e(s, t) = (dx/ds)$ where $s \geqslant (v_t \cdot t)$ and $e(s) = (dx/ds)$ where $s < (V_t \cdot t)$ (i.e. at the locked fold). Thus

$$x_r = \int_0^{s_L} e(s)\, ds + \int_{s_L}^{s_r} e(s, t)\, ds$$

$$x_r \varepsilon \text{ const} \Rightarrow \frac{dx_r}{dt} = 0$$

That is:

$$0 = e(s_L)\dot{s}_L + e(s_r, t)\dot{s}_r - e(s_L, t) + \int_{s_L}^{s_r} \frac{\partial e}{\partial t} ds.$$

Thus:

$$\dot{s}_r = -\int_{s_L}^{d_r} \frac{\partial e}{\partial t} \cdot ds.$$

We may take $\dot{s}_r = u_0$, i.e. the rate at which the Red Sea end of the plate moves towards Iran.
Thus:

$$u_0 = -\int_{s_L}^{s_r} \frac{\partial e}{\partial t} ds \qquad \text{(I)}$$

Conservation of mass

The mass of the whole system $m(t) = \int_0^{x_r} \rho y\, dx$, where the density ρ is assumed to be constant.
Therefore:

$$m(t) = \int_0^{s_L} \rho g y e(s)\, ds + \int_{s_L}^{s_r} \rho y e(s, t)\, ds.$$

Now

$dm/dt = \rho u_0 t_0$ (input of mass at the Red Sea).

Therefore:

$$\rho u_0 t_0 = \rho (ye)_{s_r} \cdot \dot{s}_r + \rho \int_{s_L}^{s_r} \frac{\partial}{\partial t}(ye)\, ds.$$

Now

$y = t_0$ and $e_{s_r} = 1[=(dx/ds)_{s_r}]$, $\dot{s}_r = u_0$.

Therefore:

$$0 = \int_{s_L}^{s_r} \frac{\partial}{\partial t}(ye)\, ds \qquad \text{(II)}$$

Conservation of rate of working

The potential energy of the system

$$\phi(t) = \tfrac{1}{2}\rho g \int_0^{s_L} (y^2 e)\, ds + \tfrac{1}{2}\rho \cdot g \int_{s_L}^{s_r} (y^2 e)\, ds.$$

Thus:

$$\left(\frac{d\phi}{dt}\right) - [\tfrac{1}{2}\rho g t_0^2 u_0] = \tfrac{1}{2}\rho g \int_{s_L}^{s_r} \frac{\partial}{\partial t}(y^2 e)\, ds.$$

Now $[\tfrac{1}{2}\rho g t_0^2 u_0]$ is the rate of work required to raise material on to the plate at the Red Sea, and thus is not involved with the horizontal thrust on the plate.
Thus:

$$F_T u_0 = \tfrac{1}{2}\rho g \int_{s_L}^{s_r} \frac{\partial}{\partial t}(y^2 e)\, ds \qquad \text{(III)}$$

Estimation of integrals

From Eqn. (II):

$$\int_{s_L}^{s_r} y \frac{\partial e}{\partial t} ds = -\int_{s_L}^{s_r} e \frac{\partial y}{\partial t} ds.$$

Subtracting $[t_0(I)]$ from this equation:

$$\int_{s_L}^{s_r} (y - t_0) \frac{\partial e}{\partial t} ds = -\int_{s_L}^{s_r} e \frac{\partial y}{\partial t} ds + t_0 u_0 \qquad \text{(IV)}$$

Now

$$e = (dx/ds) = \sqrt{\{1 - (dy/ds)^2\}} \Rightarrow$$

$$\frac{\partial e}{\partial t} = -(1/e)\left(\frac{dy}{ds}\right)^2 (\alpha V_t)$$

since, in both of the models, $(y - t_0)$ has an $e^{\alpha V_t t}$ factor.

Eqn. (IV) may thus be written (transforming from s to x):

$$-\alpha V_t \int_x^{x_r} (y - t_0)(dy/dx)^2\, dx =$$

$$t_0 u_0 - (\alpha V_t) \int_{x_L}^{x_r} (y - t_0)\, dx$$

i.e.

$$t_0 u_0 = (\alpha V_t)\int_{x_L}^{x_r} (y - t_0)\{1 - (\mathrm{d}y/\mathrm{d}x)^2\}\,\mathrm{d}x \tag{IVa}$$

Likewise:

$$u_0 = \alpha V_t \int_{x_L}^{x_r} (\mathrm{d}y/\mathrm{d}x)^2\,\mathrm{d}x \tag{Ia}$$

In similar fashion, from (III), (IV), and (I) we obtain:

$$F_t \,.\, u_0 = \tfrac{1}{2}\rho g(\alpha V_T)\int_{x_L}^{x_r} (y - t_0)^2 \times \{2 - (\mathrm{d}y/\mathrm{d}x)^2\}\,\mathrm{d}x + \tfrac{1}{2}\rho g t_0{}^2 u_0. \tag{V}$$

(The final term, $\frac{1}{2}\rho g t_0{}^2$, in (5) arises when (IV) and (I) are inserted in (III).)

For convenience x_L is taken as zero, where s is also zero. Since the integrands of (I), (IV) and (V) are functions of s (and t) it is not possible to evaluate them exactly, but by observing that:

$$\int (y - t_0)^n\,\mathrm{d}x = \int (y - t_0)^n(\mathrm{d}x/\mathrm{d}s)\,\mathrm{d}s$$

and

$$\int (y - t_0)^m(\mathrm{d}y/\mathrm{d}x)^2\,\mathrm{d}x = \int (y - t_0)^m\{(\mathrm{d}y/\mathrm{d}s)^2/(\mathrm{d}x/\mathrm{d}s)\}\,\mathrm{d}s$$

we have that:

$$\cos(\phi_m)\int (y - t_0)^n\,\mathrm{d}s \leqslant \int (y - t_0)^n\,\mathrm{d}x \leqslant 1\cdot\int (y - t_0)^n\,\mathrm{d}s,$$

and,

$$1\,.\int (y - t_0)^m(\mathrm{d}y/\mathrm{d}s)^2\,\mathrm{d}s \leqslant \int (y - t_0)^m(\mathrm{d}y/\mathrm{d}x)^2\,\mathrm{d}x \leqslant \{1/\cos(\phi_m)\}\int (y - t_0)^m(\mathrm{d}y/\mathrm{d}s)^2\,\mathrm{d}s$$

where ϕ_m is max $\{|\phi|\}$.

It is suggested that for the observed folds, $\phi_M \sim \pi/4$ and thus

$$\int (y - t_0)^n\,\mathrm{d}x \sim A\int (y - t_0)^n\,\mathrm{d}s$$

and

$$\int (y - t_0)^m(\mathrm{d}y/\mathrm{d}x)^2\,\mathrm{d}x \sim B\int (y - t_0)^m(\mathrm{d}y/\mathrm{d}s)^2\,\mathrm{d}s$$

where

$$1/\sqrt{2} \leqslant A \leqslant 1, \qquad 1 \leqslant B \leqslant \sqrt{2}.$$

Model (i) 'folding carpet'

$$\left[\frac{t_0}{h_L}\cdot\left(\frac{\alpha}{\beta}\right)^2\right]\left(\frac{v_0}{V_t}\right) = \frac{A_1}{1 + (\beta/\alpha)^2} - \tfrac{2}{3}A_2(\alpha h_L)^2 \times\left[\frac{(\beta/3\alpha)^4}{[1 + (\beta/3\alpha)^2][1 + (2\beta/3\alpha)^2]}\right] \tag{IV)(i}$$

$$\left[\frac{2F_T}{\rho g h_L^2}\right]\left(\frac{u_0}{V_t}\right) = \left(\frac{t_0}{h_L}\right)^2\left(\frac{u_0}{V_t}\right) + C_1\left[\frac{(3/8)(\beta/\alpha)^4}{[1 + (\beta/2\alpha)^2][1 + (\beta/\alpha)^2]}\right] - \frac{C_2}{32}(\alpha h_L(\beta/\alpha))^2\left[5 + \frac{95/2}{[1 + (\beta/4\alpha)^2]} + \frac{3}{[1 + (\beta/2\alpha)^2]} - \frac{1/2}{[1 + (38/4\alpha)^2]}\right] \tag{5)(i}$$

where $\beta = (2\pi/L_s)$, and $1/\sqrt{2} \leqslant A_1,\ C_1 \leqslant 1$; $1 \leqslant C_2,\ A_2,\ B_2 \leqslant \sqrt{2}$.

α is of dimension L^{-1} and thus (α/β) is non-dimensional; likewise (u_0/V_t). From Eqn (I)(i), with values for u_0 and h_L from field data, V_t estimated in the light of the field data, B_2 taken as 1.2 (likewise A_2 and C_2, with A_1, C_1 taken subsequently as 0.8), an approximate value for α is obtained. Hence we obtain t_0 from Eqn (IV)(i) and F_T from Eqn (V)(i). F_T is a force per unit length of mountain front. (Subsequently the front is taken to be 1500 km in length.)

Data (i)

$$h_L = 2.5\ \text{km}, \quad L_s = 19\ \text{km}$$

therefore

$$\beta = -0.33069.$$

It is observed that the vertical velocity of a nascent fold, of height 0.5 km adjacent to a locked fold, is estimated at 7 mm/yr (Vita-Finzi 1982).

From

$$y = t_0 + h_L e^{-\alpha s + \alpha V_t t}[1 - \cos(2\pi/L_s)]$$

we have $\dot{y}_{\text{vertex}} = \alpha V_t(y_{\text{vertex}} - t_0)$.

Thus $\alpha V_t = 7/0.5$ (V_t measured in mm/yr, α in km^{-1}).

Eqn (I)(i) thus reduces to

$$\alpha = (0.1653)[0.004556(14/\alpha) - 1]^{1/2},$$

which is satisfied by $\alpha = 0.057$. Whence $V_t = 245.6$ mm/yr. From (Eqn (IV)(i), $t_0 = 9.92$ km; From Eqn (V)(ii), $(2F_T/\rho g) = 117.6\ \text{km}^2$. Taking $\rho = 2500\ \text{kg m}^{-3}$, $g = 9.780\ \text{m s}^{-2}$, and $R = 1500$ km, the horizontal driving force is estimated to be $2.16 \times 10^{17}\ \text{kg m}^{-1}\ \text{s}^{-2}$.

Model (ii)—'pudding skin'

For this model

$$\alpha^2 = \left(\frac{2}{B_Z h_L^2}\right)\left(\frac{u_0}{V_t}\right) - \beta^2/2 \qquad \text{(I)(ii)}$$

$$\left(\frac{t_0}{h_L}\right)\left(\frac{u_0}{V_t}\right) = \left[\frac{A_1}{1 + (\beta/\alpha)^2}\right] + \frac{\alpha^2 h_L^2 A_2}{4}\left[\frac{4}{3} + \frac{(2\beta/3\alpha)^2}{[1 + (\beta/3\alpha)^2]}\right] \qquad \text{(IV)(ii)}$$

$$\left[\frac{2F_r}{\rho g h_L^2}\right]\left(\frac{u_0}{V_t}\right) = \left(\frac{t_0}{h_L}\right)^2\left(\frac{u_0}{V_t}\right) + \frac{C_1}{2} \times \left[1 + \frac{1}{[1 + (\beta/\alpha)^2]}\right] - \frac{C_2}{4}(ah_L)^2 \times \left[\frac{1 + \frac{1}{2}(\beta/\alpha)^2 + \frac{1}{2}(\beta/2\alpha)^4}{[1 + (\beta/2\alpha)^2]}\right] \qquad \text{(V)(ii)}$$

Data (ii)

1. $h_L = 2.5$ km, $L_s = 19$ km, $\alpha V_t = 14$. Equation (I)(ii) is quadratic in α, thus two possible values for α are obtainable, namely $\alpha_1 = 0.0694$ and $\alpha_2 = 0.78773$. t_0 estimated from Eqn (IV)(ii) is either 0.42 or 0.80 km. These are very small values. Moreover, $h_L > t_0$ in both cases, and thus the model is physically not viable. That there are two physically impossible solutions is unsatisfactory and thus recourse is made to the following.

2. The critical value of h_L which gives *equal* values of α^* of 0.11695 is $h_L^* = 4.79$ km. There are folds with amplitudes approaching this value.
Hence:

$$V_t^* = 119.7\ \text{mm/yr}.$$

From (IV)(ii):

$$t_0^* = 8.08\ \text{km} \quad (\text{and } t_0^* > h_L^*).$$

$$2F_T^*/\rho g = 89.55.$$

Hence the driving force $= 1.64 \times 10^{17}\ \text{kg m s}^{-2}$.

Conclusion

The estimated driving forces for the three models are, respectively, 2.24, 2.16, and 1.64×10^{17} Newton, corresponding to 0.790, 0.145, and 0.134 kbar. *Décollement* appears not to facilitate deformation. Since the models do not include elasticity and viscous effects, the true horizontal driving force will presumably exceed these values. Bird (1978) obtained a driving force of 2.8–5.5 10^{15} dyne/cm, equivalent to a compression of 0.280–0.550 kbar, distributed through 100 km of lithosphere, and showed that the smaller amount could be supplied by gravitational spreading of the Red Sea Rift. Ridge push would appear to furnish even lower values in the region of 0.2–0.3 kbar (Kusznir and Park 1984). Unless gravity glide (Price *et al.* 1988) proves acceptable, the additional problems raised by buckle initiation suggest that only the 'folding carpet' and 'pudding skin' models are appropriate.

ACKNOWLEDGEMENTS: We thank G. C P. King for critical comments on an early draft, B. Trovato for helpful discussion, M. R. Samadian for dip measurements at Tujak, B. Walker for fold data, and S. Gatsky and M. Clue for the figures.

References

Bird, P. (1978). Finite element modeling of lithosphere deformation: the Zagros collision orogeny. *Tectonophys.* **50**, 307–36.

Bird, P., Toksöz, M. N., and Sleep, N. H. (1975). Thermal and mechanical models of continent–continent convergence zones. *J. Geophys. Res.* **80**, 4405–16.

Bonatti, E. (1985). Punctiform initiation of seafloor spreading in the Red Sea during transition from a continental to an oceanic rift. *Nature* **316**, 33–7.

BP (British Petroleum) (1956). *Geological maps and sections of south-west Persia*. Stanford, London.

FALCON, N. L. (1973). Vertical and horizontal Earth movements. In *The Musandam Expedition 1971–72: Scientific results, Part I*. (P. F. Cornelius, N. L. Falcon, D. South, and C. Vita-Finzi). *Geogr. J.* **139**, 404.

FALCON, N. L. (1974). Southern Iran: Zagros Mountains. In *Mesozoic–Cenozoic orogenic belts*, (ed. A. M. Spencer), pp. 199–211, Scottish Academic Press, Edinburgh.

GIRDLER, R. W. and STYLES, P. (1978). Seafloor spreading in the western Gulf of Aden. *Nature* **271**, 615–17.

HANCOCK, P. L., AL KADHI, A., and SHA'T, N. A. (1984). Regional joint sets in the Arabian Platform as indicators of intraplate processes. *Tectonics* **3**, 27–43.

HAYNES, S. J. and MCQUILLAN, H. (1974). Evolution of the Zagros suture zone, southern Iran. *Bull. Geol. Soc. Am.* **85**, 739–44.

JACKSON, J. A. (1980). Reactivation of basement faults and crustal shortening in orogenic belts. *Nature* **283**, 343–6.

JACKSON, J. A. and MCKENZIE, D. P. (1984). Active tectonics of the Alpine–Himalayan Belt between western Turkey and Pakistan. *Geophys. J. Roy. Astron. Soc.* **77**, 185–264.

JAMES, G. A. and WYND, J. G. (1965). Stratigraphic nomenclature of Iranian Oil Consortium Agreement Area. *Bull. Geol. Soc. Am.* **49**, 2182–249.

KASSLER, P. (1973). The structural and geomorphic evolution of the Persian Gulf. In *The Persian Gulf*, (ed. B. Purser), pp. 11–32, Springer-Verlag, Berlin.

KING, G. C. P. and VITA-FINZI, C. (1981). Active folding in the Algerian earthquake of 10 October 1980. *Nature* **292**, 22–6.

KUSZNIR, N. J. and PARK, R. G. (1984). The strength of intraplate lithosphere. *Phys. Earth Planet. Int.* **36**, 224–35.

LEES, G. M. (1952). Foreland folding. *Quart. J. Geol. Soc. London* **108**, 1–34.

LE PICHON, X., FRANCHETEAU, J., and BONNIN, J. (1973). *Plate tectonics*. Elsevier, Amsterdam.

MANN, C. D. and VITA-FINZI, C. (1982). Curve interpolation and folded strata. *Tectonophysics* **88**, T7–15.

NIAZI, M., ASUDEH, I., BALLARD, G., JACKSON, J., KING, G. and MCKENZIE, D. (1978). The depth of seismicity in the Kermanshah region of the Zagros Mountains (Iran). *Earth and Plsnetary Science Letters* **40**, 270–4.

PRICE, N. J. (1975). Rates of deformation. *Journal of the Geological Society of London* **131**, 553–75.

PRICE, N. J., PRICE, G. D., and PRICE, S. L. (1988). Gravity glide and plate tectonics. This volume, p. 5.

SHEARMAN, D. J. (1976). The geological evolution of southern Iran. *Geogr. J.* **142**, 393–410.

VITA-FINZI, C. (1982). Recent coastal deformation near the Strait of Hormuz. *Proc. Roy. Soc. London* **A382**, 441–57.

VITA-FINZI, C. (1987). ^{14}C chronologies in coastal Iran, Greece and Jordan. *Journal of the Geological Society of London* **144**, 553–60.

VITA-FINZI, C. and GHORASHI, M. (1978). A recent faulting episode in the Iranian Makran. *Tectonophysics* **44**, 21–5.

WHITE, R. S. and ROSS, D. A. (1979). Tectonics of the western Gulf of Oman. *Journal of Geophysical Research* **84**, 3479–89.

C. D. MANN, University College London, Gower St, London WC1E 6BT, UK.
C. VITA-FINZI, University College London, Gower St, London WC1E 6BT, UK.

Gondwanaland and the evolution of the Indian Ocean

D. H. Tarling

ABSTRACT: In regions, such as the Indian Ocean, where geophysical data are largely lacking, reconstructions must be a compromise, allowing for the evolution of 'passive' margins and the inclusion of remanent continental blocks that are now isolated within the oceanic basins. Sea-floor spreading anomalies are re-examined in the context of the foregoing considerations, and emphasis is placed on the need to match anomalies that formed within the same transform segments. Despite subsequent attempts, it is considered that Norton and Sclater's (1979) configuration of Gondwanaland and its Mesozoic–Tertiary breakup only need small, but significant modifications. These allow for the inclusion of small, but critically located continental units between the major continental blocks. Only age differences are proposed for most of their anomaly 16 reconstruction, now largely corresponding to anomaly 20 times. No changes are proposed to Norton and Sclater's reconstruction for anomaly 34 times. Alterations to older reconstructions are mainly to allow for the inclusion of the Kerguelen, Gaussberg and Broken Ridges between India and Antarctica, and for the Agulhas–Falkland (Malvinas) Plateaux between Africa and Antarctica. The revised Gondwanaland reconstruction for early–middle Jurassic times appears to be more consistent with the geological and geophysical evidence, particularly reducing the need for differential motions of Sri Lanka and of the West Antarctic Peninsula. A review of the palaeomagnetic data indicates caution in the ability of extant data to distinguish between most reconstructions proposed during the last decade; considerable re-sampling and re-evaluation are required to adequately test between such Gondwanan models.

Introduction

The geophysical evidence relating to the previous shape and constitution of Gondwanaland can be divided into two types. (1) The definition of the continental units which once could have formed part of this supercontinent, and (2) techniques which assist in its reassembly and evolution. The identification of the different units is, perhaps, the most fundamental, as this determines the areal extent over which other techniques can be extrapolated. However, both approaches are interdependent, as evidence on the location of a specific area at a given time will also define whether or not that region was part of the Gondwanan continent.

There is general agreement on the reassembly of western Gondwanaland (South America and Africa) although there are minor differences between some recent reconstructions based on slightly differing reinterpretations of magnetic anomalies and fracture zones (Rabinowitz and Labrecque 1979; Martin *et al.* 1981; Rand and Mabesoone 1982). Such differences are significant on a local scale, and also in terms of the subsequent sea-floor spreading history of the South Atlantic—but they make little difference to the configuration of Gondwanaland itself. Similarly, the evolution of the Tasman Sea appears to have been relatively simple, although the shape of New Zealand has probably been drastically modified since its original location in Gondwanaland. There is no agreement on such a configuration, because of the complexities of more recent tectonic activity. However, neither the precise fit of South America nor the actual configuration of New Zealand have a controlling influence on the basic framework of Gondwanaland, although the location of New Zealand is important in terms of the configuration of eastern and western Antarctica. The fundamental problems, at this stage, are considered to be within the Indian Ocean area, and this paper is therefore almost entirely concerned with tectonic units now bordering the Indian Ocean (Fig. 1).

Continental units

The unambiguous identification of continental crustal fragments is difficult. This problem appeared relatively simple when the early geometric fit of bathymetric lines, such as 1000 and 2000 m (Bullard *et al.* 1965; Smith and Hallam 1979) gave similar results, and these contours

From AUDLEY-CHARLES, M. G. & HALLAM, A. (eds) *Gondwana and Tethys* Geological Society Special Publication No. 37, pp. 61–77.

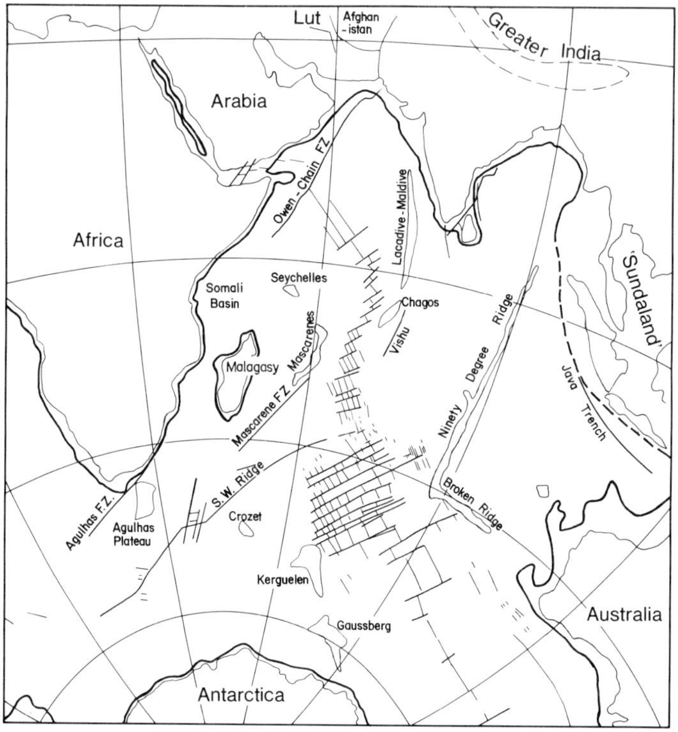

FIG. 1. The Indian Ocean region. The published sequences of anomalies, 1 to 19, are shown paralleling the north-western, south-western, and south-eastern Indian Ocean Ridges. The main mapped fracture zones are shown perpendicular to the ridge system. The main areas mentioned in the text are also shown.

mostly appeared to correspond with gravity and magnetic anomalies marking the edge of the continental crust. However, the actual evolution of 'passive' continental edges is now known to be much more complex. Furthermore, the geophysical data are not available in many crucial areas, particularly south-eastern Asia, that would enable continental and oceanic crust to be distinguished on their contrasting seismic velocity, density, magnetization, etc. In such areas, even present-day plate tectonic activity is only just becoming defined (Le Dain *et al.* 1984), although remarkable progress has been made in understanding late Cainozoic development in some areas (Ni and York 1978; Stocklin 1980) and palaeomagnetic studies are rapidly expanding this knowledge (see next section). However, the definition of specific present and past tectonic units within this complex region is becoming increasingly vital, particularly when considering their alternative locations adjacent to or as part of eastern Gondwanaland.

The significance of defining the present major continental edges is exemplified by the consequences of changing models for the crustal nature of the western Somali Basin off East Africa. This was recognized as a key area even in the earliest attempts to reconstruct Gondwanaland. Most reconstructions have followed that of du Toit (1937), as defined more precisely by Smith and Hallam (1970). This reconstruction placed Malagasy (with India, Australia, and Antarctica attached) in a position very close to the East African coastline, with Antarctica adjacent to south-eastern Africa. However, at that time the geological (Kent 1972, 1982; Kent and Perry 1973) and geophysical (Francis *et al.* 1966) evidence on the nature of the crust off East Africa were considered to show Mesozoic and older (Karroo) sediments extending for hundreds of kilometres east of the present coastline. If the western part of the Somali Basin was indeed underlain by such sediments, then that area could not simultaneously have been occupied by Malagasy (Tarling 1971). New geophysical studies of the western Somali Basin (Coffin *et al.* 1986*a*) have shown that most of the original observations, when incorporated with new data, can be reinterpreted, particularly in terms of the age of the sediments identified on seismic sections. Virtually all of the western Somali Basin comprises oceanic crust, overlain by mostly Cretaceous and younger sediments, and not Mesozoic or even Palaeozoic sediments as previously interpreted. On this basis, continental crust now only extends shortly beyond the present coastline, terminating at about 1500 m depth, beyond which only Jurassic(?) and younger sediments overlie oceanic crust. This oceanic crust is characterized by magnetic anomalies, indicative of sea-floor spreading in Jurassic times, although the location and specific identification of the anomalies varies (Segoufin 1978; Simpson *et al.* 1979; Segoufin and Patriat 1980; Rabinowitz *et al.* 1983). A young age for the Comores Archipelago (Emerick and Duncan 1982) also means that the Comores Basin can be regarded as part of the western Somali Basin until some 10 Ma ago. This revision of the location of the East African continental edge reduces the number of possible fits in this area.

Very similar arguments have also concerned the nature of the crust off southern Africa, but it is now clear that this is primarily of oceanic character, with the exception of parts of the Agulhas Plateau (Allen and Tucholke 1981; Tucholke *et al.* 1981). There now seems little doubt that the evolution of this part of the south-western Indian Ocean is fundamentally controlled by the motion of South America away from Africa, particularly the movement of the Falkland (Malvinas) Plateau along the Agulhas Fracture Zone (Dingle 1973; Dingle and Scrutton 1974; Scrutton 1973; Labrecque and Hayes 1979; Martin *et al.* 1982), with the Agulhas Plateau left behind in Cretaceous times while the South Atlantic continued to open. This evaluation also places critical constraints on the reconstruction of Gondwanaland, particularly in terms of the palaeo-position of the Antarctic Peninsula, and hence on the fit of eastern and western Gondwanaland. However, such models must be consistent with the geological evidence from such regions (de Wit 1977), the knowledge of which is continually improving (Dalziel 1980; Hawkes 1981; Dalziel and Elliot, 1983; Kristoffersen and Haugland 1986).

There are also complexities in the way in which the continental edges have evolved in several regions, although improved knowledge (Powell 1982; Williams and Corliss 1982) suggests that most of these effects are likely to be of local significance, rather than affecting global reconstructions. However, it is the complexities of southern and south-eastern Asia which, while knowledge is increasing rapidly for the more recent evolution (e.g. Molnar and Tapponnier 1978; Ridd 1980; Mitchell 1981; Curray *et al.* 1982; Hutchinson 1982; Helmcke 1982; Liu *et al.* 1983; Lyon-Caen and Molnar 1984; Pigram *et al.* 1985), are still largely unknown for the more remote past. Similarly, knowledge is gradually improving on the nature of the crust and its evolution within the western and south-western Pacific (Howell 1980; Ballance *et al.* 1982; Davey 1982; Veevers 1982; Lallemand and Jolivert 1986). (In many of these complex areas, it seems likely that palaeomagnetic techniques will be able to provide the most rapid method for placing constraints on their evolution—see next section.)

In much the same way as the previous shapes of tectonic units within south-eastern Asia are critical for making reconstructions, the positions of isolated continental fragments within the Indian Ocean are also of major importance (Fig. 1). Many shallow areas of the ocean almost certainly contain at least some fragments of continental crust, as attested by their density (assuming such isolated units to be in isostatic equilibrium). There is agreement on the location of most of these shallow features (Schlich 1982), but new studies are still defining the continental fragments in more isolated areas (Goslin *et al.* 1981; Kahle *et al.* 1981; Sinha *et al.* 1981; Mutter and Cande 1983; Coffin *et al.* 1986*b*; Ramsay *et al.* 1986).

The stretching of the continental lithosphere during initial continental separation, and the subsequent isostatic adjustment, means that geometrical fits are likely to be somewhat tighter than indicated by the simple matching of present bathymetric lines. Conversely, the existence of isolated, remanent continental fragments left within the ocean basins, mainly as a result of 'jumps' in the spreading ridge, means that reconstructions must also leave space for the location of such fragments within the original configuration. Within the Indian Ocean basin, little or no information is available on the extent of lithosphere spreading as this ocean formed. Similarly, the identification of isolated continental fragments is probably now complete, but the present dimensions of the continental crust is unclear and the original configuration is even more speculative. In the proposed Gondwana reconstruction, the edges of the major continental blocks have not been modified, although space has been made for the presence of continental fragments in key areas. While this, at first sight, might indicate that little or no thinning of the continental lithosphere occurred, it merely indicates that the scale of such distortion is small in the context of the extant reconstruction.

The palaeomagnetic evidence

Most of the geophysical data concerned with the reconstructions of past continental distributions are based on measurements of magnetic remanence of rocks from two main environments—(1) oceanic, utilizing the linearity of sea-floor spreading magnetic anomalies, and (2) continental, using the evidence for the dipolar nature of the average geomagnetic field in order to define past latitudes and the previous orientation of tectonic units relative to the rotational pole at specific times. Both systems are complementary in that identical reconstructions should be derived from each method, and they can therefore be used as independent checks. However, the sea-floor spreading anomalies offer the greatest potential for determining the precise sequence in which the Gondwanan continents separated from each other during the last 200 million years, while continental palaeomagnetic techniques, although less precise spatially, can be used to define palaeolatitudes and relative orientations, throughout geological time, and when ocean-floor evidence no longer exists—as within southern and south-eastern Asia.

Marine magnetic anomalies and fracture zones

Sea-floor spreading anomalies generally parallel the spreading ridge at which they formed. Only rarely do the anomalies depart by more than some 5° from this parallel location, and then usually only locally and for short periods. Consequently, the fracture zones which offset both the ridges and linear anomalies, are similarly constrained within some 5° of perpendicularity to the spreading ridge. At these angles, the faults may become 'leaky', but at greater angles new spreading systems generally develop (Atwater 1970).

In order to make reconstructions for different times, anomalies of identical age on both sides of the same spreading ridge, and bounded by the same fracture zones, must be identified. These are then rotated about an Euler pole until they are superimposed. If the correct pole has been used, then the fracture zones should lie on small circles relative to this pole. The ocean floor younger than the superimposed anomalies is then eliminated as not having existed at that time. The relationship between the older ocean floor and the continents can thus be determined for the time when the superimposed anomalies were actually forming at the previous ridge crest. Such a palaeogeographical reconstruction does not give absolute values for either the palaeolatitude or palaeolongitude, but very precisely defines the previous relative positions of the older units. The method is therefore relatively simple, but requires (1) the precise identification and location of the same specific anomaly on both sides of a present or past spreading ridge, and (2) a rotation that maintains the lineation of associated fracture zones.

Several anomalies are easy to identify, as they have very characteristic shapes, even when modified by their original strike orientation and latitude. Other anomalies are more difficult to identify but can usually be determined with sufficient accuracy by interpolation between identified anomalies. The age of the identifiable anomalies is now quite well established biostratigraphically, although their absolute age is still subject to errors of about 5 per cent (Tarling 1983). The identification problem is therefore only of importance when rates of sea-floor spreading are changing, or when 'jumps' of the spreading ridges have occurred, giving rise to completely new orientations of the magnetic anomaly patterns. Under such circumstances, identification by interpolation or extrapolation becomes less reliable. However, the largest source of error is the need to match anomalies within the same fracture zone segment. This can be very difficult for early times, particularly if a change in the direction of spreading has occurred, as small errors in determining the location of each successive Euler pole or rotation angle can give rise to mismatches of older fracture-zone segments.

In this review, the discussion has been largely confined to the evidence from the Indian Ocean and is an interim report of research currently in progress. The reconstructions are thus still being evaluated against a range of other evidence. It is, nonetheless, hoped that this work will at least identify the key areas where either older data needs to be re-examined, or new data needs to be acquired, in order to solve some of the outstanding problems in the breakup of Gondwanaland, and hence in deriving its original configuration. The anomaly patterns described here are based on a synthesis of published work by Johnson *et al.* (1976); Laughton (1966); Matthews (1966); McKenzie and Sclater (1971); Sclater *et al.* (1976); Bergh (1977); Heirtzler *et al.* (1978); Larson *et al.* (1978); Markl (1978); Labrecque and Hays (1979); Larson *et al.* (1979); Simpson *et al.* (1979); Goslin *et al.* (1980); Powell *et al.* (1980); Tapscott *et al.* (1980); Sclater *et al.* (1981); Martin *et al.* (1982);

Schlich (1982); Fisher and Sclater (1983); and Liu *et al.* (1983).

As the general conclusion is that the Norton and Sclater (1979) reconstructions only need minor revision, their arguments in support of specific reconstructions will not be repeated here; discussion mainly be restricted to the reasons for the proposed modifications. It has also been decided to concentrate on three intermediate ages, corresponding to anomaly 20, 34, and M1 times, as well as the original early Jurassic configuration. These are described in order of decreasing reliability, i.e. commencing with a model for anomaly 20 times, and then proceeding through to a probable Gondwanan configuration for early Jurassic and mid-Jurassic times. These reconstructions have been made relative to Africa (Tables 1 and 2), and have not

TABLE 1. *Euler poles and rotation angles for Gondwanaland.*

	A16[a] (N&S)[b]			A20 (DT)[c]			A34 (N&S and DT)		
	°N	°E	°	°N	°E	°	°N	°E	°
India	13.3	54.1	−20.8	13.3	54.1	−20.8	18.7	25.8	−56.0
Australia	13.3	54.1	−20.8	13.3	54.1	−20.8	−8.6	−112.0	30.7
Antarctica	5.8	−37.2	7.2	9.5	−20.4	8.4	19.7	−43.8	19.2
S. America	59.0	−35.0	14.9	63.0	−36.0	24.3	63.0	−36.0	33.8

	M1 (N&S)			M1 (DT)		
	°N	°E	°	°N	°E	°
India	24.5	33.5	−59.0	15.2	31.1	−68.8
Australia	−28.3	−64.6	40.0	−36.9	−74.3	34.9
Antarctica	−3.9	−27.8	45.5	−10.8	−25.8	38.0
S. America	52.6	−34.6	52.4	52.6	−34.6	52.4
Madagascar	−13.1	−41.0	8.1	0.0	0.0	0.0

	Jurassic (N&S)			Jurassic (DT)		
	°N	°E	°	°N	°E	°
India	29.6	36.1	−56.8	−23.3	−136.5	68.0
Australia	−24.6	−60.4	50.0	−28.2	−61.7	34.9
Antarctica	−2.4	−32.7	55.4	−5.4	−31.7	52.8
S. America	45.5	−32.2	57.5	45.5	−32.2	57.5
Madagascar	−16.3	−31.4	13.8	0.3	−49.9	16.0
New Zealand	—	—	—	−42.4	−45.2	55.8

[a]At all times, Arabia is rotated by −7.6° about 26.5°N 21.5°E.
[b]N&S = Norton and Sclater (1979).
[c]DT = this paper.

TABLE 2. *Intermediate Euler poles and rotation angles*

	A20 to A34			A34 to M1			M1 to closure			
	°N	°E	°	°N	°E	°		°N	°E	°
India	−26.6	−167.6	38.7	7.3	−140.8	14.3	All eastern Gondwanaland	0.3	−49.9	16.0
Australia	6.9	−90.9	11.7	−27.0	−3.4	24.8				
Antarctica	22.1	−61.8	12.1	−27.0	−3.4	24.8				
S. America	63.0	−36.0	9.5	36.7	−23.8	20.1				

yet been converted to provide palaeolatitudes for each period, although such appraisals are currently in progress. Consequently, all references to geographical orientations are in terms of the present co-ordinates of Africa.

Mid-Eocene (anomaly 20; Fig. 2)

This was a highly critical time in the evolution of the Indian Ocean. Older oceanic crust is characterized by distinctly different orientations of magnetic anomalies and fracture zones and therefore marked the cessation of spreading from the present ridge. Only in the north-western Indian Ocean is it possible that earlier spreading continued from the same spreading ridge (discussed below), although the lack of identified anomalies between anomalies 5/6 and 22 in this area prevents proper assessment of the actual situation. The significance of the change in spreading pattern is most clearly shown by comparison of the anomalies in the northern and southern Crozet Basins (Schlich 1982). The oldest anomaly parallel to the present south-eastern Indian ocean spreading ridge is anomaly 19, while the youngest definitely identified anomaly, no longer parallel to the present spreading ridge, is anomaly 21. (Sclater *et al.* 1976, identified anomalies 18, 19, and 20 near 27°S, 85°E, but these are in a tectonically disturbed area and have not been included by subsequent reviewers.) The relevance of the anomalies close to the Java Trench to the rest of the Indian Ocean is not too clear. Liu *et al.* (1983) proposed that extinct spreading ridges are present, close to the Java Trench, which are of approximately anomaly 20 age, but the precise identification of most anomalies in this area is still uncertain. However, south-eastern Asian tectonic blocks were clearly some 10 000 km away from Broken Ridge (unless separated by a subduction zone), irrespective of any revisions of the age of anomalies currently being subduc-

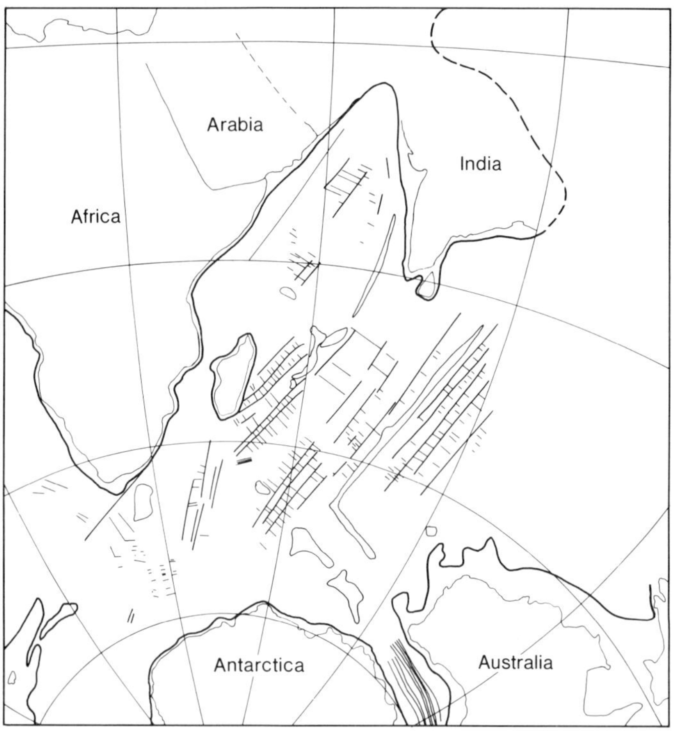

FIG. 2. The Indian Ocean in anomaly 20 (mid-Eocene) times. The location and associated fracture zones for mapped anomalies 20 to 34 are shown, but not identified. The rotations used are as given in Table 1, with the exception of the Kerguelen Plateau, which has been moved, subjectively, to become parallel to the Broken Ridge. The northern edges of India, Arabia, etc., are based on present-day geological features, but are only intended to be indicative of the possible margins in this, and also Figs. 3 to 5. The grid is for reference purposes and applies only to present-day Africa.

ted at the Java Trench. This probably marks the maximum separation of Sundaland from Gondwanaland during the last 200 million years.

There is therefore clear evidence for a major readjustment of spreading systems within the eastern Indian Ocean at anomaly 20 times. On this basis, it seems likely that spreading was occurring perpendicular to the present north-west, south-east and south-west spreading ridges until at least anomaly 19 times, and that the previous system changed to the present one after anomaly 21 times.

In view of the drastic nature of the change in spreading patterns at this time, it seems likely that the separation of the Laccadive and Mascarenes areas would also have occurred then. In Norton and Sclater's (1979) reconstruction, they calculated a new Euler pole and rotation angle (13.3°N 54.1°E –20.8°) for fitting the Laccadive and Mascarene regions. They attributed this separation to anomaly 16 times, mainly on the evidence from DSDP 238 that the oceanic basement between the present spreading ridge and the Laccadive Ridge was of Lower Oligocene age. However, they commented that this evidence could be consistent with a separation of these two areas any time between anomalies 12 and 18. However, there seems to be no reason why the separation of these units could not have been slightly older, i.e. about anomaly 20 times. Unfortunately, there are insufficient anomalies defined in the north-west Indian Ocean to test this suggested age revision, but neither does there seem to be any reason for not considering this as a probable age for their initial separation.

The movement of Arabia is constrained by the very well-defined Owen-Chain Fracture Zone (Matthews 1966), and so the Euler pole defined by this fracture zone (26.5°N 21.5°E –7.6°—McKenzie *et al.* 1970) is preferred to that based on bathymetry alone (36.5°N 18.0°E −6.1°—Smith and Hallam, 1971). The rotation angle and Euler pole for the motion of India (and Australia) relative to Africa (Table 1) is not fully consistent with the alignment of the Owen-Chain fracture zone, as it results in an angular offset of this zone. The angle involved, *c.* 19°, is similar to that between the Laccadive–Maldive Ridge and the western Indian continental edge. This similarity indicates a similar origin, or close relationship, but the tectonic complexities in north-western India and the Middle East also need to be added to the argument. Conversely, the match of the anomalies south of Australia with those off Antarctica, and the conjunction of the southern tip of the Ninetydegree East/Broken Ridges and the Kerguelen Plateau, suggests that any errors in this Euler pole are relatively small, and thus most of the observed misalignments must be due to local, rather than plate-wide disturbances. It seems probable that the opening between the Laccadive–Maldive Ridge and India was, in fact, a response to the collision with the Lut block, as this was, during anomaly 20 times, being actively rotated away from its previous position as an Arabian promontory towards its present location within Iran. However, this reconstruction still needs further examination—it is necessary, for example, to rotate the Kerguelen Plateau slightly away from Antarctica, since anomaly 20 times, in order to make it parallel to Broken Ridge.

Norton and Sclater (1979) calculated the corresponding Euler pole and rotation for Antarctica to India at anomaly 16 times (11.1°N 34.6°W 22.0°), assuming a triple junction between all three plates. This provided an excellent match between the few reported identifications of anomaly 17 in the south-eastern Indian Ocean, confirming that this Euler pole and rotation correspond to anomaly 16–17 times. The resultant Euler pole for moving Antarctica to Africa (5.8°N 37.2°W 7.2°) indicates only very small motions between Africa and Antarctica during this period—consistent with the closeness of the anomalies paralleling this ridge to the south of Africa. However, Norton and Sclater also noted that their reconstruction for anomaly 16 times was inconsistent with Bergh and Norton's (1976) identification of anomalies 21 to 23 bordering the Prince Edward Fault Zone.

The validity of the Norton and Sclater rotation for anomaly 17 times in the south-eastern Indian Ocean means that a correction is needed to compare all areas at anomaly 20 times. If a constant spreading rate is assumed in the south-eastern Indian Ocean, a slight additional rotation of Antarctica to India (2.6°) is required to make both reconstructions contemporaneous, and the Euler pole and rotation for Antarctica to Africa at anomaly 20 times becomes 9.5°N 20.4°W 8.4°. Such a rotation approximately aligns the older magnetic anomalies and fracture zones in the Crozet Basin area with those in the Mascarene and Madagascar Basins. In these basins, the youngest anomalies which are non-parallel to the present spreading ridge are 22 in the southern Crozet Basin, 20 in the Mascarene Basin, and a possible extinct spreading ridge corresponding to anomalies 27 or 28 in the Madagascar Basin. The proposed revision is now consistent with the age of anomalies previously identified south of Africa by Bergh and Norton (1976). This slightly modified version of Norton and Sclater's anomaly 16 reconstruction thus appears to be consistent with an anomaly 20 age. Its validity is vital as it constrains all older reconstructions.

Late Cretaceous (anomaly 34; Fig. 3)

The oldest anomaly so far identified in the northern Arabian Sea is anomaly 28, but this is still some distance from the Indian continental shelf, suggesting that spreading continued from earlier times. However, the counterpart of this spreading section is mostly identified within the Madagascar and Mascarene Basins. The Mascarene Basin anomalies are at a distinct angle to the previous spreading pattern, commencing at anomaly 20, and showing parallel lineation of anomalies and fracture zones as far as the continental shelf of Malagasy, where the oldest identified anomaly is 33b. It seems likely, therefore, that the west coast of India was closed against Malagasy immediately prior to anomaly 33b times—implying that anomalies 29 to 33b exist in the northern Arabian Sea, but possibly obscured by the thick sedimentary sequences now occupying parts of this area.

Such a model seems consistent with the observed patterns, with the exception of some of the anomalies in the north-eastern Madagascar Basin, where the younger anomalies have been interpreted as having originally spread from a ridge which became extinct at anomaly 29 times. It is possible to construct a spreading system that could incorporate such a spreading ridge, with subsequent jumps in spreading, but this seems unlikely in view of the clear systematic trends shown immediately to the south in the Mascarene Basin. It seems more probable that continuous spreading also occurred here and that some of the anomalies in the north-eastern section of the Madagascar Basin have been misidentified—although asymmetrical spreading may have occurred.

In the rest of the Indian Ocean, the picture is more complex. The juxtaposition of the post-anomaly 21 lineations of the southern Crozet Basin, and those of the Mascarene Basin at

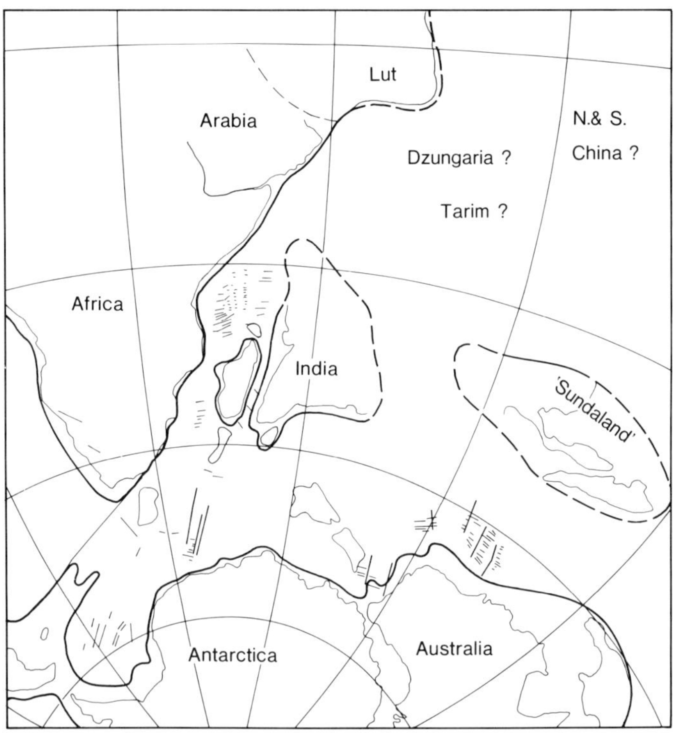

FIG. 3. The Indian Ocean in anomaly 34 (late Cretaceous) times. The location and associated fracture zones for mapped anomalies 34 to M25 are shown. The Euler poles and rotation angles are given in Tables 1 and 2, with the exception of the Kerguelen and Crozet Plateaux. The location of Sundaland is entirely schematic. Note that the grid is just for reference purposes, and applies only to present-day Africa.

anomaly 20 times, suggests that either no earlier spreading took place between these two areas, or that any spreading from the south-western ridge was so small as to be insignificant within the present level of location and identification of the magnetic anomaly patterns. It certainly seems possible that the major change in spreading at anomaly 20 times also initiated spreading from the present south-western Indian Ocean Ridge, although some spreading is indicated south and south-west of the Crozet Basin. At this stage, there seems no reason to invoke significant spreading between the Crozet and Mascarene Basins during this period.

The early identifications of the anomalies between Australia and Antarctica suggested that the oldest anomaly was anomaly 22, but this has subsequently been revised to anomaly 34 (Cande and Mutter, 1982). In the context of this series of reconstructions, the original identification would be more consistent, as it would then correspond with the initiation of spreading along the south-eastern Indian Ocean ridge by anomaly 19 times (see previous paragraph). Australia and Antarctica were certainly close by anomaly 34 times, but it is not clear whether the Kerguelen and Gaussberg areas were yet close to Antarctica. It sees probable that they were probably closed against it by anomaly 34 times.

In the East Indian Ocean, two clear provinces existed, separated by the Ninetydegree East Ridge. The Indian Plate motion, between anomaly 20 and 34 times, has been defined by assuming that the Ninetydegree East Ridge acted as a major transform fault. By anomaly 34 times, India was adjacent to Malagasy, with the Seychelles immediately north of Malagasy, and the continental parts of the Laccadive, Maldive, Chagos, and Mascarene areas just south of India and Malagasy. Spreading in the eastern East Indian Ocean between anomaly 34 and 20 times, was antiparallel to the Indian plate, so that, at anomaly 34 times, at least some parts of south-eastern Asia (Sundaland) were lying to the north-west of Australia, but their locations are not clear. Palaeomagnetic data from other south-eastern Asian blocks, and those now north of India, suggests that these were lying offshore of Gondwanaland, i.e. in positions whereby they were to collide with India and Sundaland during their Tertiary northward motion.

It seems likely that slight modifications will be required to the proposed motions south of the Crozet Basin between anomaly 20 and 34 times. At the moment, the Euler poles and rotations, as defined by Norton and Sclater (1979), result in the Crozet Plateau overlying Sri Lanka, while a position off the tip of southern India seems more likely. There are no data with which to test such modifications, other than that such a location seems more consistent with the proposed reconstruction for Jurassic times, but such reasoning is circular, as older reconstructions depend on the validity of the younger reconstructions.

Early Cretaceous (anomaly M1 Fig. 4)

The main modification proposed for this time is that Malagasy, India, Sri Lanka, Australia, New Zealand, and Antarctica were contiguous, with Malagasy still in its present location relative to Africa, i.e. eastern Gondwanaland was a single continental unit, with active spreading confined to the area of southern Africa, and possibly within the Weddell Sea. It is proposed that the actual closure of India against Antarctica was somewhat different to that used in previous reconstructions, mainly reflecting the space required for the Kerguelen Plateau and the Gaussberg and Broken Ridges. This revision also alleviates the need for a rotation of Sri Lanka relative to India (necessary in all previous reconstructions) and is thus more consistent with evidence for structural continuity between them. The Agulhas Plateau also formed part of the South American continent at this time, and was thus slightly further north-east than at present. The position of the south-eastern and Central Asian blocks is purely speculative at this stage of analysis, but they were certainly offshore from Australia.

Jurassic Gondwanaland (Fig. 5)

This reconstruction is very dependent on the validity of the younger configurations, but if the anomaly M1 reconstruction is approximately valid, then this only requires a southerly motion, along the Davie Fracture Zone, of eastern Gondwanaland relative to Africa between mid-Jurassic and anomaly M1 times. This simplicity, in itself, is persuasive and also reduces the need for a final 'twist' of Malagasy, with eastern Gondwanaland, against Africa. This 'twist' was required on previous reconstructions of this area in order to avoid an overlap of East Antarctica on to Africa and the Falkland (Malvinas) and Agulhas Plateaux. The major overlap between the Antarctic Peninsula and the Patagonian/Falkland region, which occurs in most previous reconstructions, is also reduced to the level where it can be explained in terms of local effects, such as the addition of the Rocas Verdes Complex in the Cretaceous, and late Tertiary

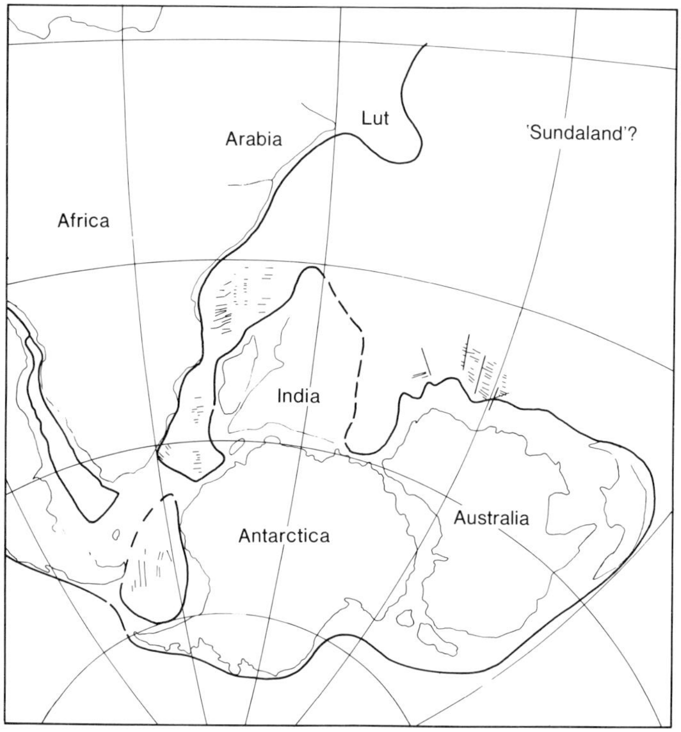

FIG. 4. The Indian Ocean in anomaly M1 (early Cretaceous) times. Mapped anomalies and fracture zones older than M1 are shown. Rotation angles and Euler poles are given in Tables 1 and 2. The locations of continental fragments (Seychelles, Mascarene, Crozet, Kerguelen, etc.), are not shown, as it is assumed that their present shapes differ from what they were at this time. The present-day bathymetry around New Zealand and New Guinea have also been strongly distorted and are purely diagrammatic. Note that the grid is just for reference purposes, and applies only to present-day Africa.

motions in at least the northern part of the Peninsula.

The problems of the locations and motions of the various Asian blocks—Tarim, Dzungaria, etc., and those of south-eastern Asia—Sundaland, South China, North China/South Korea, etc., are deliberately left unclear, as these must have had times of closure towards Gondwanaland and times of separation during this 200 Ma interval. Certainly, some of the oldest anomaly patterns around Antarctica, Australia, and East Africa suggest that several of these units were moving away from Gondwanaland in Jurassic times, but were then 'overtaken' by the motion of later fragments and the possible reversal of their own independent motions. To attempt to define these motions at the present state of knowledge does not appear to be more than speculation. It is more important, at this stage, to establish the validity of the major reconstruction.

Continental palaeomagnetism

Palaeomagnetic studies of continental rocks played a fundamental role in establishing the theory of continental drift and plate tectonics. The method is to produce a polar-wander path for each continent. Each continent is then rotated into its postulated palaeo-position, and the polar-wander curve rotated in the same way. Ideally, the correct configuration would be characterized by one clear, common polar wandering path until the time that the supercontinent began to break up. Early analyses succeeded in both confirming the existence of Gondwanaland, and placing some constraints on its configuration. Subsequent progress has resulted in a much greater understanding of the physical and chemical processes by which rocks acquire their magnetization, and this has also resulted in considerable advances in the methods of treatment and analysis of rock sam-

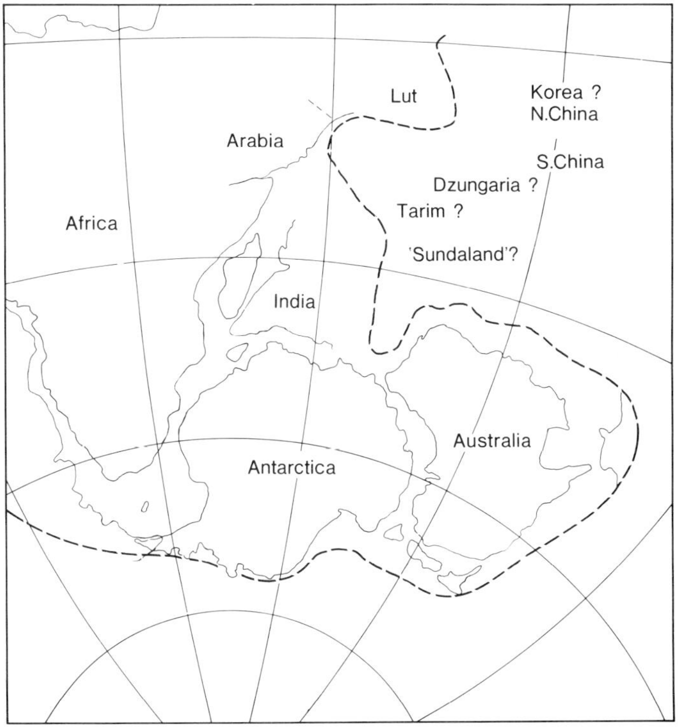

FIG. 5. Mid-Jurassic Gondwanaland. See Fig. 4 for comments on the bathymetric representations, etc. Note that the grid is for reference purposes and applies only to present-day Africa.

ples (Collinson 1983; Tarling 1983; O'Reilly 1984). As a consequence, much of the earlier continental palaeomagnetic data no longer fulfill modern requirements. During the last few years there has, unfortunately, been little improvement in the available data from the stable cratonic areas of the major Gondwanan continents, with the notable exception of Australia, for which there has been both re-sampling and re-analysis (Embleton 1981; Idnurm 1985*a*,*b* 1986). In fact, most advances have resulted in an increasing appreciation of the complexities of the magnetic processes, even when only time is involved (Moon and Merrill 1986). This has resulted in the elimination, or at least questioning, of the reliability of most previously accepted observations. In other instances, there is a much greater appreciation of the probability of local rotations having affected regions previously considered as cratonic (Steiner 1983). In this review, only the Devonian of Africa will be cited as an example of the magnitude of all such revisions.

Most reviews of the African early Phanerozoic results have considered the best-defined pole to be the Devonian pole determined from studies of the Msissi norite in Morocco (Hailwood and Mitchell 1971; Hailwood 1974). The reliability of this pole was apparently confirmed by its agreement with climatic indicators—the Llanvirn–Llandeilo glaciation of the Sahara region and the Sakmarian Dwyka glaciation in southern Africa. However, Salmon *et al.* (1986) have shown that this norite is a teschenite, and more significantly, have found radiometric ages ranging to late Jurassic or early Cretaceous. Salmon *et al.* also showed that its magnetization comprised several vectors, and Najid (1985) also found extensive remagnetization in the same region. Thus the previously well-accepted Devonian pole may represent a combination of several vectors which may, or may not, include those of Devonian age. Clearly, this pole can no longer be considered to be reliably determined, and so there are virtually no acceptable data for most of the African Palaeozoic polar-wander path.

Unfortunately this situation is typical for all Lower Palaeozoic data from the Gondwanan continents, with the possible exception of Australia. Such unreliability is not particularly surprising, but there are strong grounds for considering that similar, although less drastic, remagnetization problems exist for much of the extant African Upper Palaeozoic and Mesozoic data (Najid 1985; Ballard *et al.* 1986). Such problems have been found, and apparently overcome, in the re-examination of earlier Australia data (Embleton 1981; Idnurm 1985*a*,*b*, 1986), but virtually all extant European and North American observations must also be considered inadequate for testing all but the most divergent tectonic models. Indeed, it seems probable that many of the early postulates for the relationships between Laurentia and Gondwanaland must also be re-examined where they have been based primarily on palaeomagnetic data that should now be regarded with more caution (Irving and Irving 1982; Daly and Irving 1983; Hallam 1983; Tarling 1985*a*; Bachtadse and Van der Voo 1986). In the context of Gondwanaland, it is relevant that the palaeomagnetic data were inadequate for testing whether Malagasy was adjacent to Kenya and Somalia in Permian times, or was then still in its present position relative to Africa (Tarling and Kent 1976). There has been no significant improvement in the data base since then.

Such an assessment is *not* to dismiss all available continental palaeomagnetic data as being of no value. Many of the more 'unusual' Gondwanan reconstructions can be readily eliminated on palaeomagnetic grounds, but many variants of the original Smith and Hallam (1970) model have been proposed (e.g. Norton and Sclater, 1979; Embleton *et al.* 1980; King 1980; Powell *et al.* 1980; Rickard and Belbin 1980; Embleton and McElhinny 1982; Livermore *et al.* 1986). Each of these was found 'to be consistent with' the palaeomagnetic data. The present continental palaeomagnetic data must be considered too unreliable to evaluate Gondwanan reconstructions that do not involve drastic rotations relative to the Smith and Hallam model. Furthermore, the problem is inherent within the available data base, rather than in the procedures for comparison of polar paths. Improved techniques (Thompson and Clark 1982; Embleton *et al.* 1983), while highly desirable, can still only be utilized effectively if the original data are valid.

While there has been little recent progress in redefining continental polar-wander paths, considerable effort has been devoted to studies in structurally complex areas that had been deliberately avoided during the earlier investigations. A considerable quantity of such data are now available from southern and south-east Asia (Klootwijk and Bingham 1980; McElhinny *et al.* 1981; Wensink 1981, 1983; CCOP Project Office 1982; Achache *et al.* 1983; Klootwijk *et al.* 1983, 1985; Westphal *et al.* 1983; Besse *et al.* 1984; Otofuji and Matsuda 1984; Pozzi *et al.* 1984; Achache and Courtillot 1985; Kono and Tanaka 1985; Lin *et al.* 1985). However, while the palaeomagnetic reliability is generally much better established than for cratonic areas, most of the newer data are from areas that are frequently only poorly known geologically. Consequently, most new, and hence probably more reliable, palaeomagnetic data are from areas where effective interpretation is inhibited by the geological uncertainty or complexity. Such constraints involve:

1. The scale of the tectonic dislocation, i.e. local or regional. Individual sites provide information on the palaeolatitude and their orientation relative to the pole when they formed, but such information may only apply to a single fault-block, or it could be regional if the sites were in an extensive autochthonous block.

2. Tests to determine the actual age of magnetization may be easier to make than in relatively flat-lying areas, but information on the age of the folding may be lacking, and several phases of folding could be involved, with apparent prefolding magnetization actually having been acquired during an intermediate stage.

3. The tectonic correction for observed deformations requires separate corrections for each phase if the original direction is to be correctly calculated. Such information on the sequences and processes of deformation are often lacking and, in any case, too little is known about the magnetic effects of such processes to be clear how such corrections should be undertaken, if more than simple tilting is involved (Tarling 1983).

Such a negative assessment for even modern palaeomagnetic data does not mean that palaeomagnetic techniques do not have a vital and major role to play. In particular areas, the palaeomagnetic data can be considered to be more reliable than the poorly known geological constraints. The simple observation that virtually all of the palaeomagnetic observations from tectonic units in southern and south-eastern Asia are consistent with a Gondwanan origin is, in itself, of major significance, even though such data must be treated cautiously in trying to establish the precise tectonic evolution of each of the numerous tectonic units in this region.

Conclusions

It must be concluded, at this stage, that too few reliable palaeomagnetic data are currently available for the testing of most of the subtleties of possible Gondwanan reconstructions. This is not to say that palaeomagnetic methods do not already place major constraints on such models, e.g. Tarling (1985*b*,*c*), Livermore *et al.* (1986), etc., but at the moment it is the available geological evidence, such as that expounded by Adamia *et al.* (1981), Mitchell (1981), Audley-Charles (1983), Şengör (1984), Gatinsky (1986) and many others, that provides the most pragmatic controls on different models for the reassembly of Gondwanaland. A revision of the continental palaeomagnetic data, involving intensive and extensive re-sampling, is urgently needed for most of the continental blocks, while geological and palaeomagnetic methods, together, are likely to solve many of the outstanding problems of the changing configurations of different tectonic units within south-eastern and southern Asia, when such blocks were either loosely or firmly associated with Gondwanaland.

At this stage, it would still seem that a slightly modified Norton and Sclater (1979) series of reconstructions provides the most acceptable scenario for the Cainozoic and Mesozoic history of this ocean basin, and hence of Gondwanaland. Further work is proceeding on these proposed revisions.

ACKNOWLEDGEMENTS: This article was being researched during the chaotic process of moving to Plymouth. I am grateful to P. Hart for his assistance in modifying the FORTRAN programs to run on a different computing system.

References

Achache, J. and Courtillot, V. (1985). A preliminary Upper Triassic paleomagnetic pole for the Khorat Plateau (Thailand): consequences for the accretion of Indochina against Eurasia. *Earth Planet. Sci. Lett.* **73**, 147–57.

Achache, J., Courtillot, V., and Besse, J. (1983). Paleomagnetic constraints on the late Cretaceous and Cenozoic tectonics of southeastern Asia. *Earth Planet. Sci. Lett.* **63**, 123–36.

Adamia, S. A., Chkhotua, T., Kekelia, M., Lordkipanidze, M., Shavishvili, I., and Zakariadze, G. (1981). Tectonics of the Caucasus and adjoining regions; implications for the evolution of the Tethys ocean. *J. Struct. Geol.* **3**, 437–47.

Allen, R. B. and Tucholke, B. E. (1981). Petrography and implications of continental rocks from the Agulhas Plateau, southwest Indian Ocean. *Geology* **9**, 463–8.

Atwater, T. (1970). Implications of plate tectonics for the Cenozoic tectonic evolution of western North America. *Geol. Soc. Amer., Bull.* **81**, 1513–36.

Audley-Charles, M. G. (1983). Reconstruction of eastern Gondwanaland. *Nature* **306**, 48–50.

Bachtadse, V. and Van der Voo, R. (1986). Paleomagnetic evidence for crustal and thin-skinned rotations in the European Hercynides. *Geophys. Res. Lett.* **13**, 161–4.

Ballance, P. F., Pettinga, J. R., and Webb, C. (1982). A model of the Cenozoic evolution of northern New Zealand and adjacent areas of the Southwest Pacific. *Tectonophys.* **87**, 37–48.

Ballard, M. M., Van der Voo, R., and Halbich, I. W. (1986). Remagnetizations in Late Permian and Early Triassic rocks from southern Africa and their implications for Pangea reconstructions. *Earth Planet. Sci. Lett.* **79**, 412–18.

Bergh, H. W. (1977). Mesozoic seafloor off Droning Maud Land, Antarctica. *Nature* **269**, 686–7.

Bergh, H. W. and Norton, I. O. (1976). Prince Edward Fracture Zone and the evolution of the Mozambique Basin. *J. Geophys. Res.* **81**, 5221–39.

Besse, J., Courtillot, V., Pozzi, J. P., Westphal, M., and Zhou, Y. X. (1984). Palaeomagnetic estimates of crustal shortening in the Himalayan thrusts and Zangbo suture. *Nature* **311**, 621–6.

Bullard, E. C., Everitt, J. E., and Smith, A. G. (1965). The fit of the continents around the Atlantic. *Proc. Roy. Soc., London* **A258**, 41–51.

Caude, S. C. and Mutter J. C. (1982). A revised identification of the oldest sea-floor spreading anomalies between Australia and Antarctica. *Earth Planet. Sci. Lett.* **58**, 151–60.

CCOP Project Office. (1982). *Palaeomagnetic research in southeast and east Asia* UNDP Technical Support (RAS/81/120). (229 pp.)

Coffin, M. R., Rabinowitz, P. D., and Houtz, R. E. (1986*a*). Crustal structure in the Western Somali Basin. *Geophys. J. Roy. Astron. Soc.* **86**, 331–69.

Coffin, M. F., Davies, H. L., and Haxby, W. F. (1986*b*). Structure of the Kerguelen Plateau province from Seasat altimetry and seismic reflection data. *Nature* **324**, 134–6.

Collinson, D. W. (1983). *Methods in rock magnetism and palaeomagnetism*. Chapman & Hall, London. (503 pp.)

Curray, J. R., Emmel, F. J., Moore, D. G., and Raitt, R. W. (1982). Structure, tectonics, and geological history of the northeastern Indian Ocean. In *The ocean basins and margins, 6, the Indian Ocean* (ed. A. E. M. Nairn and F. G. Stehli), pp. 399–450, Plenum, New York.

Daly, L. and Irving, E. (1983). Paléomagnétisme des roches carbonifères du Sahara central; analyse des aimantations juxtaposées; configuration de la Pangée. *Ann. Geophys.* **1**, 207–16.

Dalziel, I. W. D. (1980). The Ellsworth Mountains as a displaced fragment of the Transantarctic Mountains: present status of the Schopf hypothesis. *Eos* **61**, p. 1197.

Dalziel, I. W. D. and Elliot, D. H. (1983). The Antarctic plate: a global perspective. *Geology* **11**, 118–20.

Davey, F. J. (1982). The structure of the South Fiji Basin. *Tectonophys.* **87**, 185–241.

de Wit, M. J. (1977). The evolution of the Scotia Arc as a key to the reconstruction of southwestern Gondwanaland. *Tectonophys.* **37**, 53–81.

Dingle, R. V. (1973). Mesozoic palaeogeography of the southern Cape, South Africa. *Palaeogeogr., Palaeoclimatol., Palaeoecol.* **13**, 203–13.

Dingle, R. V. and Scrutton, R. A. (1974). Continental breakup and the development of post-Palaeozoic sedimentary basins around Southern Africa. *Geol. Soc. Amer., Bull.* **85**, 1467–74.

du Toit, A. L. (1937). *Our wandering continents*. Oliver & Boyd, London. (366 pp.)

Embleton, B. J. J. (1981). A review of the palaeomagnetism of Australia and Antarctica. *Earth Planet. Sci. Lett.* **45**, 61–8.

Embleton, B. J. J., Veevers, J. J., Johnson, B. D. and Powell, C. McA. (1980). Palaeomagnetic comparison of a new fit of East and West Gondwanaland with the Smith and Hallam fit. *Tectonophys.* **61**, 381–90.

Embleton, B. J. J. and McElhinny, M. W. (1982). Marine magnetic anomalies, palaeomagnetism and the drift history of Gondwanaland. *Earth Planet. Sci. Lett.* **58**, 141–50.

Embleton, B. J. J., Fisher, N. I., and Schmidt, P. W. (1983). Analytical comparison of apparent polar wander paths. *Earth Planet. Sci. Lett.* **64**, 276–82.

Emerick, C. M. and Duncan, R. A. (1982). Age progressive volcanism in the Comores Archipelago, western Indian Ocean and its implications for Somali plate tectonics. *Earth Planet. Sci. Lett.* **60**, 415–28.

Fisher, R. L. and Sclater (1983). Tectonic evolution of the south-west Indian Ocean since the mid-Cretaceous: plate motions and stability of the pole and Antarctica/Africa for at least 80 Myr. *Geophys. J. Roy. Astron. Soc.* **73**, 553–76.

Francis, T. J. G., Davies, D., and Hill, M. N. (1966). Crustal structure between Kenya and the Seychelles. *Phil. Trans. Roy. Soc. Lond.* **A259**, 240–61.

Gatinsky, Y. G. (1986). Geodynamics of Southeast Asia in relation to the evolution of ocean basins. *Palaeogeogr., Palaeoclimatol., Palaeoecol.* **55**, 127–44.

Goslin, J., Segoufin, J., Schlich, R., and Fisher, R. L. (1980). Submarine topography and shallow structure of the Madagascar Ridge, western Indian Ocean. *Geol. Soc. Amer., Bull.* **91**, 741–53.

Goslin, J., Recq, M., and Schlich, R. (1981). Structure profonde du Plateau de Madagascar: relations avec le Plateau du Crozet. *Tectonophys.* **76**, 75–97.

Hailwood, E. A. (1974). Palaeomagnetism of the Msissi norite (Morocco) and the Palaeozoic reconstruction of Gondwanaland. *Earth Planet. Sci. Lett.* **23**, 376–86.

Hailwood, A. E. and Mitchell, J. G. (1971). Palaeomagnetic and radiometric results from Jurassic intrusions in south Morocco. *Geophys. J. Roy. Astron. Soc.* **24**, 351–64.

Hallam, A. (1983). Supposed Permo-Triassic megashear between Laurasia and Gondwana. *Nature* **301**, 499–502.

Hawkes, D. D. (1981). Tectonic segmentation of the northern Antarctic Peninsula. *Geology* **9**, 220–4.

Heirtzler, J. R., *et al.* (1978). The Argo Abyssal Plain. *Earth Planet. Sci. Lett.* **41**, 21–31.

Helmcke, D. (1982). On the Variscan evolution of central Mainland Southeast Asia. *Earth Evol. Sci.* **2**, 309–19.

Howell, D. G. (1980). Mesozoic accretion of exotic terranes along the New Zealand segment of Gondwanaland. *Geology* **8**, 487–91.

Hutchinson, C. W. (1982). Southeast Asia. In *The ocean basins and margins, 6, the Indian Ocean* (ed. A. E. M. Nairn and F. G. Stehli), pp. 451–512. Plenum, New York.

Idnurm, M. (1985*a*). Late Mesozoic and Cenozoic palaeomagnetism of Australia—I. A redetermined apparent polar wander path. *Geophys. J. Roy. Astron. Soc.* **83**, 399–418.

Idnurm, M. (1985*b*). Late Mesozoic and Cenozoic palaeomagnetism of Australia—II. Implications for geomagnetism and true polar wander. *Geophys. J. Roy. Astron. Soc.* **83**, 419–33.

Idnurm, M. (1986). Late Mesozoic and Cenozoic palaeomagnetism of Australia—III. Bias-corrected pole paths for Australia, Antarctica and India. *Geophys. J. Roy. Astron. Soc.* **86**, 277–87.

Irving, E. and Irving, G. A. (1982). Apparent polar wander paths Carboniferous through Cenozoic and the assembly of Gondwana. *Geophys. Surv.* **5**, 141–88.

Johnson, B. D., Powell, C. McA., and Veevers, J. J. (1976). Spreading history of the eastern Indian Ocean and Greater India's northward flight from Antarctica and Australia. *Geol. Soc. Amer., Bull.* **87**, 1560–66.

KAHLE, H.-G., NAINI, B. R., TALWANI, M., and ELDHOLM, O. (1981). Marine geophysical study of Comorin Ridge, north Central Indian Basin. *J. Geophys. Res.* **86**, 3807–14.

KENT, P. E. (1972). Mesozoic history of the East Coast of Africa. *Nature* **238**, 147–8.

KENT, P. E. (1982). The Somali Ocean Basin and the continental margin of East Africa. In *The ocean basins and margins, 6, the Indian Ocean.* (ed. A. E. M. Nairn and F. G. Stehli), pp. 185–204, Plenum, New York.

KENT, P. E. and PERRY, J. T. O'B. (1973). The development of the Indian Ocean margin in Tanzania. In *Sedimentary basins of the African coast.* (ed. G. Blant), pp. 113–31.

KING, L. C. (1980). Gondwanaland revisited. *Geology* **8**, 111–12.

KLOOTWIJK, C. T. and BINGHAM, D. K. (1980). The extent of Greater India, III. Palaeomagnetic data from the Tibetan Sedimentary Series, Thakkhola region, Nepal Himalaya. *Earth Planet. Sci. Lett.* **51**, 381–405.

KLOOTWIJK, C. T., SHAH, S. K., GERGAN, J., SHARMA, M. L., TIRKEY, B., and GUPTA, B. K. (1983). A palaeomagnetic reconnaissance of Kashmir, northwestern Himalaya, India. *Earth Planet. Sci. Lett.* **63**, 305–24.

KLOOTWIJK, C. T., CONAGHAN, P. J., and POWELL, C. McA. (1985). The Himalayan Arc: large-scale continental subduction, oroclinal bending and back-arc spreading. *Earth Planet. Sci. Lett.* **75**, 167–83.

KONO, M. and TANAKA, H. (ed.) (1985). *Rock magnetism and paleogeophysics.* DELP Publ., 4. (100 pp.)

KRISTOFFERSEN, Y. and HAUGLAND, K. (1986). Geophysical evidence for the East Antarctic plate boundary in the Weddell Sea. *Nature* **322**, 538–41.

LABRECQUE, J. L. and HAYES, D. E. (1979). Seafloor spreading history of the Agulhas Basin. *Earth Planet. Sci. Lett.* **45**, 411–28.

LALLAMEND, S. and JOLIVERT, L. (1986). Japan Sea: a pull-apart basin? *Earth Planet. Sci. Lett.* **76**, 375–89.

LARSON, R. L., CARPENTER, G. B., and DIEBOLD, J. B. (1978). A geophysical study of the Wharton Basin near the Investigator Fracture Zone. *J. Geophys. Res.* **83**, 773–82.

LARSON, R. L., MUTTER, J. C., DIEBOLD, J. B., and CARPENTER, G. B. (1979). Cuvier Basin: a product of ocean crust formation by Early Cretaceous rifting off Western Australia. *Earth Planet. Sci. Lett.* **45**, 105–14.

LAUGHTON, A. S. (1966). The Gulf of Aden. *Phil. Trans. Roy. Soc. London* **A259**, 150–71.

LE DAIN, A. Y., TAPPONNIER, P., and MOLNAR, P. (1984). Active faulting and tectonics of Burma and surrounding regions. *J. Geophys. Res.* **89**, 453–72.

LIN, J.-I., FULLER, M., and ZHANG, W.-Y. (1985). Preliminary Phanerozoic polar wander paths for the North and South China blocks. *Nature* **313**, 444–9.

LIU, C.-S., CURRAY, J. R. and MCDONALD, J. M. (1983). New constraints on the tectonic evolution of the eastern Indian Ocean. *Earth Planet. Sci. Lett.* **65**, 331–42.

LIVERMORE, R. A., SMITH, A. G., and VINE, F. J. (1986). Late Palaeozoic to early Mesozoic evolution of Pangaea. *Nature* **322**, 162–5.

LYON-CAEN, H. and MOLNAR, P. (1984). Gravity anomalies and the structure of Western Tibet and the southern Tarim Basin. *Geophys. Res. Lett.* **11**, 1251–54.

MARKL, R. G. (1978). Basement morphology and rift geometry near the former junction of India, Australia and Antarctica. *Earth Planet. Sci. Lett.* **39**, 211–25.

MARTIN, A. K., HARTNADY, C. J. H., and GOODLAD, S. W. (1981). A revised fit of South America and south Central Africa. *Earth Planet. Sci. Lett.* **54**, 293–305.

MARTIN, A. K., GOODLAD, S. W., HARTNADY, C. J. H., and DU PLESSIS, A. (1982). Cretaceous palaeopositions of the Falkland Plateau relative to southern Africa using Mesozoic seafloor spreading anomalies. *Geophys. J. Roy. Astron. Soc.* **71**, 567–79.

MATTHEWS, D. H. (1966). The Owen fracture zone and the northern end of the Carlsberg Ridge. *Phil. Trans. Roy. Soc. Lond.* **A259**, 172–86.

MCELHINNY, M. W., EMBLETON, B. J. J., MA, X. H., and ZHANG, Z. K. (1981). Fragmentation of Asia in the Permian. *Nature* **293**, 212–16.

MCKENZIE, D. P., DAVIES, D., and MOLNAR, P. (1970). Plate tectonics of the Red Sea and East Africa. *Nature* **226**, 243–8.

MCKENZIE, D. P. and SCLATER, J. G. (1971). The evolution of the Indian Ocean since the Late Cretaceous. *Geophys. J. Roy. Astron. Soc.* **25**, 437–528.

MITCHELL, A. H. G. (1981). Phanerozoic plate boundaries in mainland SE Asia, the Himalayas and Tibet. *J. Geol. Soc. Lond.* **138**, 109–22.

MOLNAR, P. and TAPPONNIER, P. (1978). Active tectonics in Tibet. *J. Geophys. Res.* **83**, 5361–75.

MOON, T. and MERRILL, R. T. (1986). A new mechanism for stable viscous remanent magnetization and overprinting during long magnetic polarity intervals. *Geophys. Res. Lett.* **13**, 737–40.

MUTTER, J. C. and CANDE, S. C. (1983). The early opening between Broken Ridge and Kerguelen Plateau. *Earth Planet. Sci. Lett.* **65**, 369–76.

NAJID, D. (1985). *Palaeomagnetic studies in Morcco.* Ph.D. Thesis, University of Newcastle-upon-Tyne.

NI, J. and YORK, J. E. (1978). Late Cenozoic tectonics of the Tibetan Plateau. *J. Geophys. Res.* **83**, 5377–84.

NORTON, I. and SCLATER, J. G. (1979). A model for the evolution of the Indian Ocean and the breakup of Gondwanaland. *J. Geophys. Res.* **84**, 6803–30.

O'Reilly, W. (1984). *Rock and mineral magnetism*. Blackie, Glasgow. (220 pp.)

Otofuji, Y.-I. and Matsuda, T. (1984). Timing of rotational motion of Southwest Japan inferred from paleomagnetism. *Earth Planet. Sci. Lett.* **70**, 373–82.

Pigram, C. J., Surono, and Supandjono, J. B. (1985). Origin of the Sula Platform, eastern Indonesia. *Geology* **13**, 246–8.

Powell, D. E. (1982). The northwest Australian continental margin. *Phil. Trans. Roy. Soc. London* **A350**, 45–62.

Powell, C. McA., Johnson, B. D., and Veevers, J. J. (1980). A revised fit of East and West Gondwanaland. *Tectonophys.* **63**, 13–29.

Pozzi, J.-P., *et al.* (1984). Paleomagnetism of the Xigaze ophiolite and flysch (Yarlung Zaigbo suture zone, southern Tibet): latitude and direction of spreading. *Earth Planet. Sci. Lett.* **70**, 383–94.

Rabinowitz, P. D. and Labrecque, J. (1979). The Mesozoic South Atlantic Ocean and evolution of continental margins. *J. Geophys. Res.* **84**, 5973–6002.

Rabinowitz, P. D., Coffin, M. L., and Falvey, D. (1983). The separation of Madagascar and Africa. *Science* **220**, 67–9.

Ramsay, D. C., Colwell, J. B., Coffin, M. F., Davies, H. L., Hill, P. J., Pigram, C. J., and Stagg, H. M. J. (1986). New findings from the Kerguelen Plateau. *Geology* **14**, 589–93.

Rand, H. M. and Mabesoone, J. M. (1982). Northeastern Brazil and the final separation of South America and Africa. *Palaeogeogr., Palaeoclimatol., Palaeoecol.* **38**, 163–83.

Rickard, M. J. and Belbin, L. (1980). A new continental assembly for Pangaea. *Tectonophys.* **63**, 1–12.

Ridd, M. F. (1980). Possible Palaeozoic drift of SE Asia and Triassic collision with China. *J. Geol. Soc. London* **137**, 635–40.

Salmon, E., Montigny, R., Edel, J. B., Pique, A., Thuizat, R., and Westphal, M. (1986). The Msissi Norite revisited: K/Ar dating, petrography and paleomagnetism. *Geophys. Res. Lett.* **13**, 741–3.

Schlich, R. (1982). The Indian Ocean: aseismic ridges, spreading centers, and ocean basins. In *The ocean basins and margins, 6, The Indian Ocean* (ed. A. E. M. Nairn and F. G. Stehli), pp. 51–147. Plenum, New York.

Sclater, J. G., Luyendyk, B. P., and Meinke, L. (1976). Magnetic lineations in the southern part of the Central Indian Basin. *Geol. Soc. Amer., Bull.* **87**, 371–8.

Sclater, J. G., Fisher, R. L., Patriat, P., Tapscott, C., and Parsons, B. (1981). Eocene to Recent developments of the South-west Indian Ridge, a consequence of the evolution of the Indian Ocean triple junction. *Geophys. J. Roy. Astron. Soc.* **64**, 587–604.

Scrutton, R. A. (1973). Structure and evolution of the sea floor south of South Africa. *Earth Planet. Sci. Lett.* **19**, 250–6.

Segoufin, J. (1978). Anomalies magnetiques mésozoiques dans le bassin du Mozambique. *C. R. Acad. Sci. Paris* **287D**, 109–11.

Segoufin, J. and Patrick, P. (1980). Existance d'anomalies mésozoiques dans le bassin du Somali. Implications pour les rélations Afrique–Antarctique–Madagascar. *C. R. Acad. Sci. Paris* **B.291**, 85–8.

Şengör, A. M. C. (1984). The Cimmeride orogenic system and the tectonics of Asia. *Geol. Soc. Amer., Spec. Pap.* **195**, 74 pp.

Simpson, E. S. W., Sclater, J. G., Parsons, B., Norton, I., and Meinke, L. (1979). Mesozoic magnetic lineations in the Mozambique Basin. *Earth Planet. Sci. Lett.* **43**, 260–4.

Sinha, M. C., Louden, K. E., and Parsons, B. (1981). The crustal structure of the Madagascar Ridge. *Geophys. J. Roy. Astron. Soc.* **66**, 351–77.

Smith, A. G. and Hallam, A. (1970). The fit of the southern continents. *Nature* **225**, 139–44.

Steiner, M. (1983). Mesozoic apparent polar wander and plate motions of North America. In *Mesozoic paleogeography of West–Central United States.* (ed. M. W. Reynolds and E. D. Dolly), pp. 1–11. Soc. Econ. Paleont. Mineral.

Stocklin, J. (1980). Geology of Nepal and its regional frame. *J. Geol. Soc., London* **137**, 1–34.

Tapscott, C., Patriat, P., Fisher, R. L., Sclater, J. G., Hoskins, H., and Parsons, B. (1980). The Indian Ocean triple junction. *J. Geophys. Res.* **85**, 4723–39.

Tarling, D. H. (1971). Gondwanaland, palaeomagnetism and continental drift. *Nature* **229**, 17–21.

Tarling, D. H. (1983). *Palaeomagnetism*. Chapman & Hall, London. (379 pp.)

Tarling, D. H. (1985*a*). Siluro-Devonian palaeogeographies based on palaeomagnetic observations. *Phil. Trans. Roy. Soc. London* **B309**, 81–3.

Tarling, D. H. (1985*b*). Problems in Palaeozoic palaeomagnetism. *J. Geodynamics* **3**, 87–103.

Tarling, D. H. (1985*c*). Carboniferous reconstructions based on Palaeomagnetism. *C.R. 10th Congrès International Stratigraphie et Géologie Carbonifère*, Madrid, **3**, 153–62.

Tarling, D. H. and Kent, P. E. (1976). The Madagascar controversy still lives. *Nature* **261**, 304–5.

Thompson, R. and Clark, R. M. (1982). A robust least-squares Gondwanan apparent polar wander path and the question of palaeomagnetic assessment of Gondwanan reconstructions. *Earth Planet. Sci. Lett.* **57**, 152–8.

Tucholke, B. E., Houtz, R. E., and Barret, D. M. (1981). Continental crust beneath the Agulhas Plateau, southwest Indian Ocean. *J. Geophys. Res.* **86**, 3791–806.

Veevers, J. J. (1982). Western and northwestern margin of Australia. In *The ocean basins and margins, 6, the Indian Ocean*. (ed. A. E. M. Nairn and F. G. Stehli), pp. 513–44, Plenum, New York.

WENSINK, H. (1981). Le contact Gondwana–Eurasie en Iran d'après les recherches paléomagnétiques. *Bull. Géol. Soc. Fr.* **23**, 547–52.

WENSINK, H. (1983). Paleomagnetism of red beds of Early Devonian age from Central Iran. *Earth Planet. Sci. Lett.* **63**, 305–34.

WESTPHAL, M., POZZI, J.-P., ZHOU, Y. X., XING, L. S., and CHEN, X. Y. (1983). Palaeomagnetic data about southern Tibet (Xijang)—2. The Cretaceous formations of the Lhasa block. *Geophys. J. Roy. Astron. Soc.* **73**, 507–22.

WILLIAMS, D. F. and CORLISS, B. H. (1982). The South Australian continental margin and the Australian–Antarctic sector of the Southern Ocean. In *The Ocean Basins and Margins, 6, the Indian Ocean*. (ed. A. E. M. Nairn and F. G. Stehli), pp. 545–84, Plenum, New York.

D. H. TARLING, Department of Geological Sciences, Plymouth Polytechnic, Plymouth PL4 8AA, UK.

Evolution of the southern margin of Tethys (North Australian region) from early Permian to late Cretaceous

M. G. Audley-Charles

ABSTRACT: A mid-Permian rifting episode appears to have removed continental blocks (now forming parts of Asia) from Australian Gondwana, but the present continental margin of Northern Australia–central New Guinea was formed by a later episode of rifting during the Jurassic. With the exception of the north-west Australian offshore region west of the Scott Plateau, this rifted continental margin has been greatly modified during the Cainozoic by collisions with arc–trench systems. Collision resulted in fold and thrust mountains, locally with ophiolite emplacements, that now characterize most of northern New Guinea and the Banda Arc. The identification of the Australia–New Guinea rifted continental margin in what is now the fold and thrust mountain belt of New Guinea, islands of the Outer Banda Arc, East Sulawesi, and Buton, rests largely on stratigraphic correlation and the recognition of the pre-rift sequence, breakup unconformity, and the post-breakup marine transgression and prolonged subsidence in these now highly deformed rocks. One implication of the Jurassic continental rifting is that the pre-rift rocks of the region accumulated in an intracratonic basin of eastern Gondwana. Another important implication is that the continental blocks rifted from the Northern Australia–central New Guinea margin must have drifted northwards towards Asia as the Mesozoic Tethys ocean spread across the tropics and equatorial region. These continental fragments of Gondwana collided with parts of south-east Asia that had been derived from Gondwana in an earlier phase of rifting.

Introduction

Since the recognition of the late Jurassic magnetic ocean-floor spreading lineations in the north-east Indian Ocean adjacent to the continental margin of north-western Australia (Veevers and Heirtzler, 1974), it has been obvious that a very large continental block (or several blocks) must have been rifted from Northern Australia and central New Guinea (Figure 1). Furthermore, the petroleum company seismic-reflection surveys of the north-west Australian shelf revealed abundant indications of extensional tectonics during the Permian and Mesozoic, with a major breakup unconformity of late Jurassic–early Cretaceous age (Powell 1976; Falvey and Mutter 1981, and many others). Hamilton (1979) drew attention to the indications that a rifted continental margin must be present in central New Guinea, where its age appeared to fall within the range of late Triassic to mid-Jurassic. Pigram and Panggabean (1984), calling upon the available stratigraphic data in New Guinea and some of the islands of the Banda Arc, were able to recognize the breakup unconformity, and subsequent marine transgression with subsidence, in New Guinea and part of the Banda Arc. However, Pigram and Panggabean (1984) built their interpretation on the concept of microcontinents, which they postulated wherever they recognized indications of distinct crustal blocks (Fig. 2). In doing this they were not concerned to demonstrate the evidence for oceanic crust being created or destroyed between their postulated microcontinents, and neither were they aiming to identify the usual diagnostic continental margin sequences and structures of shelf, slope, and rise. Some criticism of the details of their interpretation and methodology has already been made by Dow and Sukamto (1986). The stratigraphy and structure of the region seem to offer little support for these postulated microcontinents, and in some cases these data appear to conflict strongly with the concept. This applies particularly to the Banda Arc.

This paper presents a summary of the stratigraphic information available from the complexly deformed region of the Banda Arc and Sulawesi, and attempts a correlation with the data from the Australian shelf and New Guinea (Figs 3, 4, 5, 6, and 7). From this, the breakup unconformity and subsequent subsidence with marine transgression have been interpreted. The series of palaeogeographical maps (Figs 8, 9, 10, 11, 12) follow an interpretation of the

From AUDLEY-CHARLES, M. G. & HALLAM, A. (eds) *Gondwana and Tethys* Geological Society Special Publication No. 37, pp. 79–100.

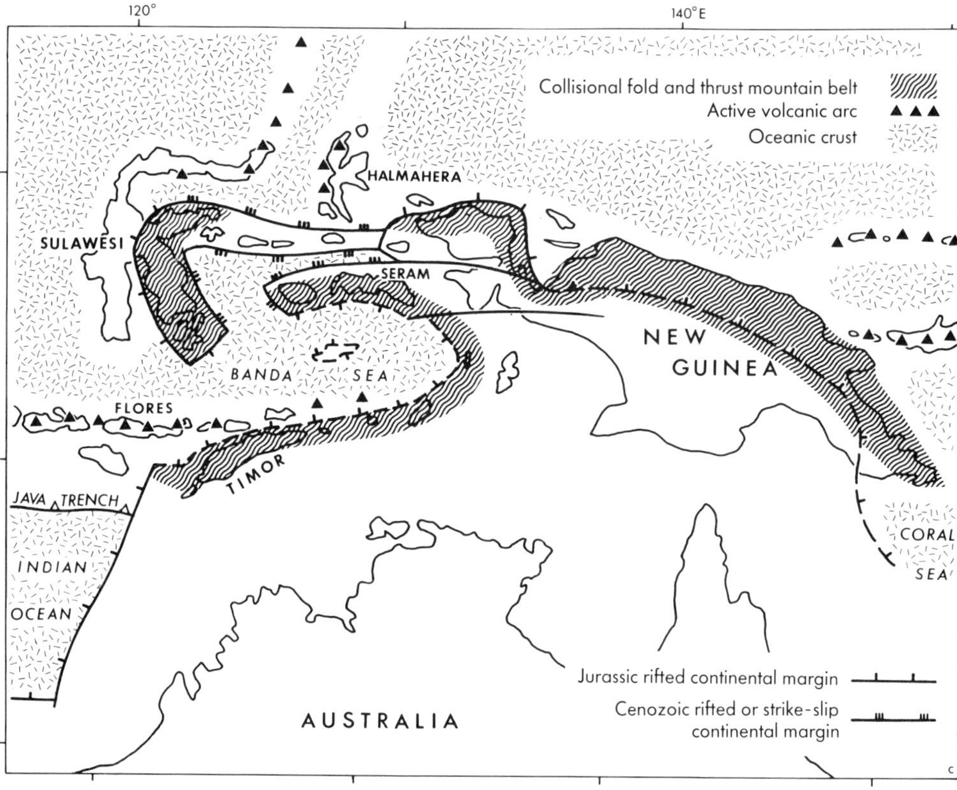

FIG. 1. Location of the Jurassic rifted continental margin of Australian Gondwana and its modification by Cainozoic tectonic collision.

evolution of this part of Gondwana put forward by Audley-Charles (1983*b*, 1984). This may be summarized as the development of a rifted continental margin in the middle Permian, as Iran, North Tibet and Indo-China were removed by the creation of the new ocean Tethys II, followed by another episode of continental rifting in the Jurassic which, by spreading of Tethys ocean III, removed South Tibet, Burma, western Thailand, Malaya and Sumatra. It was this Jurassic rifting that produced the present continental margin of northern Australia–central New Guinea.

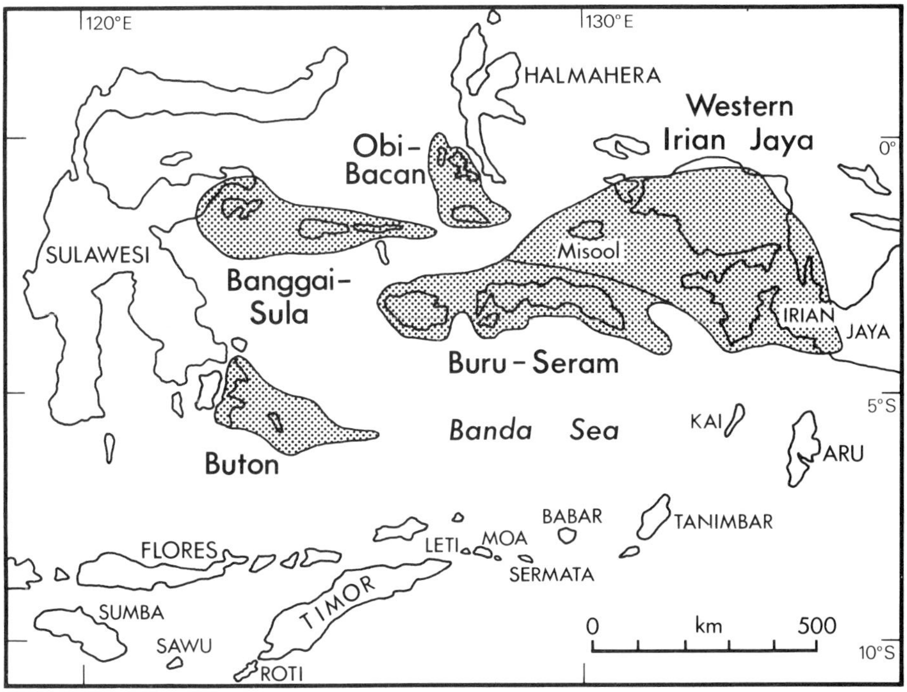

FIG. 2. Concept of microcontinents, applied to breakup of Australian Gondwana by Pigram and Panggabean (1984). Their proposed microcontinents are ornamented.

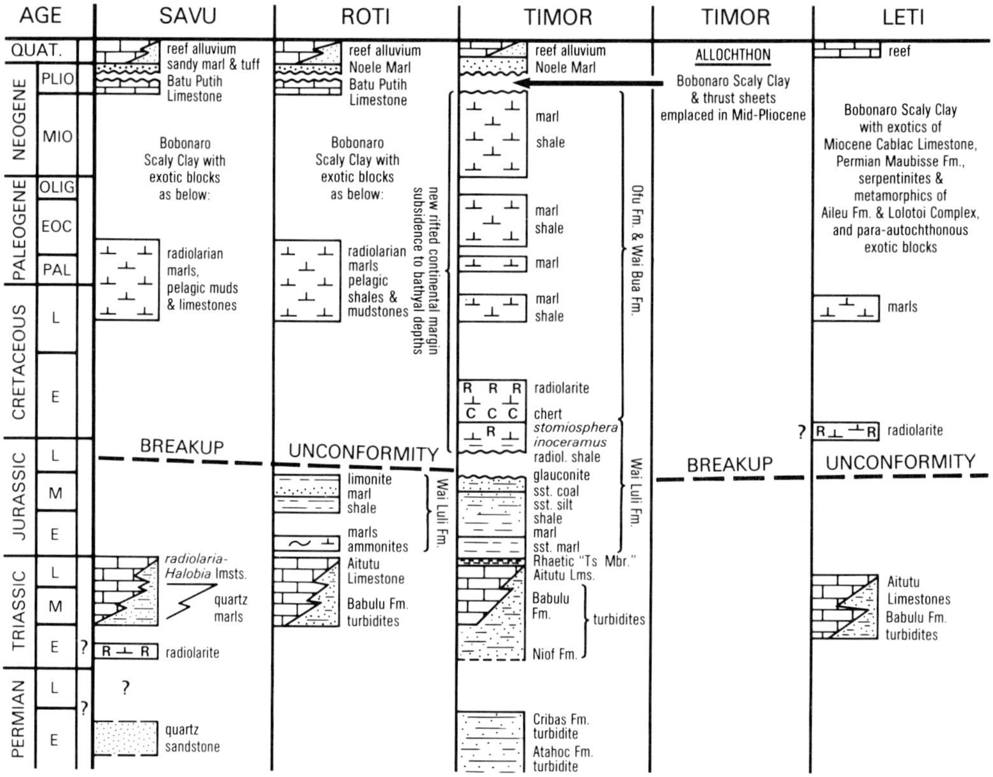

FIG. 3. Proposed correlation of sections in some Outer Banda Arc islands. Data from mapping in Savu, Roti, and Timor by Audley-Charles and others, for Leti from Van Bemmelen (1949). Stratigraphy of Timor partly after Cook (1987) and Bird (1987).

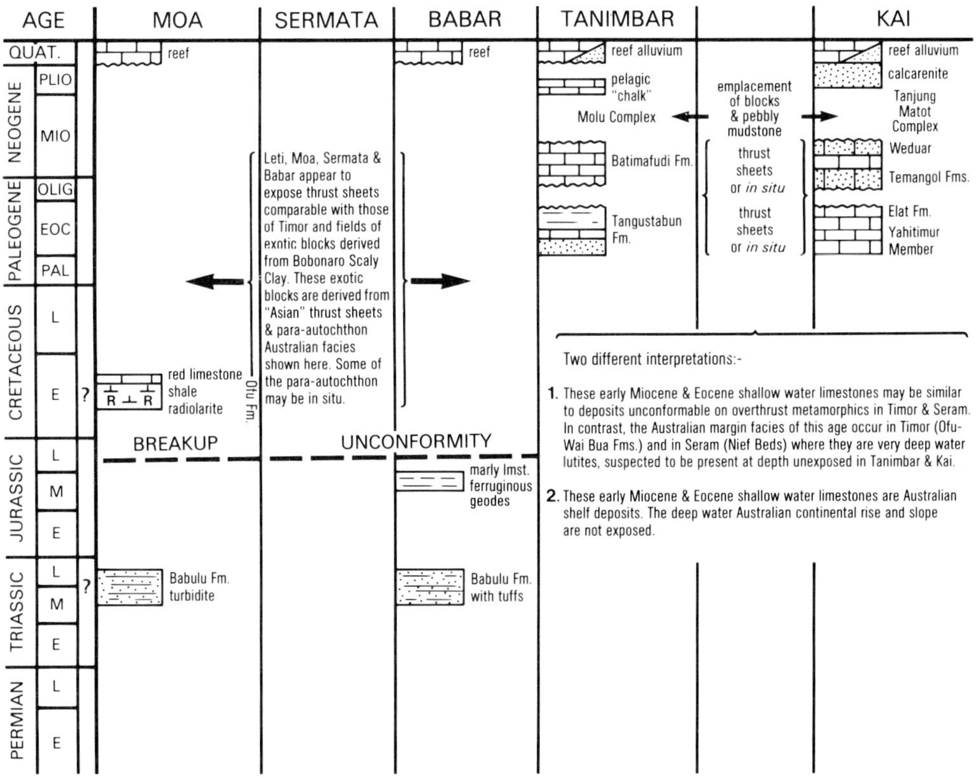

FIG. 4. Proposed correlation of sections in some Outer Banda Arc islands, based on data from Van Bemmelen (1949), Sukardi and Sutrisno (1981), and Achdan and Turkandi (1982).

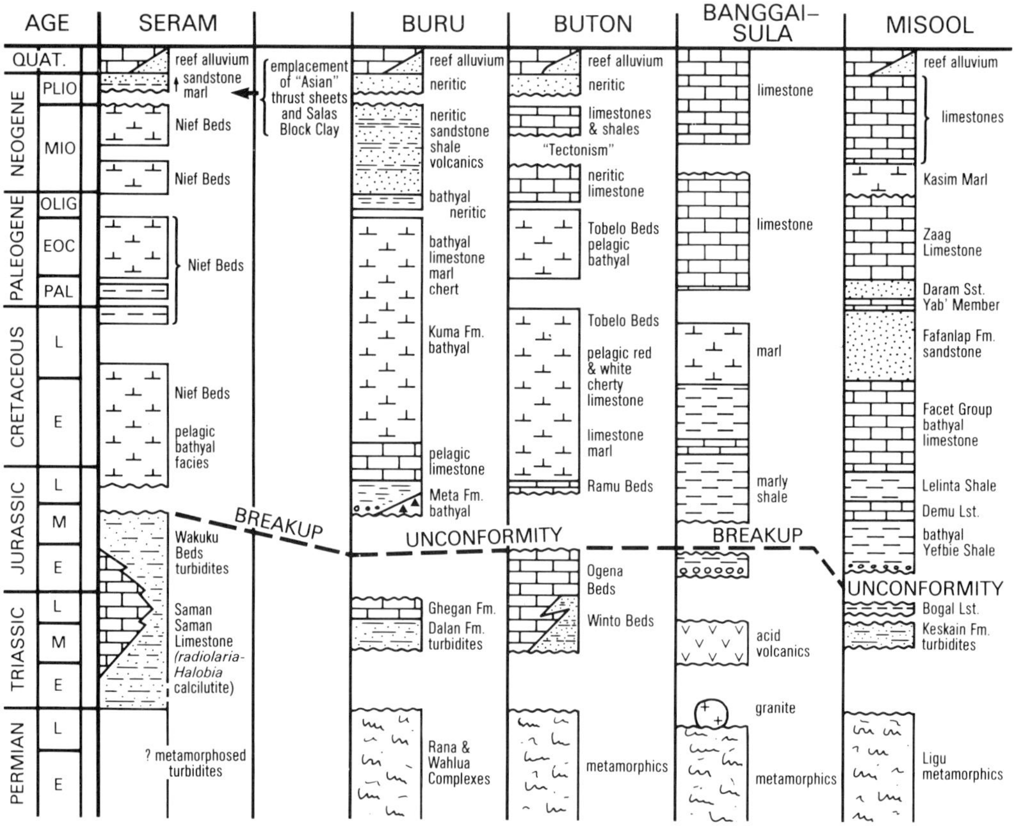

Fig. 5. Proposed correlation in some Outer Banda Arc islands based on data from Van Bemmelen (1949), Audley-Charles (1978), Audley-Charles *et al.* (1979), Tjokrosapoetro and Budhitrisna (1982), Smith (1983), and Pigram and Panggabean (1984).

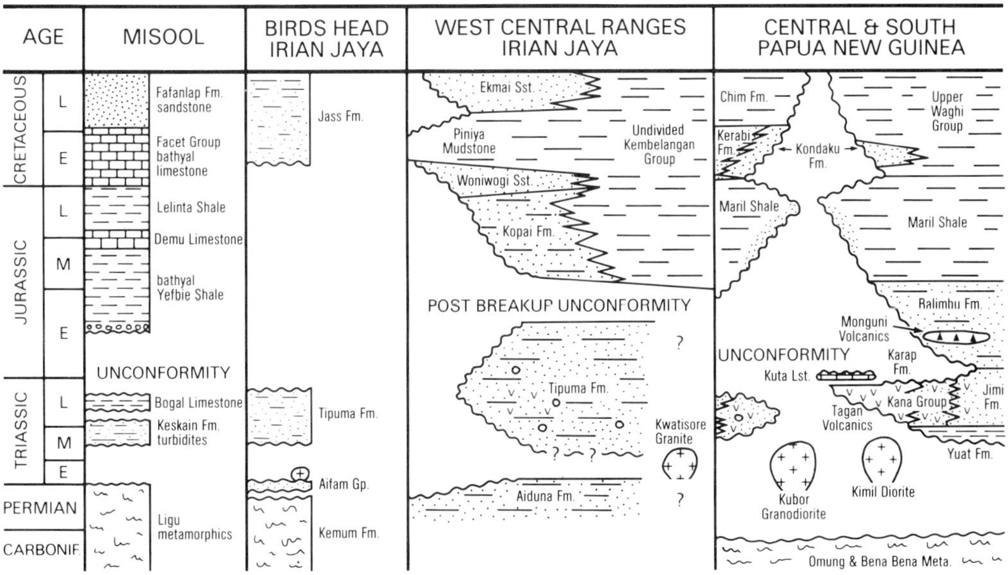

FIG. 6. Proposed correlation of Misool and New Guinea from Pigram and Panggabean (1984).

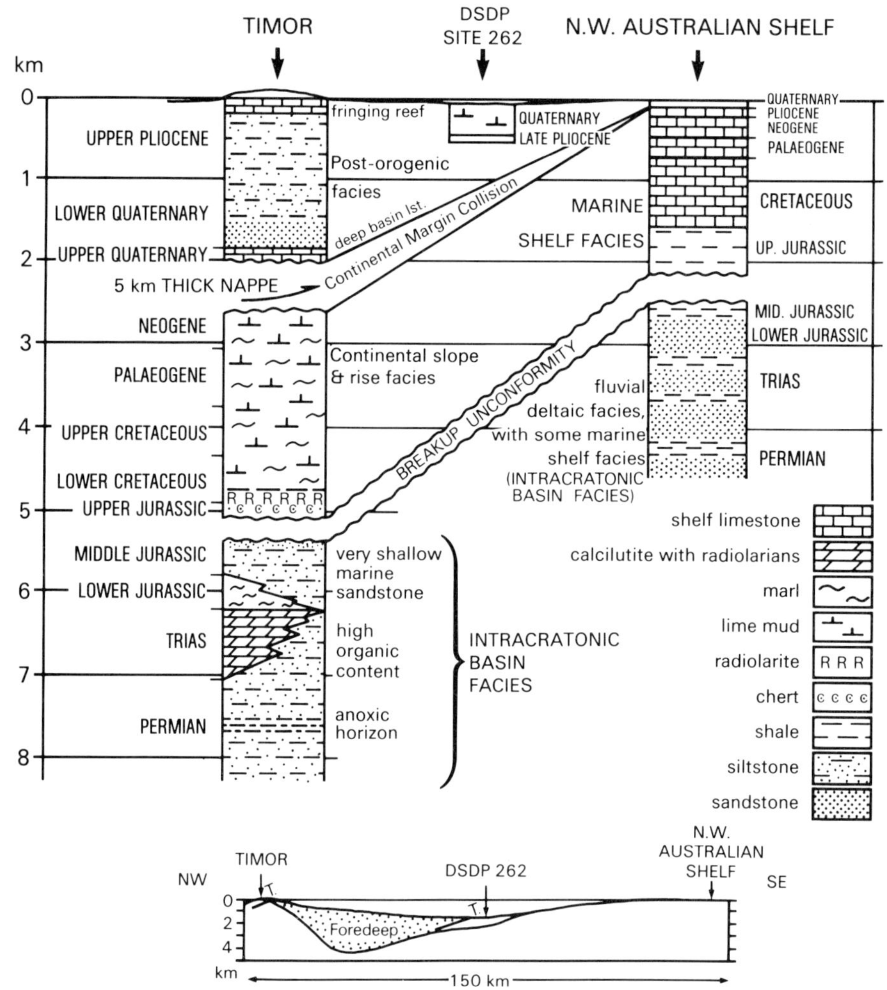

FIG. 7. Correlation of stratigraphic succession in Timor and north-west Australian shelf. (After Audley-Charles 1986*b*.) Note that both pre-breakup sections accumulated in an intracratonic basin. See Fig. 14 for pre-collision sections.

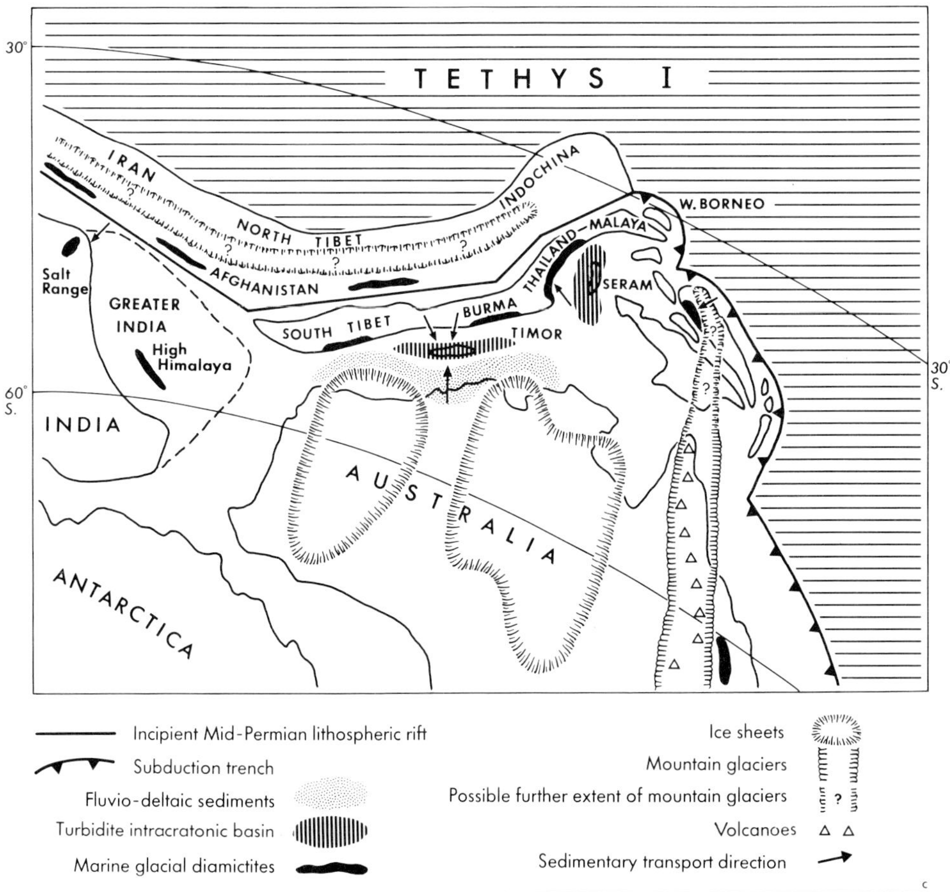

FIG. 8. Palaeogeographical sketch map of Australian Gondwana during late Carboniferous to early Permian times (modified after Audley-Charles (1984)); Permian sediment transport directions from Bird (1987). Outlines are for reference only.

FIG. 9. Palaeogeographical sketch map of Australian Gondwana during late Permian to early Triassic times. The warm current was proposed by Robinson (1973). Outlines for reference only.

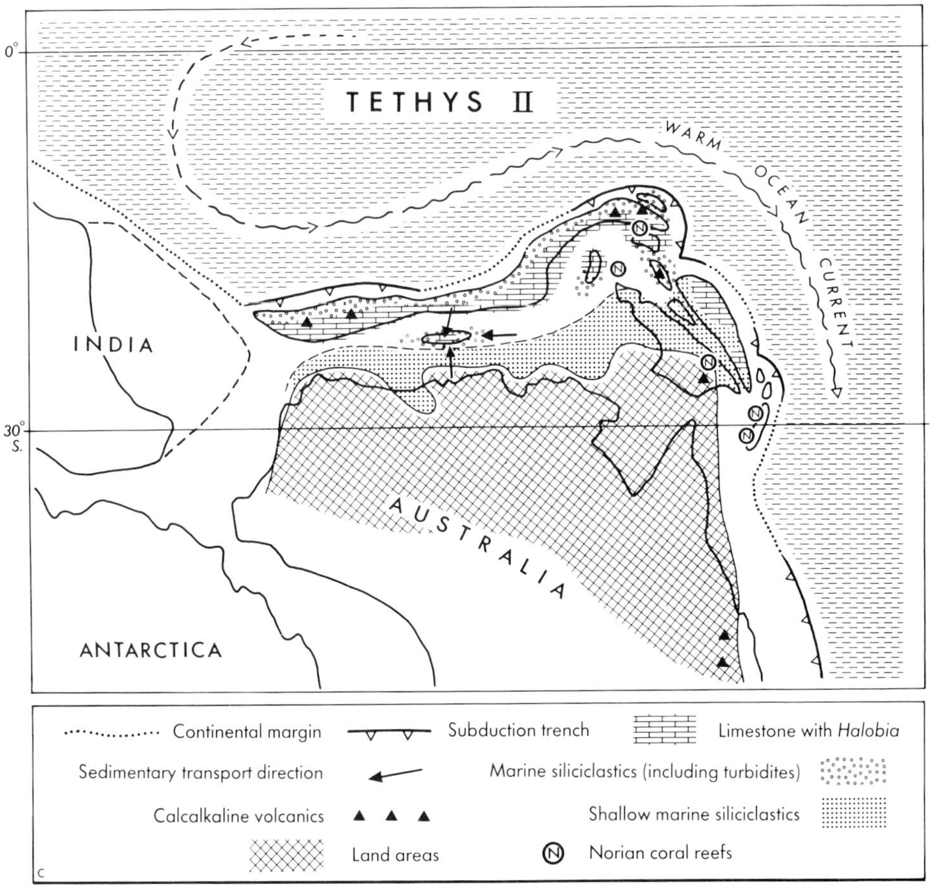

FIG. 10. Palaeogeographical sketch map of Australian Gondwana during late Triassic times. The warm current was proposed by Tollmann and Kristan-Tollmann (1985). Sediment transport directions from Cook (1986). Outlines for reference only. (Partly after Audley-Charles, Ballantyne and Hall (In press).)

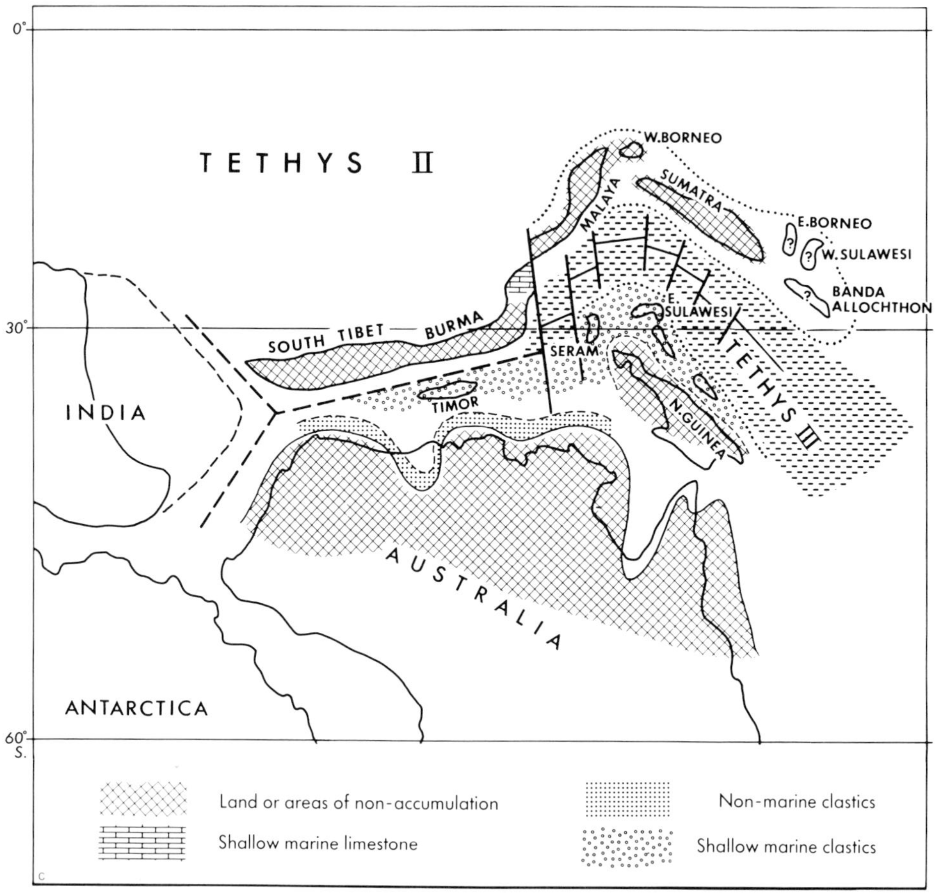

FIG. 11. Palaeogeographical sketch map of Australian Gondwana during late Jurassic times. Outlines for reference only. (Partly after Audley-Charles, Ballantyne, and Hall (In press).)

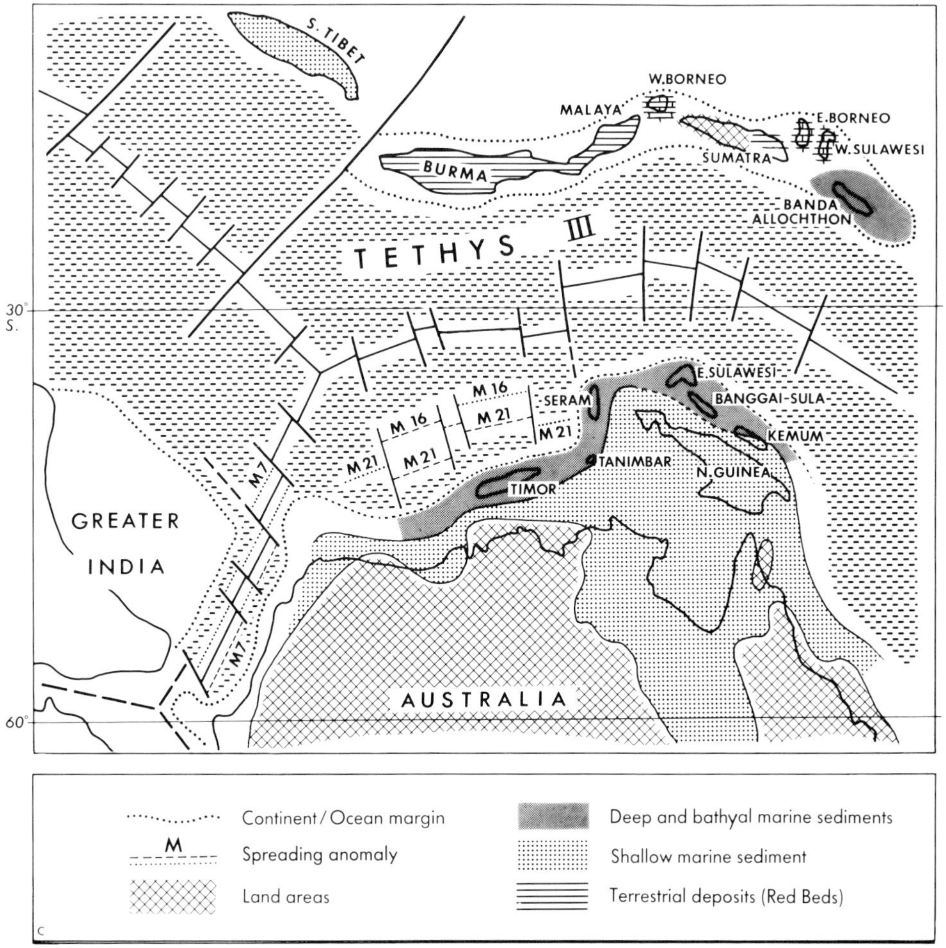

FIG. 12. Palaeogeographical sketch map of Australian Gondwana during early Cretaceous times. Outlines for reference only. (Partly after Audley-Charles, Ballantyne and Hall (In press).)

Identification of the northern Australia–New Guinea continental margin

The gravity survey data for the northern Australian shelf and Timor region of the Banda Arc indicate that the Australian continental crust continues below the 2–3-km deep Timor Trough to the north coast of Timor, where it ends abruptly (Chamalaun *et al.* 1976). Seismic refraction surveys in the Timor–Tanimbar Trough and over the north-east Australian shelf (Jacobson *et al.* 1978; Bowin *et al.* 1980) have shown that the Australian continental crust extends into the islands of the Outer Banda Arc (Timor via Tanimbar, Kai to Seram and Buru). Similarly, geophysical data have demonstrated that the Australian continental basement continues from Cape York Peninsula below the Torres Strait into southern New Guinea (O'Brien *et al.* 1961; Vind and Harwood, 1965).

The lithofacies, faunas, and floras of the Permian and Triassic rocks in the Outer Banda Arc islands appear to be closely related to those of the northern Australian shelf (Fig. 14), as seen in seismic-reflection surveys (Powell 1976), and in the various wells on the shelf (Bird 1987 and Cook 1986). Similarly, the Cretaceous and pre-collision Cainozoic rocks of the Outer Banda Arc islands can be correlated with those of the north Australian shelf (Audley-Charles *et al.* 1979, Schluter and Fritsch, 1985). It has also been shown that the Mesozoic and Cainozoic deposits of the Carpentaria Basin (east of Cape

York) can be traced from northern Australia into the southern part of Papua New Guinea (Smart and Senior 1980; Smart *et al.* 1980; and Veevers 1984, pp. 127–9). It has been argued by Pigram and Panggabean (1984) that the post-breakup unconformity can be traced from central and southern Papua New Guinea (PNG), where it is of early Jurassic or possibly even late Triassic age, to western PNG, where it is of Early Jurassic age, to the western central ranges of Irian Jaya, where it is early Mid-Jurassic in age. This transgressive unconformity reflects the marine waters flooding over the terrestrial sediments of the New Guinea hinterland, as a consequence of subsidence of the cooling lithosphere. This subsidence followed the thermal uplift associated with lithospheric rifting and the initiation of spreading of new Tethys oceanic crust.

Hamilton (1979) and Pigram and Panggabean (1984) used stratigraphic and geophysical indications to draw the line, marking the edge of the rifted continental margin in New Guinea, as it formed in the Jurassic. Similarly, it is possible to draw this line in the region of the Outer Banda Arc although, as in New Guinea, the Australian continental margin deposits and the underlying pre-rift intracratonic basin deposits have been strongly deformed. They are now found in Seram, and in the southern Banda Arc islands such as Timor, strongly folded and thrust with a vergence towards the Australian continent. Although the present continental margin of Australia can be recognized on geophysical and geological criteria on the Banda Sea side of the Outer Banda Arc islands, it might be argued that the Jurassic rifted margin does not correspond closely with the present margin. This is because the late Cainozoic collision of the rifted Australian continental margin involved this margin underthrusting the forearc, so that the more distal part of the margin may have been subducted. However, the distal continental rise deposits of rifted margins accumulate on oceanic, not continental, crust. Thus, the position of the Jurassic rifted continental margin may correspond quite closely with the present margin, unless the continental crust has been subducted. The Chamalaun *et al.* (1976) view, based on the gravity data and supported by deep magnetic sounding (Chamalaun and White 1975)—that the present limit of the Australian continent corresponds to the northern edge of the southern Banda Arc, appears justified on the basis of the strata exposed in the Outer Banda Arc islands, for example in Timor, where Australian siliciclastic facies of Permian and Triassic age are exposed along the north coast.

Pigram and Panggabean (1984) interpreted the edge of the Australian continental rifted margin, in the Banda Arc region, as corresponding to the present axis of the Timor Trough. That view appears to be based on the interpretation of the earthquake data, i.e. that these data indicate Timor Trough marks the trace of the Benioff zone (McCaffrey *et al.* 1984; McCaffrey *et al.* 1985, and McCaffrey and Nabelek, 1986). This has been challenged by Chamalaun *et al.* (1981), Audley-Charles (1983*a*), and Price and Audley-Charles (1983). The Pigram and Panggabean (1984) interpretation ignores the gravity and deep magnetic sounding data, and takes no account of the exposed stratigraphy on the island of Timor.

The northern arm of the Outer Banda Arc, which includes the archipelago of small islands between Kai Ketjil and Seram, and also includes the island of Seram, is designated by Hamilton (1979) as part of the Australian continental block. Hamilton drew the continental margin at the inner (Asian) boundary of these islands. The geology and gravity field of Seram appear to be a mirror image of those of Timor, thus supporting the suggestion that the continental margin of Australia corresponds with the inner (Asian) margin of the islands of the Outer Banda Arc (Fig. 1). The reported presence of continental metamorphic basement exposed in Buru (Tjokrosapoetro and Budhitrisna, 1982), with only gently deformed Mesozoic and Cainozoic strata unconformable above it, suggests that Buru represents a detached continental block from Irian Jaya (Hamilton, 1979). The case for considering Buru as part of Mesozoic Australia is strongly supported by the stratigraphic sequence; its correlation (Fig. 5) has been discussed by Audley-Charles (1978), Pigram and Panggabean (1984), and others.

Hamilton (1979), and Pigram and Panggabean (1984), followed Klompe (1956) in regarding Buton and Sula–Banggai islands as detached parts of the Australia–New Guinea continent. The stratigraphic evidence (Figs 5, 6, and 13) for this has been discussed by Audley-Charles (1978), Smith (1983), and Pigram and Panggabean (1984).

The east and south-east arms of Sulawesi are usually regarded as integral parts of Sulawesi (Silver *et al.* 1978; Hamilton 1979; Pigram and Panggabean 1984). However, the presence of Australian-type shallow-marine facies of Triassic and Jurassic age, and the presence of deep-marine Cretaceous pelagic facies—imbricated with shallow-marine Eocene carbonates and fault slivers of ophiolite—in the east arm (Kundig, 1956) led Audley-Charles (1978) to

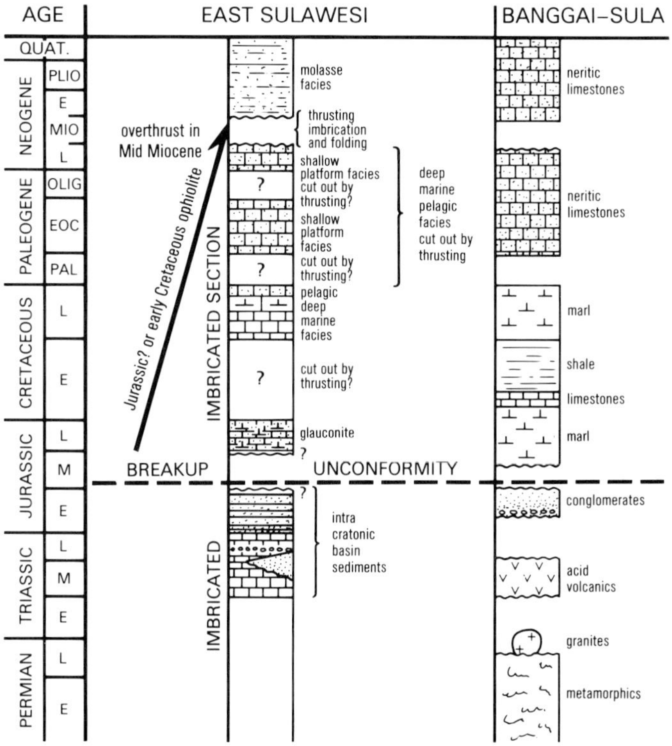

FIG. 13. Proposed correlation of Sulawesi and Banggai-Sula from data in Westermann *et al.* (1978), Sukamto and Simandjuntak (1983), and Simandjuntak (1986). Note that the Mesozoic ophiolite may have first overthrust in the Palaeocene. The Miocene thrusting appears to be related to the collisional tectonics.

suggest that eastern Sulawesi was detached from Australia–New Guinea by rifting (Fig. 1). Recent work by Sukamto and Simandjuntak (1983), and by Simandjuntak (1986), suggests that the central Sulawesi province of blueschist and other metamorphic rocks represents a major tectonic suture. It separates the western Sulawesi Cainozoic carbonate platform and volcanic arc from the eastern province composed of Australian continental margin sediments and slices of ophiolitic ocean floor. If we look at the detail of this Australian continental margin, we see a 'basement' of Triassic and early Jurassic shallow-marine strata, which can be correlated with the intracratonic deposits of that age exposed in the other islands of the outer Banda Arc, such as Timor. These are overlain by late Jurassic and Cretaceous deep-marine, pelagic sediments deposited on the Australian continental slope and rise. All these sedimentary rocks are imbricated with Australian shelf carbonates and related deposits of Eocene to early Miocene age, and also with slices of ophiolite, probably representing part of the late Mesozoic ocean floor. Thus, eastern Sulawesi is interpreted here as an imbricate zone composed of the Australian continental margin with slices of ocean floor and associated metamorphic rocks. This imbricate zone occurs between the blueschists and other metamorphic rocks of the tectonic suture of central Sulawesi, and the relatively undeformed Australian platform deposits of Mesozoic and Cainozoic age that crop out in the Banggai-Sula islands—where they are unconformable upon eroded Permo-Triassic granites (Pigram and Panggabean, 1984).

One outstanding feature of the Sulawesi collision zone, is the apparent absence of the forearc accretionary wedge between the volcanic arc of western Sulawesi, and the collided fragment of Australia–New Guinea—represented in eastern Sulawesi. This absence of the forearc is remarkably similar to the Timor collision zone, where the forearc accretionary wedge is also missing between the volcanic arc and the Australian continental margin exposed in Timor. According to Price and Audley-Charles (1983 and 1987), its absence from Timor can be explained by the overriding of the forearc by the ruptured Australian lithospheric plate; perhaps a similar mechanism has operated in the central Sulawesi collision zone.

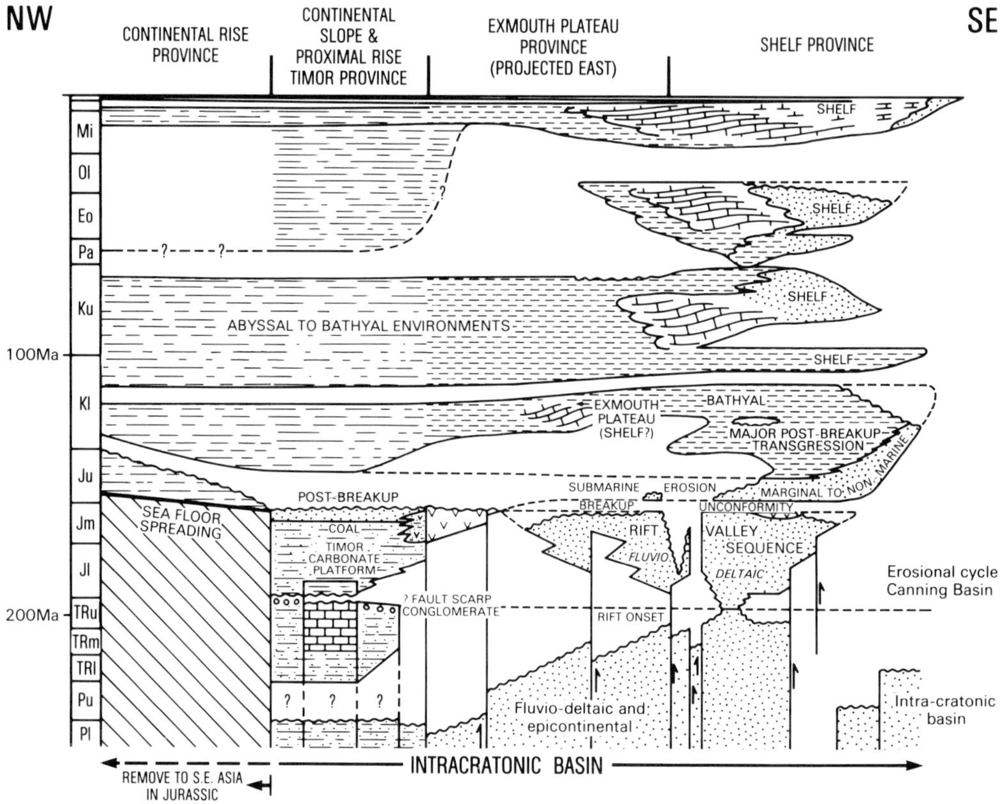

FIG. 14. Generalized time-stratigraphic cross-section of the northern Australian continental margin (after Falvey and Mutter 1981). Compare it with Fig. 7 to see the effect of collision of this rifted margin with the Sunda arc–trench system.

Stratigraphic correlation of Northern Australia, Banda Arc, Sulawesi, Buton, and New Guinea

Summaries of the stratigraphic successions are presented in Figs 3–6, 13, and 14. These demonstrate that the characteristic pre-breakup, breakup, and post-breakup successions can be recognized and correlated throughout this huge region.

The basis for the stratigraphic classification of the rocks of the Outer Banda Arc, into allochthon, para-autochthon and autochthon has been discussed in detail in Audley-Charles (1968), and revised in Audley-Charles (1986*a*). Here we will just note that the para-autochthonous strata of the Outer Banda Arc and eastern Sulawesi are correlated with the successions encountered below the shelf of northern Australia and their outcrop onshore. The allochthonous rocks of the Outer Banda Arc were emplaced in the late Cainozoic collision with the volcanic island arc, and hence do not concern us here. Similarly, the autochthonous rocks of the Outer Banda Arc which represent the late Cainozoic post-orogenic deposits are not considered in this paper.

Two outstanding features to emerge from these correlation tables is the predominance of siliciclastic facies in the pre-breakup sequences of the Banda Arc and Northern Australia, and the dramatic diminution in importance of siliciclastic deposits, particularly in the Banda Arc (and to a lesser extent in the outer part of the Australian shelf) after the breakup unconformity. The late Jurassic and early Cretaceous post-breakup transgression is characterized by muds, in which radiolarians and chert are important. They give way to carbonate mud with relatively little dilution by terrigenous mud, while silt and sand are extremely rare, particularly in the Banda Arc region, until the late Cainozoic, when the effects of collision with the Banda forearc result in the accumulation of siliciclastics.

In central and southern New Guinea, the proximity to a source of terrigenous detritus possibly led to the accumulation of siliciclastic deposits, including sand, after the post-breakup transgression. The more distal outer shelf and continental slope and rise deposits, comparable with those found in the islands of the Outer Banda Arc, probably occur at depth below the overthrusts in the mountain belt of the northern New Guinea province.

One notable feature of the pre-breakup sequence of the Outer Banda Arc is the palaeocurrent evidence for the derivation of some of the siliciclastic sediment, from the northern 'Asian' side of what was to become the new rifted continental margin (Bird 1987). The high woody-fragment content of the Permian and Triassic sandstones in Timor (Bird 1987, and Cook 1986) and Seram (Audley-Charles *et al.* 1979) imply their derivation from large deltaic complexes. Such deltas have been postulated to characterize much of the region that now lies below the northern Australian shelf, which during the Permo-Triassic was part of the large intracratonic basin (Crostella and Barter 1980; Bhatia *et al.* 1984; Cook *et al.* 1985).

Rifting of the Northern Australian continental margin

Most of the late Carboniferous and Permian sediment under the present north-western shelf area is predominantly non-marine, with marine incursions (Powell 1976; Falvey and Mutter 1981). On both seismic reflection and well data, the late Carboniferous and Permian sediments amount to about 4 km thickness locally. There is also abundant evidence for tensional movements, block faulting, and considerable local subsidence to accommodate this thickness, much of which appears to be deltaic. These large-scale regional changes may be related to the major lithospheric rifting episode at the northern margin of this part of Gondwana, which has been postulated to explain the rifting of continental blocks tentatively identified as Iran, Indo-China, and northern Tibet (Audley-Charles 1983*b*, 1984). The presence (Metcalfe, this volume) of glacial-marine diamictites of late Carboniferous–early Permian age in northern Tibet (Changtang), Afghanistan, and Iran, indicates that northern Tibet remained part of Gondwana, at least until the end of the early Permian. This further suggests a mid–late Permian rifting.

It may be important that, on the Outer Banda Arc island of Timor, Bird (1987) has identified sedimentological and faunal indications of significant changes in water depth in the intracratonic basin, which he suggests may result from Permian block faulting. This, together with the evidence of a major phase of Permian extensional faulting, from seismic reflection data of the northern Australian shelf (Powell 1976), and the regional arguments put forward by Audley-Charles (1983*b*, 1984), appear to support the rifting of continental blocks from this part of Gondwana in mid-Permian times—with the spreading of a new Tethys ocean carrying these blocks to collide with Asia (Metcalfe, this volume). Metcalfe argued that because the Carboniferous and Permian floras of Indo-China do not appear to be related to those of north-east Gondwana, we should therefore regard Indo-China as having rifted from Gondwana in the pre-Carboniferous. This apparent conflict might be resolved, if the warming influence of the Tethys ocean—at subtropical latitudes—adjacent to this part of north-east Gondwana, was sufficient to sustain a different flora along the seaboard of the continent. This is in contrast to the colder and much more severely continental climate of the vast interior of late Palaeozoic Gondwana. Robinson (1973) argued the case for a warm, equatorial, Tethys ocean current, turning southward, and sweeping eastward, along the northern shores of Australian Gondwana during the Permian. Tollmann and Kristan-Tollmann (1985) applied the same idea to the Triassic.

Seismic-reflection surveys over the northern Australian shelf have provided abundant indications of extensional faulting during the Triassic and Jurassic (Powell 1976; Veenstra 1985). Spreading of new oceanic floor, creating the north-east Indian Ocean adjacent to north-west Australia, occurred in the Callovian (Veevers and Heirtzler, 1974). It was the culmination of the extension tectonics that produced block faulting during the Triassic and Jurassic, and which resulted in the formation of the rifted continental margin of northern Australia during the Jurassic.

Seismic sections and drilling on the present shelf (Fig. 14) have found that some fault blocks moving in the late Triassic and early and middle Jurassic were strongly eroded, while elsewhere, half-grabens subsided (Laws and Kraus 1974; Veenstra 1985). The large-scale slumping and the slump folding in the Triassic and early Jurassic rocks of Timor (Bird 1987, and Cook 1986)

FIG. 15. Breccio-conglomerate of clasts of Middle and Upper Triassic Aitutu Limestone, found at the margin of that formation, and interpreted as a late Triassic contemporaneous submarine fault-scarp breccia. Near Tilomar, east Timor. (See Fig. 14.)

may be regarded as the result of block faulting during the late Triassic and early to middle Jurassic. Indications of strong erosion in Timor during the Triassic are illustrated in Figs 15 and 16. Perhaps the most spectacular indication is that of the intraformational breccio-conglomerates (Fig. 15), interpreted as a fault-scarp deposit associated with the block faulting affecting the pelagic carbonate platforms (Aitutu Formation). These platforms must have stood above the middle and late Triassic basin floor, where turbidites and siliciclastics of the Babulu Formation (Giani 1971) accumulated.

The principal indication of a regional structural event, that has a major influence on sedimentation and stratigraphy, may be seen in the exposed sections of Oxfordian–Callovian age in southern Timor. Whereas most of the Permo-Triassic and Lower Jurassic para-autochthonous deposits of Timor are siliciclastic sediments—much of them turbidites deposited in deep basinal environments, the Middle Jurassic deposits exposed in southern Timor (e.g. in Kolbano and Aliambata) comprise glauconitic sandstones with belemnites and a thick-shelled coiled serpulid. At Aliambata a notable bed of lignite is exposed. Thus these Middle Jurassic deposits provide a striking indication of very shallow-marine conditions, in contrast to the Permo-Triassic and early Jurassic deposits, and in even greater contrast to the late Jurassic–Pliocene deep-marine deposits above. These glauconitic sandstones, exposed at Kolbano and Aliambata, are overlain by pink, grey, and violet marls and calcilutites with *Stomiosphaera*, rare radiolaria, *Cadosina* and *Inoceramus* prisms. These sections appear to record the breakup unconformity marked by the glauconitic sandstones and the post-breakup transgression preserved in the fine-grained lutites rich in radiolaria and calcareous plankton, which represent deposits of the newly formed continental slope and proximal rise. Remarkably similar sequences are exposed in Seram (Audley-Charles *et al.* 1979). Furthermore, the manganese nodules in red clay, sharks' teeth, and radiolaria from the Niki Niki region of central West Timor (where they occur as exotic blocks within the Bobonaro Scaly Clay), represent deposits of the distal continental rise or even the bathyal floor of the spreading Cretaceous Tethys ocean (Margolis *et al.* 1978). These nodules are now located in central Timor, as a consequence of structural telescoping associated with the collision of this continental margin with the Banda volcanic arc in the middle Pliocene.

FIG. 16. Intraformational conglomerate in the Middle and Upper Triassic Aitutu Limestone of West Timor. This indicates an important phase of relative uplift of the Aitutu carbonate platform during late Triassic times, resulting in the erosion of the lithified Aitutu Limestone. (See Fig. 14.)

Indication of Jurassic rifting in New Guinea

The evidence for a phase of lithospheric rifting, that had the effect of removing a continental block from the northern margin of what is now central New Guinea, has been discussed by Hamilton (1979), and Pigram and Panggabean (1984). The principal difference between the interpretation adopted in this paper and that of Pigram and Panggabean (1984) is that the latter take the view that western Irian Jaya (i.e. the Bird's Head and Neck including the Onin and Kumawa peninsulas), together with Misool, Buru and Seram, formed a separate microcontinent. In their view this microcontinent was located far from the rest of New Guinea. This idea, and the stratigraphic arguments on which it based, have been challenged by Dow and Sukamto (1986). The very close similarity in stratigraphic succession throughout the Mesozoic and Cainozoic in Timor and Seram, and other islands of the Outer Banda Arc, argue strongly against Seram having been a microcontinent during that time.

Permian to Cretaceous palaeogeography

A series of palaeogeographical maps of the Northern Australian continental margin (Figures 8–12) have been based on my recent interpretation of the rifting of this continental margin in the mid- to late Jurassic, and the identification of the continental blocks that were adjacent to Australian Gondwana until this mid-Mesozoic rifting event—as South Tibet, Burma, Western Thailand, Malaya, Sumatra, and the Banda allochthon. Aspects of this contentious interpretation have been discussed in detail elsewhere (Audley-Charles, 1983*b*, 1984, 1987, Audley-Charles *et al.*, in press, and Metcalfe, this volume). One vital piece of evidence concerns the age of the major orogenic compressional event that strongly folded, thrust, and cleaved the Triassic rocks of the Malay Peninsula. The general view (Metcalfe, this volume), recently supported by Şengör (1986), regards this as a Triassic event. An expedition to investigate this problem in the field was undertaken in 1986. We found that the Carboniferous and older rocks of the Peninsula display multiphase penetrative deformation distinctly different from the Permian and Mesozoic rocks. We regard this as indicating a mid-late Cretaceous collision event. We found no evidence of a Triassic orogenic phase in Peninsula Malaya (paper in preparation by Harbury *et al.*).

ACKNOWLEDGEMENTS: I thank Janet Baker and Colin Stuart for the art work.

References

Achdan, A. and Turkandi, T. (1982). Preliminary geologic map of the Kai (Tayandi and Tual) quadrangle, Maluku (1:250,000). Geological Research and Development Centre, Indonesia.

Audley-Charles, M. G. (1968). The Geology of Portuguese Timor. *Mem. Geol. Soc. London* **4**, 1–76.

Audley-Charles, M. G. (1978). Indonesian and Philippine Archipelagoes. In *The Phanerozoic geology of the world II. The Mesozoic A*. (ed. M. Moullade and A. E. M. Nairn), pp. 165–207, Elsevier, Amsterdam.

Audley-Charles, M. G. (1983*a*). Comments on 'Analogous tectonic evolution of the Ordovician foredeeps, southern and central Appalachians'. *Geology* **11**, 490–3.

Audley-Charles, M. G. (1983*b*). Reconstruction of eastern Gondwanaland. *Nature* **306**, 48–50.

Audley-Charles, M. G. (1984). Cold Gondwana, warm Tethys and the Tibetan Lhasa block. *Nature* **310**, 165–6.

Audley-Charles, M. G. (1986*a*). Rates of Neogene and Quaternary tectonic movements in the southern Banda Arc based on micropalaeontology. *J. Geol. Soc. London* **143**, 161–75.

Audley-Charles, M. G. (1986*b*). Timor–Tanimbar Trough: the foreland basin of the evolving Banda orogen. In *Foreland Basins*. (ed. P. A. Allen and P. Homewood). Internat. Assoc. Sedimentologists Spec. Pub. **8**, 91–102.

Audley-Charles, M. G. (1987). Dispersal of Gondwanaland: relevance to the evolution of the angiosperms. In *Biogeography of the Malay Archipelago*. (ed. T. C. Whitmore), Clarendon Press, Oxford, pp. 5–25.

Audley-Charles, M. G., Carter, D. J., Barber, A. J., Norvick, M. S., and Tjokrosapoetro, S. (1979). Reinterpretation of the geology of Seram: implications for the Banda Arcs and northern Australia. *J. Geo. Soc. London* **136**, 547–68.

Audley-Charles, M. G., Ballantyne, P. D., and Hall, R. (1988). Mesozoic–Cenozoic rift–drift sequence of Asian fragments from Gondwanaland. *Tectonophys*. (In press).

Bhatia, M. R., Thomas, M., and Boirie, J. M. (1984). Depositional framework and diagenesis of the Late Permian gas reservoirs of the Bonaparte basin. *Austral. Petrol. Explor. Assoc. J.* **24**, 299–313.

Bird, P. (1987). The Geology of the Permo-Triassic rocks of Kekneno, west Timor. Unpublished Ph.D. Thesis, University of London.

Bowin, C., *et al.* (1980). Arc–continent collision in Banda Sea region. *Am. Assoc. Petrol. Geol. Bull.* **64**, 868–915.

Chamalaun, F. H. and White, A. (1975). Electromagnetic induction at Dili, Portuguese Timor. *J. Geophy.* **41**, 537–40.

Chamalaun, F. H., Lockwood, K., and White, A. (1976). The Bouguer gravity field and crustal structure of Timor. *Tectonophys.* **30**, 241–59.

Chamalaun, F. H., Grady, A. E., von der Borch, C. C., and Hartono, H. M. S. (1981). The tectonic significance of Sumba. *Bull. Geol. Res. & Dev. Centre, Bandung* **5**, 1–20.

Cook, S. (1986). Triassic sediments from east Kekneno, west Timor. Unpublished Ph.D Thesis, University of London.

Cook, A. C., Smyth, M., and Vos, R. G. (1985). Source potential of Upper Triassic fluvio-deltaic systems of the Exmouth Plateau. *J. Austral. Petr. Explor. Assoc.* **25**, 204–15.

Crostella, A. and Barter, T. (1980). Triassic–Jurassic depositional history of the Dampier and Beagle sub-basins, northwest shelf of Australia. *J. Austral. Petrol. Explor. Assoc.* **20**, 25–33.

Dow, D. B. and Sukamto, R. (1986). Western Irian Jaya: the end product of oblique plate convergence in the Late Tertiary—Reply. *Tectonophys.* **121**, 348–50.

FALVEY, D. A. and MUTTER, J. C. (1981). Regional plate tectonics and the evolution of Australia's passive continental margins. *BMR J. Austral. Geol. Geophys.* **6**, 1–29.

GIANI, L. (1971). The geology of the Belu district of Indonesian Timor. Unpublished M.Phil. Thesis, University of London.

HAMILTON, W. (1979). Tectonics of the Indonesian region. *US Geol. Surv. Prof. Pap.* **1078,** 345 pp.

JACOBSON, R. S., SHOR, G. G., KIECKHEFFER, R. M., and PURDY, G. M. (1978). Seismic refraction and reflection studies in the Timor–Aru Trough system and Australian continental shelf. *Mem. Am. Assoc. Petrol. Geol.* **29**, 209–22.

KLOMPE, T. H. F. (1956). The structural importance of the Sula Spur (Indonesia). *Pacific Science Congress, 8th Quezon City Philippines, 1955 Proceedings 2A*, 869–89.

KUNDIG, E. (1956). Geology and ophiolite problems of East Celebes. *Koninkl. Nederlandsch Geol.-Mijnbouw Genoot Verh. Geol.* **16**, 210–35.

LAWS, R. A. and KRAUS, G. P. (1974). The regional geology of the Bonaparte Gulf Timor Sea area. *J. Austral. Petrol. Explor. Assoc.* **14**, 77–84.

MARGOLIS, S. V., KU, T. L., GLASBY, G. P., FEIN, C. D., and AUDLEY-CHARLES, M. G. (1978). Fossil manganese nodules from Timor: geochemical and radiochemical evidence for deep-sea origin. *Chem. Geol.* **21**, 185–98.

MCCAFFREY, R. and NABELEK, J. (1986). Seismological evidence for shallow thrusting north of the Timor Trough. *Geophys. J. Roy. Astron. Soc.* **85**, 365–81.

MCCAFFREY, R., *et al.* (1984). Preliminary results of the 1982 microearthquake survey in Nusa Tenggara, Timor, Indonesia. *Bull. Geol. Res. & Dev. Centre, Indonesia* **10**, 1–9.

MCCAFFREY, R., MOLNAR, P., and ROECKER, S. (1985). Microearthquake seismicity and fault plane solutions related to arc–continent collision in the eastern Sunda Arc, Indonesia. *J. Geophys. Res.* **90**, 4511–28.

METCALFE, I. (1988). Origin and assembly of southeast Asian continental terranes. This volume, p. 101.

O'BRIEN, C. A. E., GRAY, K. W., and GILLESPIE, I. (1961). The geological results of petroleum exploration in western Papua. *Geol. Soc. Austral. J.* **8**, 1–133.

PIGRAM, C. J. and PANGGABEAN, H. (1984). Rifting of the northern margin of the Australian continent and the origin of some microcontinents in eastern Indonesia. *Tectonophys.* **107**, 331–53.

POWELL, D. E. (1976). The geological evolution of the continental margin off northwest Australia. *J. Austral. Petrol. Explor. Assoc.* **16**, 13–23.

PRICE, N. J. and AUDLEY-CHARLES, M. G. (1983). Plate rupture by hydraulic fracture resulting in overthrusting. *Nature* **306**, 572–5.

PRICE, N. J. and AUDLEY-CHARLES, M. G. (1987). Tectonic collision processes after plate rupture. *Tectonophys.* **140**, 121–9.

ROBINSON, P. L. (1973). Palaeoclimatology and continental drift. In *Implications of continental drift to the Earth sciences*. Vol. 1. (ed. D. H. Tarling, and S. K. Runcorn), pp. 451–76, Academic Press, London.

SCHLUTER, H. U. and FRITSCH, J. (1985). Geology and tectonics of the Banda Arc between Tanimbar island and Aru island. *Geol. Jb.* **30**, 3–41.

ŞENGÖR, A. M. C. (1986). The dual nature of the Alpine–Himalayan system: progress, problems and prospects. *Tectonophys.* **127**, 177–95.

SILVER, E. A., JOYODIWIRYO, Y., and MCCAFFREY, R. (1978). Gravity results and emplacement geometry of the Sulawesi ultramafic belt, Indonesia. *Geology* **6**, 572–81.

SIMANDJUNTAK, T. U. (1986). Sedimentology and the tectonics of the collision complex of the east arm of Sulawesi, Indonesia. Unpublished Ph.D. Thesis, University of London.

SMART, J. and SENIOR, B. R. (1980). Jurassic–Cretaceous basins of northern Australia. In *The geology and geophysics of northeastern Australia*. (ed. R. A. Henderson and P. J. Stephenson), pp. 315–28, Geological Society of Australia.

SMART, J., GRIMES, K. G., DOUTCH, H. F., and PIN CHIN, J. (1980). The Carpentaria and Karumba basins, north Queensland. *Austr. BMR Geol. & Geophys. Bull.* **202**.

SMITH, R. B. (1983). Sedimentology and tectonics of a Miocene collision complex and overlying late orogenic clastic strata, Buton Island, eastern Indonesia. Unpublished Ph.D. Thesis, University of California, Santa Cruz.

SUKAMTO, R. and SIMANDJUNTAK, T. U. (1983). Tectonic relationship between geological provinces of western Sulawesi, eastern Sulawesi and Banggai-Sula in the light of sedimentological aspects. *Bull. Geol. Res. & Dev. Centre, Indonesia* **7**, 1–12.

SUKARDI and SUTRISNO (1981). Preliminary geological map of the Tanimbar Islands quadrangle, Maluku (1:250,000). Geological Research and Development Centre, Indonesia.

TJOKROSAPOETRO, S. and BUDHITRISNA, T. (1982). Geology and tectonics of the northern Banda Arc. *Bull. Geol. Res. & Dev. Centre, Indonesia* **6**, 1–17.

TOLLMANN, A. and KRISTAN-TOLLMANN, E. (1985). Paleogeography of the European Tethys from Paleozoic to Mesozoic and the Triassic relations of the eastern part of Tethys and Panthalassa. In *The Tethys*. (ed. K. Nakazawa and J. M. Dickins), pp. 3–22, Tokai University Press, Tokyo.

VAN BEMMELEN, R. W. (1949). *The Geology of Indonesia*. Government Printing Office, The Hague.

VEENSTRA, E. (1985). Rift and drift in the Dampier sub-basin, a seismic and structural interpretation. *J. Austral. Petrol. Explor. Assoc.* **25**, 177–89.

Veevers, J. J. (1984). *Phanerozoic Earth history of Australia*. Clarendon Press, Oxford. (418 pp.)

Veevers, J. J. and Heirtzler, J. R. (1974). Tectonic and palaeogeographic synthesis of Leg 27. In *Initial Rep. DSDP* **27**, pp. 1049–54.

Vind, E. W. and Harwood, C. R. (1965). Geophysical exploration in Torres Strait. *Austr. Petrol. Explor. Assoc. J.* **5**, 188–90.

Westermann, G. E. G., Sato, T., and Skwarko, S. K. (1978). Brief report on the Jurassic biostratigraphy of the Sula Islands, Indonesia. *Newsl. Stratigr.* **7** (2), 96–101.

M. G. Audley-Charles, Department of Geological Sciences, University College London, Gower Street, London WC1E 6BT, UK.

Origin and assembly of south-east Asian continental terranes

I. Metcalfe

ABSTRACT: The pre-Tertiary continental core of south-east Asia is a composite of at least four tectonostratigraphic terranes; Sibumasu (Shan States of Burma, north-west Thailand, Peninsular Burma and Thailand, Western Malaya, and north-western Sumatra), Indo-China (east Thailand, Laos, Vietnam south-west of the Song Ma–Song Da zone and Kapuchea), East Malaya (including south-east Sumatra and Natuna) and south-west Borneo. These terranes are separated from each other by sutures, and are now joined to South China along the Song Ma–Song Da zone.

Stratigraphic, palaeontological, and palaeomagnetic evidence suggests that all of these crustal blocks probably had their origin on the north-eastern margin of Gondwanaland where they formed part of a complex continental margin.

The Indo-China and East Malaya terranes both have typical late Palaeozoic Cathaysian floras, which indicate that they were close to South China in the Permian. Stratigraphic studies in north-east Vietnam suggest that Indo-China and South China may well have sutured as early as the late Devonian to early Carboniferous. Carboniferous and Permian faunas and floras of these blocks do not appear to be related to those of north-eastern Gondwanaland, indicating that they rifted away at some earlier time.

The Sibumasu block remained on the north-western Australian margin of Gondwanaland until the late Lower Permian, and was in continuity with the Tibetan Lhasa and Changtang blocks (south of the Langcangjiang Fracture Zone) and the proto-Kreios microcontinent further to the west. Late Carboniferous and Lower Permian sediments of the Sibumasu, Lhasa, and Changtang blocks, and the Iranian and Afghan portions of proto-Kreios include extensive glacial-marine deposits. Early Permian shallow-marine faunas were of cold-water type until a major change occurred in late Lower Permian times, with Middle Permian faunas becoming typically warm-water Tethyan in aspect. It seems most likely that Sibumasu (still attached to the Lhasa and Changtang blocks and proto-Kreios) rifted from Gondwanaland in the late Lower Permian.

Amalgamation of the south-east Asian blocks was probably completed by the end of the Triassic. This is indicated by terrestrial vertebrate biogeography and palaeomagnetism. It is still a matter of debate whether Indo-China sutured to South China in the Devonian–Carboniferous along the Song Ma line, or in the late Triassic along the Song Da zone.

Evidence seems stronger for a late Triassic suturing of Sibumasu to the other 'Cathaysian' south-east Asian blocks, although a Permian closure of 'palaeo-Tethys' has recently been proposed for the Shan-Thai–Indo-China region.

The south-west Borneo block probably rifted from the margin of Indo-China in the Cretaceous, and travelled south to its present position during the opening of the proto-South China Sea.

Introduction

Continental south-east Asia is a composite of allochthonous terranes which accreted to one another during the Palaeozoic and Mesozoic. The exact number of individual terranes or blocks present is not yet clear, but it seems probable that at least four major ones, now sutured to South China, comprise continental mainland south-east Asia. These are: Sibumasu, Indo-China, East Malaya, and south-west Borneo, now separated by sutures representing former oceans that are now recognized by mobile belts, ophiolites, volcano-plutonic arcs and major tectonic lineaments (Fig. 1).

These tectonostratigraphic terranes generally have internally consistent but mutually different stratigraphies, structural styles, and volcano-plutonic histories. They had accreted to each other by the late Triassic, although some component blocks had sutured together at earlier times. Gatinsky *et al.* (1978), Gatinsky *et al.* (1984), and Gatinsky and Hutchison (in press) suggested that the South China and Indo-China terranes (Cathaysia and Indosinia in their terminology) had sutured together by late early Carboniferous times to form the 'East Asian Continent'. Helmke and Kraikhong (1982), Helmke and Lindenberg (1983) and Helmke (1983, 1985*a*) have suggested a Permian suturing of the 'Shan-Thai' craton (Sibumasu) to Indo-

From AUDLEY-CHARLES, M. G. & HALLAM, A. (eds) *Gondwana and Tethys* Geological Society Special Publication No. 37, pp. 101–18.

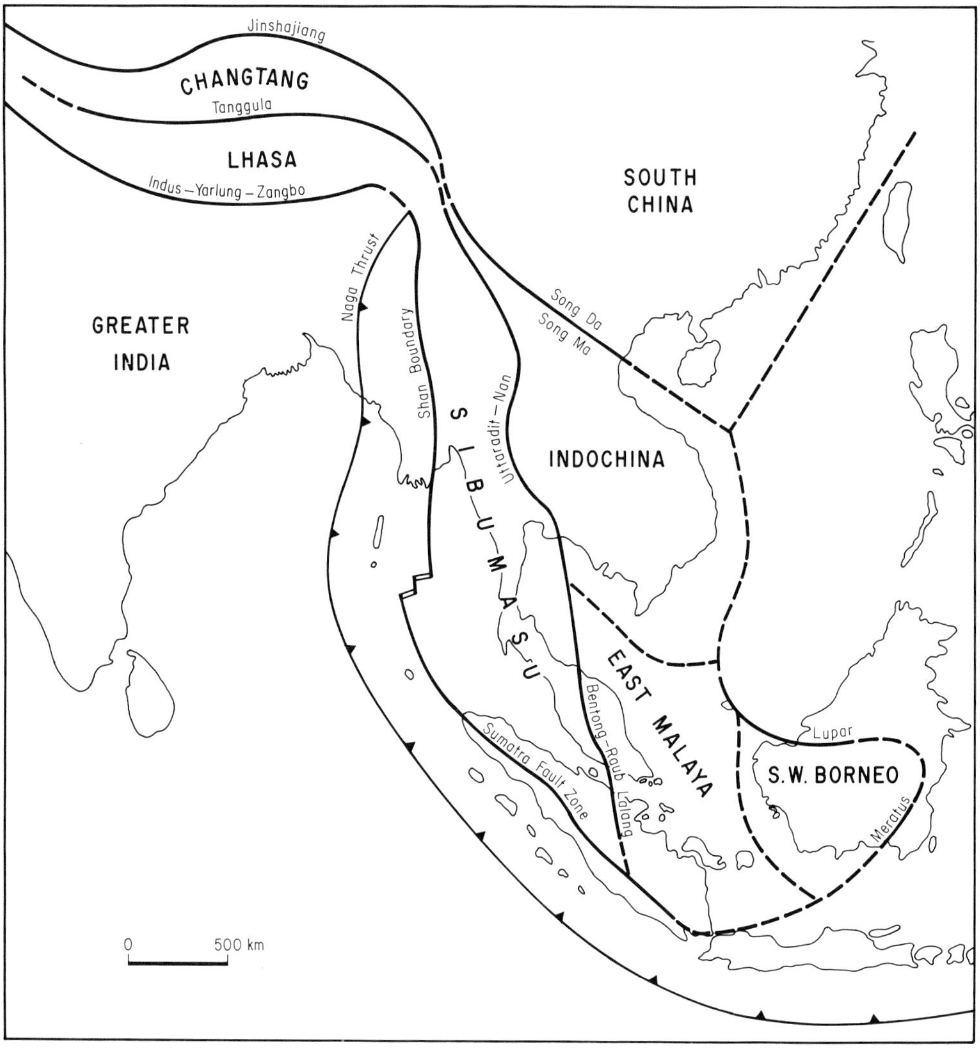

FIG. 1. The tectonostratigraphic terranes which now constitute continental mainland south-east Asia.

China, and Helmke (1985*b*) has also suggested that this suturing may have occurred as early as Carboniferous times. Other authors (e.g. Ridd 1980; Bunopas 1982; Mitchell 1981; Şengör 1984, 1985*a*) regard a late Triassic suturing of South China to Indo-China and 'Shan-Thai' to Indo-China most probable. A more extreme view was taken by Audley-Charles (1983, 1984) who suggested that Sibumasu remained on the north-eastern margin of Gondwanaland until the Jurassic (Oxfordian), thus proposing a post-Oxfordian suturing to Indo-China.

During the last decade there seems to have been a general consensus that large parts, if not all, of south-east Asia had its origin on the margin of Gondwanaland, and that various fragments of the Gondwanaland margin have been consistently drifting northwards during the Palaeozoic and Mesozoic. The precise original locations of these fragments, their times of rifting from Gondwanaland, and their subsequent drift and assembly has seen much recent debate, not only for fragments which now comprise south-east Asia, but also other Gondwanaland fragments which drifted across the Tethys and accreted to Laurasia.

Any scenario of the Palaeozoic and Mesozoic history of the south-east Asian continental terranes must be consistent with the known stratigraphic, palaeontological, and palaeomagnetic data. An attempt is made here to assess the evidence for the origins of these terranes, their

Palaeozoic and Mesozoic drift histories, and subsequent assembly into south-east Asia, within currently known geological and geophysical constraints.

The terranes and their stratigraphic and palaeontological features

Sibumasu

The Sibumasu terrane comprises the Shan States of Burma, north-west Thailand, Peninsular Burma and Thailand, Western Malaya, and north-west Sumatra. It is bounded on the east by the Nan-Uttaradit suture in Thailand—which extends down through Peninsular Malaysia as the Bentong–Raub suture and into Sumatra as the Lalang Line (Howell *et al.* 1983; Tjia 1985; Tjia and Zaiton 1985). The western and south-western boundary is formed by the Shan Boundary and Sumatran fault zones (Fig. 1). On stratigraphic and palaeontological grounds it is most probable that Sibumasu extends into Tibet as the Lhasa block and part of the Changtang block (see next paragraph), but the lateral connection is difficult to trace because of the tectonic complexity in the region of the north-east India syntaxis.

The Sibumasu block basement consists of Precambrian (Proterozoic) gneiss, marble, schist, and phyllite. The overlying Palaeozoic sequence comprises Upper Cambrian deltaic sediments, Ordovician and Silurian peritidal and subtidal limestones and shales, Devonian shelf carbonates and clastics, and Carboniferous and Permian shallow-marine limestones and clastics (Fig. 2). Within this Palaeozoic sequence, there occurs an extensive (2000 km) belt of late Carboniferous–early Permian diamictites extending from Sumatra (Bohorok Formation) through north-west Malaya (Singa Formation), Peninsular Thailand (Kaeng Krachan Group), Peninsular Burma (Mergui Group) to north-west Burma (Martaban Series and Lebyin Group), see Fig. 2. Two hypotheses as to the origin of these diamictites have been proposed, glacial-marine (Stauffer and Mantajit 1981; Stauffer 1983; Stauffer and Lee 1984) and a mass-flow/turbidite origin on a continental slope or in a rift environment (Burton, in press; Gatinsky and Hutchison, in press). The evidence for a glacial-marine origin seems stronger, in view of the great lateral extent, shallow-marine nature, types of clasts, presence of dropstones, and age (corresponding with the age of glaciation on Gondwanaland). Glacial-marine sediments of the same age are also found on the probable extension of Sibumasu in Tibet, where they are found on the Lhasa block and on the Changtang block (south of the Langcangjiang Fracture Zone), see Fig. 3. Interpretation of these diamictites as glacial-marine necessitates that Sibumasu was marginal to Gondwanaland in the late Palaeozoic.

The Mesozoic stratigraphy is characterized by Triassic shallow-marine limestones and deeper-marine radiolarian cherts and rhythmites (turbidites), and Jurassic–Cretaceous continental red beds and shallow-marine strata, with the exception of Sumatra, where there are Jurassic marine volcaniclastics and Cretaceous ophiolites.

Within the Sibumasu stratigraphic succession there are five stratigraphic breaks recognized: in the early–Mid-Cambrian, mid–late Ordovician, late Devonian–early Carboniferous, latest Permian–earliest Triassic, and latest Triassic–early Jurassic (Fig. 2).

An examination of the faunal affinities and biogeographical provinces of Sibumasu faunas is most revealing. Early Palaeozoic (Cambrian and Ordovician) marine faunas show close affinities to comparable faunas of north-west Australia, Tibet, Argentina and northern China (Stait and Burrett 1982; Jell *et al.* 1984; Webby *et al.* 1985; Burrett and Stait 1986). Silurian and Devonian faunal affinities are as yet uncertain, but Lower Carboniferous foraminifera also show north-west Australian affinities (Fontaine and Vachard 1981).

Early Permian brachiopods and foraminifera indicate cool- or cold-water conditions on Sibumasu (Ingavat and Douglass 1981; Waterhouse 1982; Rao 1984 and pers. comm.), which is consistent with the presence of glacial-marine sediments of that age (see Fig. 3). The early Permian brachiopods also show affinities with north-west Australia, New Guinea, and Timor (Archbold *et al.* 1982). By mid-Permian times, faunas had become warm-water, eastern Tethyan types (Ishii *et al.* 1981, Nakamura *et al.* 1985), and the Triassic and Jurassic faunas continue to represent the eastern Tethyan province (Matsuda 1985; Hayami 1984). There are no post-early Permian faunas with Gondwanaland affinities on Sibumasu. See Fig. 2 for summary of faunal affinities and palaeoclimatic interpretations.

East Malaya

This terrane includes eastern Malaya, south-east Sumatra, Natuna and the pre-Tertiary basement area beneath the Malay, west Natuna, and

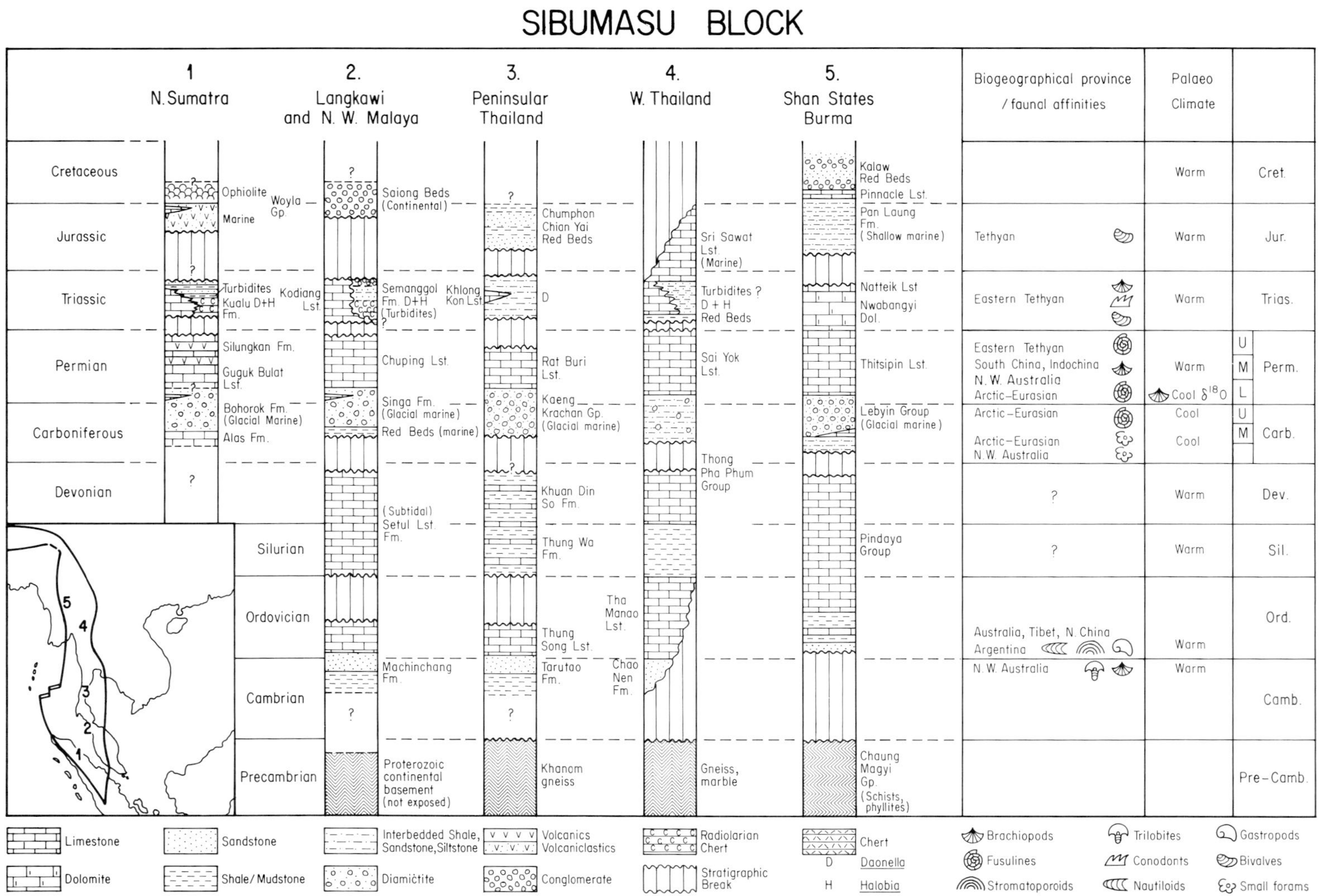

FIG. 2. Representative stratigraphic sections, faunal affinities, and palaeoclimates for the Sibumasu block.

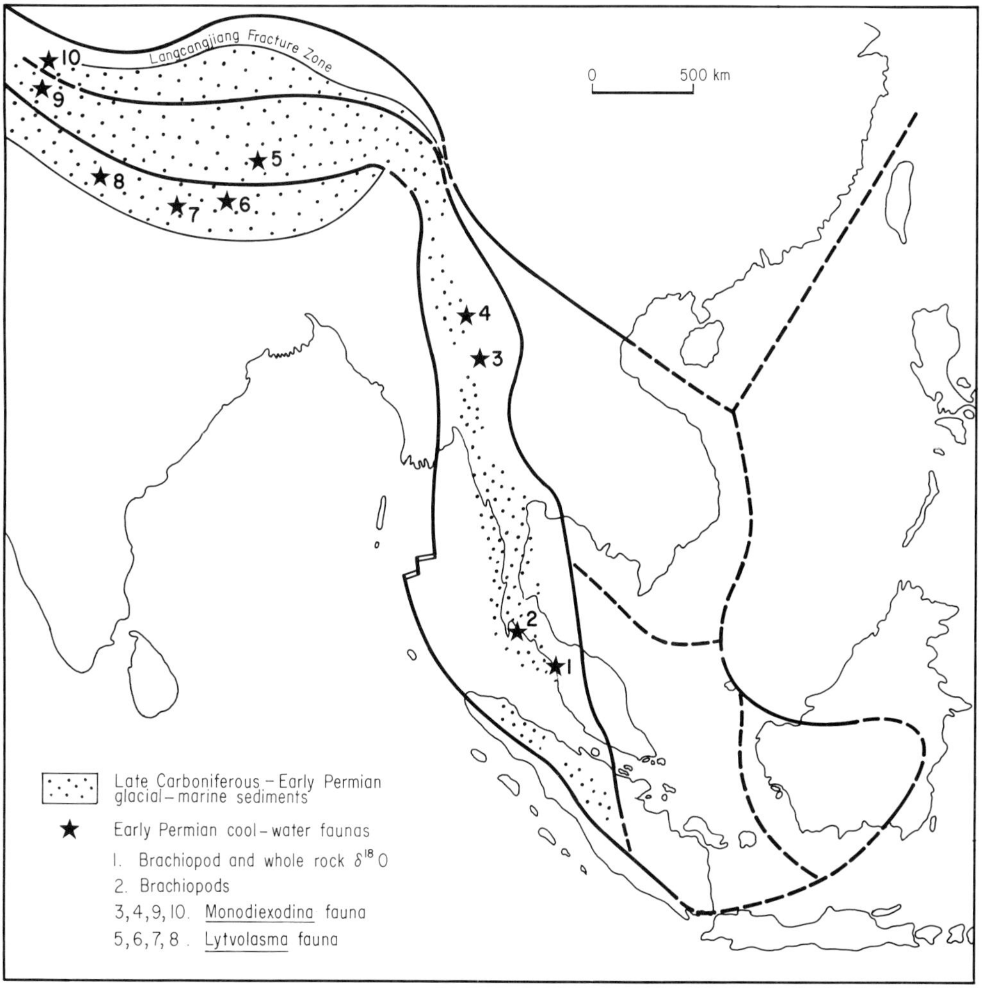

FIG. 3. Distribution of late Carboniferous–early Permian glacial-marine sediments and early Permian cool-water faunas on Sibumasu, and the probable extension of Sibumasu into Tibet.

Penyu Basins and the Sunda Shelf north-east of Sumatra. It is bounded on the west and south-west by the Bentong–Raub suture and its extension north along the west margin of the Thai Basin and south along the Lalang Line and Sumatra Fault Zone. On the east it is bounded by the Natuna extension of the Lupar Suture, and north and south by major pre-Tertiary lineaments.

A Precambrian (Proterozoic) basement is indicated by U–Pb zircon inheritance ages and Nd model ages of granitoids (Liew and McCulloch 1985).

The Palaeozoic and Triassic succession on this block comprises deep- to shallow-marine, predominantly slope-deposited clastics and limestones (some of which occur in carbonate build-ups), with abundant intermediate and acidic volcaniclastics and volcanics. The large proportion of volcaniclastics on this block is in sharp contrast to Sibumasu, where they are virtually absent. Continental red beds of late Jurassic to early Cretaceous age unconformably overlie this sequence, except in Sumatra where marine conditions persisted into the Upper Jurassic.

Immediately east of the Bentong–Raub line, Silurian marine and Devonian deep-marine (with radiolarian cherts) sediments occur with associated serpentinite bodies. These pass up into shallow-marine volcaniclastics and limestones of Carboniferous to Triassic age. The Raub Red Beds, previously interpreted as Carboniferous in age on stratigraphic grounds

(Haile *et al.* 1977) are here regarded as equivalents to the Jurassic–Cretaceous Tembeling Formation. This is based on lithological similarity, and the fact that although the Raub Red Beds are steeply dipping, they appear to be relatively undeformed internally (i.e. conglomerate clasts show little or no tectonic flattening and mudstone layers show only weak or no cleavage), whereas the Palaeozoic, and to some extent the Triassic rocks of the same area show much more deformation, with well-developed cleavages and tectonically deformed macrofossils. Attempts to obtain palynomorphs from these red beds have so far proved negative.

In the same area, *mélange* with Lower Triassic and Permian limestones and volcanic clasts has recently been discovered near Raub (Chakraborty and Metcalfe 1987). It is not yet clear whether this *mélange* is of sedimentary or tectonic origin, but initial studies indicate that it is an olistostrome type deposit.

Slightly further to the east, in the Central Belt of Peninsular Malaysia, Triassic volcaniclastic turbidites (with minor limestones) appear to contain large blocks of Permian and possibly older shallow marine strata (mainly limestones), and these are overlain unconformably by late Jurassic–Cretaceous continental red beds.

In the Eastern Belt of Peninsular Malaysia, Triassic sediments appear to be largely absent and the Upper Jurassic continental rocks lie unconformably on the Carboniferous–Permian sequence (Fig. 4).

The late Palaeozoic and Mesozoic faunas of this terrane are of warm-water type, and of Eurasian and Tethyan aspect (Yanagida and Aw 1979; Kobayashi and Tamura 1984; Matsuda 1985). There are no known Gondwanaland faunas on this block. Carboniferous floras are Euramerican (Metcalfe 1983), and Permian floras belong to the tropical Cathaysian province (Kon'no and Asama 1970; Kon'no *et al.* 1970; Asama 1984). Jurassic–Cretaceous floras are of the east Eurasian Ryoseki type (Kimura 1984), see Fig. 4.

Indo-China

The Indo-China terrane comprises eastern Thailand, Laos, Kampuchea, and Vietnam (except the portion north-east of the Song Ma–Song Da zone). It has also been referred to as Indosinia (e.g. Helmke, 1985*a*), and Annamia (e.g. Şengör 1984). It is bounded to the south-west by the Nan-Uttaradit suture and its extensions; it is bounded to the north by the Trung Song–Song Ma–Song Da suture zone, and to the south and east by tectonic lineaments in the pre-Tertiary basement.

The stratigraphy comprises Palaeozoic and Mesozoic shallow-marine, epicontinental and continental sediments which overlie a Precambrian metamorphic basement of mainly gneiss, schist, amphibolite, and quartzite. Widespread stratigraphic breaks occur in the mid–late Ordovician, late Devonian–early Carboniferous, late Permian–early Triassic, and early–middle Jurassic (Fontaine and Workman 1978; Phan 1981; Tran 1979), see Fig. 5.

Both Palaeozoic and Mesozoic faunas and floras are of warm climate type. Late Palaeozoic and Mesozoic marine faunas are of eastern Tethyan type and have strong affinities with those of South and North China (Hill 1973; Fontaine and Workman 1978; Fontaine *et al.* 1982). Late Palaeozoic floras are tropical Cathaysian (Asama 1984), and Triassic floras have affinities with Yunnan (Kimura 1984). Many Mesozoic terrestrial vertebrates are now known from Indo-China, and they all show, with one exception, Laurasian affinities (Buffetaut 1981, 1982, 1983, 1985; Buffetaut *et al.* 1981; Buffetaut and Ingavat 1980, 1982, 1983, 1984; Ingavat and Janvier 1981; Martin and Ingavat 1982; Martin, 1985; see Fig. 6). The one exception is the occurrence of a dicynodont at Luang Prabang, Laos, which would indicate Gondwanaland connections. Unfortunately, the specimen's age is in doubt, its taxonomy questionable, and the specimen is apparently lost (Buffetaut 1985).

South-west Borneo

This terrane comprises west Sarawak, and south-west Kalimantan, and is bounded to the north by the Lupar suture, to the south-east by the Meratus suture, and to the south-west by the limits to the east Malaya block. This terrane presumably has a Precambrian basement, but there is little direct or indirect evidence for this, except for the presence of diamonds in south-west Kalimantan.

Palaeozoic rocks are poorly known in south-west Borneo. The oldest dated rocks are shallow-marine Devonian limestones (Rutten 1940) which form part of the 'Old Slates Formation' (a probable *mélange*). The late Carboniferous and early Permian is represented by shallow-marine limestones with fusulinids and conodonts. The middle and late Permian, and early Triassic, appear to be absent. The middle and late Triassic are represented by basal conglomerates over-

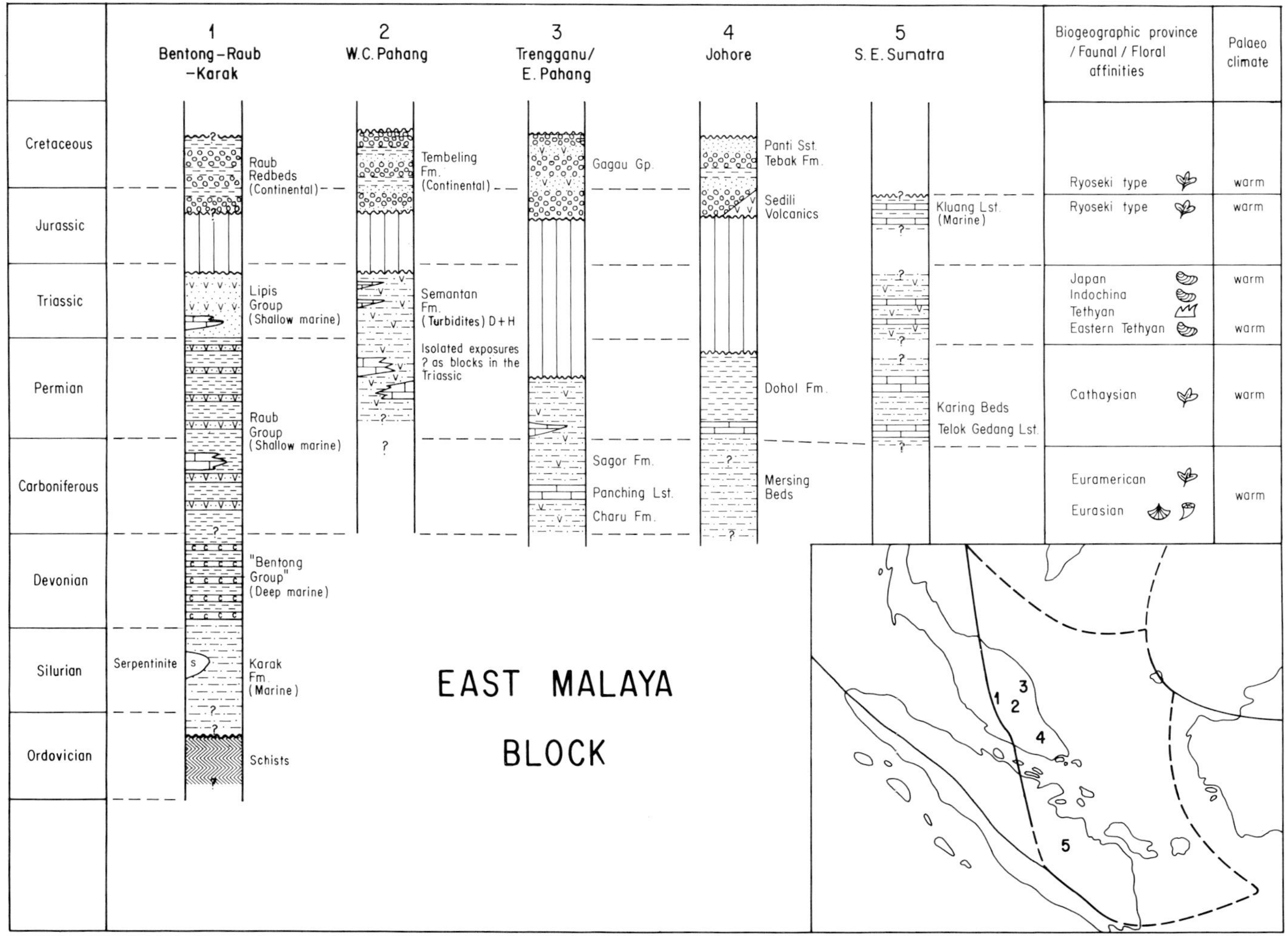

FIG. 4. Representative stratigraphic sections, faunal affinities, and palaeoclimates for the East Malaya block. Symbols are as used in Fig. 2.

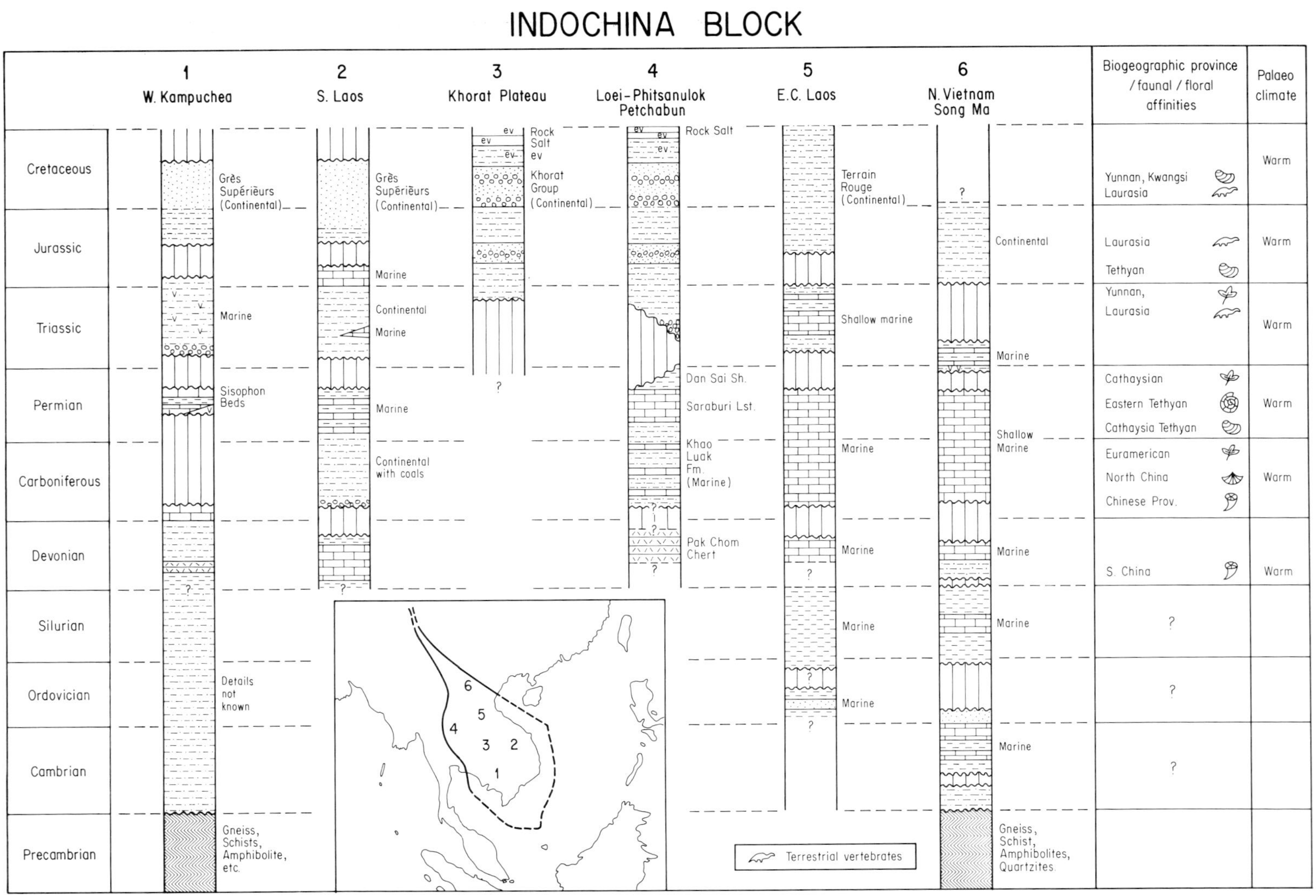

FIG. 5. Representative stratigraphic sections, faunal affinities, and palaeoclimates for the Indo-China block. Symbols are as used in Fig. 2.

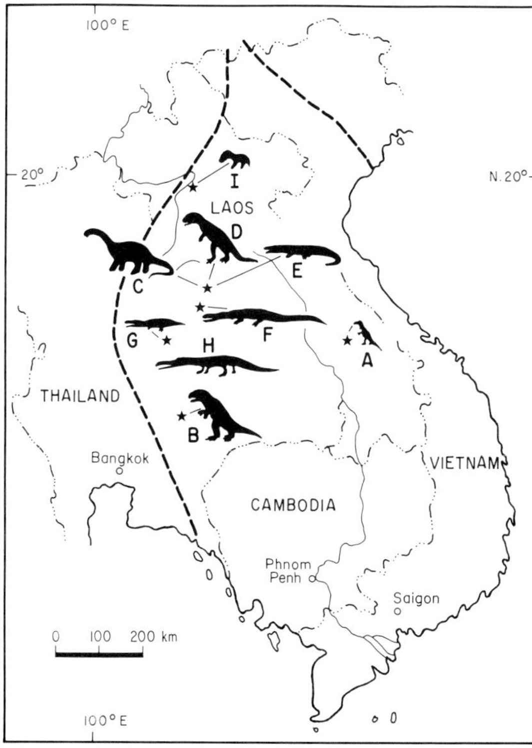

	Vertebrates	Age	Affinities
A.	Sauropods, hadrosaurs	U. Cretaceous	Europe China
B.	Carnosaur, hybodont shark	M. Cretaceous	Tibet (Lhasa Block)
C.	Sauropod	U. Jurassic	Cosmop.
D.	Carnosaur		
E.	*Goniopholis* (crocodilian)		
F.	*Sunosuchus* (crocodilian)	L. Jurassic	N. China
G.	*Cyclotosaurus*	U. Triassic	Laurasia
H.	Phytosaur		
I	Dicynodont	L. Triassic ?	Gondwana

FIG. 6. Mesozoic vertebrates of south-east Asia and their affinities. Compiled from the papers by Buffetaut, Ingavat, Martin, and Janvier. Symbols are as used in Fig. 2.

lain by marine volcaniclastics with *Daonella* and *Halobia*. The latest Triassic and early and middle Jurassic appear to be absent. The late Jurassic and early to middle Cretaceous are represented by shallow-marine limestones, clastics, and volcaniclastics. These are overlain by latest Cretaceous to Tertiary continental sandstones.

Both Palaeozoic and Mesozoic faunas and floras appear to represent warm climatic types. Triassic bivalves are typical Tethyan forms and Triassic plants show affinities with South China, Indo-China, and Japan (Kimura 1984; Kobayashi and Tamura 1984). Jurassic bivalves show close affinities with those of East Asia, Japan and the Philippines (Hayami 1984), see Fig. 7.

Palaeomagnetic data

Palaeomagnetic data from south-east Asia is very limited, and most of it comes from the late Mesozoic and Tertiary (Haile and Briden 1982). Palaeozoic data is extremely limited and its quality is highly suspect, due to poor age control, lack of fold tests, evidence of thermal heating and deformation. Mesozoic data in general appears to be more reliable but is still at the reconnaissance stage.

It is as yet impossible to establish apparent polar-wander paths for the south-east Asian tectonic blocks. These data are a prerequisite for comparison with the established apparent polar-wander paths (APWPs) for Eurasia and Gondwanaland. However, there are sufficient data to allow some constraints to be placed on the geodynamic history of the region.

Late Triassic and Jurassic data (Barr *et al.* 1978; Bunopas *et al.* 1978; Bunopas 1982; Achache and Courtillot 1985) indicate that the Sibumasu, Indo-China and South China blocks were united at that time. Data from the Cretaceous and Tertiary (Achache *et al.* 1983) indicate a clockwise rotation of Indo-China and South China since the middle Cretaceous, probably related to the collision of India with Eurasia, and consequent 'extrusion' of south-east Asia. Achache *et al.* (1983), using the data of Haile (now published in Haile *et al.* 1985) postulated that 'south-east Malaya' lay at a palaeolatitude

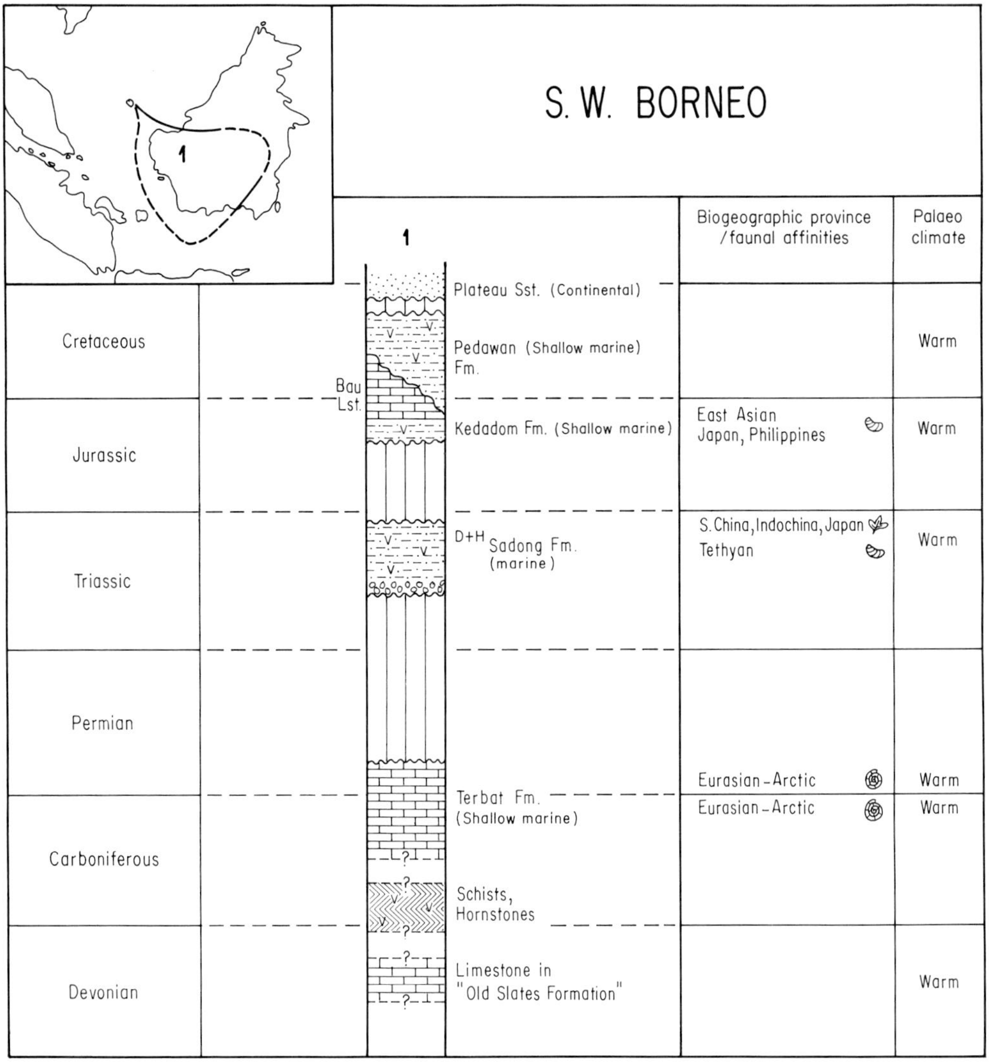

FIG. 7. Representative section, faunal affinities, and palaeoclimate for south-west Borneo. Symbols are as used in Fig. 2.

of 23°N in the late Cretaceous, and therefore separated from 'north-west Malaya' by more than 10° of latitude. This conclusion is at variance with the geology, which indicates that the Malay Peninsula was a single rigid unit in the late Triassic.

Haile (1979), using Triassic, Cretaceous and Tertiary data, postulated a 12° northwards drift and 40° clockwise rotation of Sumatra since the late Triassic. He also suggested an anticlockwise rotation of West Borneo, Malay Peninsula and south-west Sulawesi, and stated that Sumatra was not coupled to 'Sundaland' until the middle Tertiary. These conclusions are inconsistent with the geology of the region, which suggests that north-west Sumatra, the western part of the Malay Peninsula, and other parts of Sibumasu have acted as a single unit since the late Palaeozoic (Cameron *et al.* 1980; Metcalfe 1983, 1985).

McElhinny *et al.* (1974) presented data from the Malay Peninsula (treated by them as a single unit) which, they suggested, showed that the Peninsula was not a part of Gondwanaland, since it lay at 15°N in the late Palaeozoic. This result was based on measurements of rocks of uncertain age, which show a pronounced development of cleavage, and is at strong variance with stratigraphic and palaeontological data. Haile (1980) reported data from the Ordovician and Silurian limestones in north-west Malaya, and suggested a palaeolatitude of 43°N in the Ordovician. This result is not constrained by a fold test and the limestones have undergone regional metamorphism, with temperatures in excess of 300°C (as indicated by conodonts).

Bunopas (1982) also reported Ordovician data from Peninsular Thailand, which give a palaeolatitude of 12°N. This result is also suspect for the same reasons. Recent petrographic and palaeobiogeographical studies (Wongwanich *et al.* 1983; Burrett and Stait 1984) indicate low palaeolatitudes for north-west Malaya in the Ordovician and proximity to north-west Australia. Carboniferous and Permian data for Sibumasu (McElhinny *et al.* 1974; Bunopas 1982) give conflicting results, and their reliability is here considered doubtful (see Fig. 8).

Palaeomagnetic data from the South China block indicate low palaeolatitudes during the Carboniferous (Lin 1987) and Permian (McElhinny *et al.* 1981), which are consistent with stratigraphic and palaeontological indications.

Origin of the terranes

Sibumasu

Palaeozoic provenance studies indicate that Sibumasu was attached to a Precambrian craton during the Palaeozoic, and that the attachment was on its present western side (Jones 1968; Stauffer 1974, 1983; Stauffer and Mantajit 1981). The extensive belt of late Carboniferous–early Permian glacial-marine diamictites on this block, with associated cool-water faunas, indicate that it lay marginal to a late Carboniferous–early Permian glacial area. The only glacial region of this age is found in Gondwanaland, and the Sibumasu block must therefore have originated on the margin of Gondwanaland. The precise area of former attachment is indicated by stratigraphy and palaeobiogeography. Stratigraphic comparisons between the Palaeozoic of Sibumasu and north-west Australia reveal striking similarities. In particular, the identification of a stratigraphic break within the Ordovician of Sibumasu, cutting out Caradocian strata, corresponds with an identical break over the whole of Northern Australia (Burrett and Stait 1986). Ordovician faunas of Sibumasu include

	SIBUMASU	INDOCHINA	EAST MALAYA	S.W. BORNEO
JURASSIC	15° N (2)	25° N (2) 20° N (4)	15° N (7)	——
TRIASSIC	Equatorial (2) (5) (6)	25° N (4) 11° N (2)	16° N/S (3)	——
PERMIAN	19° S (2)*	37° N/S (2)*	——	——
CARBONIFEROUS	13° S (2)* 15° N (3)*	——	——	——
DEVONIAN	——	——	——	——
SILURIAN	——	——	——	——
ORDOVICIAN	12° N/S (2)* 43° N/S (1)*	——	——	——
CAMBRIAN	——	——	——	——

FIG. 8. Palaeolatitude data for the south-east Asian tectonic blocks. Numbers in brackets refer to publications listed below. Starred data is considered unreliable, due to poor dating of the sampled rocks, possible remagnetization effects and lack of fold tests (see text).
1. Haile (1980), *2*. Bunopas (1982), *3*. McElhinny *et al.* (1974), *4*. Achache and Courtillot (1985), *5*. Sasajima *et al.* (1978), *6*. Haile (1979), *7*. Haile and Khoo (1980).

nautiloids endemic to Sibumasu, Tibet, and Australia (Stait and Burrett 1983; 1984*b*), the Gondwanaland snail *Peelerophon oehlerti* (Bergeron), and stromatoporoids, polyplacophorans and rostroconchs of Australian affinities (Jell *et al.* 1984; Webby *et al.* 1985; Stait and Burrett 1984*a*).

In addition, Sibumasu late Cambrian brachiopods and trilobites (Shergold *et al.* in press) show north-west Australian affinities. Lower Carboniferous foraminifera from the Sumatran part of Sibumasu have been closely compared with faunas of the same age from the Bonaparte Gulf Basin of north-west Australia (Fontaine and Vachard 1981). Early Permian brachiopods from Thailand are said to have affinities with those of New Guinea, Timor, and north-west Australia (Archbold *et al.* 1982), whereas middle Permian brachiopods show affinities with Indo-China and South China (Nakamura *et al.* 1985).

There is therefore strong evidence from stratigraphic, sedimentological, and palaeontological data that Sibumasu had its origin on the north-west Australian margin of Gondwanaland. Palaeomagnetic data are of little help in confirming this, but the available late Palaeozoic data are not in conflict with this conclusion.

Indo-China

There is no unequivocal evidence that Indo-China had Gondwanaland origins, except perhaps the unestablished occurrence of a Gondwanaland dicynodont in Laos. Audley-Charles (1983, 1984) places Indo-China on the north-east margin of Australian Gondwanaland in the late Palaeozoic. Further biogeographical studies of the Indo-China Palaeozoic faunas are required, together with reliable palaeomagnetic data, to confirm such a position. If Indo-China did have its origin on the margin of Gondwana, it must have rifted away by the early Carboniferous at the latest (see next paragraph).

East Malaya

Lower Carboniferous faunas of this block contain forms which are in general Eurasian, but which also show Australian affinities. However, Middle Carboniferous forms show relationships with Chinese, Russian, and European faunas, and also have no Australian affinities. Permian floras are Cathaysian and Permian faunas are of typical Tethyan types. This terrane may also have originated in Gondwanaland, but like Indo-China, must have drifted north at a fairly early time (pre-middle Carboniferous).

South-west Borneo

This terrane appears to have rifted from the north-eastern portion of Indo-China sometime in the late Cretaceous (Gatinsky *et al.* 1984; Gatinsky and Hutchison, in press), during the opening of the proto-South China Sea. Affinities with East Asia, Japan and the Philippines are indicated by Jurassic bivalves (Hayami 1984). It may well have formed a single tectonic block with Indo-China in pre-Cretaceous times, but more data from the Palaeozoic and Triassic are required to confirm this.

Rift, drift, and assembly into south-east Asia

There has been much discussion and debate in recent years, regarding the evolution of the Tethys ocean and the histories of fragments now forming part of Eurasia, which appear to have had their origins in the Southern Hemisphere, where they formed part of the supercontinent Gondwanaland. Much of the debate has concerned the western part of Tethys, but increasing interest in the eastern Tethys has led to a greater knowledge of this part of the globe. Much information has recently come out of Tibet mainly the result of joint French and Chinese work.

With regard to Southeast Asia, three schools of thought have evolved as to the geodynamic history of the region. All three groups generally agree on the Gondwanaland origins of the Southeast Asian Blocks but disagree markedly on the timing of their amalgamation.

Şengör (1984, 1985*a*, 1985*b*, 1985*c*) invokes the rifting of an elongate sliver like continent, the Cimmerian continent (which includes Sibumasu), during the Triassic from the north-eastern margin of Gondwanaland and the formation of a new ocean, Neotethys. As the Cimmerian continent drifted across Tethys, rather like a wind shield wiper, the older Palaeotethys was destroyed and the Neotethys opened up. During the Cimmerian continent's drift across the Tethys, it also split into two slivers at its eastern end dividing north and south Tibet and opening the 'Tanggula-Waser Ocean'. The suturing of Indochina (Annamia) to South China and Sibumasu (Thai Malay Peninsular Block) to Indochina was considered to have occurred in the late Triassic (a view shared by Mitchell 1981, 1982, etc.). Tollmann and Tollmann (1985) invoke a similar scenario and term the western part of the Cimmerian continent Kreios.

Helmke and KraiKhong (1982), Helmke and Lindenberg (1983), and Helmke (1983, 1985*a*, 1985*b*) propose an earlier amalgamation of the Southeast Asian blocks during the late Palaeozoic, perhaps as early as Carboniferous.

A third possibility has been proposed by Audley-Charles (1983, 1984), which involves late Carboniferous–early Permian rifting of Indo-China and 'Central Tibet' (attached to Turkey and Iran) from Gondwana, followed by the rifting of the other south-east Asian blocks together with 'South Tibet' in the middle to late Jurassic.

Other authors have proposed combinations of these views involving Palaeozoic suturing of Indo-China and South China, and a later Triassic suturing of Sibumasu to Indo-China and eastern Malaya (Gatinsky *et al.* 1978; Gatinsky *et al.* 1984; Metcalfe 1986) or even vice versa (Hahn 1985). An attempt is made here to put constraints on the rifting of the south-east Asian terranes from Gondwanaland, on their subsequent drift northwards, and on their assembly into their current configuration.

Rifting from Gondwanaland

Sibumasu

This terrane remained on the margins of Gondwanaland until early Permian, as indicated by the late Carboniferous–early Permian glacial-marine diamictites, cool-water faunas, absence of warm climate floras, and presence of faunas of Gondwanaland (north-west Australian) affinities. A remarkable change occurs in the shallow-marine biotas of Sibumasu in the late early Permian and middle Permian, and subsequent faunas are of eastern Tethyan types and have strong affinities with South China and Indo-China (indicating proximity to those blocks already in the middle Permian). Since South China and Indo-China were at low northern palaeolatitudes in the Permian, Sibumasu must have already rifted from Gondwana by latest early Permian times. It is considered here that Sibumasu, Tibet (between the Indus–Yarlung–Zangbo suture and the Langcangjiang Fracture Zone) and proto-Kreios were in continuity and rifted together at the same time from Gondwanaland. This is consistent with the data from western Tethys (Tollmann and Tollmann 1985) and Tibet (Allègre *et al.* 1984; Wang 1984; Xiao and Gao 1984).

Indo-China

It seems most probable that Indo-China and South China had already sutured together by mid-Carboniferous times (Gatinsky *et al.* 1978; Tran 1979; Gatinsky *et al.* 1984). This is evident from both stratigraphy and palaeontology; palaeomagnetism and stratigraphic and palaeontological climatic indicators suggest that South China was in equatorial latitudes at that time. Indo-China must therefore have rifted from Gondwanaland (assuming that it had its origin there) at an earlier time. The presence of a widespread late Devonian–early Carboniferous stratigraphic break on Sibumasu, Indo-China, and north-west Australia, indicates that an early Carboniferous rifting is most probable.

East Malaya

The Middle Carboniferous and Permian faunas and floras of East Malaya are Eurasian and Cathaysian respectively, and do not exhibit any cool climate or Gondwanaland forms. This terrane, therefore, if originally attached to Gondwana, must have rifted in pre-middle Carboniferous times. Again, an early Carboniferous rifting seems likely, and Indo-China and East Malaya may have acted as a single unit since that time.

South-west Borneo

Late Carboniferous and early Permian fusulinids and conodonts indicate a warm palaeoclimate for south-west Borneo at these times, and a position adjacent to Indo-China (as suggested by Jurassic bivalves) indicating that south-west Borneo formed part of the Indo-China terrane until late Mesozoic times. Rifting of the south-west Borneo block from Indo-China probably occurred sometime in the Cretaceous, during the opening of the proto-South China Sea.

Drift and assembly into south-east Asia

Early Palaeozoic global reconstructions are still highly conjectural, and the lack of palaeomagnetic data for south-east Asia makes it impossible to construct a map showing the relative positions of the south-east Asian blocks at that time. However, Cambrian and Ordovician faunas indicate proximity to north-west Australia (Burrett and Stait 1984).

Data for the Carboniferous and Permian is

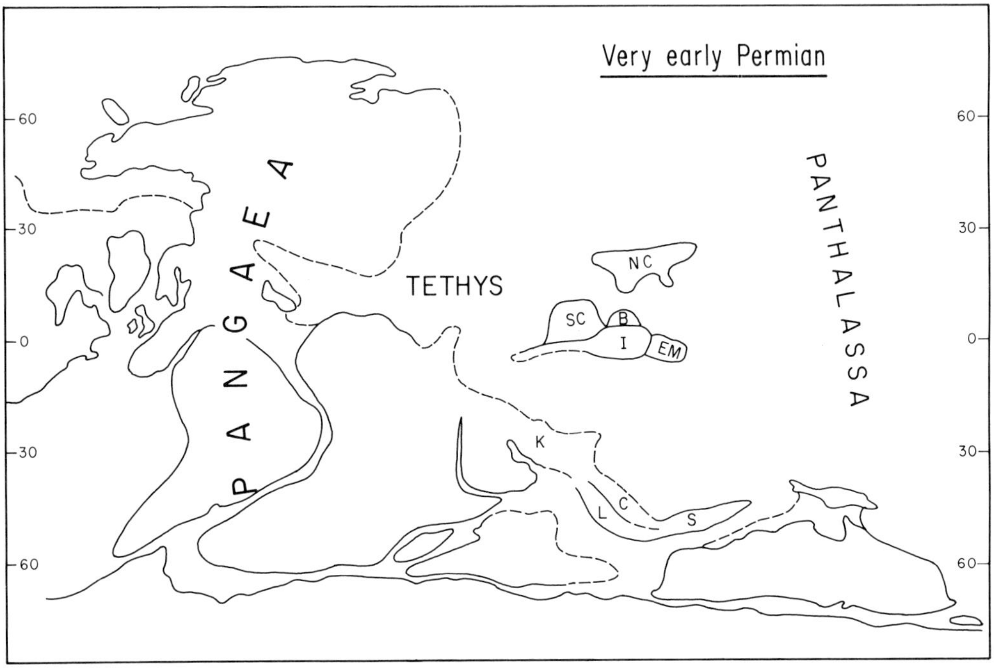

FIG. 9. Palaeogeographical reconstruction showing the positions of the south-east Asian, Tibetan, and Chinese tectonic blocks in the earliest Permian. *NC* = North China, *SC* = South China, *B* = Borneo, *I* = Indo-China, *EM* = east Malaya, *S* = Sibumasu, *L* = Lhasa block, *C* = Changtang block, *K* = Kreios. Other continental positions after Smith *et al.* (1981).

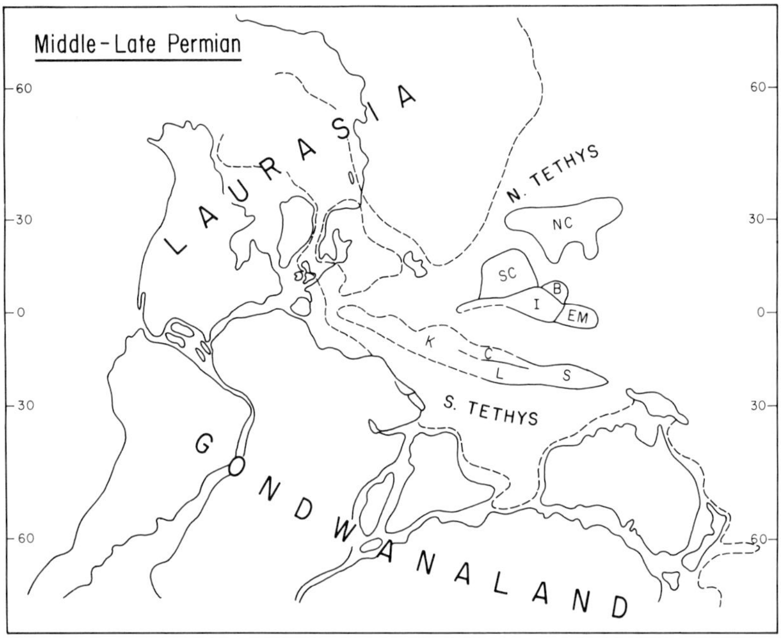

FIG. 10. Palaeogeographical reconstruction showing the relative positions of the south-east Asian, Tibetan, and Chinese blocks in the middle–late Permian. Other continental positions after Smith *et al.* (1981). Abbreviations as in Fig. 9.

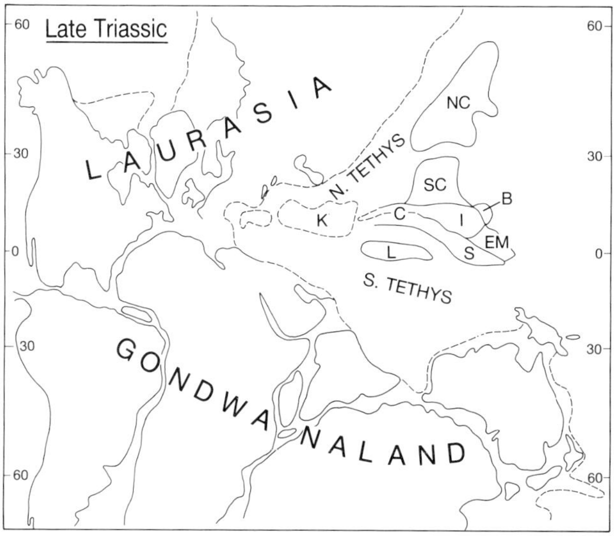

FIG. 11. Palaeogeographical reconstruction showing the relative positions of the south-east Asian, Tibetan, and Chinese blocks in the late Triassic. Other continental positions after Smith *et al.* (1981). Abbreviations as in Fig. 9.

somewhat better. Indo-China, East Malaya and south-west Borneo appear to have rifted from Gondwanaland in the early Carboniferous and drifted northwards, colliding with South China, which was situated in equatorial latitudes, in late-early Carboniferous times. This suturing formed a unified microcontinent, broadly equivalent tq what Gatinsky *et al.* (1984) called the 'East Asian Continent', on which developed the tropical *Gigantopteris* flora. The northern part of the Changtang block of Tibet (which does not have glacial-marine sediments, and has Cathaysian floras and Tethyan faunas in the Permian) is considered to have formed part of this microcontinent. North China must also have been relatively close by, as indicated by its Cathaysian flora and faunal similarities.

Sibumasu was still attached to north-west Australia until late-early Permian times. Fig. 9 shows the probable positions of the south-east Asian blocks in earliest Permian times.

During the rest of the Permian, South China, Indo-China, East Malaya, and Borneo lay in low northern palaeolatitudes, and Sibumasu—attached to Tibet and Kreios—left Gondwanaland in late-early Permian times and travelled northwards. Sometime during its northward travel (probably in the Permian and early Triassic) the elongate continent broke up into several fragments, and the Tibetan Lhasa block was left behind to follow at a later time (Fig. 10). Collision of Sibumasu and the Changtang block with Indo-China and South China probably occurred in the late Triassic, although a late Permian suturing is also possible (see papers by Helmke and co-workers). In the latest Triassic, South China sutured to North China along the Qinling suture line, and the northern branch of Tethys ('Palaeo-Tethys') was closed along its western part, allowing continental communication for vertebrates between the newly formed 'south-east Asia' and Laurasia (see Fig. 11).

Subsequent modifications to this proto-south-east Asia involve the south-eastward migration of south-west Borneo in the Cretaceous and the clockwise rotation and south-eastward 'extrusion' of south-east Asia during the collision of India and Eurasia.

References

ACHACHE, J. and COURTILLOT, V. (1985). A preliminary Upper Triassic palaeomagnetic pole for the Khorat plateau (Thailand): consequences for the accretion of Indochina against Eurasia. *Earth Planet. Sci. Lett.* **73**, 147–57.

ACHACHE, J., COURTILLOT, V., and BESSE, J. (1983). Palaeomagnetic constraints on the late Cretaceous and Cenozoic tectonics of Southeastern Asia. *Earth Planet. Sci. Lett.* **63**, 123–36.

ALLÈGRE, C. J. *et al.* (1984). Structure and evolution of the Himalaya–Tibet orogenic belt. *Nature* **307**, 17–22.

ARCHBOLD, N. W., PIGRAM, C. J., RATMAN, N., and HAKIM, S. (1982). Indonesian Permian brachiopod fauna and Gondwana–South-East Asia relationships. *Nature* **296**, 556–8.

ASAMA, K. (1984). *Gigantopteris* flora in China and southeast Asia. *Geol. Palaeont. Southeast Asia* **25**, 311–23.

AUDLEY-CHARLES, M. G. (1983). Reconstruction of eastern Gondwanaland. *Nature* **306**, 48–50.

AUDLEY-CHARLES, M. G. (1984). Cold Gondwana, warm Tethys and the Tibetan Lhasa block. *Nature* **310**, p. 165.

BARR, S. M., MACDONALD, A. S., and HAILE, N. S. (1978). Reconnaissance palaeomagnetic measurements on Triassic and Jurassic sedimentary rocks from Thailand. *Bull. Geol. Soc. Malaysia* **10**, 53–62.

BUFFETAUT, E. (1981). Éléments pour une histoire paléobiogéographique du Sud-Est asiatique: l'apport des Vertèbres fossiles continentaux. *Bull. Soc. Géol. Fr.* **23**, 587–93.

BUFFETAUT, E. (1982). Mesozoic vertebrates from Thailand and their palaeobiological significance. *Terra Cognita* **2**, 27–34.

BUFFETAUT, E. (1983). Dating the collision of the Indochina block with Asia: the contribution of vertebrate palaeontology. *Terra Cognita* **3**, p. 239.

BUFFETAUT, E. (1985). The palaeobiogeographical significance of the Mesozoic continental vertebrates from South-East Asia. *Mem. Soc. Geol. Fr. N. S.* **147**, 37–42.

BUFFETAUT, E. and INGAVAT, R. (1980). A new crocodilian from the Jurassic of Thailand, *Sunosuchus thailandicus* n.sp. (Mesosuchia, Goniopholididae) and the palaeogeographical history of Southeast Asia in the Mesozoic. *Geobios* **13**, 879–89.

BUFFETAUT, E. and INGAVAT, R. (1982). Phytosaur remains (Reptilia, Thecodontia) from the Upper Triassic of north-eastern Thailand. *Geobios* **15**, 7–17.

BUFFETAUT, E. and INGAVAT, R. (1983). *Goniopholis phywiangensis* nov. sp., a new mesosuchian crocodile from the Mesozoic of north-eastern Thailand. *Geobios* **16**, 79–91.

BUFFETAUT, E. and INGAVAT, R. (1984). The lower jaw of *Sunosuchus thailandicus*, a mesosuchian crocodilian from the Jurassic of Thailand. *Palaeont.* **27**, 199–206.

BUFFETAUT, E., GINSBURG, L. INGAVAT, R., JANVIER, P., MARTIN, M. and SUTTETORN, V. (1981). A new continental vertebrate fauna from the Upper Triassic of Thailand. *Terra Cognita* (special issue) Spring 1981, 116.

BUNOPAS, S. (1982). Palaeogeographic history of Western Thailand and adjacent parts of Southeast Asia—a plate-tectonics interpretation. *Geol. Surv. Pap. No. 5* Department of Mineral Resources, Thailand. (810 pp.)

BUNOPAS, S., VELLA, P., PITAKPAIVAN, K., and SUKROO, T. (1978). Preliminary palaeomagnetic results from Thailand sedimentary rocks. *Third Reg. Conf. Geol. Min. Resources Southeast Asia, Bangkok, Proc.* pp. 25–32.

BURRETT, C. and STAIT, B. (1984). South East Asia as a part of an Early Paleozoic Gondwanaland. *Abstracts of Papers, GEOSEA V Congress*, Kuala Lumpur, p. 3.

BURRETT, C. and STAIT, B. (1986). South East Asia as a part of an Early Palaeozoic Gondwanaland. *Bull. Geol. Soc. Malaysia,* **19**, 103–7.

BURTON, C. K. *in press*. The Kanchanaburi Supergroup of Peninsular and Western Thailand. *Bull. Geol. Soc. Malaysia,* **20**.

CAMERON, N. R., CLARKE, N. C. G., ALDISS, D. T., ASPDEN, T. A., and DJUNUDDIN, A. (1980). *The geological evolution of northern Sumatra.* Ninth Indonesian Petroleum Assoc. Convention, Jakarta. (53 pp.)

CHAKRABARTY, K. R. and METCALFE, I. (1987). Occurrence of sheared diamictite in the Ramb area its possible extensions and tectonic implications. *Scol. Soc. Malaysia Ann. Conf. Abstracts* 5.

FONTAINE, H. and VACHARD, D. (1981). A note on the discovery of the Lower Carboniferous (Middle Viséan) in Central Sumatra. *CCOP Newsl.* **8**, 14–18.

FONTAINE, H. AND WORKMAN, D. R. (1978). Review of the geology and mineral resources of Kampuchea, Laos and Vietnam. *Proc. GEOSEA III*, Bangkok, 541–603.

FONTAINE, H., LOVACHALASUPAPORN, S., and SEKTHEERA, B. (1982). Distribution of corals and coral reefs in the Permian of Thailand. *CCOP Newsl.* **9**, 13–19.

GATINSKY, Y. G. and HUTCHISON, C. S. *in press.* Cathaysia, Gondwanaland and the Paleotethys in the evolution of continental Southeast Asia. *Bull. Geol. Soc. Malaysia*, **20**.

GATINSKY, Y. G., MISHINA, A. V., VINOGRADOV, I. V., and KOVALEV, A. A. (1978). The main metallogenic belts of Southeast Asia as a result of different geodynamic conditions interference. *Third Regional Conf. Geol. and Min. Res. Southeast Asia*, Bangkok, pp. 313–18.

GATINSKY, Y. G., HUTCHISON, C. S., MINH, N. N., and TRI, T. V. (1984). Tectonic evolution of Southeast Asia. *27th Int. Geol. Congress, Moscow, Rep.* **5**, 225–39.

HAHN, L. (1985). The Indosinian orogeny in Thailand and adjacent areas. *Mem. Soc. geol. Fr. N. S.* **147**, 71–82.

HAILE, N. S. (1979). Palaeomagnetic evidence for rotation and northward drift of Sumatra. *J. Geol. Soc. London* **136**, 541–5.

HAILE, N. S. (1980). Palaeomagnetic evidence from the Ordovician and Silurian of northwest Peninsular Malaysia. *Earth Planet. Sci. Lett* **48**, 233–6.

HAILE, N. S and BRIDEN, J. C. (1982). Past and future paleomagnetic research and the tectonic history of East and Southeast Asia. *Proc., CCOP Workshop Palaeomagnet. Res. Southeast & East Asia*, Kuala Lumpur.

Haile, N. S. and Khoo, H. P. (1980). Palaeomagnetic measurements on Upper Jurassic to Lower Cretaceous sedimentary rocks from Peninsular Malaysia. *Bull. Geol. Soc. Malaysia* **12**, 75–8.

Haile, N. S., Stauffer, P. H., Krishnan, D., Lim, T. P., and Ong, G. B. (1977). Palaeozoic redbeds and radiolarian chert: reinterpretation of their relationships in the Bentong and Raub areas, West Pahang, Peninsular Malaysia. *Bull. Geol. Soc. Malaysia* **8**, 45–60.

Haile, N. S., Beckinsale, R. D., Chakraborty, K. R., Abdul, H. H., and Tjahjo, H. (1985). Palaeomagnetism, geochronology and petrology of the dolerite dykes and basaltic lavas from Kuantan, West Malaysia. *Bull. Geol. Soc. Malaysia* **16**, 71–85.

Hayami, I. (1984). Jurassic marine bivalve faunas and biogeography in Southeast Asia. *Geol. Palaeont. Southeast Asia* **25**, 229–37.

Helmke, D. (1983). On the Variscan evolution of central mainland Southeast Asia. *Earth Evol. Sci.* **4**, 309–19.

Helmke, D. (1985*a*). The Permo-Triassic 'Paleotethys' in mainland Southeast-Asia and adjacent parts of China. *Geol. Rdsch.* **74**, 215–28.

Helmke, D. (1985*b*). The orogenic evolution (Permian–Triassic) of central Thailand. Implications on paleogeographic models for mainland SE-Asia. *Me. Soc. Geol. Fr., N. S.* **147**, 83–91.

Helmke, D. and Kraikhong, C. (1982). On the geosynclinal and orogenic evolution of central and northeastern Thailand. *J. Geol. Soc. Thailand* **5**, 52–74.

Helmke, D. and Lindenberg, H. G. (1983). New data on the 'Indosinian' orogeny from Central Thailand. *Geol. Rdsch.* **72**, 317–28.

Hill, D. (1973). Lower Carboniferous corals. In *Atlas of palaeobiogeography*. (ed. A. Hallam), Elsevier, Amsterdam.

Howell, D. G., Schermer, E. R., Jones, D. L., Ben-Avraham, Z., and Scheibner, E. (1983). Tectanostratigraphic terrane map of the Circum–Pacific region. *US Geol. Surv.* Open File Map 83–716.

Ingavat, R. and Douglass, R. (1981). Fusuline fossils from Thailand, Part XIV. The fusulinid genus *Monodiexodina* from Northwest Thailand. *Geol. Palaeont. Southeast Asia* **22**, 23–34.

Ingavat, R. and Janvier, P. (1981). *Cyclotosaurus* cf *posthumus* Fraas (Capitosaudidae, Stereospondyli) from the Huai Hin Lat Formation (Norian, Upper Triassic) of northeastern Thailand, with a note on capitosaurid biogeography. *Geobios* **14**, 711–23.

Ishii, K., Okimura, Y., and Ichikawa, K. (1985). Notes on Tethys biogeography with reference to Middle Permian fusulinaceans. In *The Tethys*. (ed. K. Nakazawa and J. M. Dickins), pp. 139–55, Tokai University Press, Tokyo.

Jell, P. A., Burrett, C. F., Stait, B., and Yochelson, E. L. (1984). The Early Ordovician bellerophontoid *Peelerophon oehlerti* (Bergeron) from Argentina, Australia and Thailand. *Alcheringer* **8**, 169–76.

Jones, C. R. (1968). Lower Palaeozoic rocks of Malay Peninsula. *Bull. Amer. Assoc. Petrol. Geol.* **52**, 1259–78.

Kimura, T. (1984). Mesozoic floras of East and Southeast Asia, with a short note on the Cenozoic floras of Southeast Asia and China. *Geol. Palaeont. Southeast Asia* **25**, 325–50.

Kobayashi, T. and Tamura, M. (1984). The Triassic Bivalvia of Malaysia, Thailand and adjacent areas. *Geol. Palaeont. Southeast Asia* **25**, 201–28.

Kon'no, E. and Asama, K. (1970). Some Permian plants from the Jengka Pass, Pahang, West Malaysia. *Geol. Palaeont. Southeast Asia* **8**, 94–132.

Kon'no, E., Asama, K., and Rajah, S. S. (1970). The Late Permian Linggiu flora from the Gunong Blumut area, Johore, Malaysia. *Bull. Nat. Sci. Museum* **13**, 491–580.

Liew, T. C. and McCulloch, M. T. (1985). Genesis of granitoid batholiths of Peninsular Malaysia and implications for models of crustal evolution: Evidence from a Nd–Sr isotopic and U–Pb zircon study. *Geochim. Cosmochim. Acta* **49**, 587–600.

Lin, J. (1987). Late Carboniferous palaeogeographic reconstruction. International Congress of Carboniferous Stratigraphy and Geology. Beijing, China. Abstracts of Papers **1**, 283.

Martin, M. (1985). The actinopterygian scales and teeth (Pisces) from the continental Upper Triassic of Thailand, their paleogeographical significance. *Mem. Soc. Geol. Fr., N. S.* **147**, 101–5.

Martin, M. and Ingavat, R. (1982). First record of an Upper Triassic ceratodontid (Dipnoi, Ceratodontiformes) in Thailand and its paleogeographical significance. *Geobios* **15**, 111–14.

Matsuda, T. (1985). Late Permian to Early Triassic conodont paleobiogeography in the 'Tethys Realm'. In *The Tethys*. (ed. K. Nakazawa and J. M. Dickins), Tokai University Press, Tokyo.

McElhinny, M. W., Haile, N. S., and Crawford, A. R. (1974). Palaeomagnetic evidence shows Malay Peninsula was not a part of Gondwanaland. *Nature* **252**, 641–5.

McElhinny, M. W., Embleton, B. J. J., Ma, X. H., and Zhang, Z. K. (1981). Fragmentation of Asia in the Permian. *Nature* **293**, 212–16.

Metcalfe, I. (1983). Southeast Asia. In *The Carboniferous of the world*. Vol. 1 (ed. P. H. Wagner, C. F. Winkler Prins, and L. F. Granados), pp. 213–43, IUGS Publ. No. 16.

Metcalfe, I. (1985). Stratigraphy, palaeontology and palaeogeography of the Carboniferous of Southeast Asia. *Mem. Soc. Geol. Fr. N. S.* **147**, 107–18.

Metcalfe, I. (1986). Late Palaeozoic palaeogeography of Southeast Asia: some stratigraphical, palaeontological and palaeomagnetic constraints. *Bull. Geol. Soc. Malaysia* **19**, 153–64.

Mitchell, A. H. G. (1981). Phanerozoic plate boundaries in mainland Southeast Asia, the Himalayas and Tibet. *Jour. Geol. Soc. London* **138**, 109–22.

Nakamura, K., Shimizu, D., and Liao, Z. (1985). Permian palaeobiogeography of brachiopods based on faunal provinces. In *The Tethys*. (ed. K. Nakazawa and J. M. Dickins), pp. 185–98, Tokai University Press, Tokyo.

Phan, C. T. (1981). The limits of the natural division in the late Palaeozoic in North Vietnam. *Add Proc. GEOSEA IV, Manila*, pp. 40–7, Inst. Geoscientific Information, Vietnam.

Rao, P. C. (1984). Sedimentology of some Permo-Triassic carbonates of Malaysia. *GEOSEA V, Kuala Lumpur, Abstr.*, p. 28.

Ridd, M. F. (1980). Possible Palaeozoic drift of S.E. Asia and Triassic collision with China. *J. Geol. Soc. London* **137**, 635–40.

Rutten, M. G. (1940). On Devonian limestones with *Clathrodictyon* cf *spatiosum* and *Heliolites porosus* from Eastern Borneo. *Proc. Kon. Nederl. Akad. Wetensch.* **43**, 1061–64.

Sasajima, S., Otofuji, Y., Hirook, K., Suparka, Suwijanto, and Hehuwat, F. (1978). Palaeomagnetic studies on Sumatra Island: on the possibility of Sumatra being part of Gondwanaland. *Rock Magn. Palaeogeophys.* **5**, 104–10.

Şengör, A. M. C. (1984). The Cimmeride orogenic system and the tectonics of Eurasia. *Geol. Soc. Am. Spec. Pap.* **195**, 82 pp.

Şengör, A. M. C. (1985*a*). The story of Tethys: how many wives did Okeanos have? *Episodes* **8**, 3–12.

Şengör, A. M. C. (1985*b*). Die Alpiden und die Kimmeriden: die verdoppelte Geschichte der Tethys. *Geol. Rdsch.* **74**, 181–213.

Şengör, A. M. C. and Hsü, K. J. (1985). The Cimmerides of Eastern Asia: history of the eastern end of Palaeo-Tethys. *Mem. Soc. Geol. Fr. N. S.* **147**, 139–67.

Shergold, J., Burrett, C., Laurie, J., and Stait, B. Upper Cambrian trilobites of the Tarutao Formation, Southern Thailand (in press).

Smith, A. G., Hurley, A. M., and Briden, J. C. (1981). *Phanerozoic palaeocontinental world maps*. Cambridge University Press. (102 pp.)

Stait, B. and Burrett, C. (1982). *Wutinoceras* (Nautiloidea) from the Setul Limestone (Ordovician) of Malaysia. *Alcheringa* **16**, 193–6.

Stait, B. and Burrett, C. (1983). Biogeography of Ordovician nautiloids and trilobites from Australia, Malaysia and Thailand and their tectonic implications. *Geol. Soc. Austral. Abstr.* **9**, 221–2.

Stait, B. and Burrett, C. (1984*a*). Lower Ordovician polyplacophoran *Chelodes whitehousei* from Tarutao Island, southern Thailand. *Alchieringa* **8**, p. 112.

Stait, B. and Burrett, C. (1984*b*). Ordovician nautiloid faunas of Central and Southern Thailand. *Geol. Mag.* **121**, 115–24.

Stauffer, P. H. (1974). Malaya and Southeast Asia in the pattern of continental drift. *Bull. Geol. Soc. Malaysia* **7**, 89–138.

Stauffer, P. H. (1983). Unravelling the mosaic of Palaeozoic crustal blocks in Southeast Asia. *Geol. Rdsch.* **72**, 1061–80.

Stauffer, P. H. and Lee, C. P. (1984). Late Palaeozoic glacial-marine facies in Southeast Asia and its implications. *Abstr. Pap., GEOSEA V, Kuala Lumpur* pp. 31–2.

Stauffer, P.H. and Mantajit, J. (1981). Late Palaeozoic tilloids of Malaya, Thailand and Burma. In *Earth's pre-Pleistocene glacial record.* (ed. M. J. Hambrey and W. H. Harland), pp. 331–7, Cambridge.

Tjia, H. D. (1985). Gaya struktur rantau selat Melaka: suatu pembandingan. *Sains Malaysiana* **14**, 37–64.

Tjia, H. D and Zaiton, H. (1985). Regional structures of Peninsular Malaysia. *Sains Malaysiana* **14**, 95–107.

Tollmann, A. and Tollmann, E. K. (1985). Palaeogeography of the European Tethys from Paleozoic to Mesozoic and the Triassic relations of the eastern part of Tethys and Panthalassa. In *The Tethys*. (ed. K. Nakazawa and J. M. Dickins), pp. 3–22, Tokai University Press, Tokyo.

Tran, V. T. (1979). *Geology of Vietnam. (North part).* General Geological Department, Hanoi, Vietnam.

Wang, N. (1984). Le paléocontinent qingzangindien et sa suturation avec le paléocontinent cathaysien. In *Mission Franco-Chinoise au Tibet, 1980.* (ed. J. L. Mercier and Guangcen Li), pp. 33–54, CNRS, Paris.

Waterhouse, J. B. (1982). An early Permian cool-water fauna from pebbly mudstones in south Thailand. *Geol. Mag.* **119**, 337–54.

Webby, B. D., Wyatt, D., and Burrett, C. (1985). Ordovician stromatoporoids from the Langkawi Islands, Malaysia. *Alcheringa* **9**, 159–66.

Wongwanich, T., Wyatt, D., Stait, B., and Burrett, C. (1983). The Ordovician System in southern Thailand and northern Malaysia. *Proc. Workshop stratigr. correl. Thailand & Malaysia, Haad Yai, 1983*, pp. 77–95.

Yanagida, J. and Aw, P. C. (1979). Upper Carboniferous, Upper Permian and Triassic brachiopods from Kelantan, Malaysia. *Geol. Palaeont. Southeast Asia* **20**, 119–41.

Xiao, X. and Gao, Y. (1984). Tectonic evolution of the Tethys-Himalayas of China. *27th Intern. Geo. Congr., Moscow, Rep.* Vol. 5, pp. 181–9.

I. Metcalfe, Department of Geology, Universiti Kebangsaan Malaysia, Bangi, Selangor, Malaysia.

Origin and assembly of the Tethyside orogenic collage at the expense of Gondwana Land

A. M. C. Şengör, Demir Altıner, Altan Cin, Timur Ustaömer, and K. J. Hsü

ABSTRACT: The *Tethysides* are a superorogenic complex flanking the Eurasian continent to the south and consisting of the *Cimmerides* and *Alpides*, products of *Palaeo-* and *Neo-Tethys* respectively. We here review their evolution, mainly on the basis of new maps showing the distribution of sutures, magmatic rocks, certain palaeobiogeographically and palaeoclimatologically significant taxa and facies, and fragments of Pan-African (900–450 Ma) orogenic system forming the basement of many Tethyside blocks. These are supplemented by palaeomagnetic data reported in the literature.

A fundamental tenet of this paper is that major sutures which contain ophiolite fragments, represent tectonic sections between continental blocks where oceanic crust has been subducted. Palaeo-Tethys came into existence largely in late Carboniferous time. Coevally, it began to be consumed by both internal and peripheral subduction zones, which continued into the Permian; some of these had been inherited from pre-Tethyan times. In the later Permian, rifting subparallel with the northern margin of Gondwana Land began between the Zagros and Malaysia, separating a Cimmerian continent from N. Gondwana Land, and thus heralding the opening of Neo-Tethys and other smaller oceans that were back-arc basins of Palaeo-Tethys. This rifting possibly also extended farther west into Crete and mainland Greece. However, the North China block, Yangtze block, Huanan block, the eastern moity of the Qangtang block (North Tibet), and Annamia, all originally pieces of the end-Proterozoic–early Palaeozoic Gondwana Land, had already separated from it in pre-late Carboniferous times, possibly during the Devonian. All of these blocks, and the Cimmerian continent, were characterized by Cathaysian floral elements in late Palaeozoic time. Palaeomagnetic and palaeontological data showing the original Gondwana Land affinity of these continental blocks are supplemented by correlating late Proterozoic–early Palaeozoic Pan-African sutures, orogenic belts, and sedimentary basin fragments across Tethyside sutures. Late Permian foraminiferal provinces are related to this palaeogeographical interpretation.

By Triassic times, most Cimmeride subduction zones were already in existence. The Cimmerian Continent accelerated its separation from Gondwana Land and—locally in the late Permian—began disintegrating internally along the Waser/Rushan–Pshart/Banggong Co–Nu Jiang/Mandalay ocean. By late Triassic time all of the Chinese blocks—except Lhasa—and Annamia had collided with each other and with Laurasia. The resulting enormous orogenic collage had a 'soft cushion' between itself and Laurasia, in the form of the enormous accretionary complex of the Songpan–Ganzi. This connection enabled Laurasian land vertebrates to reach south-east Asia by late Triassic time. In late Triassic to middle Jurassic times, most major Cimmeride collisions were completed. Widespread aridity in Central Asia occurred in late Jurassic time, probably in the rain shadow of the newly formed Cimmeride mountain wall.

Neo-Tethyan subduction systems formed along the S. margin of the Cimmerides or within Neo-Tethyan oceanic lithosphere during the Jurassic. Most, if not all, were north- or east-dipping. They continued the northerly migration of the Tethyside blocks.

Evolution of the Tethysides influenced the distribution of marine and terrestrial organisms, and affected sea-level changes and patterns of atmospheric circulation during much of the Mesozoic and Cainozoic. It is likely to have reflected the surface expression of a persistent trend in the large-scale convective circulation in the mantle, that continuously transported material northward into the Tethyan domain.

Within the Alpine–Himalayan–Indonesian mountain ranges, two distinct but largely superimposed orogenic systems exist, parts of which correspond with Suess' (1909, 1911) original *Cimmerian Mountains* and the *Alpides* (Şengör 1984). Respecting Suess' priority, Şengör (1984) called the older of these two systems *Cimmerides*, and the younger *Alpides*, together forming the superorogenic complex *Tethysides*. Both of these systems have been shown to include a very large number of sutures, but two dominant times of ocean closure along

From AUDLEY-CHARLES, M. G. & HALLAM, A. (eds) *Gondwana and Tethys* Geological Society Special Publication No. 37, pp. 119–81.

them are discernible: the earlier interval from the late middle Triassic to earliest middle Jurassic, corresponds with the formation of the Cimmerides, whereas the later episode, the Alpide, was dominantly from the late Palaeocene to late Eocene. Although in both systems a number of suture segments closed outside the intervals indicated, these intervals appear valid for such long stretches that it may be possible to schematize them into a 'northern group' of Cimmeride sutures and a 'southern group' of Alpide sutures, with a long Cimmerian continent (or a continental archipelago), and a number of 'exotic' continental fragments in between (Fig. 1) (Şengör 1979*a*, 1984). This grossly simplified scheme makes a rapid overview easy, and also provides a practical nomenclature for large-scale palaeogeographical and palaeotectonic elements of Pangaea in the Old World. We have, thus, Laurasia in the north (Permian to Cretaceous), Palaeo-Tethys to its south (early Carboniferous to middle Jurassic), the Cimmerian continent in the middle (Triassic to middle Jurassic), Neo-Tethys between the Cimmerian continent and northern Gondwanaland (? latest Permian or Triassic to largely Eocene, locally still extant) and finally Gondwana Land itself in the south (*c.* Ordovician to Jurassic) (Fig. 1) (Şengör 1984, Şengör and Hsü 1984).

The closure of these two sets of 'Tethyan' sutures generated the Cimmeride and Alpide orogenic systems (Fig. 2), with their associated wide and complex areas of cratonic disruption that now dominate the architecture of nearly the whole of Eurasia. Palaeo- and Neo-Tethys, plus the continental portions separating them or contained within them may be designated as the *Tethyan domain* or the *Tethyan realm*.

Although the simple classification given in Fig. 1 serves as a useful terminological basis for further research, the suture map displayed in Fig. 3 emphasizes how much more complicated the real situation is. The Tethysides consist of an enormous orogenic collage, containing a large number of continental and otherwise buoyant blocks, stitched together by an anastomosing suture network. Following the dual division of the Tethysides, this *Tethyside collage* also con-

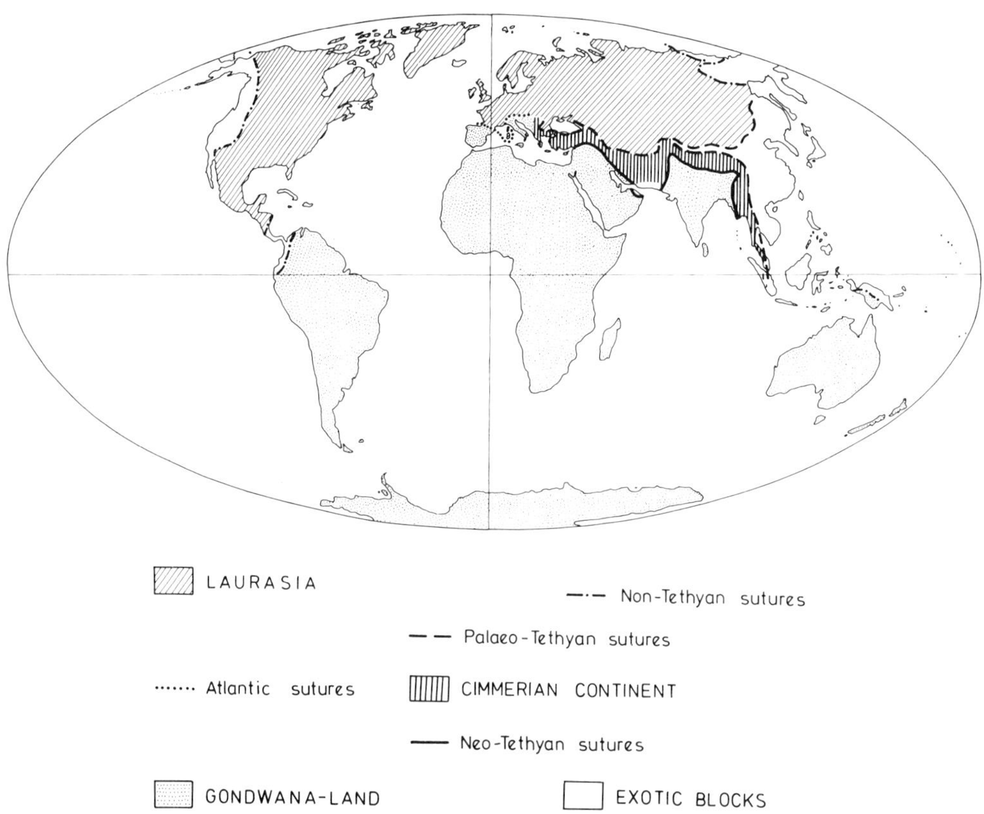

FIG. 1. First-order palaeogeographical/palaeotectonic elements taking part in the architecture of the Tethysides, and their place in the present structure of our planet.

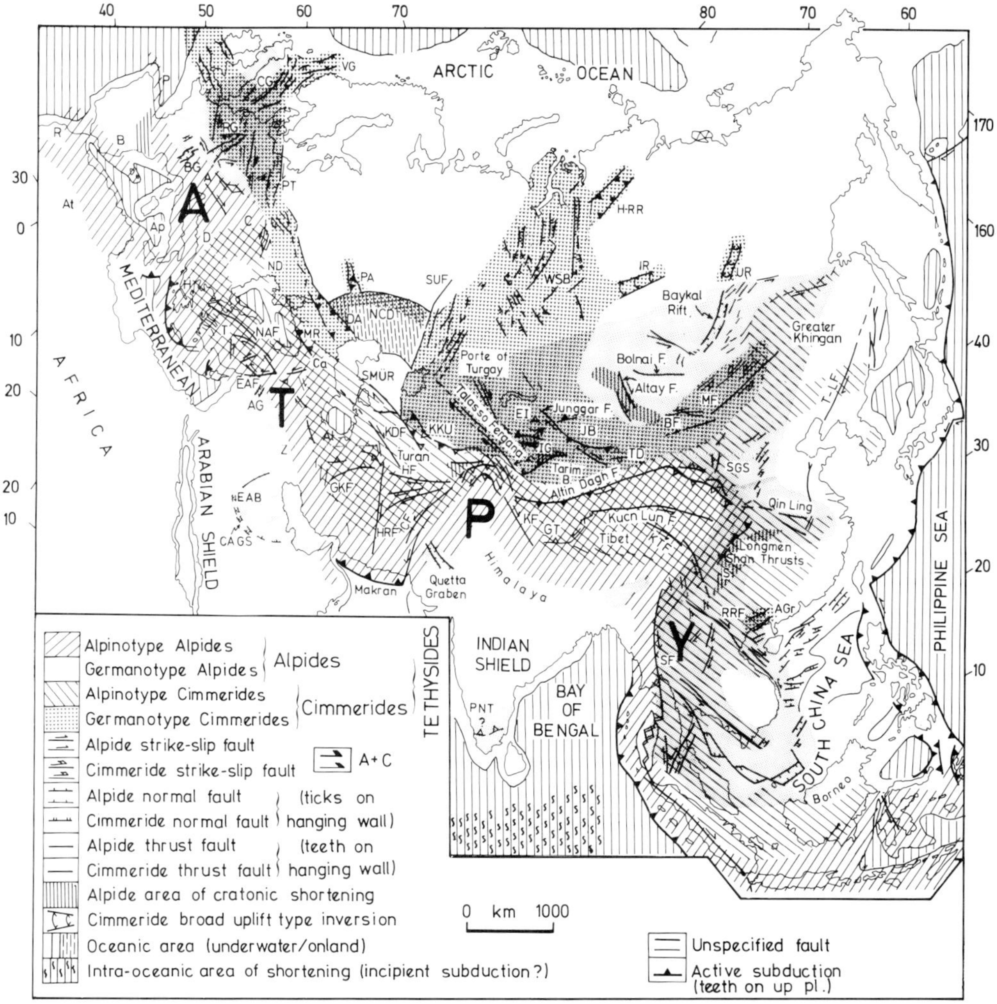

FIG. 2. The Tethyside superorogenic complex, showing both the Cimmerides and the Alpides and their associated areas of fore- and hinterland deformation. Notice the degree of overprinting of the Cimmerides by the Alpides, especially between the Carpathians and China, which has long hindered the recognition of the former as an independent orogenic belt. Key to lettering: larger bold letters *A*, *T*, *P*, and *Y* are the Alpine, Turkish, Pamir, and Yunnan syntaxes respectively. Smaller letters: *A*—Alps, *AG*—Akçakale Graben, *AGr*—An Chau Graben, *Al*—Alborz, *Ap*—Apennines, *At*—Atlas mountains (*sensu lato*), *B*—Betics, *BF*—Bogdo Fault, *BG*—Bresse Graben, *C*—Carpathians, *Ca*—Caucasus, *CAGS*—Central Arabian graben system, *CF*—Chaman Fault, *CG*—Central Graben, *D*—Dinarides, *DA*—Dnyepr–Donetz aulacogen, *EAB*—East Arabian block, *EAF*—East Anatolian Fault, *EI*—East Ili Basin, *GKF*—Great Kavir Fault, *GT*—Gerze thrust, *H*—Hellenides, *HF*—Herat Fault, *HRF*—Harirud Fault, *H-RR*—Hantaj–Rybninsk Rift, *IG*—Issyk Gol Basin, *IR*—Irkineev Rift, *KDF*—Kopet Dagh Fault, *KF*—Karakorum Fault, *KKU*—Kizil Kum uplift, *KTF*—Kang Ting Fault, *MF*—Mongolian Faults, *MR*—Main Range of the Greater Caucasus, *NAF*—North Anatolian Fault, *NCD*—North Caspian depression, *ND*—North Dobrudja, *PA*—Pachelma aulacogen, *PNT*—Palni–Nilgiri Hills Thrust, *PT*—Polish Trough, *RG*—Upper Rhine Graben, *RRF*—Red River Fault, *S*—Sichuan Basin, *SF*—Sagain Fault, *SGS*—Shanxi Graben System, *SMÜR*—South Mangyshlak–Üst Yurt Ridge, *SUF*—South Ural Faults, *T*—Turkish Ranges, *TD*—Turfan depression, *T-LF*—Tan-Lu Fault, *UR*—Ura Rift, *VG*—Viking Graben, *WSB*—West Siberian Basin, *Z*—Zagrides.

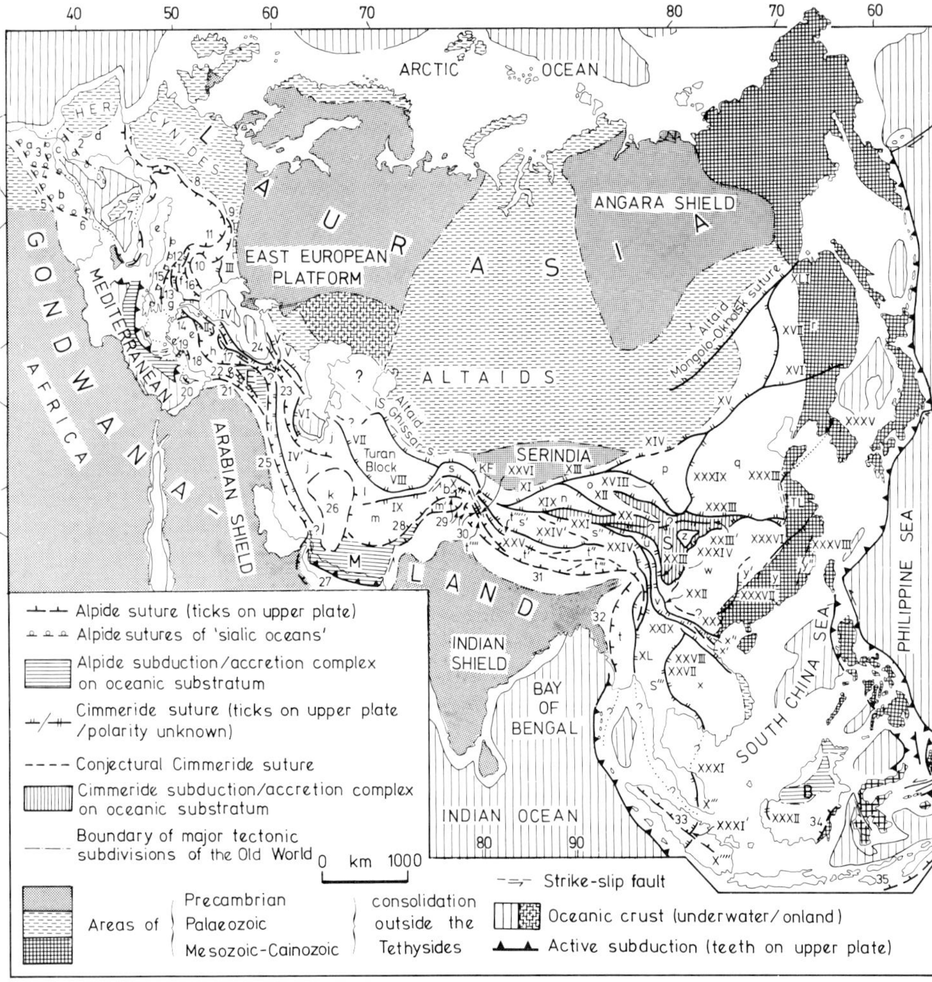

FIG. 3. Tectonic map of Eurasia and North Africa, showing their major tectonic subdivisions and the distribution of sutures within the Tethysides. Tethysides correspond with white areas on land. Cimmeride sutures: *I*—Palaeo-Tethyan suture in Greece and Yugoslavia, *II*—Karakaya, *III*—Luncavita–Consul, *IV*—North Turkish, *IV′*—Sanandaj–Sirjan suture, *V*—Svanetia, *V′*—Chorchana–Utslevi, *VI*—Talesh, *VII*—Kopet Dagh/Mashhad, *VIII*—Paropamisus/Hindu Kush/North Pamir, *IX*—Waser, *X*—Rushan–Pshart, *XI*—Northern Kuen-Lun, *XII*—Qimandag, *XIII*—Altin Dagh, *XIV*—Suelun–Hegen Mountains, *XV*—Inner Mongolian, *XVI*—Suolun–Xilamulun, *XVII*—Greater Khingan, *XVIII*—Tergun Daba Shan/Qinghai Nanshan, *XIX*—Southern Kuen-Lun, *XX*—Burhan Budai Shan/Anyemaqen Shan, *XXI*—Hoh Xil Shan/Jinsha Jiang, *XXII*—Maniganggo, *XXIII*—Litang, *XXIII′*—Luochou 'arc-trench belt', *XXIV*—Banggong Co–Nu Jiang, *XXIV′*—'Mid-Qangtang', *XXV*—Shiquan He, *XXVI*—south-west Karakorum, *XXVII*—Nan-Uttaradit/Sra-Kaeo, *XXVIII*—Tamky–Phueson, *XXIX*—Song Ma, *XXX*—Song Da, *XXXI*—Bentong–Raub, *XXXI′*—Mid-Sumatra, *XXXII*—Serabang (West Borneo), *XXXIII*—Qin-Ling/Dabie Shan, *XXXIII′*—northernmost Jiangsu, *XXXIV*—Longmen Shan/Qionglai Shan, *XXXV*—Chugareong, *XXXVI*—Shaoxing–Pingxiang, *XXXVII*—Tianyang, *XXXVIII*—Lishui–Haifeng, *XXXIX*—Helan Shan, *XL*—Mandalay, *XLI*—Shilka. Alpide sutures: *1*—Pyrenean, *2*—Betic, *3*—Riff, *4*—High Atlas, *5*—Saharan Atlas, *6*—Kabylian, *7*—Apennine, *8*—Alpine, *9*—Pieniny Klippen Belt, *10*—circum-Moesian, *11*—Mureş, *12*—Vardar, *13*—Peonias/Intra-Pontide, *14*—Almopias/İzmir–Ankara, *15*—Pindos–Budva–Bükk, *16*—Srednogorie, *17*—Ilgaz–Erzincan, *18*—Inner Tauride, *19*—Antalya, *20*—Cyprus, *21*—Bitlis (Miocene) and Assyrian (Senonian), *22*—Maden, *23*—Sevan/Akera/Qaradagh, *24*—Slate-Diabase zone, *25*—Zagros, *26*—circum-Central Iranian Microcontinent (including Sabzevar and Sistan), *27*—Oman, *28*—Waziristan, *29*—Kohistan (*29* North—Chalt, *29* South—Main

sists of two parts of unequal size. The older and much larger part is formed from fragments accreted to Laurasia during the Cimmeride evolution, and is called the *Cimmeride collage*. The younger and much smaller part was accreted to Eurasia during the Alpide evolution, and is called the *Alpide collage* (Şengör 1986*a*). In Fig. 3, the Cimmeride and Alpide sutures are identified with separate symbols to distinguish their components from one another.

In the following paragraphs we first comment on our improved suture map of the Tethysides (Fig. 3), and establish the timing of events along them. We then explore the provenance of the major components of the Cimmeride and Alpide collages, on the basis of palaeontological, palaeomagnetic, and other geological data, showing that most, if not all, of them originated in Gondwana Land. Following a highly schematized and tentative evolutionary history of the Tethysides, we point out the extremely important role of multiphase strike-slip motion in complicating the evolution of the Tethysides. We conclude by emphasizing the vast complexities and the tremendous mobility inherent in the evolution of orogenic complexes in general, and underline the misleading role of such simplistic methods as the 'terrane concept' that are employed for their analysis.

Suture zones of the Tethyside orogenic collage

Fig. 3 illustrates all of the known major suture zones within the Tethyside superorogenic complex. The most conspicuous feature of this map is the nearly complete independence of the Tethyside sutures from those of older orogenic complexes such as the Hercynides and the Altaids. Other than two exceptions (the Ghissar and the Mongolo–Okhotsk sutures: Fig. 3) none of the Tethyside sutures seem to connect with sutures outside the Tethyside suture network in Eurasia. One implication of this is that a few long and thin (present geometries) continental objects had collided with the older Altaid orogenic collage to 'seal it off' before the Tethyside collisions with Asia commenced. In the following sections we review first the Cimmeride, and then the Alpide sutures, mainly with respect to the ages of ocean opening and closing along them. Their orogenic polarities (i.e. pre-collision facing of main subduction zone) are shown in Fig. 3. Figs 4 through 7 display the evolution of magmatism along these suture zones, and provide a major basis for their identification.

Cimmeride sutures[1]

Palaeozoic Cimmeride sutures

A few of the earliest 'Tethyside' collisions had taken place between individual components of the Cimmeride collage (e.g. the pre- or intra-Devonian Tamky–Phueson suture in North Vietnam (*XXVIII*): Hutchison, in press, or the coeval Tergun Daba Shan–Qinghai Nanshan suture (*XVIII*): Yang *et al.* 1986) even before the triangular Pangaean gap came into existence. Thus, these sutures are not Tethyside in the strict sense. The reason why they are so considered here is that their closure seems to have triggered the onset of subduction activity along the nearby Tethyside sutures, thus exercising a

[1] In this paper Roman numerals in parentheses refer to Cimmeride sutures shown in Fig. 3, Arabic numerals refer to Alpide sutures, and lower case letters refer to Tethyside blocks. Hereinafter they are cited without repeated reference to Fig. 3.

Mantle Thrust), *30*—Ladakh (*30* North (N)—Shyok, *30* South (S)—Indus), *31*—Indus/Yarlung–Zangbo, *32*—Indo-Burman, *33*—Woyla, *34*—Meratus, *35*—Timor. Tethyside blocks: *a*—Moroccan Meseta, *b*—Oran Meseta, *c*—Alboran, *d*—Iberian, *e*—African Promontory (including the Menderes–Taurus block; *e'*—Alanya Massif), *f*—Rhodope–Pontide, *g*—Sakarya, *h*—Kırşehir, *i*—Sanandaj–Sirjan zone, *j*—north-west Iran, *k*—central Iranian microcontinent, *l*—Farah, *m*—Helmand, *m'*—Kohistan, *n*—western Kuen-Lun 'Central Meganticlinorium', *o*—Qaidam, *p*—Alxa (Ala Shan), *q*—North China ('Sino-Korean'), *r*—North China Foldbelt, *s*—Central Pamir–Qangtang–Sibumasu (*s'*—Central Pamir/West Qangtang, *s''*—East Qangtang, *s'''*—Sibumasu), *t*—Lhasa–Central Burma (*t'*—Bongthol Tangla, *t''*—Nagqu, *t'''*—Lhasa proper: A. Gansser, verbal comm. 1985, *t''''*—Ladakh, *t'''''*—west Lhasa–south Pamir), *u*—Shaluli Shan, *v*—Chola Shan, *w*—Yangtze, *x*—Annamia (possibly with two exotic island arcs of early (*x'*) and late (*x''*) Palaeozoic age and eastern Malay Peninsula (*x'''*) and south Sumatra (*x''''*)), *y* (*sensu lato*)—Huanan (*y* (*sensu stricto*)—Huanan, *y'*—Hunan, *y''*—coastal block), *z*—Songpan Massif. Key to lettering: *B*—Sarawak accretionary complex, *b*—South-east Pamir black slates, *E*—East Anatolian accretionary complex, *KF*—Karakorum Fault, *M*—Makran accretionary complex, *TL*—Tan-Lu fault.

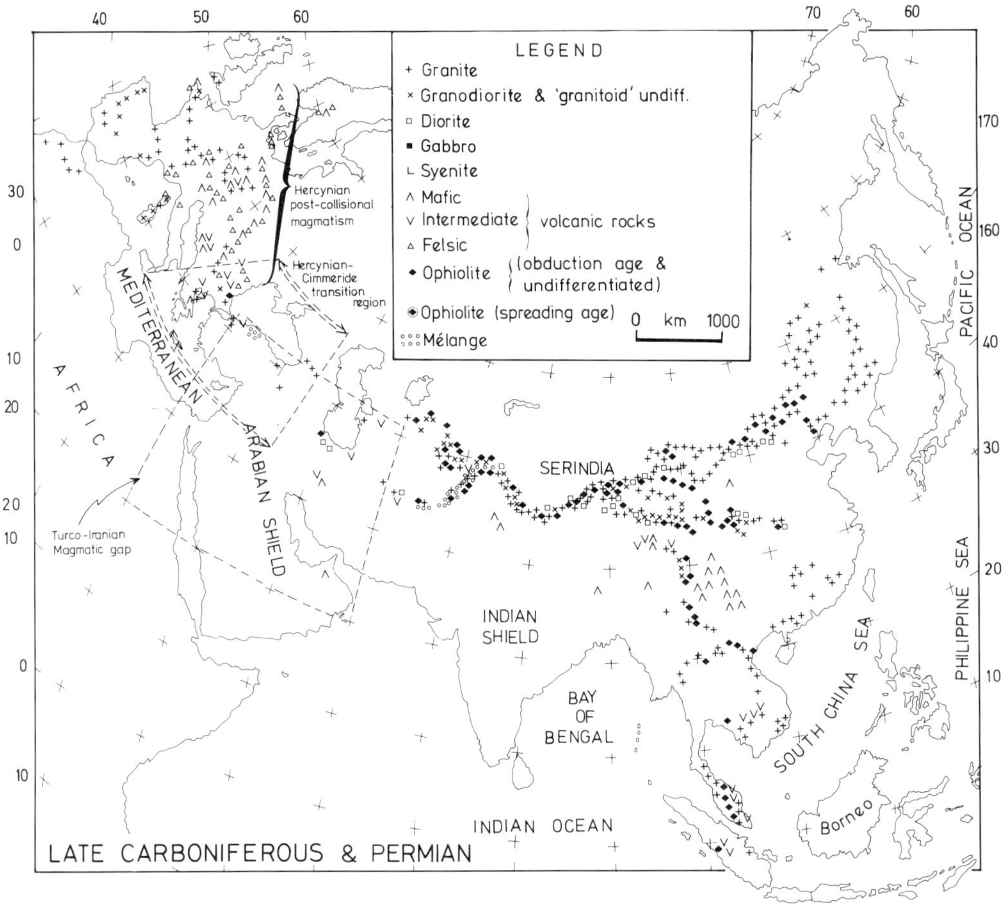

FIG. 4. Late Carboniferous and Permian magmatism in the Tethysides and in the Hercynides. Notice the conspicuous contrast between the irregular distribution of collision-related magmatism of the Hercynides, and the linear/arcuate subduction-related magmatism of the Cimmerides. Notice also that there seems to be no natural break between the two in the Balkan Peninsula and in the Caucasus. This is one reason why the Hercynides and the Cimmerides have been confused for so long in these regions. The list of references, from which this map was compiled, may be obtained from the Geological Society of London for the price of photocopying.

direct influence on the future Tethyside evolution.

The earliest 'Tethyside' (i.e. post-Devonian within the Tethyside architecture) sutures are also those between individual 'exotic' blocks of the Cimmeride collage. The earliest of these was probably the Song Ma (Red River) suture (*XXIX*) in northern Vietnam, that closed between the already accreted (?arc) fragment *x′* and the new (?arc) fragment *x″* along a south-west-dipping subduction zone during the Tournaisian–Viséan interval (Fontaine and Workman 1978). Morgunov (1970) indicated a somewhat later time for the onset of intense south-west-directed thrusting, namely during the Viséan to Namurian transition. Recently, both Gatinsky (1985) and Hutchison (in press) argued that the closure along not only the Song Ma suture, but also the final collision between Annamia (*x*) and the 'South China block' *sensu lato* (*w* or *y*) had already taken place during the Devonian. The basis for this interpretation is the observation that an *c.* 10-km thick section south of the suture, consisting of Cambrian through Silurian shallow-water sediments, was deformed and uplifted before the beginning of the Devonian. A regional unconformity is believed to exist at the base of the Devonian (Fontaine and Workman 1978). However, the sea returned quickly, and the Devonian through early Carboniferous marine sedimentation was finally terminated by intense south-west-vergent folding and large-scale thrusting, during the early to middle Carboniferous, forming such major structures as the Nam Nhuong nappe (with a 220 km along-strike length and some 60 km of minimum displacement; Fromaget 1927). The presence of late Palaeozoic ophiolites along the Song Ma suture, and granites to the south of the suture (Fig. 4) further strengthens the interpretation of an early to middle Carboniferous collision along it. Farther to the north-west, along the Ailao-Shan segment of the same suture (just under the letter *v* in Fig. 3) the collision may have happened later, during the Triassic, as evidenced by a complex terrane of sedimentary and volcanic rocks plus some ophiolite slivers with ages ranging Devonian–Triassic (Bally *et al.* 1980; Şengör and Hsü 1984).

Another Carboniferous suture between two 'exotic' Cimmeride collage components is the Suolun–Xiamulun (*XVI*) suture between the North China block (*q*), and what is known as the 'North China Foldbelt' (*r*—Klimetz 1983). Yang *et al.* (1986) ascribe a middle Carboniferous age to this suture.

Finally, the Qimandag (*XII*) suture united the 'Central Meganticlinorium of the Western Kuen-Lun' block of Belyaevsky (1976) (*n*) with the Qaidam block (*o*), sometime during the late Palaeozoic, as judged from the abundant arc-type intrusive rocks and ophiolites of 'Hercynian' age along it (Fig. 4).

Whether the formation of the Qimandag suture post-dated the sutures connecting the Central Meganticlinorium of the Western Kuen-Lun (*n*) and the Qaidam (*o*) blocks with Serindia[2] or Tarim (the 'Northern Synclinorium of the Western Kuen-Lun' of Belyaevsky 1976 (*XI*) and the Arjin suture of Yang *et al.* 1986, (*XIII*) respectively), is difficult to tell so far. All are unconformably covered by Permian and younger terrestrial molasse sequences, including local Lower Triassic limestones and Middle Triassic coal (Şengör and Hsü 1984). Şengör (1984) and Şengör and Hsü (1984) interpreted the late Palaeozoic tectonics of the Western Kuen-Lun in terms of the accretion of a Central Meganticlinorium block (*n*) to the south-east margin of Tarim along suture *XI*, because of the ophiolites and granites of this age along it (Fig. 4). Recent field investigations, as summarized by Chang *et al.* (in press), have been interpreted by these authors to indicate the presence of a geometry and evolution in the Western Kuen-Lun similar to the situation encountered along the Tergun Daba Shan–Qinghai Nan Shan suture (*XVIII*), and allegedly different from the Northern Pamirs. Chang *et al.* (in press) point out that much of the northern part of the Western Kuen-Lun consists mainly of a thick sequence of marine clastic rocks, and metamorphosed intermediate and mafic submarine volcanics of Sinian–Cambrian age. This assemblage is found in association with plagiogranite-bearing ophiolites. A K–Ar age of 517 Ma was obtained from a hornblende in the plagiogranite. The metavolcanic rocks contain jadeite and quartz, indicating blueschist-facies conditions, and have led Chang *et al.* (in press) to interpret this entire rock suite as a former subduction complex. All of these rocks are unconformably overlain by fossil-bearing Ordovician rocks, which in turn are covered by a widespread Devonian molasse. This pre-Ordovician subduction complex is largely equivalent to what Norin (1946) calls 'the crystalline rocks' of the extreme Western Kuen-Lun (between 77°E, and 82°E longitudes), and what Belyaevsky (1976) designates as part of the 'Central Meganticlinorium'.

[2] The term 'Serindia' was used by Argand (1924) to denote the continental fragment forming the basement of the Tarim Basin. Herrmann (1923), however, suggested that *Serindia* was a misspelt form of *Serinda* which might be synonymous with the Arabic *Serendib*, and presented arguments to show that it was probably Ceylon.

Using the same historical data as Herrmann, however, von Richthofen (1877, pp. 528–9, 550–1) came to the conclusion that Serinda must be identical with Khotan. Since the term Serindia had a priority in the geological literature, and has been repeatedly used, by von Richthofen (as Serinda), and Stille, among others, we adopted the term (as did Şengör (1987) in an earlier draft of the manuscript), and the illustrations have been drafted with this geographical designation.

Huang (1945), called, however, the block in question Tarim. He was of the opinion that the name Tarim is better known among the Chinese. He believed that the term Serindia was an abbreviation of the expression 'Silk Roads to India', which used to run north and south of the Taklamakan Desert. The literal Chinese translation of Serindia would be *Shi Indu*, carrying the meaning Silky India. Such a translation would be incomprehensible to Chinese readers, and it would also carry the wrong implication that the area was once a part of India. Finally, Serindia as originally suggested by Argand (1924) implies that the area is underlain by a craton. Yet this is no longer a certainty, as we used to believe, now that one of us has suggested that the Tarim Desert is underlain in part by oceanic crust of a relict Palaeo-Tethyan back-arc basin (Hsü in press). For all those reasons, we now have changed all references to Serindia in the text into Tarim, except that we have left the figures as they were originally drafted.

However, the early Palaeozoic metasediments, metavolcanics, and ophiolites reported from the Western Kuen-Lun are not dissimilar to correlative assemblages known from the Northern Pamirs and the Western Hindu Kush (Boulin and Bouyx 1977, Kravchenko 1979). Farther westwards, from the Paropamisus, Wolfart and Wittekindt (1980) report similar early Palaeozoic rocks.

Thus, from the Paropamisus to the Western Kuen-Lun, and then all the way to the Tergun Daba Shan–Qilian Shan system (i.e. in regions just north of the suture *VIII*, and along the sutures *XI*, *XIII*, and *XVIII* themselves) we see clear evidence of an early Palaeozoic orogenic event, involving possibly subduction-related volcanism, ophiolite obduction and, locally, blueschist metamorphism. Whether it involved continental collision in Afghanistan, in the Northern Pamirs, and along the sutures *XI* and *XIII* is unclear, but in the east it seems to have involved the collision of a Qaidam 'block' (*o*: a collection of arcs?) with the Alxa (Ala Shan) block (*p*). Traces of this early Palaeozoic event are not encountered south of the Qaidam block in the Burhan Budai Shan and in the Anyemaqen Shan (Chang *et al.* 1986, Chang *et al.* in press).

Chang *et al.* (in press) interpret the early Palaeozoic record along suture *XI* as indicating the closure of a marginal basin, and the late Palaeozoic record as the growth of a south-facing continental margin arc/subduction–accretion complex. Although they seem unable to identify the accreted early Palaeozoic arc fragment in the Central Kuen-Lun, late Proterozoic rocks shown on the Geological Map of China (1976), just to the south-west of the Qimandag suture (*XII*) may in fact represent the basement of such a fragment.

The extant sparse data could be interpreted in one of two ways: (1) either the fragment *n* is viewed as a Palaeozoic accretionary complex, consisting essentially of a flysch/*mélange* wedge, with local, Devonian–Permian shallow-water forearc sediments, all of which function as the 'basement' to a late Palaeozoic continental-margin magmatic arc along suture *XIX*, or (2) the fragment *n* is viewed as a mini-continent—carrying a record of early Palaeozoic subduction—which collided with Tarim in late Palaeozoic times and then slid left-laterally along suture *XI* to close the Qimandag suture (*XII*) somewhat later. In this paper we adhere to the second possibility, mainly because of the presence of large masses of late Proterozoic rocks south of Qimandag and also because a well-defined belt of late Palaeozoic arc-type magmatic rocks and ophiolites everywhere separates the fragments *n* and *o* from Tarim. This belt implies active subduction, and therefore the presence of an ocean between these objects (Fig. 4) at least until the middle Carboniferous. One of us (K. J. Hsü) thinks that this subduction activity opened a major back-arc basin behind the northern Kuen-Lun-Altyn Dagh Range during the Palaeozoic. This basin is now preserved in the Tarim area, which is thought to be underlain by Palaeozoic oceanic crust (Hsü in press). This is an interesting possibility, but whether Tarim is a microcontinent or a relict back-arc basin, the Palaeo-Tethyan suture geometry and evolution around it, as presented here, remains unaffected.

The following picture thus emerges concerning the ages of the sutures in the Kuen-Lun Mountains as a whole: by middle Palaeozoic time, the Qaidam block (*o*) collided with the Alxa Massif (*p*). Later, during the late Palaeozoic, the combined Qaidam/Alxa block ($o + p$), and the block *n* independently collided with Tarim along the sutures *XI* and *XIII*, and with the Altaid Asia along the suture *XIV*, as clearly indicated by the abrupt cessation of subduction-related magmatic activity along all of these sutures after the Palaeozoic (Fig. 5). These collisions were followed shortly by the closure of the Qimandag remnant ocean.

The last sutures we discuss in the category of Cimmeride Palaeozoic sutures is the Helan Shan suture (*XXXIX*), and its apparent prolongation northward along sutures *XV* and *XVIII*. Klimetz (1982, 1983) and, following him, Şengör (1984) and Şengör and Hsü (1984), considered the Helan Shan as a middle Mesozoic suture zone, mainly because of an erroneous identification of blueschists in the Jurassic Siasouchan Series (Lexique Stratigraphique International 1963). The main—and a very strong—argument against the presence of a middle Mesozoic suture along the Helan Shan, is the absence of any Mesozoic marine sediments. It may possibly represent a late Palaeozoic suture (or late Proterozoic according to some, e.g. Sobolev 1984) that was later reactivated to accommodate further intracontinental shortening—probably a substantial amount, as can be judged from the intensity of folding in the Jurassic sediments (Teilhard de Chardin and Licent 1924).

A problem similar to the one encountered in the Helan Shan case, also exists along the only major Palaeozoic marginal suture belt separating Altaid Asia from the larger fragments of the Cimmeride collage, namely the Inner Mongolian–Da Hingan (Greater Khingan) suture (*XV*, *XVII*) (Wang and Liu 1986), which

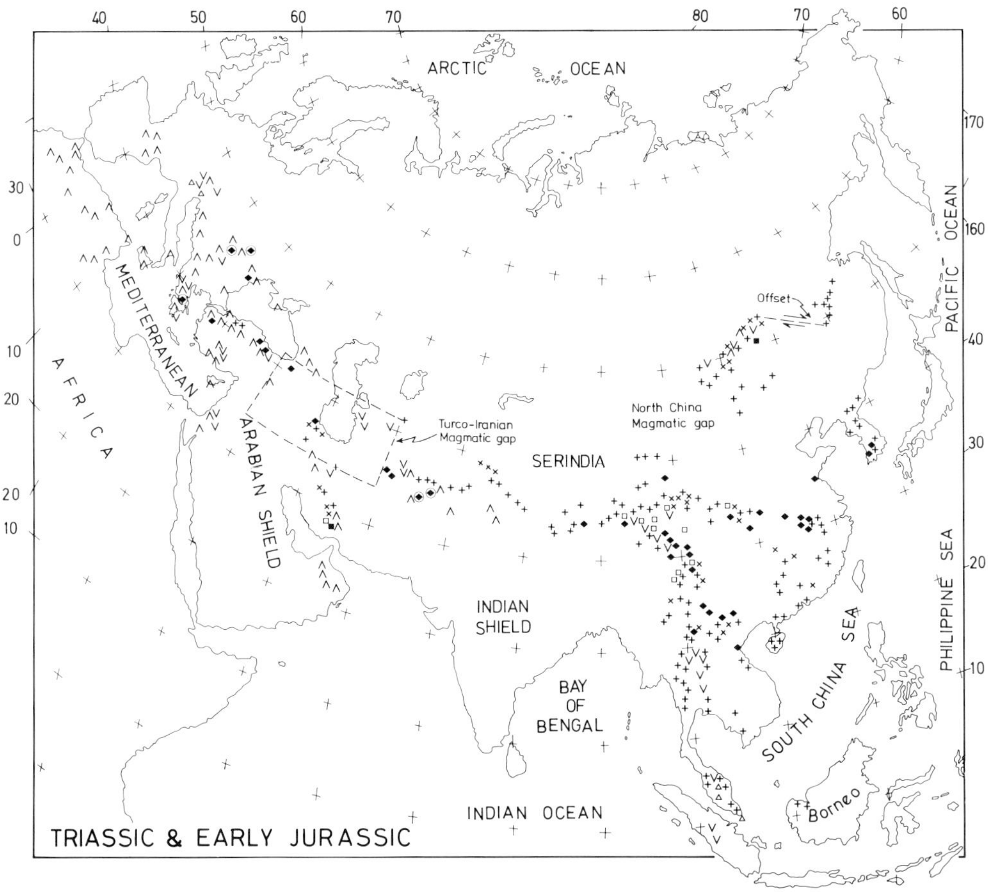

FIG. 5. Triassic and early Jurassic ('Indosinian' in China) magmatism in the Tethysides. The list of references, from which this map was compiled, may be obtained from the Geological Society of London for the price of photocopying. For legend see Fig. 4.

continues directly into the Shilka suture (*XLI*). A well-developed early Cretaceous magmatic arc, and continental-margin sediments along the Shilka suture (Kosygin and Parfenov 1981), leave no room for doubt for the existence of a late Mesozoic suture here. Farther south, however, despite some Triassic and early Jurassic, and very intensive and widespread late Jurassic to early Cretaceous intermediate and felsic magmatism (Figs 6 and 7), neither coeval marine sediments nor any ophiolites, are seen. Although Şengör (1984) interpreted the ultramafics located generally to the east of the extensive Mesozoic felsic magmatic rocks, as indicating the existence of a coeval suture zone, this interpretation can no longer be upheld, because the associated ultramafics all seem to be late Palaeozoic in age.

Convergent plate-margin magmatism appears to be confined to the Greater Khingan–Inner Mongolian suture zone during the late Palaeozoic (Fig. 4). This suture probably closed at this time, as the late Palaeozoic ophiolites and the nearly complete cessation of all magmatism bear witness (Fig. 5) (see also Wang and Liu 1986). Immediately following this closure, the locus of convergence seems to have jumped to the Mongolo-Okhotsk ocean and to have begun consuming it. Following the final elimination of the Mongolo-Okhotsk ocean in the late Jurassic, continuing intracontinental convergence generated the wide felsic magmatic area, and probably also triggered subduction along the Ussuri suture in the Sikhote Alin (Fig. 6). In this episode, the Greater Khingan suture appears to have been reactivated in a compressional mode, and this reactivation extended its effects probably as far south as the Helan Shan (Wong 1929).

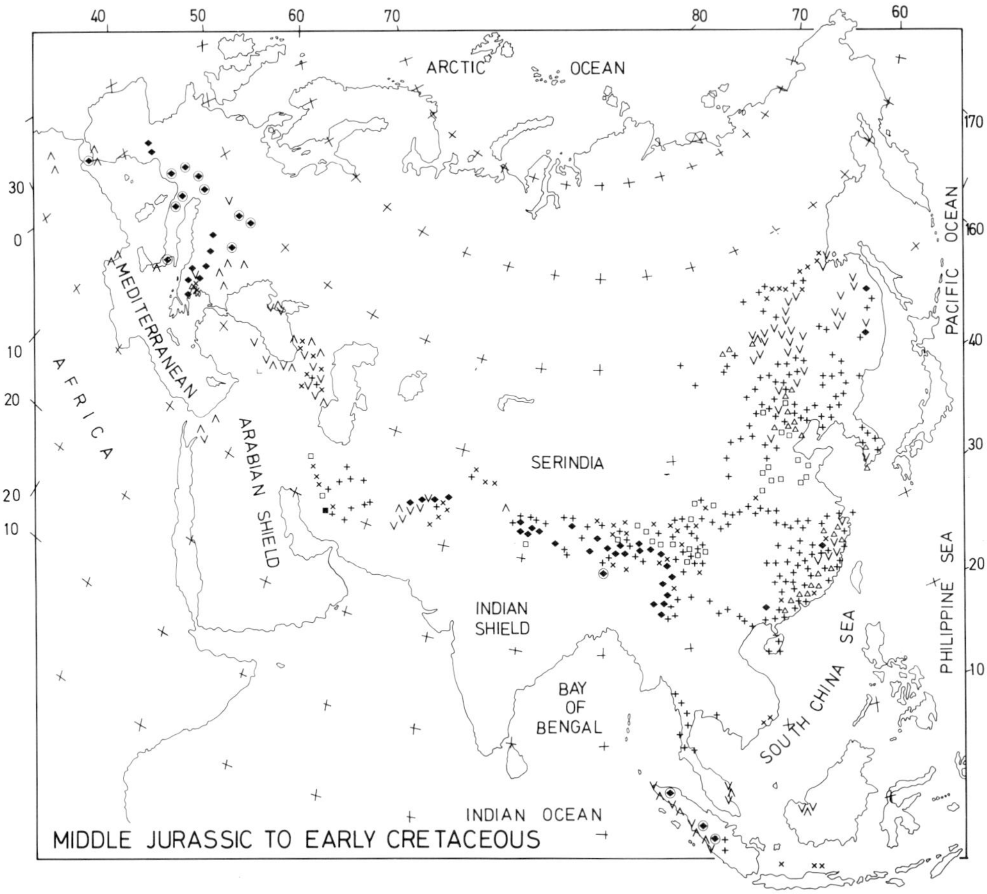

FIG. 6. Middle Jurassic to early Cretaceous ('early Yenshenian' in China) magmatism in the Tethysides. The list of references, from which this map was compiled, may be obtained from the Geological Society of London for the price of photocopying. For legend see Fig. 4.

Mesozoic Cimmeride sutures

As the 'North China magmatic gap' in Fig. 5 also shows, all oceanic plate boundaries that connect north-east China with the rest of the Tethyside system, had become extinct by the beginning of the Mesozoic, as a consequence of late Palaeozoic continental collisions in this part of the world (Wang and Liu 1986). The Mesozoic Pacific-type, active continental margin thus became more uniformly south-facing (present geographical orientation), along which magmatic arc activity continued without break, and very much in the same places as in the late Palaeozoic (compare Figs 4 and 5). The only significant exception was in the Central Kuen-Lun, where the magmatic arc axis jumped a few hundred kilometres southwards, presumably after the accretion of the Central Meganticlinorium (*n*) to Tarim.

The earliest Mesozoic continental collisions again appear to have been between individual components of the Cimmeride collage. The first of these was probably the collision along the Jinsha Jiang suture (*XXI*), between the eastern segment of the North Tibetan Qangtang block (*s″*) and the Chola Shan sliver (*v*). Along the Hoh Xil Shan Arc (to the west of the Jinsha Jiang suture) magmatic activity lasted until the Carnian, and then the suture zone was uplifted. The marine Triassic section along it commonly ends with the Carnian (Şengör and Hsü 1984).

A probable coeval suture is the Song Da (Black River) suture (*XXX*) in North Vietnam. Along the early Carboniferous Song Ma suture, which closely parallels the later Song Da suture,

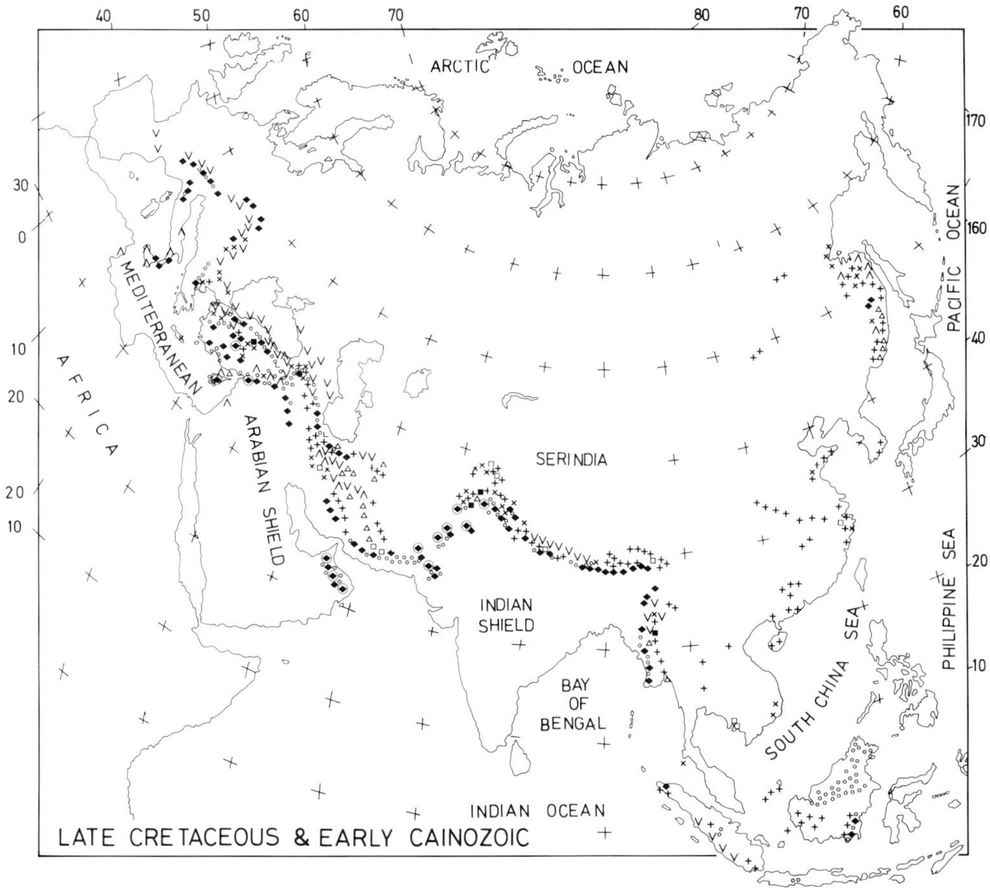

FIG. 7. Late Cretaceous and early Cainozoic ('late Yenshenian' and partly 'Himalayan' in China) magmatism in the Tethysides. References, from which this map was compiled, may be obtained from the Geological Society of London for the price of photocopying. For legend see Fig. 4.

the post-deformation rocks are almost exclusively middle Carboniferous to Permian shallow-water limestones, similar to the Devonian rocks deposited following the arc–continent collision along the Tamky–Phueson suture (*XXVIII*) to its south. This facies suggests that the early Carboniferous deformation along the Song Ma suture was likely due to the collision with Annamia of yet another island arc, now perhaps buried under the younger deposits of the Cainozoic Song Ma rift zones. That the Song Ma collision did not represent the terminal welding of Annamia to South China, is suggested by the supposedly 10 000-metre thick, pre-Norian Triassic section, which contains abundant volcanic material such as spilites, albitophyres, basalts, plus *ultramafics* (Salun *et al.* 1975) along the Song Da suture. These rocks are now extremely deformed and imbricated, with a clear north-east vergence under the unconformable coal-bearing Norian–Rhaetian molasse (Fromaget 1934; Burrett 1974). Şengör (1984) interpreted them as a subduction–accretion complex, in which measured stratigraphic thicknesses would have little meaning.

To the north-east of the Song Da suture, on its foreland, the Hagian–Bac Kan nappe carried Precambrian metamorphic basement rocks on to folded late Permian limestones, during the late Triassic. This episode ended with the 'intrusion' of the Viet Bac mantled gneiss dome into the Hagian–Bac Kan nappe (Fontaine and Workman 1978; Şengör and Hsü 1984). Triassic and Jurassic convergent-margin magmatism, in the form of dominantly granitic intrusions (Figs 5 and 6) also indicates that the final collision along the Song Da suture could not have been much earlier than the late Carnian.

Additional support for this view comes from the Ailo Shan, in the Chinese province of Yun-

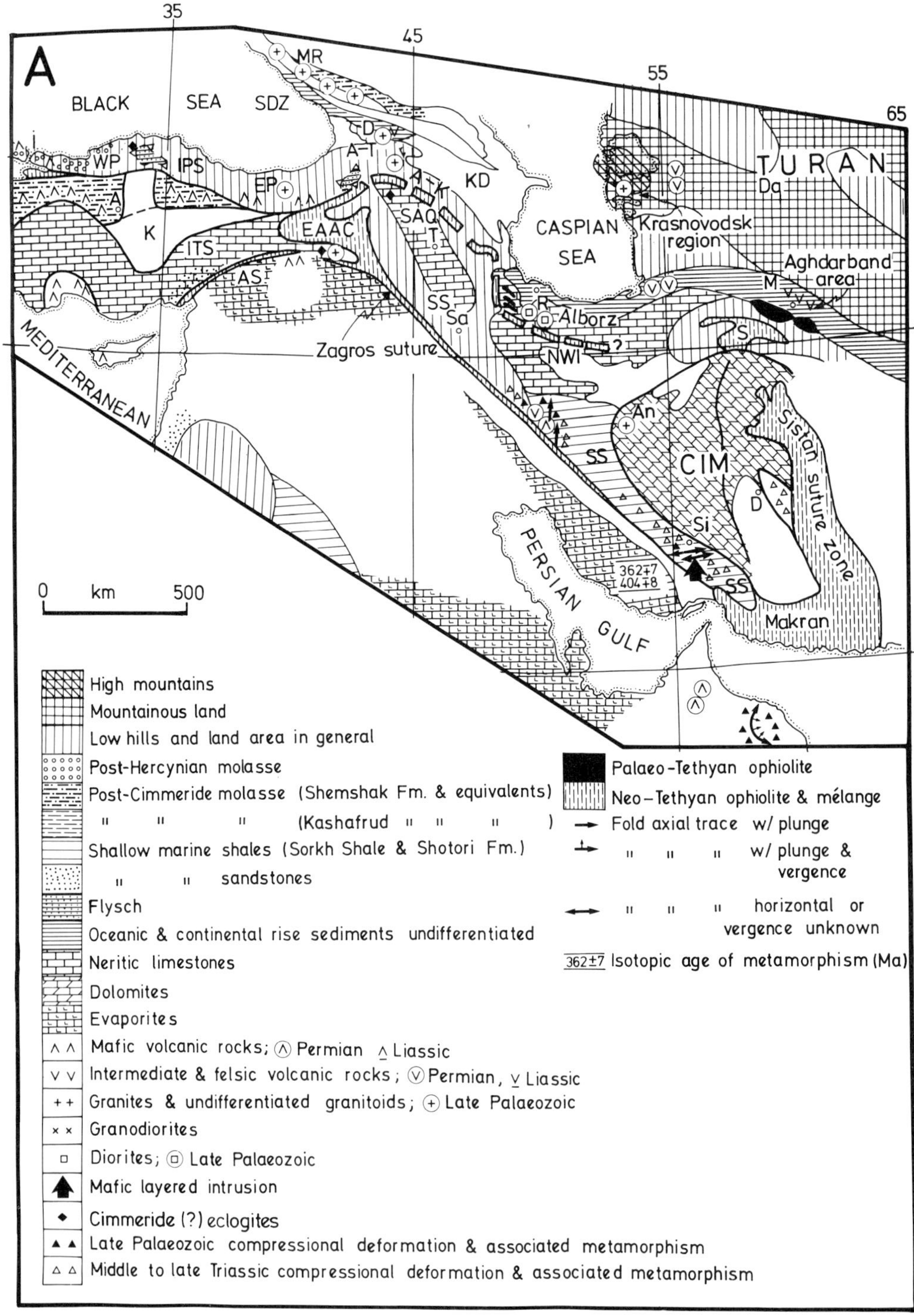

FIG. 8. (A) Distribution of major early and middle Triassic sedimentary facies, igneous rocks, metamorphism, and deformation in Iran and surrounding regions. Compiled from Thiele *et al.* (1968), Vereshchagin and Ronov (1968), Lunnov (1971), Vialon *et al.* (1972), Khain (1975), Stämpfli (1978), Berberian and King (1981), Şengör and Yılmaz (1981), Michard (1983), Roman'ko and Morozov

nan (see suture segment *XXII*). In these mountains there is a 10- to 15-km wide ophiolitic *mélange* zone of middle Triassic age. Late Triassic red beds and felsic volcanics unconformably overlie the *mélange* (Bally *et al.* 1980).

The initial collisional contact established between Laurasia and the Cimmerian Continent occurred in Iran, and represents the next oldest Cimmeride suturing event. Following Stöcklin (1974), both Hsü (1977) and Şengör (1979*a*, 1984, 1985*a*, *b*; 1986*a*) argued that the ophiolite slivers encountered under the unconformable Rhaeto-Liassic terrestrial Shemshak Formation, in the Alborz/Binalud ranges in northern Iran, near the towns of Rasht (Davies *et al.* 1972; Clark *et al.* 1975) and Mashhad (Alavi 1979), were remnants of Palaeo-Tethys. They further argued that the Palaeo-Tethys in Iran had closed before the Rhaetian and after the Carnian, because the platform deposits, on which the Mashhad ophiolites had moved, reached as far upwards as the Carnian (Alavi 1979).

None of these authors, however, could account for the strong, coeval compressional deformation, kyanite-grade metamorphism and arc-related volcanic and intrusive activity of Devonian to late Triassic age along the *south-eastern* border of central Iran, in the Sanandaj–Sirjan zone (*i*) under unconformable Liassic rocks of the Shemshak Formation equivalents (*SS* in Figs 8a and b).

Berberian and King (1981) recognized the necessity of assuming convergent plate-margin tectonics along the Sanandaj–Sirjan zone in the late Palaeozoic Triassic interval, and postulated the subduction *northward* under the Sanandaj–Sirjan zone of a Palaeozoic 'Zagros ocean'. However, they also admit that stratigraphic, palaeobiogeographical and palaeomagnetic data unequivocally demand that Iran be kept as an

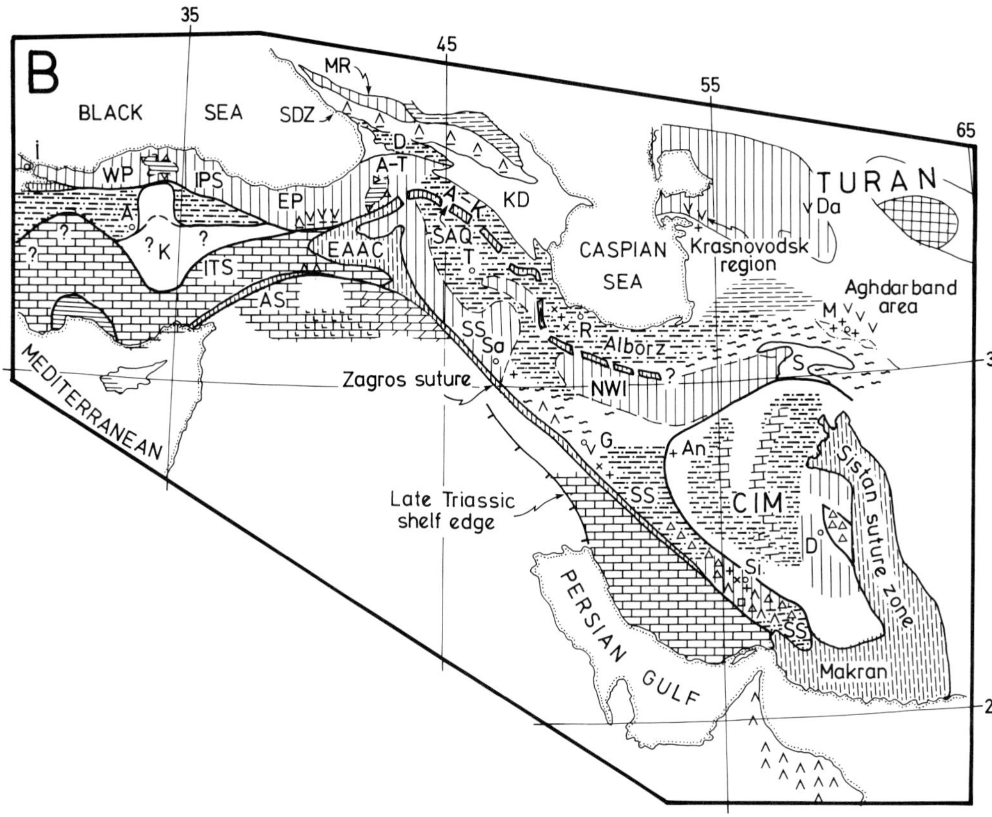

(B) Distribution of major late Triassic and earliest Jurassic sedimentary facies, igneous rocks, metamorphism, and deformation in Iran and surrounding regions. Sources as in Fig. 8a. Symbols are as used in Fig. 8a.

(1983), Adamia and Belov (1984), Davoudzadeh and Schmidt (1984), Meliksetyan *et al.* (1984), Ruttner (1984), Şengör *et al.* (1984*b*), Okay *et al.* (1985).

integral part of Gondwana Land until the early Triassic, so that no ocean could have existed between it and Arabia, prior to at least the middle Triassic. Moreover, Michard (1982, 1983) described folded and slightly metamorphosed terrains, constituting the pre-Permian basement of the Oman Mountains, and ascribed their origin to a late Palaeozoic orogeny (Fig. 8a). Thus, deformation and metamorphism similar to those known from the pre-Liassic basement of the Sanandaj–Sirjan zone, also took place on the 'Arabian' side of Berberian and King's (1981) Palaeozoic Zagros ocean, which cannot be explained by northward subduction under the Sanandaj–Sirjan zone.

There is also abundant evidence that the Zagros segment of the Neo-Tethys, between Iran and Arabia, began rifting during the late Permian (e.g. Szabo and Kheradpir 1978), with continuing stretching and abundant basaltic volcanism in the early and middle Triassic (Searle *et al.* 1980, Searle and Graham 1982), and finally the onset of ocean-floor spreading in the late Triassic.

The late Palaeozoic and Triassic (locally later) tectonism in the Sanandaj–Sirjan zone is very reminiscent of the events known from the Aghdarband/Kopet Dagh regions of north-east Iran and south-west USSR (Turkmenia SSR; along suture *VII*). In the Krasnovodsk uplift of the Kopet Dagh trend, there was both late Palaeozoic (mainly Permian), and Triassic arc-type magmatism, whereas in the Aghdarband uplift, this activity is seen only in the Triassic (Figs 4, 5, and 8a). On the basis of borehole and geophysical data, Bazhenov and Burtman (1986) traced a continuous late Palaeozoic magmatic arc, composed mainly of andesites and dacites, between Krasnovodsk and Aghdarband. Stämpfli (1978) reported Permian volcanics from the Gonbad–ı Qabus area, near the south-east corner of the Caspian Sea (Figs 4 and 8a). Between the Sanandaj–Sirjan zone and the Alborz Mountains, the entire Iranian area contains no comparable deformations or magmatism, except in the Anarek area (Davoudzadeh *et al.* 1981, Roman'ko and Morozov 1983) (Fig. 8a and 8b).

Within the framework of the present-day geography it is thus difficult to find a suitable convergent plate boundary of Permo-Triassic (and older) age, to create the Sanandaj–Sirjan magmatic arc belt. An examination of Figs 4 and 5 shows the continuous late Palaeozoic–early Triassic magmatic arc belt—related to the consumption of Palaeo-Tethys in Central Asia—to stop rather abruptly in the Aghdarband area. This interruption, the Turco–Iranian magmatic gap (Fig. 4), is partly related to younger cover in the Turkmenian SSR, hiding the segment of the arc between Aghdarband and Krasnovodsk (see Bazhenov and Burtman 1986; Fig. 2). But farther west from the Krasnovodsk, in the Greater Caucasus, we see no trace of the Triassic arc. Instead, a long but seemingly isolated arc segment is present in the Sanandaj–Sirjan zone, that continues into the Pontides in northern Turkey (Figs 5, and 8a and b). The apparent interruption in eastern Turkey, visible in Figs 4 and 5, is due to the presence of a late Cainozoic volcanic cover (Şengör and Kidd 1979).

This haphazard geometry, involving abrupt termination of major fossil magmatic arc systems (as in Krasnovodsk), presence of fragments of others, apparently with no associated sutures (as in the Sanandaj–Sirjan zone), and very small pieces of former arcs lying in complete isolation amidst anomalous palaeogeographical zones (as in the Anarek region of Central Iran; Figs 4, 5, 8a, and 8b) suggests the disruption and redistribution, by strike-slip faulting, of a formerly more 'conventionally' organized orogenic belt. This view is strengthened by the 135° anticlockwise rotation of the Central Iranian Microcontinent (*k*) between the late Triassic and the Cretaceous (Davoudzadeh *et al.* 1981; Soffel and Förster 1984). As a result of this rotation, the originally south-east end-segment of the Sanandaj–Sirjan zone was torn away and transferred to eastern Iran, to the east of Deh Salm (Davoudzadeh and Schmidt 1984), whereas the Anarek region, originally close to the northern Palaeo-Tethyan suture near Mashhad, was brought westward to its present location (Davoudzadeh *et al.* 1981) (Figs 8a and 8b). This is also documented by the rotation of Cambrian (Zarkhov 1984), Devonian (Weddige 1984), and Mesozoic sedimentary facies (Davoudzadeh and Schmidt 1984) within the Central Iranian Microcontinent (see Fig. 8a and b for the rotation of the middle Triassic to earliest Jurassic facies realms).

Geological information from Iran alone is insufficient to reconstruct with confidence the pre-disruption geometry of the Iranian portion of the Cimmeride orogenic collage. We therefore defer the presentation of our reconstruction of how Iran evolved during the Mesozoic, until the end of our discussion on the distribution of Tethyside sutures and blocks.

A probably coeval collision occurred sometime between the middle Carnian and the Rhaetian in Thailand, along the Nan-Uttaradit–Sra-Kaeo suture (*XXVII*) (Bunopas 1985; Şengör 1986*b*), which continues southwards into the Malay Peninsula where it becomes the Bentong–Raub suture (*XXXI*) which closed after the Ladinian and before the deposition of the late Triassic Tembeling Formation (Şengör 1984, 1986*b*). Farther south, this suture crosses the Makassar Strait and separates northern Sumatra from southern Sumatra (*XXXI'*) (Hamilton 1979, Hutchison in press).

Another late Triassic suture belt within the Cimmeride collage is the Qin-Ling (*XXIII*)/Dabie Shan (*XXXIII*)/northernmost Jiangsu (*XXXIII'*)/Chugareong (*XXXV*)/system, that united the North China platform (*q*) with the Yangtze block (*w*). Of this long suture belt, the Qin-Ling/Dabie Shan segment (*XXXIII*) is the best-known part.

Controversy however persists, both on the timing of suturing (e.g. Şengör 1985*c*) and on the predominant vergence (e.g. compare Xu *et al.* 1984, with Zhu *et al.* 1983 and Liu and Li 1984) of this complex and much disrupted orogenic belt. During the last two decades, opinions on the age of the main orogeny along the Qin-Ling have varied from Devonian (*c.*400 Ma) (e.g. Mattauer *et al.* 1985) to late Triassic (*c.*220 Ma) (e.g. Şengör and Hsü 1984; Şengör 1985*c*) or even early Jurassic (*c.*200 Ma) (Opdyke *et al.* 1986). Şengör (1985*c*) summarized the observations supporting a late Triassic collision.

Further evidence for a Triassic collision is also found farther east along the strike, in the Dabie Shan. Fig. 9 shows a cross-section across the Dabie Shan near its eastern end, where the entire orogenic belt is cut and offset some 500 km northward by the Tan-Lu Fault (Fig. 10) (Xu and Lu in press). This cross-section shows that most of the Dabie Shan consists of large slabs of Precambrian basement rocks (Dabie Group, with an age ranging from early Proterozoic to late Proterozoic, and with scattered K–Ar cooling ages as young as 532 Ma: Xu *et al.* 1984*a* and Xu Shutong pers. comm. 1986) separated by *mélange* zones, where ultramafic rocks, mafic volcanics, and marble blocks are seen scattered in a metapelitic matrix. The metamorphism of these *mélange* zones ranges from greenschist to high amphibolite grade, with local anatectic melting. Both within the Dabie Group and in the *mélange* belts, eclogites of different ages have been encountered (see Xu *et al.* 1984*a* for a map). Recently, Li *et al.* in press reported Sm–Nd isotopic ages from the eclogites and from the ultramafic rocks found in the Northern Ultramafic Belt of the Dabie Shan (Fig. 9). These ages are 243.9 ± 5.6 Ma for the eclogite, and 230.6 ± 30.7 Ma and 402.6 ± 17.4 Ma for the ultramafics, suggesting that the ultramafics formed at two different intervals (Triassic and early Devonian), and that subduction was under way already during the earliest Triassic. Moreover, Sang *et al.* (1986) have dated the metavolcanics of the Susong Group in S. Dabie Shan (Figs 9 and 11) and have concluded that the volcanics were formed during the Ordovician, and that following a later episode of metamorphism they were cooling by 231 Ma ago. Sang *et al.* also obtained younger cooling ages around 210 Ma. Stratigraphic indication of a post-Palaeozoic continental collision in the Dabie Shan, comes from the presence of a fairly thick Palaeozoic continental margin sequence: the Fozhilin Group, which was strongly deformed, with a northerly vergence, after the ?Permian and before the Jurassic (Zhu *et al.* 1983).

Fig. 11 shows our interpretation of the evolution of the Dabie Shan segment of suture *XXXIII*. This suture probably continues through the widespread ultramafic and mafic rocks of northernmost Jiangsu (XXXIII') (Chinese Academy of Geological Sciences 1973) and the 'Chugareong zone' of Lee (1984), located south of the Permo-Triassic Pyeongan Basin (Lee 1974).

The Qin-Ling/Dabie Shan mountain range delimits a continental region to the north, known as the South China platform or the South China block in the literature (Fig. 10; e.g. Klimetz 1983, Lin *et al.* 1985*a*). Huang (1978) separated the South China block into a Yangtze paraplatform in the north and east, and a South China fold system in the south and west. A much smaller tectonic unit, called the 'south-east maritime fold system', was distinguished by Huang (1978), east of the South China fold system and separated from it by the Lishui–Haifeng 'depth fracture' of Palaeozoic age (Fig. 10). Huang views the South China fold system as a unit that formed during the early Palaeozoic ('Caledonian' as he calls it), and the south-east maritime fold system he regards as being late Palaeozoic ('Variscan') in age. In 1981, Hsü pointed out that the notion of a 'South China Platform' is erroneous and misleading, owing to extensive Mesozoic deformation of the 'platform sediments' and coeval convergent-margin magmatic activity. Hsü (1981) also indicated that South China was greatly similar to the Central and Southern Appalachians, comparing the flat-lying Mesozoic strata of the Sichuan Basin (Fig. 10) with the Alleghany Plateau, and the narrow folds involving Mesozoic strata (between SP and TY in Fig. 10) with those of the Valley and Ridge Province. Farther to the south-east of these folds, are extensive outcrops of a thick unfossiliferous sequence of slates and turbidites, forming generally mountainous terrain. Blocks of metamorphosed mafic volcanics

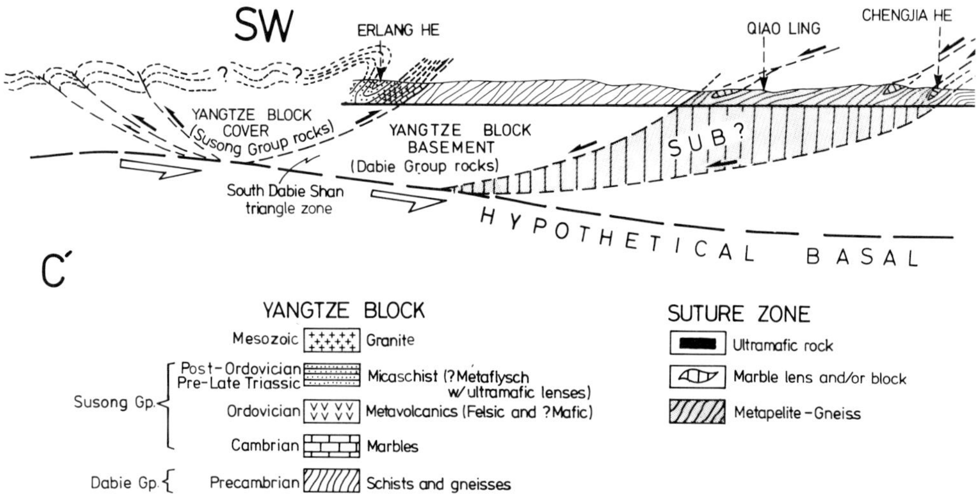

FIG. 9. An interpretive structural cross-section across the eastern end of the Dabie Shan, between Songshan Zai and Erlang He (for location see Fig. 10). Data were reinterpreted from the Taihu (H–50–XV, 1970), Lioan (H–50–III, 1974), and the Yuexi (H–50–IX, 1974) sheets of the 1:200 000

('greenstones'), submarine tuffs, gabbro, and serpentinite are locally present within the dominant slate matrix, with few chert lenses. In the provinces of Hunan and Guangxi, this complex is known as the Banxi Group (Fig. 10), and has been assigned an Upper Proterozoic age (Chinese Academy of Geological Sciences 1973). Hsü *et al.* (in press) extended this appellation to all such rocks in the South China Platform, and interpreted them as lying over the autochthonous Yangtze carbonate rocks and clastics above a thrust contact (Fig. 10). Hsü *et al.* (in press) argue that these allochthonous flysch/*mélange* masses were expelled from what they term the Xinagganzhe suture (from the abbreviated Chinese expressions for Hunan, Jiangxi, and Zhejiang), that coincides with the boundary between Huang's (1978) Yangtze paraplatform and the South China fold system tectonic units (*XXXVI* and *XXXVII*).

The Xiangganzhe suture itself actually consists of two separate segments (*XXXVI* and *XXXVII*), whose relations to each other are as yet unclear. In the north, it starts near the town of Shaoxing and extends south-westward to Jiangshan (Shui *et al.* 1986). This north-west-vergent segment—contained entirely within the Zhejiang province—continues directly into the Pingxiang–Yushan segment in Jiangxi. We will call these two segments (collectively) the Shaoxing–Pingxiang suture (*XXXVI*; *SP* in Fig. 10). To the north-west of this suture, large outcrops of the Banxi are themselves tectonically overlain by Precambrian plutonic and metamorphic basement rocks forming prominent topographic highs.

In the province of Hunan the continuity of the suture zone is interrupted; and it jumps some 500 km westwards to continue as the Tianyang suture zone (*XXXVII*; *TY* in Fig. 10). To the west of the Tianyang suture we see again an enormous mass of Banxi Group rocks thrust westward.

That the Xiangganze structure represents a major suture zone in its various segments is now generally acknowledged (e.g. Jin and Fu 1986, Shui *et al.* 1986), but the timing of suturing and the expulsion of the Banxi allochthons remains in dispute. Although the Banxi rocks have long been considered unfossiliferous, Hsü *et al.* (in press) quote unpublished information gathered by the Jiangxi Geological Survey about Permian spilites and radiolarites included in the Banxi rocks in the Jiangxi province (i.e. north-west of the Shaoxing–Pingxiang suture: SP in Fig. 10). To the west of the Tianyang segment, Yang *et al.* (1981) reported the existence of Permo-Triassic radiolarian cherts and spilitic volcanics (as in Jiangxi farther to the north-east) the former containing a 'Tethyan' fauna, including such forms as *Otoceras* and *Ophiceras*. These rocks are seen to be associated with flysch deposits that form part of the Banxi allochthon in eastern Guizhou and northern Guangxi.

The sparse biostratigraphic information thus suggests that the Banxi rocks are at least partly

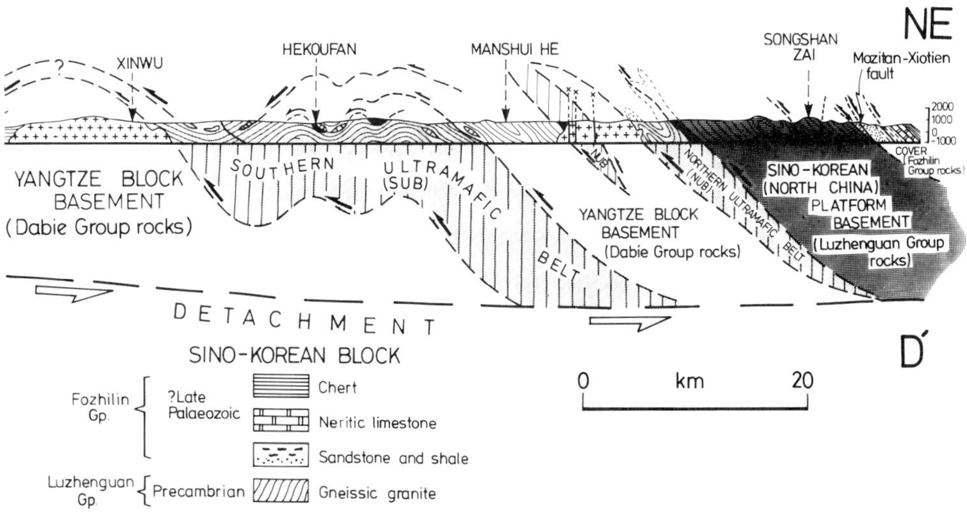

geological map of the Anhui province and Xu *et al.* (1984, and in press), locally supplemented by our own observations.

as young as the early Triassic, and therefore their thrusting westwards and the suturing of the Xiangganzhe branch of Palaeo-Tethys could not be earlier than that date, at least in its south-western one-third.

In order to establish the upper limit for the age of closure of the Xiangganzhe ocean, Şengör and Hsü (1984) plotted the distribution of Palaeozoic and Mesozoic granites to the east of its suture zone. In south-east China there seem to exist *three* Mesozoic intrusive, and subordinate volcanic, belts parallel with the Xiangganzhe suture: two I-type granite belts flanking a much wider S-type granite belt. Şengör and Hsü (1984) interpreted the western I-type granite belt and a considerable portion of the S-type belt as manifestations of a south-east-dipping subduction under the Huanan block (*y* at large). On the Yangtze block (*w*) and on the Huanan block the post-orogenic molasse sediments are late Triassic or younger (Chinese Academy of Geological Sciences 1973; Hsü *et al.* in press) and we therefore place the suturing between these two blocks in the middle Triassic, with intracontinental convergence continuing until early Cretaceous times, and causing both the formation of the wide, S-type, 'Tibet-style' intrusive belt and the widespread middle Jurassic to middle Cretaceous compressional intramontane basins.

Although major suturing along the entire length of the Xiangganzhe suture zone was a middle Triassic affair, earlier deformation, and the deposition of terrestrial 'molasse' sediments are known from the Huanan block in Devonian and pre-Devonian times. A Devonian 'clastic wedge' thins to the west, suggesting that it was fed from an easterly source (Hsü *et al.* in press). Grabau was the first to notice this, and he called the hypothesized source area the 'Cathaysia' landmass. Hsü *et al.* (in press) followed Li *et al.* (1982) in interpreting the Lishui–Haifeng depth fracture of Huang (1978) as a pre-Devonian suture zone (*XXXVIII*) that united the 'southeast maritime fold system' (*y″*: 'coastal block' of Hsü *et al.* in prep.) with the Huanan block proper (*y* in the restricted sense).[3] Although Devonian clastic rocks also exist on the Yangtze block, they are not unconformable on older rocks. Here also, they do not have a typically molasse character. A possible early Palaeozoic deformation, affecting the Yangtze block, may be indicated by middle Ordovician turbidites with a greywacke composition, in the Lin'an county of the Zhejiang province to the north of the Shaoxing–Jiangshan suture zone.

Fig. 12 represents the interpretation of Hsü *et al.* (in press), of the evolution of what Hsü *et al.* (in prep.) call the 'Cathaysian orogenic belt' and we adopt it here also. The Palaeozoic granites, shown intruding the Huanan block in Fig. 12, have been partly dated recently and are located in the provinces of Hunan, Fujiang, Guangdong

[3] Recent observations by Professor J. L. Li in the Fujiang province suggest that at least parts of the Lishui–Haifeng structure may represent Mesozoic sutures.

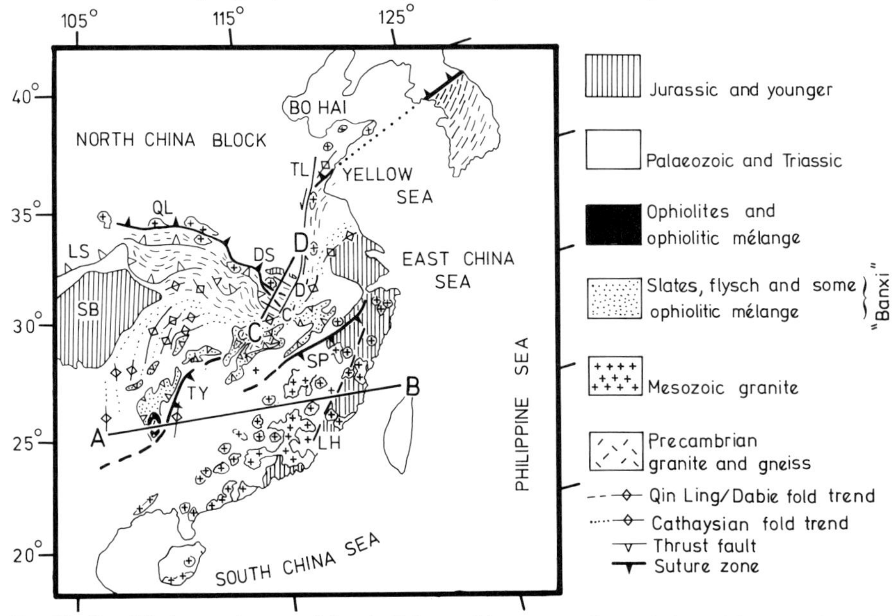

FIG. 10. Simplified tectonic map of South China and its proposed correlation with Korea (compiled from the Geological Map of China 1976, Lee 1984, and Hsü *et al.* in press). Key to lettering: *C*—Chugareong suture zone, *DS*—Dabie Shan, *LH*—Lishui–Haifeng suture zone, *LS*—Longmen Shan, *QL*—Qin-Ling, *SP*—Shaoxing–Pingxiang suture zone, TL—Tan-Lu Fault, TY—Tianyang suture zone. *C′–D′*, C–D, and A–B show the locations of the cross-sections shown in Figs 9, 11 and 12 respectively.

and eastern Guangxi (Fig. 4; Institute of Geochemistry 1980).

Before we close our discussion of the Cimmeride tectonics of South China we should mention the possibility of the presence of an independent Hunan block (*y′*) in the architecture of South China as suggested in Fig. 3. Alternatively, the free ends of the Tianyang and the Shaoxing–Pingxiang sutures may have been originally contiguous and later offset by a post-collisional, right-lateral, strike-slip fault (Jin and Fu 1986). This latter interpretation seems also to be favoured by the presence of several north-north-east trending, narrow, fault-controlled Cretaceous depressions.

With the discussion of the Cimmeride assembly of South China, we have come to the end of our description of the Triassic suture zones within the Tethysides. However, there is one more major region within the Cimmerides where a large number of sutures are present and where marine conditions largely ended by about the end of the Triassic. This region, which has been interpreted as an 'Indosinian' zone of consolidation, because of the retreat of the sea from it during the late Triassic (e.g. Chen 1977; Huang 1978; Li 1980; Li *et al.* 1982), is the Songpan–Ganzi System (Fig. 4), the giant suture knot of the East Asian Cimmerides (Şengör 1981, 1984).

Şengör (1981, 1984, 1986*b*) and Şengör and Hsü (1984) discussed the geology of this extremely complex and little-known region. As shown in Fig. 3, the south-west margin of the Songpan–Ganzi System consists of a north-east-convex bundle of four suture zones. In reality, the outermost of these, the Litang *mélange* zone (*XXIII*), is not a suture but simply a structural boundary, marked by a few 'Indosinian' ultramafics separating the Shaluli Shan Arc system (*u*) from the thick Triassic flysch fill of the main mass of the Songpan–Ganzi System. This entire flysch fill, the Bayanhar Group, represents a suture knot itself (see Şengör 1981, 1984, 1985*a*, 1986*b*).

A second belt of Indosinian ultramafics (largely Triassic) representing the Maniganggo suture (*XXII*), separate the Shaluli Shan Arc from the Hoh Xil Shan Arc constructed directly on top of the basement of the western part of the western segment of the Qangtang block (*s″*) between 88°E and about 97°E; farther east a new zone of dominantly Jurassic–early Cretaceous granitic and granodioritic plutonic activity begins inserting itself between the Shaluli Shan Arc and the Hoh Xil Shan continental-margin arc. This new zone is termed the Chola Shan Arc (*v*) by Şengör (1984). The Maniganggo suture also separates the Chola Shan from the Shaluli Shan. Between the Hoh Xil Shan Arc and the

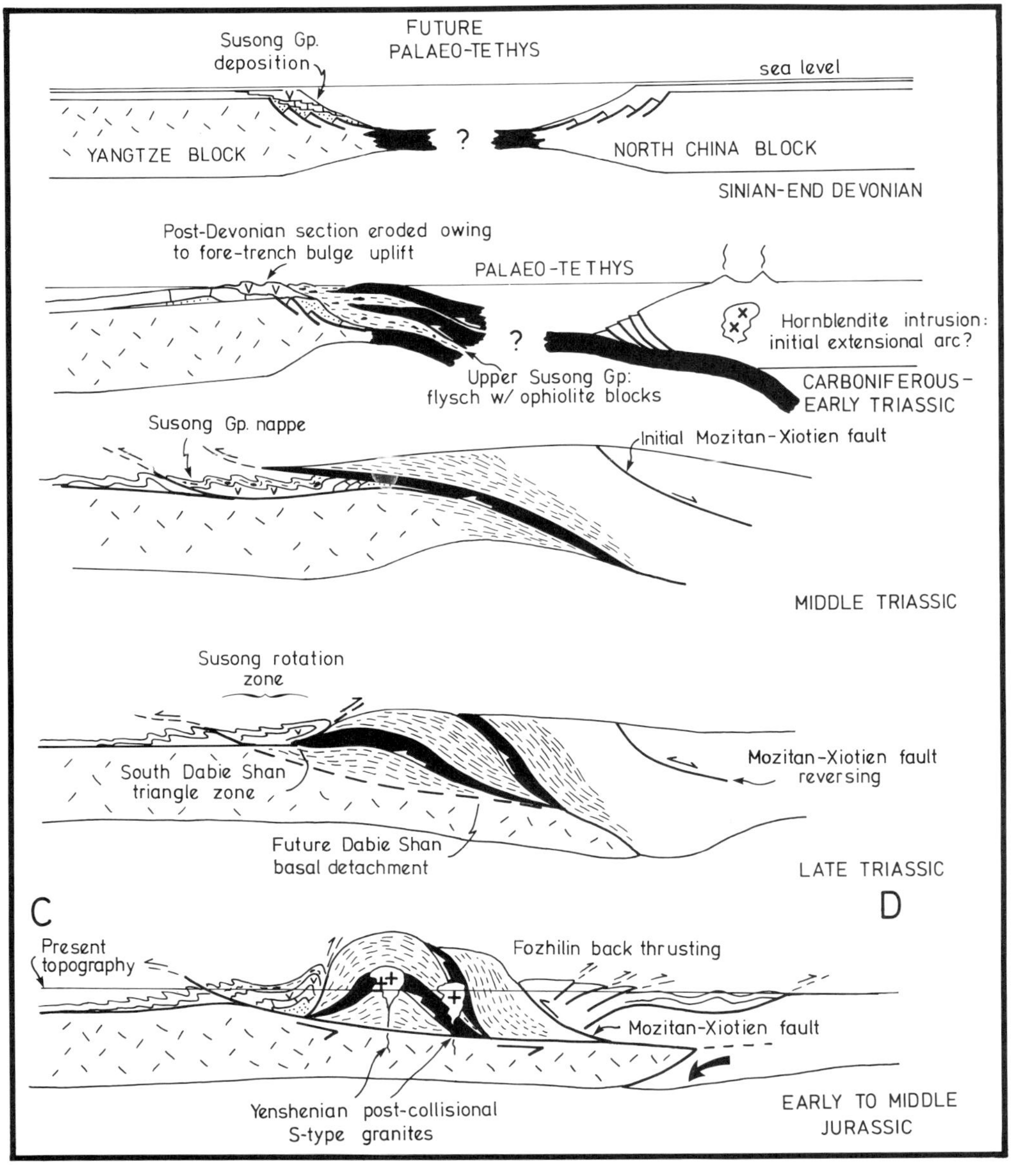

FIG. 11. Schematic, sequential, evolutionary cross-sections, showing the development of the Dabie Shan between the Sinian (late Proterozoic) and the middle Jurassic. For location see Fig. 10.

Chola Shan is the middle Triassic Jinsha Jiang suture (*XXI*).

In all of these suture zones, the Norian and later stratigraphic sections are represented by terrestrial sediments, despite the fact that *arc* magmatism in both the Chola Shan and in the Shaluli Shan Arcs continued generally into the Jurassic, and locally even into the early Cretaceous. Şengör (1984, 1985*a*) proposed that 'hidden subduction' beneath the overthickened flysch/*mélange* fill of the Songpan–Ganzi System may have been responsible for this peculiar observation. Thus, although marine conditions ended by the late Triassic in the entire Songpan–Ganzi system, subcutaneous 'hidden' subduction lasted until the early Cretaceous.

Although the structural style of the eastern margin of the Songpan–Ganzi System is vastly different from the south-west margin, characterized by enormous cuspate allochthons carrying thick Triassic clastics of the Bayanhar Group on to the Sichuan Basin, there too, hidden

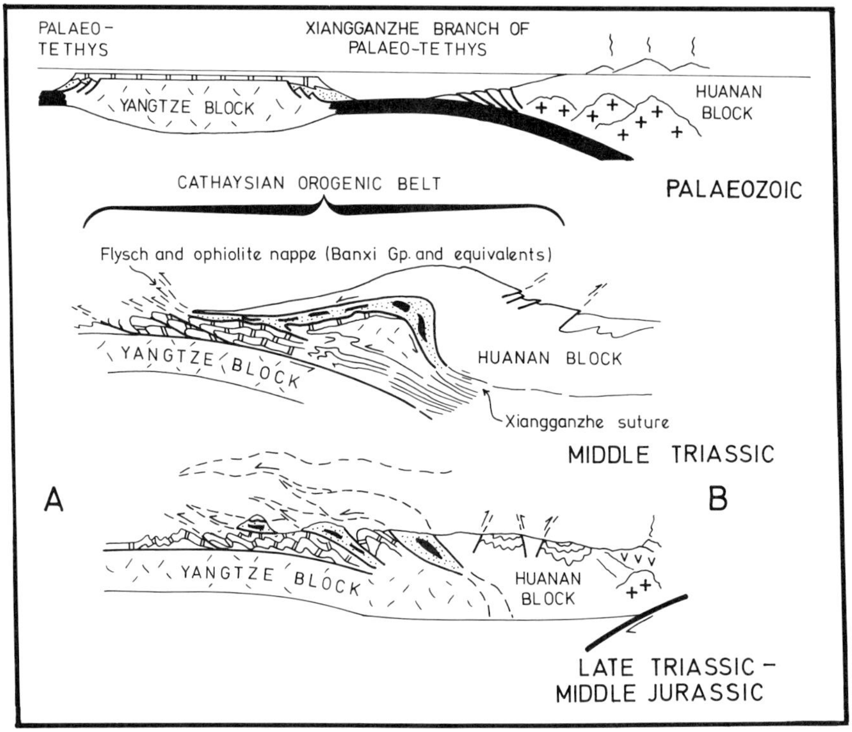

FIG. 12. Schematic, sequential, evolutionary cross-sections, showing the development of the Cathaysian orogenic belt between the late Palaeozoic and the middle Jurassic. For location see Fig. 10 (redrawn after Hsü *et al.* in press).

subduction may have played an important role in the Jurassic–early Cretaceous magmatic development of the Longmen Shan and Qionglai Shan, together forming the 'suture' *XXXIV*. But here a *line* of suturing between two continents is not present (except perhaps along the eastern margin of the Songpan Massif (*z*): Yang *et al.* 1986): 'suture' *XXXIV* is nothing more than a thrust-front, carrying the spilled-over flysch content of the constricted Songpan–Ganzi oceanic hole on to the Sichuan Basin (Figs 3 and 10).

Along the northern margin of the Songpan–Ganzi system, in the Anyemaqen Shan and Burhan Budai Shan, magmatic activity largely ended during the Triassic with very few and isolated Jurassic and Cretaceous plutons. Along the strike, westwards from the Anyemaqen Shan/Burhan Budai Shan 'suture' (*XX*) is the Southern Kuen-Lun suture (*XIX*). Along this suture, Şengör (1984) and Şengör and Hsü (1984), reviewed the available field evidence and concluded that collision probably took place by the early Jurassic.

Within the triangle defined by the Burhan Budai Shan (*XIX*) in the west, the Qin-Ling in the east (*XXXIII*) and the north-facing hairpin curve of the Litang 'suture' (*XXIII*), in the south, the Songpan–Ganzi System is filled with the enormous volume of the Triassic Bayanhar Group (which comprises dominantly turbiditic clastic rocks and some Permian deep-sea mudstones and radiolarian cherts, amounting to some five million cubic kilometres; Zhang 1981). The joint 1985 Royal Society–Academia Sinica Tibetan Geotraverse Team encountered these deposits between the Jinsha Jiang suture and the Kuen-Lun Pass on the Lhasa-Golmud highway. Along this stretch, they are made up of fine-grained, deep-water Triassic mudstones with graded wackes. The Geotraverse Team followed Şengör (1981, 1984) in interpreting these as parts of an accretionary prism made up of turbidites and contourites, but they also indicated that part of the sequence may have been deposited on a passive-margin continental rise. Because the Burhan Budai Shan/Anyemaqen Shan margin of Palaeo-Tethys was already a consuming margin, as judged from the abundant but very narrowly disposed late Palaeozoic arc-magmatic rocks,

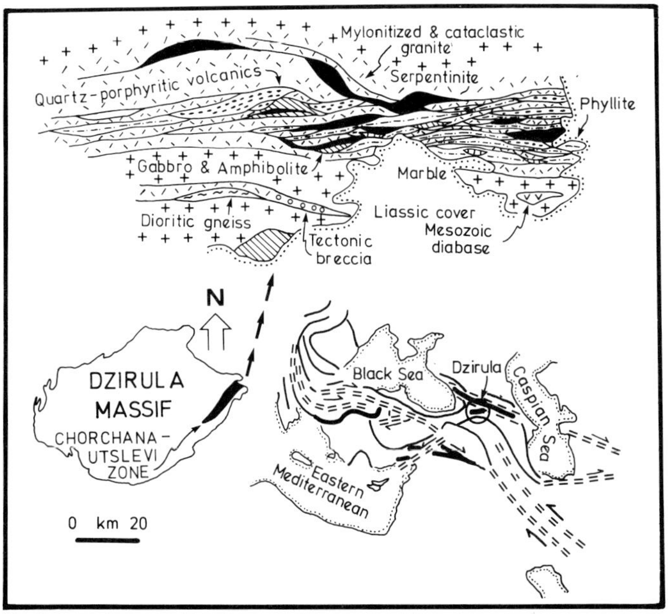

FIG. 13. Simplified geological map of the Chorchana–Utslevi shear zone of the Dzirula Massif. The location of the zone within the Dzirula Massif, and the location of the Dzirula area within the tectonic framework of the Tethysides, are shown in the two lower sketch maps. The lower right-hand map is taken from Fig. 22.

this conjecture seems unlikely. A forearc setting would be more likely for some of the Bayanhar lithologies.

A number of Cimmeride sutures are encountered under unconformable Liassic sequences in the western part of the orogenic zone. Although some of these may have already closed in the late Triassic, the available evidence seems to allow an intra-Liassic age for most of them. For only one of these, the Karakaya suture (*I*) in N. Turkey, related to a peri-Gondwanian back-arc basin of Palaeo-Tethys (Şengör and Yılmaz 1981; Şengör *et al.* 1984*a*), a late Triassic age seems at least locally well established (Tekeli 1981). Present data suggest that it probably began opening in the late Permian, at least in north-west Turkey, but may have been older farther east, near Erzincan (?Carboniferous) (Şengör *et al.* 1984*a*). Westwards this opening may have continued into the Maliac zone in Greece (Vergely 1984) or, in the early Triassic, into the Pindos–Budva Trough—probably also a Palaeo-Tethyan marginal basin (Şengör 1985*b*). Although it is not known whether the Karakaya suture extends any farther east, Y. Yılmaz (pers. comm. 1986) mapped sub-Liassic zones of intense deformation within the Malatya metamorphics (to the north-east of *22* in Fig. 3), thus extending the zone of sub-Liassic deformation previously known from the Menderes–Taurus block (known from the Lycian Nappes—north of *19* in Fig. 3—the Beyşehir autochthon—north of 18—and the Bolkardağ—north-west of *22*: Şengör *et al.* (1984*a*)) farther east. Şengör *et al.* (1984*a*) suggested that this deformation may represent the effects of the closure of the Karakaya marginal basin.

The Palaeo-Tethyan suture fragments in the Caucasian region all lie under unconformable Liassic sequences. The best outcrop of a Palaeo-Tethyan suture fragment in this region is in the Trans-Caucasian Dzirula Massif, where it is known as the Chorchana–Utslevi zone (V′; Fig. 13; Adamia and Belov 1984). It outcrops from under unconformable Lias, and consists of extremely sheared oceanic rocks and remnants of (?one) former magmatic arc. Their ages range from early Cambrian, through late Devonian to the late Palaeozoic (Adamia and Belov 1984). The present geometry of the Chorchana–Utslevi zone and its internal structures, suggests important amounts of strike-slip faulting following suturing. Both the process of suturing and the strike-slip faulting must have ended before the deposition of the overlying Liassic rocks. Coeval ophiolites also outcrop in the Khrami Massif along the Chochiani River gorge, within the

Artvin–Karabagh zone (*K* in Fig. 8a; Adamia and Belov 1984). Peive *et al.* (1980), Şengör (1984, 1985*a*), Şengör *et al.* (1984*a*), and Belov *et al.* (1986) argued that yet another, but cryptic, segment of the main Palaeo-Tethyan suture must be present in the Svanetia region of the south flank of the Greater Caucasus.

It is clear from the foregoing paragraphs that Palaeo-Tethyan suture segments in the Caucasian region occur in several different tectonic zones, whose formation post-dated Cimmeride tectonism. This suggests, as in Iran, that a formerly more regular pattern was later disrupted and led to suture tripling (from north to south: Svanetia, Dzirula, Khrami) in one cross-section in the Caucasus. We shall see in the following paragraph that this disruption of the Palaeo-Tethyan suture in the Caucasus was related to the same events that also disrupted the Palaeo-Tethyan suture in Iran.

Westwards from the Caucasus, the main Palaeo-Tethyan suture is younger. Eastwards, it is older in Turkmenia SSR and in northern Iran (along *VII*), but becomes younger again in Afghanistan and in the Northern Pamirs (along *VIII*). In these regions the Herat–Wanch–Akbaytal lineament (Stöcklin 1977) marks the main Palaeo-Tethyan suture (*VIII*), north of which was a large early Carboniferous to late Triassic subduction–accretion complex, that once formed the southern margin of Laurasia (Şengör 1984). The Lias witnessed final emergence, and in the northern parts of the Paropamisus and the Western Hindu Kush, plant-fossil-bearing Liassic clastics, equivalents of the Shemshak Formation in Iran, were laid down. Although the Triassic and older sediments were metamorphosed in these regions, younger sedimentary rocks were generally not metamorphosed (Shareq 1981).

North of the Herat–Wanch–Akbaytal lineament, Triassic to Jurassic granodiorites and tonalites (ages between 230 and 210 Ma) intruded the highly deformed late Palaeozoic and Triassic clastic and volcaniclastic rocks all along the Paropamisus (Berberian and Berberian 1981; Debon *et al.* 1986), and suggest an earliest Jurassic collision along the Herat–Wanch–Akbaytal lineament. Kravchenko (1979) describes a similar picture from the Northern Pamirs. In all these regions the first post-orogenic sedimentary rocks that unconformably overlie the suture are early Cretaceous red beds (Şengör 1984).

Palaeo-Tethyan suture zones younger than the Lias are known from the two extremities of the Cimmerides (not counting the Songpan–Ganzi System). In northern Turkey, the westerly continuation of the Palaeo-Tethyan suture (*IV*) from the Caucasus is represented by the ophiolites, deep-sea muds, and cherts, and the flysch plus shallow-water limestone knockers now making up the Küre Nappe (Şengör 1984; Şengör *et al.* 1984*a*; Yılmaz and Şengör 1985; Tüysüz 1986). Earlier workers assigned ages, ranging from questionable Permian to the earliest Dogger, to the lithologies of the Küre Nappe (see Şengör *et al.* 1980) a range confirmed by the TPAO Arama Grubu (1986). The TPAO Arama Grubu also found some late Carboniferous palynomorphs in the Küre lithologies (M. Aydın verbal comm. 1986; see also Yılmaz and Şengör 1985) thus suggesting that the age of the Küre Nappe oceanic assemblage might extend down into the Carboniferous.

Şengör *et al.* (1984*a*) interpreted the Küre Nappe as a large subduction–accretion complex, and correlated it with the diabase–phyllitoid association and the Lipacka flysch from the Strandja Mountains in Bulgaria (to the immediate south of *16* in Fig. 3; Chatalov 1979), with the Tauridian flysch in the Crimea (Beznosov *et al.* 1978; Kotansky 1978), and with the Nalbantian flysch in the North Dobrudja (Sandulescu 1978, 1980). Both in northern Turkey and in Bulgaria, sedimentation within the Küre Nappe largely ended by the end of the Liassic (the latest Triassic in Bulgaria) and it was tectonially covered by the Kırklareli, Istanbul, and the Bayburt Nappes, representing the northern margin of the Cimmerian continent before the (?middle) Dogger. This overthrusting also extensively disrupted an already complex internal structure within the Küre Nappe (see, for example Tüysüz 1985 for the details of this event).

Judging from the record of arc-type igneous activity within the Kırklareli, Istanbul, and the Bayburt Nappes, Şengör *et al.* (1984*a*) concluded that Palaeo-Tethyan subduction southwards under northern Turkey commenced during the late Carboniferous (Westphalian) and lasted until the early Jurassic. The latest representatives of this regime may have been the so-called Kastamonu granitoid belt, containing allegedly subduction-derived plutons (Boztuğ *et al.* 1984; Yılmaz and Boztuğ 1986). Both Boztuğ *et al.* (1984) and Yılmaz and Boztuğ (1986) considered the Kastamonu belt a product of *northward* subduction because, they observed, these plutons occurred to the north of some of the Küre Nappe ophiolitic rocks and high-grade metamorphics. However, because the Küre Nappe as a whole represents an accretionary complex, the tectonic position of the Kastamonu belt appears very similar to the late Triassic–early Jurassic magmatic arc in the Paropamisus and Western Hindu Kush; alternatively, it may be related to post-collisional thickening and partial melting of the Küre rocks, as originally suggested by Şengör *et al.* (1980). This latter interpretation is supported by the observation that the partial melting of subduction–accretion complexes commonly produces granodioritic magmas, as in the Chugach Mountains of southern Alaska (Hudson and Plafker 1982).

Where the main Palaeo-Tethyan suture in northern Turkey and in south-east Bulgaria extends to farther west is still debated. We here adhere to the scheme proposed in Şengör *et al.* (1984*a*), who suggested, following S. Dimitriadis' unpublished results, that the Therma–Volvi–Gomati ophiolites may mark the location of a pre-Tithonian Palaeo-Tethyan suture in Greece. They traced this suture farther north into eastern Serbia, into the region of Jastrebac, a suggestion followed here (northern part of suture *I* in Fig. 3).

III shows the Luncavita–Consul suture in the North Dobrudja. It represents a northerly-vergent thrust complex, bringing the Triassic Niculitel ophiolite on to the Tulcean Massif and uniting the latter with the Macin zone to the south of the suture. This suture probably closed in the early Jurassic. Şengör *et al.* (1984*a*) suggested that the whole of the North Dobrudja may have been a segment of the northern Turkish Cimmerides, fitting into the interruption of the suture *IV* (indicated by a dotted line). The North Dobrudja probably left this locality, and was transferred to its present position by strike-slip faulting during the earliest Cretaceous (Şengör *et al.* 1984*a*).

Farther north and west of eastern Serbia, indications of a Palaeo-Tethyan suture are reduced to slivers of undated ultramafic rocks, located within poorly-dated, high-grade metamorphic terranes within the Southern and Eastern Carpathians. Şengör (1984) discussed these in some detail, but it is now strongly doubted whether they really represent remnants of Palaeo-Tethys (M. Sandulescu, pers. comm. 1985). Thus, Greece and Yugoslavia seem, so far, to contain the westernmost Palaeo-Tethyan suture zones. In both of these countries, however, both the precise location and the age of Palaeo-Tethyan closure are still matters of dispute (e.g. Robertson and Dixon 1984; Vergely 1984; Mountrakis 1986; Şengör 1986*a*).

A probably early to mid-Jurassic Palaeo-Tethyan suture fragment in the eastern extremity of the Cimmerides is probably represented by the Serabang Formation in western Sarawak, and the equivalent rocks of immediately adjoining regions of Kalimantan (*XXXII*). These consist of strongly folded slaty flysch with chert and ophiolites (Haile 1976). Both Haile (1976) and Audley-Charles (1978) consider the Serabang Formation and its equivalents to be older than Jurassic. This general region is also characterized by early Mesozoic granite intrusions and coeval and younger felsic, intermediate, and mafic, lavas and tuffs (Fig. 5 and 6). Şengör (1984) argued that this region represented a pre-late Jurassic Palaeo-Tethyan suture. From here the suture probably goes through the West Borneo basement. Murphy (in press) suggested that a Carboniferous to late Triassic magmatic arc, extending along the hypothesized Serabang suture, is probably truncated at the late Cretaceous Meratus suture (*34*) which here defines the boundary between the Tethysides and the Circum-Pacific superorogenic complex (Murphy in press). Murphy (in press) correlates the Bornean Palaeo-Tethyan suture with the Nan Uttaraddit/Sra-Kaeo suture in Thailand. It seems, however, more likely that the latter suture continues directly into the Bentong–Raub line in the Malay Peninsula as argued previously, and that the Serabang/West Borneo suture is a prolongation of the Song Da suture in Vietnam (Şengör 1987*a*).

With the single exception of the Shilka suture in the Soviet Far East (*XLI*), the only Cretaceous sutures are related to the closure of the Waser/Rushan–Pshart/Banggong Co–Nu Jiang/Mandalay ocean, the *Meso-Tethys* of Shvolman (1978) and Belov (1981). This substantial ocean split a considerable portion of the Cimmerian Continent into at least two halves, and may have caused the disintegration of the half lying to its north into two pieces in northern Tibet (*s′* and *s″* separated by *XXIV′*) (Montenat *et al.* 1986).

So far the most abundant information concerning this suture zone has come from the Afghan (*IX*) and the Soviet (*X*) sectors (for the most recent review of data from the Afghan and Tibetan parts see Montenat *et al.* 1986). Shvolman (1978) described the evolution of what he termed the Meso-Tethys, in terms of late Permian rifting, early (?to late) Triassic intracontinental stretching with abundant basaltic volcanism, later Triassic–Jurassic sea-floor spreading and Cretaceous closure along the Rushan–Pshart zone. A fairly complete ophiolite is described, involving a serpentinized harzburgite, containing dykes of gabbro and gabbro-diabase. The upper 200–300 metres of this peculiar gabbro–harzburgite complex are brecciated and contain, in addition to gabbros, also diorite and plagiogranites. This igneous section is overlain by sheared and differently metamorphosed calcareous schists and cherty greywackes and siltstones. A pile of variously spilitized tholeiitic lavas caps the metasedimentary section.

From Shvolman's (1978) descriptions one obtains the impression of a fossil fracture zone. Shvolman (1978) further argued that the tholeiite pile was the initial product of an ensimatic island arc, whose age reaches into the Cretaceous. We believe that this arc may have nucleated on a fracture zone.

The vergence of the present structure in the central and south-eastern Pamirs, along the boundary of which the Rushan–Pshart zone is located, is clearly northward (Shvolman 1978). From the location of most of the Cretaceous granites to the south of the Rushan–Pshart zone (see, for example, fig. 1 in Desio 1977) one obtains the impression of a north-facing orogenic polarity which also fits the structural vergence. The presence of Permian diabases and andesitic basalts on *both* sides of the Rushan–Pshart zone (Shvolman 1978) suggests that the Rushan–Pshart ocean may have opened as a back-arc basin disrupting a continental margin arc during the late Permian, much like the Karakaya marginal basin in northern Turkey (*II*; see also Fig. 20a).

Both westwards and eastwards the Rushan–Pshart zone is cut and offset by large, Cainozoic, strike-slip faults (Fig. 3). The Jurm, Zebak–Mun'ya, and Harirud Faults offset it south-westwards by some 400 km (Shvolman 1977, fig. 52; Bazhenov and Burtman 1986; fig. 1) where it continues westwards as the Waser (or the Farah–Rud) suture (*IX*). The Waser suture (*IX*) is also a north-vergent structure (Blaise *et al.* 1978). Within it, the fossils recovered from the volcano-sedimentary carapace of the Waras ophiolite indicate a middle to late Triassic spreading age. Although no direct evidence exists for the timing of initial rifting of the Waser ocean in Afghanistan, analogy with the Rushan–Pshart zone makes it likely that this too began rifting during the late Permian or the early Triassic at the latest. The closure of the Waser ocean took place during the early Cretaceous with a northerly vergence. Cretaceous granodiorites and arc volcanics indicate the southerly subduction of its floor, but Debon *et al.* (1986) argue that the late Cretaceous ages abundant in these magmatic rocks may indicate a later age of closure here than hitherto considered (Fig. 7), although its suture is unconformably covered by *Orbitolina*-bearing limestones of Aptian–Albian age (Blaise *et al.* 1978).

In Tibet, the Banggong Co–Nu Jiang suture (*XXIV*) has been studied mainly in its eastern half. In its western part, observations are sparse, but Norin (1976, 1979) identified the south-east continuation of the Rushan–Pshart zone in the Aghil–east Loqzung Mountains of north-west Tibet. The Aghil Ranges are offset to the south-east from the Rushan–Pshart zone by the Karakorum Fault for about 100 km or less. The Loqzung Mountains are characterized by Permian limestone, quartzites, quartz sandstones, and shales of marine origin, with which mafic porphyries seem to be associated (Norin 1946). Towards the south, north of the Banggong Co, the possible along-strike continuations of Norin's 'porphyries' are seen as ophiolites. In the best-studied segment of the Banggong Co–Nu Jiang suture zone, around the Donqiao–Gyanco area of eastern Xizang (to the north-east of *t″* in Fig. 3), Upper Palaeozoic low-grade metamorphic grey limestone, quartzites, and shales, similar to the Loqzung facies, are seen to be tectonically interleaved with well-developed ophiolites. Both the Upper Palaeozoic metasediments and the ophiolites were thrust southwards above a mid- to late Jurassic flysch and all were then unconformably overlain by Aptian–Albian red clastic rocks and volcanics (Girardeau *et al.* 1984*a*). The Banggong Co–Nu Jiang suture was closed before the *Orbitolina*-bearing late Cretaceous limestones were deposited upon it in a wide region (Bally *et al.* 1980).

Recently, Srimal (1986) suggested that the Banggong Co suture should be correlated westwards with the Shyok suture north of the Ladakh batholith. This suggestion is based on comparisons between poorly-dated ophiolite slivers from the northern Saltoro Range and Banggong Co, and does not take the geology of the Central and south-eastern Pamirs into account. We shall return to the question of the age and significance of the northern Saltoro ophiolites in the following paragraphs.

Farther east, the suture zone follows the course of the Nu Jiang and turns abruptly southwards around the eastern syntaxis of the Himalaya (Fig. 3). From there it goes to the Mandalay region (to the north-west of *XXIX*), where it is doubled back on to itself and on to the southern continuation of the Indus–Yarlung Zangbo suture (*31*) along a north-south right-lateral strike-slip fault, whose offset is about 400 km in the Mandalay region (Mitchell in press, Curray *et al.* 1982).

From Burma southwards, we follow Şengör's (1984) correlation of the earliest Cretaceous (?latest Jurassic) deformation, metamorphism, and granitic intrusions in the basement of the eastern Sumatra Basin (Katili 1976) with the

Mandalay zone. Murphy (in press) recently also adopted this interpretation.

With the discussion of the long suture zone of the Waser/Rushan–Pshart/Banggong Co–Nu Jiang/Mandalay/east Sumatra oceanic realm we have come to the end of our discussion of the known Cimmeride sutures. We should perhaps mention three other sutures in this connection, although neither their age, nor their relationship to the Cimmeride evolution is known with any certainty, nor are they generally considered as Cimmeride sutures by the majority of the workers active in this field. One of these is the Shiquan He suture (*XXV*) reported by Mei *et al.* (1981) and Baud (1985). Baud (1985) encountered ophiolites along the Shiquan He during a traverse from Ladakh to Yarkand. Previously, Norin (1976) had indicated the presence of what he called the Upper Shyok(?)–Shaksgam 'rift zone' (Norin called all major sutures 'rifts' including the one along the Indus) and implied a major zone of disturbance characterized by a distinctive black shale facies. This zone may possibly be identical with, and represent the north-west continuation of, the Shiquan He suture. How the Shiquan He–Shaksgam zone may continue into the Pamirs to the west is unknown, but we will mention the possibility of its being represented by the late Palaeozoic–early Mesozoic black slate troughs of the south-eastern Pamirs, south of the Rushan–Pshart suture zone (see fig. 1 in Norin 1976). This possibility is indicated schematically in Fig. 3 by the conjectured suture fragment *b* (for 'black slates') to the south-east of the Rushan–Pshart zone (*X*). If our tentative correlation across the Karakorum fault is correct, then the Shiquan He suture should have closed before the Bajocian (see Norin 1976, 1979). Mei *et al.* (1981, p. 551), however, indicated that 'Cretaceous continental basalt-andesite-rhyolite formations are developed . . . on the . . . Nganglung Kangri Range' north of the Shiquan He ophiolites (see Mei *et al.* 1981, fig. 1), implying that the Shiquan He ocean was still open during the Cretaceous. This is the interpretation preferred by Mei *et al.* (1981), although, curiously, Srimal (1986) considers the Shiquan He ophiolites to be late Palaeozoic, for which there is no evidence. The arc-type magmatic rocks of the Nganglung Kangri Range may, on the other hand, belong to a south-dipping subduction zone associated with the Banggong Co–Nu Jiang ocean, and thus be unrelated to the Shiquan He suture.

A. Baud (verbal comm. 1985) and Srimal (1986) suggest that the Shiquan He suture zone may continue westward into the southern Tibetan Lhasa block, and in fact may have separated the Trans-Himalayan late Mesozoic–early Cainozoic magmatic arc from regions lying to the north of it. They indicate that the Xainxa ultramafic rocks located within the Lhasa block (near 31°N lat. and 88°E long.; Girardeau *et al.* 1985) may be the eastern prolongation of the Shiquan He suture zone. A. Gansser (verbal comm. 1985) further suggested in support of this view, that the entire Lhasa block may have consisted of four separate blocks (t', t'', t''' and t''''). However, the recent field observations of the 1985 Royal Society–Academia Sinica Tibetan Geotraverse Team seem to suggest that the ophiolitic outcrops within the Lhasa block actually represent remnants of a single, northerly-derived sheet of ophiolite, that roots into the Banggong Co–Nu Jiang suture zone, and that became disrupted by post-obduction deformation of the Lhasa block itself. The geological and geochemical studies of the Lhasa block ophiolites by Tang and Wang (1986) also corroborate this conclusion.

Chang *et al.* (in press) recently reported total sediment thicknesses in excess of 15 km from within the Qangtang block (*s*). This observation, suggesting at least substantial crustal thinning, may indicate the presence of a major suture zone within the Qangtang block, when we combine it with earlier reports of glaucophane schists from within the same block (Hennig 1915). Suture *XXIV'* in Fig. 3 was drawn by connecting the observation points of Chang *et al.* (in press) and Hennig (1915). This suture probably closed by the end of the Triassic, because middle Jurassic sediments lie above the Triassic rocks across an important unconformity around its southern end (Chang *et al.* in press).

The third possibly Cimmeride suture is the south-west Karakorum suture (*XXVI*), reported by M. P. Searle (verbal comm. 1985). Neither the age nor the along-strike continuations of this zone are known in any detail, but the northern Saltoro ophiolites, reported to be of late Palaeozoic age by Srimal (1986), represent the direct east–south-east prolongation of the metamorphosed harzburgites, serpentinized dunites, and garnet pyroxenites reported by Searle *et al.* (1987) and Bertrand and Kienast (1987) from within the Karakorum metamorphics south of the Axial Batholith, between the Masherbrum Glacier in the south-east and the Panmak Valley in the north-west.

Bertrand and Kienast (1987) indicate the presence of a large amount of metagreywackes within the Karakorum metamorphics, in which the ultramafic rock slivers are found. The probably Carboniferous 'flysch-like' successions, reported from along the Yasin and Ishkuman Valleys in the Hindu Raj region by Casnedi (1980) may be correlative deposits, and may imply the presence of a small late Palaeozoic oceanic trough in this region, that may have stretched from the present Saltoro range to the Hindu Raj region. When this trough closed is unknown, but its closure and the metamorphism of the Karakorum metamorphics must have predated the Mango Gusar granite intrusion (36.4 ± 1 Ma U–Pb zircon age) (Searle *et al.* 1987). Casnedi (1980) pointed out that shallow-water Permian sediments followed the ?Carboniferous deep-sea deposits and implied the possible presence of a pre-Permian tectonism here. On the basis of regional considerations, we do not believe that the age of closure of this trough was much younger than the early Cretaceous.

Alpide sutures

Compared with the broad, eastward-widening net of the Cimmeride sutures, the better-known Alpide sutures are much simpler, commonly consisting of single strands for long stretches, with the notable exception of a braid of sutures between Yugoslavia and Iran, which is still small compared with the Cimmeride net. In the following account we briefly review their geology and emphasize only the more controversial points.

Mesozoic Alpide sutures

In the entire Alpide orogenic system there are very few sutures that closed terminally during the Mesozoic Era, the majority having formed during the early Cainozoic. The earliest Alpide suture is the southern and eastern sections (the Nish–Trojan suture: Hsü *et al.* 1977) of the Circum-Moesian suture zone (*10*) that closed by south-dipping subduction beneath the Rhodope–Pontide fragment (*f*) in the Albian (Hsü *et al.* 1977, Burchfiel 1980).

The next youngest Alpide suture is the so-called Northern Suture (*29N*) north of the Kohistan island arc (*m'*) and its eastern equivalent, the Shyok suture (*30N*) in northern Ladakh (*t''''*). Petterson and Windley (1985) have shown that the Kohistan segment of this suture must have closed between about 100 Ma and 75 Ma ago, that is, roughly between the Albian and Campanian.

Geological relationships and the timing of events are similar in Ladakh (*t''''*). We agree with Petterson and Windley (1985) that both Kohistan and Ladakh used to be parts of a single ensimatic arc system that formed just offshore of the southern margin of Eurasia about 100 Ma ago. This united arc collided with Eurasia during the middle Cretaceous and became a continental-margin arc.

Two distinct suture belts of probably pre-Maastrichtian age exist in Turkey and in northern Greece. Şengör and Yılmaz (1981) have argued that the Peonias–Intra-Pontide suture (*13*) must be of pre-Maastrichtian age, because the associated arc along the Black Sea coast terminated its activity at this time. However, continuing arc volcanism farther south, and the fact that the oldest sediments unconformably covering the deformed rocks of the suture in Turkey are of Lutetian age (Tokay 1973), may be indicative of a later closure.

In south-east Turkey, along the Assyride suture (not exposed; covered by the products of *21*), the Bitlis–Pötürge–Malatya–Keban portion of the Menderes–Taurus block (the eastern end of the African promontory, *e*) is believed to have collided with the northern margin of Arabia in pre-Maastrichtian time (Şengör and Yılmaz 1981). Widespread ophiolite obduction southward on to the Arabian platform was associated with this collision. A nearly coeval collision was postulated by Gealey (1977) to obduct the giant Semail ophiolite nappe in Oman (*27*). Recent geochemical studies by Searle *et al.* (1980) and Pearce *et al.* (1981) confirm Gealey's (1977) arc collision hypothesis.

Both of these obduction events took place along the southernmost continental margin of the Alpides in the Middle East facing an ocean that had already begun rifting in the Permian (in the Zagros) and the early Triassic (southern Turkey). Until the middle Triassic, intracontinental extension continued in both places, followed by sea-floor spreading in Oman, and also probably in the Zagros in the late Triassic. In southern Turkey and around the eastern Mediterranean (southern branch of Neo-Tethys) the geological record of the continental margins suggests that sea-floor spreading probably commenced in the Jurassic. A side-branch of the eastern Mediterranean, in the form of a small embayment, formed the Pamphylian Basin in south-west Turkey in the middle to late Triassic (Şengör and Yılmaz 1981) by rifting and rotation of a small fragment of the Menderes–Taurus block, the Alanya Massif (*e'*).

In Sumatra, Cameron *et al.* (1980) interpreted the record of the late Mesozoic Woyla Group as indicating the latest Jurassic–early Cretaceous opening, and late Cretaceous (probably between 110 and 85 Ma ago) compressional obliteration of a marginal basin, whose suture is today represented by the Woyla Group (*33*). This suture may continue eastwards into the island of Java and then turn to the north-east, to re-emerge eventually in the Meratus Mountains of south-east Borneo (*34*), where ophiolite obduction occurred between the Cenomanian and Turonian (Murphy in press).

Cretaceous collisions have been proposed for other Alpide sutures as well (e.g. Tapponnier 1977, for the Alps (*8*), Bergougnan and Fourquin 1980, for the İzmir–Ankara–Erzincan suture (*14* and *17*), Okay 1984, for the İzmir–Ankara suture (*14*), Berberian and King 1981 for the Zagros suture (*25*)), but most of these are based on overinterpretation of the available data, or result from erroneous conceptions of tectonic processes, or are based on negative evidence. Some of these authors have later retracted their original erroneous interpretations (e.g. Berberian *et al.* 1982). One reason why so many workers have favoured late Cretaceous continental collisions along the Alpides has been that the late Cretaceous was a time of very widespread ophiolite obduction and blueschist metamorphism—from Corsica to Oman. More recently, Searle (1983) has argued that late Cretaceous ophiolite obduction also must have taken place in the Himalaya, on to the north-west margin of the Indian Shield in the Zanskar Range (immediately south-east of *30* in Fig. 3) but there is little evidence to support this conjecture (see Şengör 1987*b*).

It is difficult to prove that all of the ophiolite obduction events in the Alpides were not associated with continental collisions (cf. Ben-Avraham *et al.* 1982), but so far no evidence has been found to show that the late Jurassic–early Cretaceous ophiolites in Yugoslavia and Greece (Spray *et al.* 1984), late Cretaceous ophiolites in Corsica (Gruppo di Lavoro sulle Ofioliti Mediterranee 1977), in Turkey (Şengör and Yılmaz 1981), and in the Zagrides (Ricou 1976) were associated with any collision. They all appear to have been pre-terminal obduction events (cf. Dewey 1976), mostly caused by the interaction of oceanic fracture zones with passive continental margins.

Cainozoic Alpide sutures

It is possible to divide the Cainozoic suturing events in the Alpides into two large classes: Palaeogene and Neogene sutures.

The Vardar zone and the İzmir–Ankara suture appear to have closed during the Palaeocene (Burchfiel 1980; Şengör and Yılmaz 1981; Vergely 1984). The easterly continuations of both the Intra-Pontide suture and the İzmir–Ankara suture (*13* and *17* respectively) closed later, during the Lutetian and Priabonian. Along all of these sutures, ocean opening took place mainly during the Hettangian–Sinemurian interval (Görür *et al.* 1983), with the possible exception of the İzmir–Ankara and the Vardar sutures, along which rifting was probably earlier—in the Triassic.

In north-east Turkey, in a number of places it is difficult to draw a line of demarcation between the Karakaya and the İzmir–Ankara sutures. In such places one obtains the impression that the Karakaya ocean did not completely close, small remnants of it having survived between sutured segments, and that these remnants were later incorporated into the İzmir–Ankara ocean.

The small Pamphylian oceanic basin in south-west Turkey began closing in the late Cretaceous and closed terminally during the latest Palaeocene–Eocene by north-vergent thrusting along the Antalya suture (*19*; Şengör and Yılmaz 1981; Ozgül 1985).

To the south of Ankara (immediately south of *11* in Fig. 3) the Karakaya ocean was probably in direct communication with the Inner Tauride ocean that began opening in the early Triassic (Özgül 1984). This Triassic opening appears to have been a local intracontinental rifting event only, because regions lying to its south—the entire Menderes–Taurus block—seem to have been affected by the closure of the Karakaya Basin to the north (Şengör *et al.* 1984*a*, Y. Yılmaz, pers. comm. 1986). The Inner Tauride ocean closed during the middle Eocene (Görür *et al.* 1984).

Another latest Palaeocene–earliest Eocene closure was that of the Waziristan (*28*)/Main Mantle Thrust (*29S*)/Indus–Yarlung Zangbo (*30S* and *31*) closure that welded India on to Eurasia (Fuchs 1984; Petterson and Windley 1985). Collision along this suture seems to have progressed from west to east during the Eocene, as a result of which the entire Indian subcontinent rotated in an anticlockwise fashion for about 15° (Klootwijk 1981).

The timing of opening of the Indus–Yarlung Zangbo ocean, i.e. the Himalayan Neo-Tethys, is fairly well-established by observations on both the Lhasa block and on the northern margin of the Indian Shield. In the south-west part of the Lhasa block (*t''''*), an overall shallowing and local emergence is seen above a 500-metre thick Lower Permian limestone horizon. Near the Indus–Yarlung Zangbo suture zone the same overall shallowing is observed. During the middle to late Triassic a mixed volcanism with andesites and basalts is seen, accompanied by clastic sedimentation and normal faulting (Bally *et al.* 1980).

On the Indian side, a rapid subsidence during the earliest Triassic is accompanied by extensive normal faulting and fissuring (Bassoullet *et al.* 1978) in Ladakh (along *30S*). Widespread Triassic turbidites flank almost the entire northern margin of the Indian Shield, and indicate the establishment of a north-facing passive continental margin. It thus seems clear that the Himalayan Neo-Tethys came into being by the middle to late Triassic at the latest.

The consumption of the floor of Neo-Tethys was already under way by north-dipping subduction under the Lhasa block at large, during the Portlandian (late Jurassic). Initially, this subduction probably began within the oceanic floor of Neo-Tethys. (Giardeau *et al.* 1984*b*). Chang *et al.* (in press) pointed out the presence of small patches of mainly Lower and Middle Cretaceous and partly Upper Jurassic andesite–dacite–keratophyre associations, forming part of the basement of the later Trans-Himalayan magmatic arc. These volcanic rocks show the presence of calc-alkaline volcanism during the late Jurassic–early Cretaceous interval along the southern margin of the Lhasa block. Thus, models involving a late Jurassic rifting of this block away from north-west Australia (e.g. Rowley *et al.* 1986) do not appear to be compatible with data from it.

The onset of the consumption of the Neo-Tethyan ocean floor in the Himalayan segment, post-dates the obduction of the Banggong Co–Nu Jiang ophiolites by about 20 Ma only (obduction *c.*170 Ma, onset of magmatism *c.*150 Ma), although the major, widespread Trans-Himalayan plutonism did not begin until about 110 Ma ago (Xu *et al.* 1985).

Another latest Palaeocene to earliest Eocene collision was the encounter of the African Promontory with the European south-facing continental margin in the Alps, along the Alpine suture (*8*). Ongoing controversy still characterizes discussions on the original palaeogeography, polarity of subduction zone(s), and the timing of events in the Alps. We here adopt the palaeogeographical scheme proposed by Trümpy (1971, 1975, 1980; Trümpy and Haccard 1969) and refined by Lemoine and Trümpy (1987), because other models (e.g. Dercourt *et al.* 1986) do not account for the available observations nearly as fully as does Trümpy's scheme. Late Cretaceous southerly-vergent structures in the Pennine zone (e.g. Milnes 1978) and in the Austro-Alpine nappes of eastern Switzerland (e.g. Şengör 1982*a*) seem to be related to reactivation of north-dipping normal faults of the southern continental margin of the Alpine ocean in a dominantly north-facing subduction geometry, as recently confirmed by the presence of Turonian to earliest Coniacian trench deposits in the Walsertal zone along the boundary of the Penninic and Austroalpine units (Winkler and Bernoulli 1986). However, the possibility of a south-facing subduction zone along the southern margin of the Briançonnais mini-continent (fragment contained between the sutures *7* and *8* where they meet in Fig. 3), as originally proposed by Hsü and Schlanger (1971), remains open.

The age of opening of the Alpine ocean, especially the Piedmont–eastern Valais segments (northernmost part of *7* and the part of *8* lying to the east of its junction with *7*) is fairly well established, that is: Hettangian–Sinemurian rifting (with a Rhaetian prelude) and subsidence (Bernoulli and Jenkyns 1974; Lemoine and Trümpy 1987) followed by the middle Jurassic onset of ocean-floor spreading (Isler and Pantiç 1980; Lemoine and Trümpy 1987).

Although the age of rifting remains remarkably uniform on both sides of the Alps proper—both westwards through the Apennines and into the Tellian Atlas/Riff/Betic sutures (*7*, *6*, *3*, and *2* respectively) and eastwards into the Pieniny Klippen belt (*9*), the times of ocean closure vary greatly. Moreover, ocean-floor spreading took place west of the Alps, in the Apennines (Bigazzi *et al.* 1972), probably in the Pyrenees (Boillot and Capdevila 1977) and also possibly in the Betic chain (*ophiolites bétiques*? in Wildi 1983). Between Sicily and Gibralter no ophiolites have yet been confirmed. The ultramafic nappes of Ronda in southern Spain (Wildi 1983) and Beni Bouchera in the Riff (Reuber *et al.* 1982) are subcontinental ultramafic rocks exposed by extreme extension during the middle Mesozoic. In the Apennines, ocean closure occurred mainly in the Oligocene (Amodio-Morelli *et al.* 1976, Kligfield 1979), whereas along the westernmost Mediterranean chains it was later, mainly in the early to middle Miocene (Wildi 1983). As Şengör (1984) pointed out, because the opening of the Alpine ocean (and its westerly continuation into the Apennines and beyond) was a part of the opening history of the Central Atlantic ocean, it is more correct to consider these mountain belts as extra-Tethyan and parts of a separate system: the *Atlantides*, products of the closure of a part of the Atlantic Ocean (see Fig. 1). However, because the extant data are not sufficient to separate the Tethysides and the Atlantides with confidence, we continue to include the latter into the Alpides.

The Pyrenees represent the evolution of a small ocean basin that opened in the early Cretaceous and closed in the middle Eocene (Boillot and Capdevila 1977). An earlier, Triassic rifting nearly coincident with the future Pyrenean axis, failed to produce oceanic lithosphere, and became dormant by the Jurassic (Ziegler 1982). We do not follow the models portraying the Pyrenees as a dominantly strike-slip-related orogen not involving a Wilson cycle (e.g. Choukroune *et al.* 1973, Daignieres *et al.* 1982), simply because the Triassic salt lying on the Landais marginal plateau at the southern apex of the Bay of Biscay (Valéry *et al.* 1971) makes it geometrically impossible to slip the Iberian Peninsula into its present place solely along a strike-slip fault, without involving a significant extension both in the Bay of Biscay and in the Pyrenean region. Moreover, convergent structures north of the Iberian Peninsula (Montadert *et al.* 1974), showing the former existence of a north-facing subduction zone, support the model of Boillot and Capdevila (1977), and suggest that the Pyrenean orogen is the product of the subductive removal of a small ocean and subsequent continental collision.

Eastwards, in the Carpathians, the Neo-Tethyan suture becomes progressively younger around the Carpathian bend east of the Rhodope–Pontide fragment, in the prolongation east and south-east of the Pieniny Klippen belt zone (*9*), and then in the so-called 'external flysch trough' of the eastern Carpathians (Sandulescu 1975; Burchfiel and Royden 1982).

Collision along the southerly prolongation of the Pieniny Klippen Belt, in the Apuşeni Mountains (*11*), and in the Vardar zone of eastern Serbia (*12*) was nearly coeval with that in the Alps (Burchfiel 1980, Sandulescu 1980, Grubiç 1980). This more 'internal zone' extends southward to connect with the Vardar zone *sensu lato* in northern Greece (*12*, *13*).

Thus an inspection of the suturing episodes around the African promontory in the circum-Adriatic regions reveals a progression of suturing in both directions away from the Alps. As Caire (1977) has shown, this progressive closure of the westernmost Tethyan–Atlantic branch resulted in an accentuation of the curvature of the orogenic belts around the African promontory in the course of the Tertiary, a phenomenon further intensified in the west by the opening of the western Mediterranean basins (Cohen 1980).

Another event that narrowed the African promontory in the south, and accentuated the Z-shape of the Carpathian–Hellenide segment of the Alpides was the latest Eocene–Oligocene closure by westward-thrusting of the Pindos–Budva Trough. Deformation since this time has migrated farther west in Greece and Yugoslavia (Richter *et al.* 1978) and active folding and thrusting now characterize the coastal regions of these countries facing the Adriatic Sea (McKenzie 1972, 1978).

In the Balkan Ranges, the Srednogorie tectonic province represents a former marginal basin that began opening during the Cenomanian–Senonian transition, behind the north-dipping Peonias–Intra-Pontide suture zone (Antonijevič *et al.* 1974, Aiello *et al.* 1977, Hsü *et al.* 1977). This marginal basin was originally the western continuation of the Black Sea marginal basin (Görür, in press), but closed along the suture *16* during the Eocene, with a northerly vergence (Hsü *et al.* 1977).

Along the southern slope of the Greater Caucasus the Slate-Diabase zone ocean (*24*) also closed in the Eocene by south-vergent thrusting (Khain 1975).

Two other suture belts of Oligocene age, in addition to the Apennines and the Pindos–Budva zone, are the Sistan suture in eastern Iran (*26*; Tirrul *et al.* 1983) and the Indo-Burman suture (*32*) along the north-eastern border of the Indian Shield (Brunnschweiler 1966). The Sistan suture ('East Iranian Flysch Trough' of Stöcklin 1974, and Berberian and King 1981) represents an ocean that probably opened during the early Mesozoic as a continuation of the Waser ocean in Afghanistan, and closed by east-dipping subduction between the Campanian and the Oligocene (Tirrul *et al.* 1983).

We close this survey of the Tethyside sutures, with the Neogene sutures of the Alpides which are found at two extremities of the Tethyside superorogenic system around the loop of Gibraltar (*2* and *4*), in the island of Timor in the extreme south-east of the Tethysides (*35*), and around the Arabian Peninsula (*21*,*25*).

Ricou (1973), Şengör and Kidd (1979), Şengör and Yılmaz (1981) and Berberian *et al.* (1982) documented the late-middle Miocene to Pliocene ocean closure along the Bitlis–Zagros suture zones (*21*,*25*), that finally welded Arabia to Eurasia.

Cimmeride and Alpide sutures: a comparison

As the earlier discussion shows, convergence along many of the Alpide sutures is still in progress, and thus not only the Alpide orogen is being enlarged continuously in such regions, but also its fore- and hinterland areas are being disrupted extensively by germanotype structures

(Fig. 2). The Alpides thus represent a still-active system. In fact, in a number of places, Neogene sutures pass along the strike into still-active subduction zones (e.g. *21* westwards into the Cyprus subduction zone, *25* and *28* into the Makran subduction zone, Indo-Burman sutures and the Timor suture into the Andaman–Sumatra–Java subduction zone).

Mini-oceans opened as a consequence of the destruction of the Neo-Tethys, mostly as back-arc basins and/or pull-apart structures along large strike-slip belts, such as the western Mediterranean basins (Cohen 1980), the Black Sea (Görür in press), and the southern Caspian Depression (Berberian 1983) and these also still survive within the Alpide edifice.

None of the Cimmeride sutures today continue into active subduction zones. The mini-oceans generated by the Cimmeride tectonics no longer remain within the Cimmeride edifice, with the possible exception of the Tarim, Junggar, and Qaidam Basins (Hsü in press). This shows that, in contrast to the Alpides, the Cimmerides represent a fossil orogen, whose ruins served as a foundation to the later construction of the Alpides. Like the Alpides, the Cimmerides also generated a vast field of germanotype cratonic structures in Eurasia (Fig. 2) that developed between the late Permian and the early Cretaceous, with the greatest intensity and areal extent being in the late Triassic–Jurassic (Şengör 1984, 1985*a*). Many of these structures also terminated their activity by the middle Cretaceous, but some remained active, being inherited by the Alpides (Şengör 1984).

A few of the sutures within the Tethyside edifice represent oceans that *opened as a part of the Cimmeride evolution, but closed during the growth of the Alpides*. For example, the Pindos–Budva Trough opened as a Palaeo-Tethyan marginal basin, possibly as the westerly continuation of the Karakaya mini-ocean (Şengör 1985*a*, 1986*a*), but closed in the latest Eocene–Oligocene, as a result of suture progradation from the Almopias–İzmir–Ankara suture zone. Similarly, the Sistan ocean in eastern Iran opened as a part of the Waser–Banggong Co–Nu Jiang ocean and closed in the Oligocene by the westward expulsion of the Helmand block. In all these cases we classified the associated sutures as Alpide, because their *closure* was a part of the Alpide tectonics, independent of any Cimmeride influence. After all, significant stretches of Neo-Tethys itself—in eastern Turkey, Iran, Pamirs, south-east Asia—seem to have opened initially as Palaeo-Tethyan back-arc basins, and only during the Jurassic did their united descendants acquire an independent life of their own as the Neo-Tethys (Şengör 1979*a*, 1984). Such genetic relationships, more than spatial coincidence, induced Şengör (1984) to combine the Cimmerides and the Alpides under one common system, the Tethyside superorogenic complex. Many of the blocks contained within the Tethyside orogenic collage are bounded by both Cimmeride and Alpide suture segments. In order to understand the evolution of these sutures and the growth of the collage, it is necessary to know the original provenance and the route of migration of these Tethyside blocks. In the next section we briefly review the relevant information to establish these two parameters.

Blocks of the Tethyside collage

Fig. 3 shows all of the known Tethyside blocks ('block' is here used in the sense of a primary collage component, as defined by Şengör, 1987, and in press). Four major lines of evidence are currently available to establish the original provenance and the path of migration of the Tethyside blocks. In the order in which they are commonly employed, these are palaeobiogeography, palaeoclimatology, stratigraphy and structural evolution, and palaeomagnetism.

Palaeobiogeographical and palaeoclimatological data from the Tethyside blocks

In Figs 14 through 17 we have plotted the distribution of various palaeobiogeographical and related palaeoclimatological indicators in and around the Tethysides, to see to what palaeobiogeographical and palaeoclimatic provinces the various Tethyside blocks might have belonged at various time intervals. The data we report are not exhaustive, but were chosen mainly to identify Laurasia versus Gondwana Land affinities. For more comprehensive summaries of palaeobiogeographical and palaeoclimatological data the reader is refered to Ziegler *et al.* (1979), Parrish *et al.* (1982), Parrish and Curtis (1982), Buffetaut *et al.* (1984), Nakazawa and Dickins (1985), Rowley *et al.* (1985), and the papers in this volume.

Late Carboniferous through early Permian

Fig. 14 displays the distribution of late Carboniferous to early Permian floras, cold-water faunas, and evidence for glaciation. Evidence for glaciation both from south of the Indus–Yarlung Zangbo suture in India (*31*) and west of

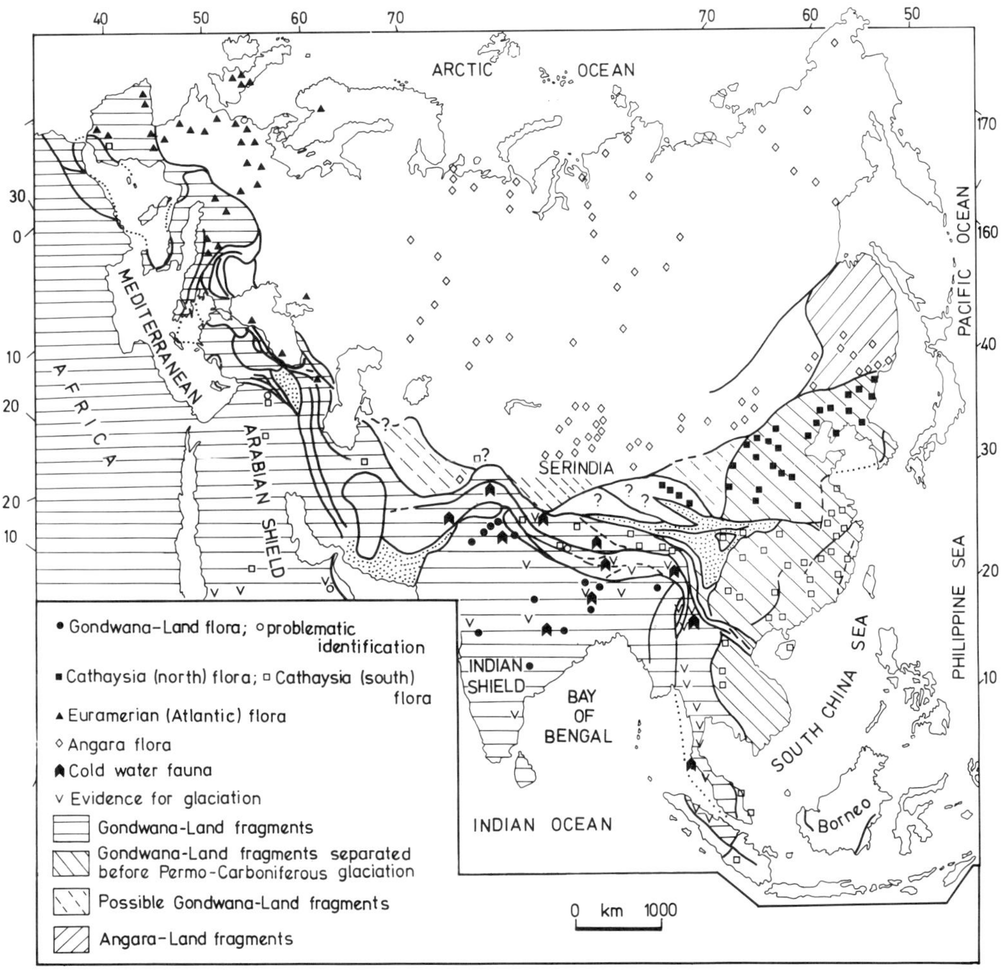

FIG. 14. Distribution of late Carboniferous–early Permian floras, cold-water faunas, and traces of glaciation in and around the Tethysides. Modified from Şengör (1987*a*), mainly using Kerey (1984), Belov *et al.* (1986), Cao (1986), and W. Chaloner (written comm. 1986).

the Zagros–Oman (*25*,*27*) suture on the Arabian Peninsula, and from the north of the Neo-Tethyan sutures (Fig. 14) belong to the 'direct evidence' category, as classified by Crowell (1983). Although the glacial nature of the Damxung–Linzhu 'tillites' located in the eastern part of the Lhasa block (in the northern part of t''') was questioned by Allègre *et al.* (1984), there is at least one locality reported earlier from farther east in the same block (near Baxoi, in the extreme east part of t''), that has unquestionable Middle and Upper Carboniferous tilloids (Chen 1981). Later, Lin (1983) reported Upper Carboniferous to Lower Permian glacio-marine sediments from the Xainza County near the central part of the Lhasa block. Most recently, Chang *et al.* (1986) not only confirmed the glacio-marine tilloid nature of the Upper Carboniferous Damxung mixtite, but they also reported the presence of striated pebbles from at least two localities on the Lhasa block, plus dropstones. This, combined with earlier reports, now puts the Gondwanian affinity of the Lhasa block, and its close proximity to glaciated terrains, as late as the early Permian, beyond dispute. Similarly, the presence of both glacio-marine deposits in north-west Qangtang (s') (Norin 1946, 1976), and the discovery of rich *Eurydesma* faunas in three separate localities within the same block (Liu and Cui 1983), indicate that the *western part* of the Qangtang block (s') was also a part of Gondwanaland until at least the early Permian. Evidence from the eastern Qangtang block (s'') indicates that this block has a Cathaysian flora.

No evidence for glaciation has yet been reported from this block, and Chang *et al.* (1986) conclude that it was probably no longer part of Gondwanaland by the Permian.

Recently, Altermann (1986) questioned the tilloid aspect of the Carboniferous to early Permian pebbly mudstones of Thailand, located on the Sibumasu block (s'''). His arguments, however, are based on unsupported estimates of the age span of these deposits, insufficient attention to relationships between pebble (and boulder) roundness and lithology, misunderstanding of the significance of the included diamonds, and neglect of the palaeontological evidence. A general review of these deposits by Stauffer and Mantajit (1981), and a detailed analysis of the outcrops in the Langkawi islands by Stauffer and Lee (in press) strongly support the glacio-marine origin of these deposits. Moreover, Cao (1986) recently reported the presence of striated pebbles and the cold-water forms *Stepanoviella* and *Eurydesma*, in glacio-marine deposits of late Carboniferous age from Baoshan, located in the northernmost extremity of the Sibumasu block (s''') in western Yunnan.

From the data shown in Fig. 14 the following conclusions emerge. Until mid-Permian times, almost the entire Asiatic part of the Cimmerian Continent (s', t', t'', t''', m) constituted a part of Gondwanaland. In three localities, Euramerian late Carboniferous floras have been found on the Cimmerian continent. Two of these are in northern Turkey, both from the Istanbul and Bayburt nappes, and one is from the Khrami Massif in Transcaucasia. Belov *et al.* (1986) argue that this may be because of the westward narrowing of the Palaeo-Tethyan triangle, enabling some floras to 'cross the suture'. The 'North China Foldbelt' block was already a part of Laurasia in the late Palaeozoic, as was the Turan block, whereas the other Chinese blocks (q, z, w, y at large), Annamia (x), and eastern Malaya, were away from both Gondwana Land and Laurasia, and were probably located somewhere near the equatorial latitudes. A Cathaysian flora occurs on the Arabian Peninsula (Gondwana Land!) and in Iran. The Iranian occurrence was probably similar to those on the Arabian Peninsula, i.e. located on Gondwana Land.

Because both fossils of land plants and marine macro-organisms are not as frequently encountered in the Tethysides as those of micro-organisms, we have also plotted the distribution of certain foraminifera within and around the Tethysides, for the late Permian, early Jurassic, and late Jurassic to Neocomian times (Figs 15, 16, and 17). Similar studies recently undertaken by others (e.g. Ross and Ross 1981; Ross 1982*a*, *b*; Ishii *et al.* 1985; Okimura *et al.* 1985, for the Permian; Cherchi and Schroeder 1973; Schroeder *et al.* 1978; Moullade *et al.* 1980, 1985; Bassoullet and Bergougnan 1981; Pélissié *et al.* 1982; and Bassoullet *et al.* 1985) have contributed to the palaeobiogeographical analysis of the Tethyan domain mainly in two aspects. Firstly, diagnostic forms for palaeobiogeographical provinces have been identified from among cosmopolitan taxa. Secondly, their distributions have been plotted on, and checked against, continental reconstructions.

Many ecological factors, such as temperature, salinity, depth, or pH of water, control the distribution of benthonic foraminifera. A change in one or more of these, might at times form ecological barriers. These barriers may be generated by the presence of oceans between different centres of populations, and by ecological changes within these oceans and their peripheral shelf seas, caused by plate motions. The benthonic foraminifera constitute excellent tools in palaeobiogeographical studies to check continental reconstructions.

The three maps we present here (Figs 15, 16, and 17) have been distilled from a large data base. Both the details of the individual fossil localities and the relevant references will appear in a separate palaeontological paper, including a discussion on different biozonation schemes of the three selected time intervals (Altıner, in prep.), to which we refer the interested reader.

Late Permian

From mid-Permian times onwards, the glacially influenced sediments and biotas disappear, and the biogeographically significant marine faunas gain in importance for purposes of continental reconstruction. In the evolution of the late Permian foraminifera, two main trends are observed towards the end of this epoch. The large fusulines (*Eopolydiexodina*, *Neoschwagerina*, *Yabeina*, *Lepidolina*, etc.) disappeared before the upper limit of the 'Midian' (Leven 1981; see Altıner 1984 for a discussion) and the number of taxa of the smaller foraminifera (*Shanita*, *Paraglobivalvulina*, *Paradagmarita*, *Colaniella*, etc.) notably increased.

In Murghabian time, *Neoschwagerina* spread over the entire Tethyan domain, with the exception of central, south-central, and eastern Turkey, large parts of the Zagros shelf, Oman, and also possibly western Yunnan in China, Burma, and western Thailand. Although not shown in Fig. 15, the genus *Verbeekina* seems to follow very much the distribution pattern of *Neoschwagerina*, albeit not as frequently encountered. Both of these genera are therefore very largely confined to Cimmerian continent, the Chinese blocks south and east of the sutures *XIX*, *XX*, *XXXIX*, *XV* and *XVI*, and to a few spots on the northern Gondwanian shelf as the southern and central parts of the African promontory (*e*), the Abadeh area in the Zagros, and the Tibetan Himalaya. Neoschwagerinids thus seem to have come in contact with Laurasia during the late Carboniferous collision of the North China block with this continent. Otherwise, they seem confined entirely to Tethyside blocks derived from Gondwana Land, and to Gondwana Land itself.

Eopolydiexodina, usually accompanied by *Rugososchwagerina* (latter not shown in Fig. 15) is known from Yugoslavia, Turkey, Transcaucasia, Afghanistan, the Darvas region in Turan, Burma, and western Thailand. This form has not been encountered in China, Indo-China east of the Nan-Uttaradit/Bentong Raub suture (*XXVII*, *XXXI*) and Japan, making it highly significant for palaeobiogeographical analysis. *Eopolydiexodina* thus seems confined only to the Cimmerian Continent, and to the north and north-east shelf of the Arabian Peninsula (Gondwana Land), with the exception of Darvas in Turan, whose tectonic position is still uncertain.

Among the smaller foraminifera of the late Permian, *Dagmarita* (not shown in Fig. 15) is found to be a cosmopolitan form and *Abadahella* spreads over the regions also occupied by the genus *Neoschwagerina*.

During the 'Midian', *Lepidolina* spread over Japan, southern China, and Indo-China, and thus formed an important province in eastern Asia, including only those blocks of Gondwana Land origin that had left the supercontinent before the Permian. The only exception is the

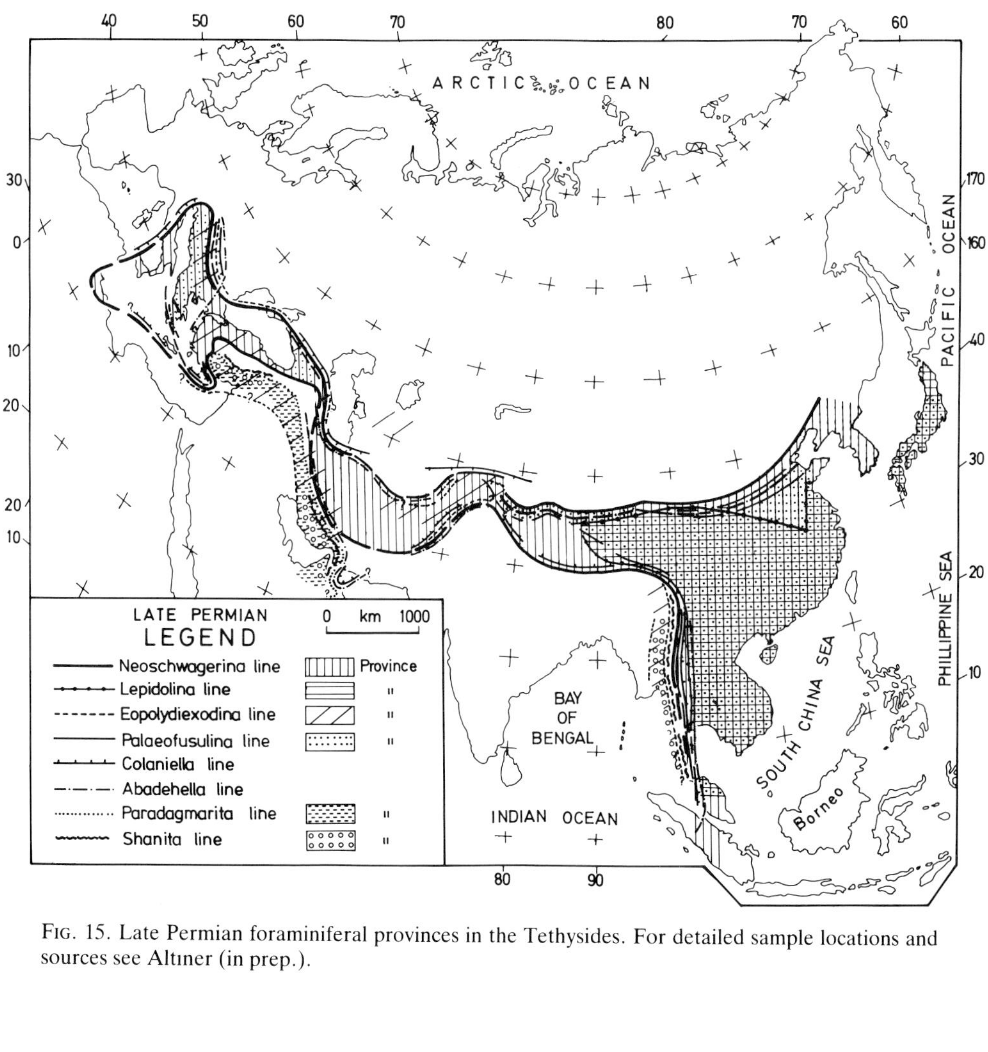

FIG. 15. Late Permian foraminiferal provinces in the Tethysides. For detailed sample locations and sources see Altıner (in prep.).

Lhasa block where *Lepidolina* also occurs. *Shanita* is also a good palaeobiogeographical marker, being found only in southern and south-east Turkey, in the Zagros shelf, in Oman, and in western Yunnan, Burma, and Thailand. It seems also confined to the Zagros margin of Gondwanaland and to the future Neo-Tethyan margin of the Cimmerian Continent.

The latest two stages of the late Permian epoch, the Djulfian and Dorashamian, are palaeobiogeographically characterized by three genera, namely: *Palaeofusulina*, *Colaniella*, and *Paradagmarita*, which are also excellent index fossils. The aberrant *Palaeofusulina* is a typical eastern Asian form, being abundant in Japan, China, and Indo-China, but it is also found in the western Tethyan domain (Caucasia, Greece, and Yugoslavia). This genus is absent in Afghanistan and probably also in Iran. It probably had a very narrow path of migration along the northern margin of the Cimmerian continent. Interestingly, the places where it is absent are marked by south-dipping subduction north of the Cimmerian continent, where quiet platform sedimentation was disturbed, as in northern Turkey, Iran, and the Pamirs.

The genus *Colaniella* has a large distribution and commonly follows the distribution pattern of *Neoschwagerina*. *Paradagmarita*, on the other hand, has a very restricted distribution and is found only in southern and south-east Turkey, and along the Zagros shelf and in Oman.

In summary, the genus *Neoschwagerina* spreads over vast areas of the Tethyan domain but is generally limited to the Cimmerian continent, except in China and Indo-China. In the central Mediterranean region, where the Cimmerian continent was hinged to Pangaea, and in the Tethyan Himalaya, it spills over on to Gondwana Land. Genera such as *Yabeina*, *Abadahella*, and *Colaniella* follow the same distribution. Two well-defined provinces are formed by the genera *Paradagmarita* and *Lepidolina*, in the Middle East and Far East, respectively. The distribution of *Shanita* corresponds with that of *Paradagmarita* in the Middle East, whereas in Burma it occurs unaccompanied by this form where *Neoschwagerina* and *Lepidolina* are also absent. *Eopolydiexodina* and *Lepidolina* never spatially coincide. In the Far East, *Eopolydiexodina* occurs together with *Shanita*. Finally, the genus *Palaeofusulina* seems to have spread over into the western Tethyan regions from its main region of distribution in the Far East, via a narrow path possibly along the northern margin of the Cimmerian continent.

Early Jurassic

In the early Jurassic, the most spectacular development in the complex agglutinated foraminifera is seen in the populations of the genus *Orbitopsella (O. primaeva, O. praecursor, O. dubari)*, and the other phylogenetically related form, *Cyclorbitopsella*. In the Pliensbachian–Domerian, the genus *Orbitopsella* spread over the Arabian Platform, Morocco (High and Middle Atlas), Tunisia, northern Egypt, Israel, Syria, south-east Turkey, the Zagros shelf, and Oman (Fig. 16). In Europe, its distribution follows the Betic chains of southern Spain, Balearic islands, Italy, Dinarides in Yugoslavia, Greece west of the Vardar zone (*12*) and southern Turkey. In Central and East Asia, the data on the distribution of *Orbitopsella* and the related genus *Cyclorbitopsella* are fragmentary. They have been reported from the Lhasa block (*t*), from the Markam region in the extreme eastern end of the eastern part of the Qangtang block (*s″*), and from western Thailand. This pattern of distribution shows that, although in the Mediterranean and Middle Eastern regions *Orbitopsella* was confined to regions south of the northern branch of Neo-Tethys, in Asia it was found both on independent blocks within the Tethyan domain (such as the Lhasa block) and along the Laurasian southern margin in Markam and western Thailand. This Asiatic anomaly is now difficult to explain.

The palaeobiogeographical distributions of many taxa follow the limits of the *Orbitopsella* province. These are *Mayncina termieri*, *Pseudocyclammina liassica*, *Lituosepta* (*L. recorensis*, *L. compressa*), *Pseudopfenderina butterlini*, and *Haurania*. The last genus however has been recorded in western France and in north-west Tibet, indicating an anomalous excursion in these regions, far into Laurasia outside the *Orbitopsella* province (Fig. 16).

Late Jurassic to Neocomian

Many complex and agglutinated late Jurassic larger foraminifera, such as *Kurnubia palastiniensis*, *Parurgonia caelinensis*, *Kilianina lata-rahonensis*, *Labyrinthina mirabilis*, *Alveosepta jaccardi*, and *Anchispirocyclina lusitanica*, show a cosmopolitan nature. These taxa spread over a large band of terrain, limited in the north along a line from northern France to Kopet Dagh, and in the south from Morocco to Senegal. The occurrence of late Jurassic foraminifera in Central and East Asia has been reported from only a few localities, and the

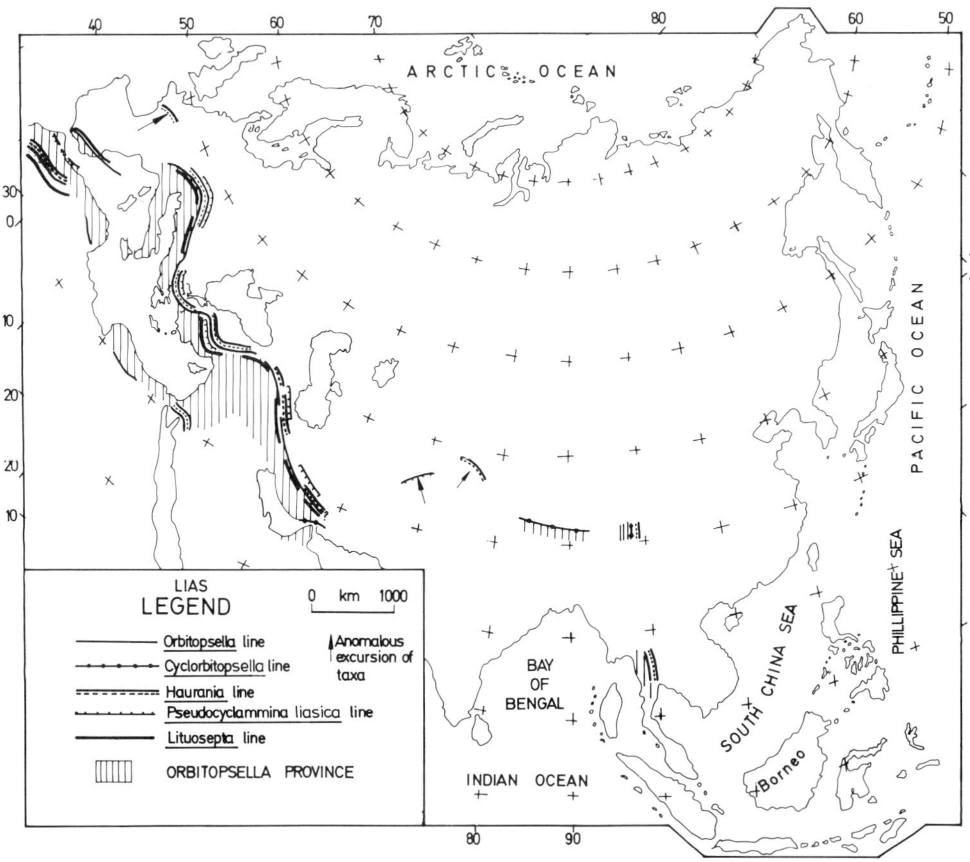

FIG. 16. Early Jurassic (Lias) foraminiferal provinces in the Tethysides. For detailed sample locations and sources see Altıner (in prep.).

easternmost occurrence is known from the Philippines (Fig. 17).

Within the broader realm of the late Jurassic larger foraminifera it is possible, however, to define a narrower and more useful palaeobiogeographical zone. Algeria, Tunisia, Italy, Yugoslavia, and Greece south-west of the Vardar zone (*12*), southern and south-eastern Turkey, Iraq, Syria, Israel, Egypt, Saudi Arabia, Qatar, Oman, and the Zagros Shelf are regions where the genus *Kurnubia* exhibits a tremendous plexus in its evolution. The various steps in the evolution from *K. palastiniensis* to *K. wellingsi*, with the other related form '*Conicokurnubia*', are not observed in the *Kurnubia* populations found in the regions to the north of this narrower biospace. This *Kurnubia* plexus province (Fig. 17) coincides with regions lying to the south of the northern branch of Neo-Tethys and the Zagros/Oman suture zone (*6*, *7*, *8*, *11*, *12*, *13*, *14*, *17*, *25*) and seems to be delimited rather sharply by this major oceanic space.

In the Neocomian, several species of agglutinated foraminifera described from Western Europe constitute important but small palaeobiogeographical provinces, and their tectonic usefulness is thus limited (Fig. 17). One exception to this is *Orbitolinopsis capuensis*, distributed in northern Algeria, Tunisia, Italy (mainly in the Apennines), Yugoslavia, Greece west of the Vardar zone (*12*) and the Menderes–Taurus block (Fig. 17). This form thus appears to have been confined to the African Promontory and to its extension in Turkey—the Menderes–Taurus block. In the Oran Meseta (*b*) the *Orbitolinopsis capuensis* and the *Choffatella pyrenaica* provinces overlap (Fig. 17).

Many other Neocomian larger foraminifera, such as *Protopeneroplis trochoangulata*, *Pseudotextulariella salevensis*, and the long-ranging forms recorded also from the Upper Jurassic or above the Neocomian (e.g. *Conicospirillina*, *Trocholina alpina*, *Trocholina elongata*, *Nautiloculina cretacea*) are cosmopoli-

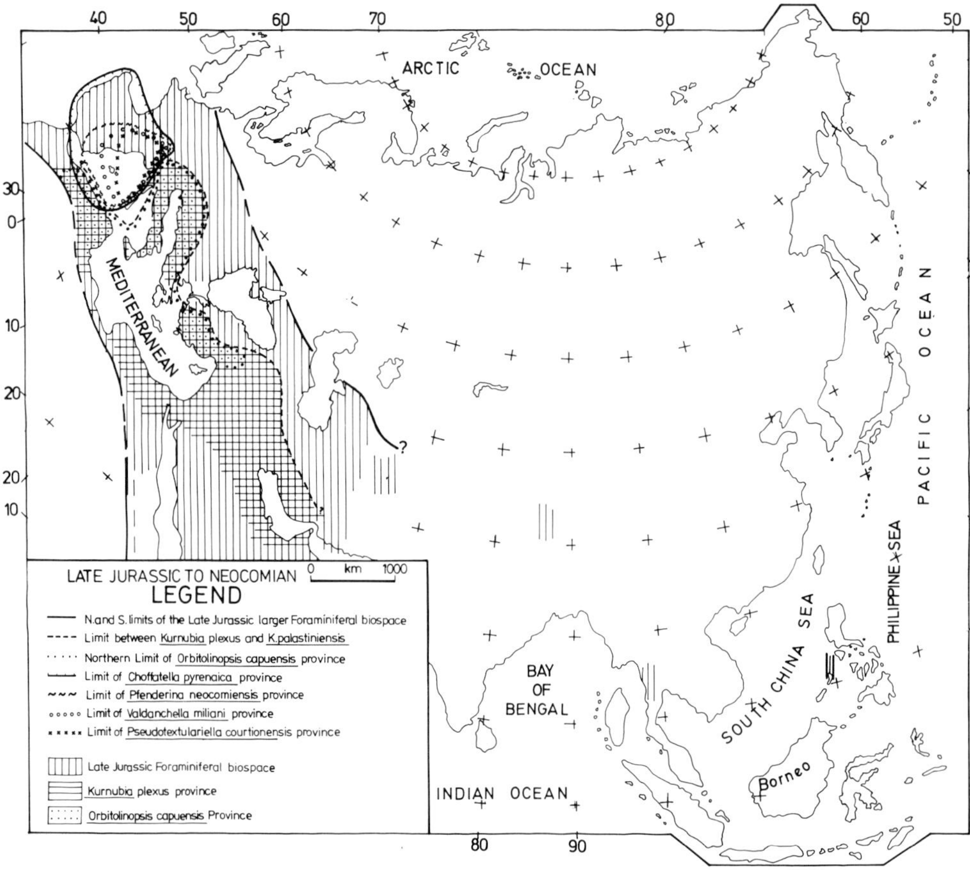

FIG. 17. Late Jurassic to Neocomian foraminiferal provinces in the Tethysides. For detailed sample locations and sources see Altıner (in prep.).

tan forms, and their palaeobiogeographical distributions coincide more or less with those of the late Jurassic forms.

The palaeobiogeography of certain foraminiferal genera presented here, shows the following features from a tectonic viewpoint. In the Permian, *Neoschwagerina* and the associated genus *Verbeekina* define the components of the Tethyside collage, except for blocks *a*, *b*, *c*, *d*, *n*, *o*, *p*, and *r*, and spill over on to Gondwanaland only in a few places along its northern shelf south of the Neo-Tethyan sutures. These forms did not everywhere cross the future Neo-Tethyan sutures in Asia east of Turkey. In these regions the southern boundary of the *Neoschwagerina* and *Verbeekina* province is mostly coincident with, or close to, the future Neo-Tethyan rift axis. This province crosses the future northern Turkish Neo-Tethyan sutures, and those in the Apennines, indiscriminately, corroborating their post-late Permian age.

The sharp S-shaped *Neoschwagerina* province boundary in western Turkey and Cyprus may be a result of lack of detailed information from the central Turkish regions where the facies are not exposed, but it may also be indicative of a left-lateral motion in post-Permian times.

Both the early Jurassic *Orbitopsella* and the late Jurassic *Kurnubia* plexus provinces are delimited to the north by the northern branch of Neo-Tethys and the Zagros suture, except in Asia where *Orbitopsella* has been reported both from an intra-Tethyan block (Lhasa) and from the southern margin of Laurasia (Markam, Thailand). Neither of these two provinces seem to be affected by the presence of the southern branch of Neo-Tethys (now represented by the sutures *19*, *20*, the buried Assyrian suture under *21*, and the eastern Mediterranean itself), which formed a boundary only for the *Orbitolinopsis capuensis* province.

Palaeobiogeographical and palaeoclimatological data of late Carboniferous to early Permian age thus largely show that the Cimmerian Continent almost entirely constituted the northern and north-east margin of Gondwana Land at this time, with the exception of the small eastern Qangtang fragment. The North China (*q*), Yangtze (*w*), Huanan (*y*), and Annamia (*x*) major blocks, and the eastern Qangtang (*s''*) fragment plus the Shaluli Shan (*u*) and the Chola Shan (*v*) fragments, were somewhere in the tropics. They all share a Cathaysian flora, but the North China block is distinct in having a northern Cathaysian flora, as opposed to a southern Cathaysian flora common to all the others. Both the Turan block and the North China Foldbelt (*r*) were by this time parts of Laurasia. Although both northern Turkey (*f*) and Transcaucasia (north-western parts of *i* and *j*) have Euramerian floras, we consider them to be parts of Gondwana Land, because all other data become unintelligible otherwise. These localities were not very far away from other Euramerian localities such as Yugoslavia.

In the later Permian, the foraminifera may suggest the beginning of separation of the Cimmerian continent from northern Gondwana Land in Asia east of Turkey, and may underline the unity of the Cimmerian continent, again with the exception of eastern Qangtang (*s''*).

In the Mesozoic, the Neo-Tethyan northern branch in the Mediterranean region, and the Zagros ocean, emerge as important barriers to foraminiferal dispersion, the southern branch of Neo-Tethys in the present eastern Mediterranean being important only for the *Orbitolinopsis capuensis* province. Both in the early Jurassic and in the late Jurassic–Neocomian interval, the Saharan Atlas was a southern boundary for foraminiferal distribution in North Africa.

At no time have the Pyrenean (*1*), Pindos–Budva (*15*), and the Sevan–Akera–Qaradagh (*23*) sutures been important barriers in agreement with the small size of the oceans involved (e.g. Pindos–Budva), or their short life-span (e.g. Pyrenees).

The distribution of Mesozoic land quadrupeds within the Tethysides has been a difficult topic to study, because of the dominantly marine nature of most of the available record. Recently, however, Buffetaut (1981) reported a number of Mesozoic quadrupeds from Thailand indicating a clear 'Laurasian' affinity since the latest Triassic.

Palaeomagnetic data from the Tethyside blocks

Independent support for most of the conclusions drawn in the last section is supplied by palaeomagnetic studies which we do not review here. For a good overview of the available sources of information the reader may consult Şengör (1987*a*). Newer data from the Yangtze block have been recently reported by Huang *et al.* (1986) and Opdyke *et al.* (1986).

Basement geology of the Tethyside blocks

Once palaeobiogeography, palaeoclimatology, and palaeomagnetic studies inform us in what latitudinal range, in what orientation, and possibly on what side of a major continent our object of interest must have been located, it is once more regional geology that will enable us to make a more detailed reconstruction by matching geological markers across suture zones.

Within the Tethyside orogenic collage, one of the best markers is the now-disrupted Pan-African orogenic system, making up the basement of many of the Tethyside blocks. Şengör (1986*a*) pointed out that the Pan-African 'event', stretching between about 900 Ma and 450 Ma (Kröner 1984), was responsible for the assembly of Gondwana Land—age of orogeny becoming progressively younger towards the margins of the supercontinent, as seen especially well in the Arabian Peninsula, India, and Australia. He further argued, that in the Asiatic Tethysides, the Tethyside rifting and orogenic events were commonly the first significant tectonic events to follow the terminal Pan-African events. Şengör *et al.* (1984*b*) showed that in Turkey, for example, events previously called 'Caledonian' actually constituted a continuation and the termination of the Pan-African orogenic phenomena. Although, unfortunately, still called 'Caledonian', the same is true for early Palaeozoic events in the Alps, (*e*: Frisch *et al.* 1984) in Iran (*i*, *j*, and *k*: Davoudzadeh *et al.* 1986), in Afghanistan (*m*: Montenat *et al.* 1981), in the Lhasa block (*t*: Chang *et al.* in press), in Annamia (*x*: Workman 1977) and in the Huanan block (*y*: Shui *et al.* 1986). Fig. 18 shows the distribution of possible Pan-African-related structures and their broad age groups. If we could reconstruct the original Pan-African geology of these fragments we might be able to reconstruct the Pan-African orogenic system, and thus at least the northern and north-east part of Gondwana Land, which would be equivalent to reconstructing the pre-dispersal geometry of the majority of the Tethyside collage components. Unfortunately, the existing data base is not nearly sufficient to do this, most of the available information from the Asiatic Tethyside blocks being merely in the form of ages of deformation and/or ages of unconformable successions lying on the Pan-African terrains. Less frequently, grade of metamorphism is crudely indicated, but the kind of metamorphism, structural geometry, type of associated magmatism, or the protoliths of the metamorphic rocks, are only rarely reported. With the present information, the best that one can do is to point out an overall trend of younging towards the periphery of Gondwana Land, from about 900 Ma to Ordovician (*c.*450 Ma), and to expect that when the Tethyside blocks are returned to their original positions in Gondwana Land this geometry would be reproduced.

As Fig. 18 shows, within the various blocks making up the Cimmerian continent, this geometry is preserved even today, with the exception of Iran (*i*, *j*, and *k*), south-east (*e*) and north-west Turkey (*f*). In southern China and Indochina this logic would demand that the Yangtze block (*w*) and the coastal block (*y″*) be placed farther towards the interior of Gondwana Land than the Huanan block (*y* and *y′*) and Annamia (*x*). This speculation is supported by independent Cambrian palaeomagnetic data from the Yangtze block, and stratigraphic data from both the Yangtze and the Huanan blocks (Lin *et al.* 1985*b*, fig. 6). The exceptional situation in Iran arises because the younger (early Palaeozoic) Pan-African terrain of the Sanandaj–Sirjan zone (*i*) is now located between older (latest Precambrian) Pan-African terrains of Arabia and Central Iran (*j*). This anomaly is automatically resolved to produce the expected

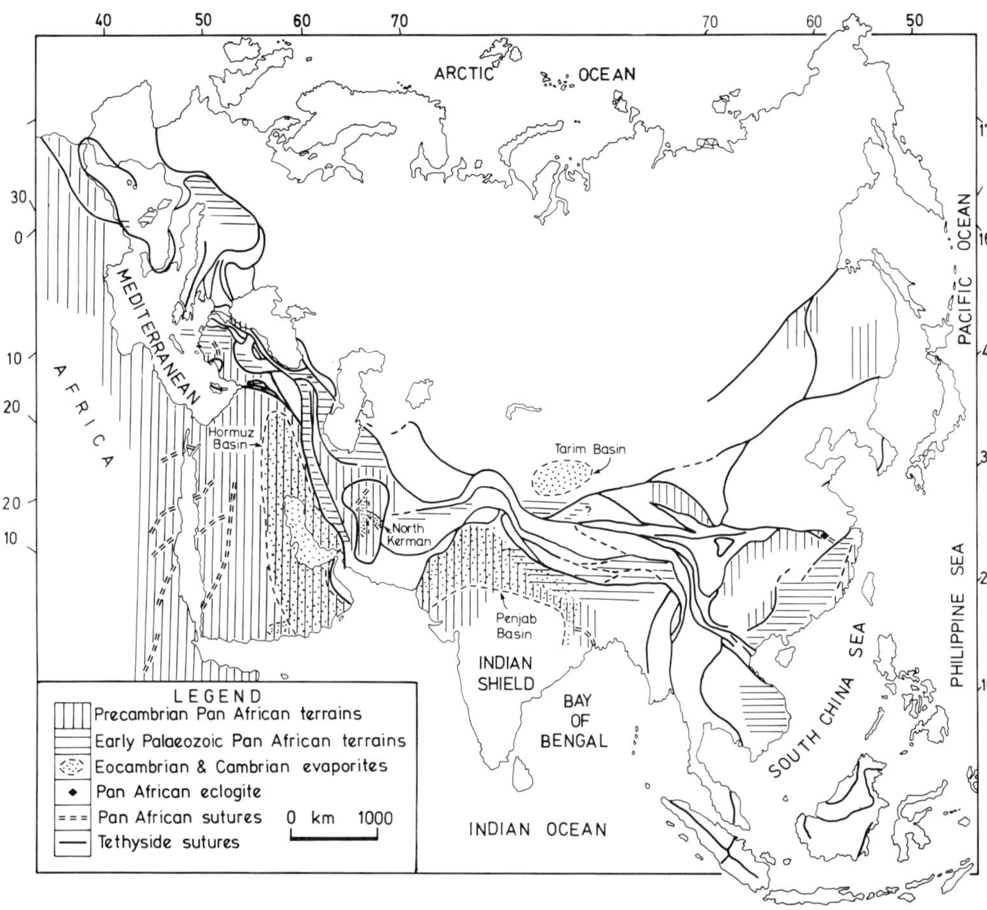

FIG. 18. Areas of known and presumed Pan-African deformation, metamorphism, and magmatism, within and around the Tethysides. Data were taken from: Workman (1977), Burke *et al.* (1978), Berberian and King (1981), Montenat *et al.* (1981), Frisch *et al.* (1984), Şengör *et al.* (1984*a*), Xu *et al.* (1984*a*), Zharkov (1984), Şengör (1986*a* and in press), Shui *et al.* (1986), Yang *et al.* (1986), Chang *et al.* (in press). Also shown are late Precambrian–Eocambrian evaporite basins after Zharkov (1984).

'younging outwards' geometry when the disrupted Palaeo-Tethyan suture and arc fragments in Iran are reconstructed to their original geometry (see next paragraph). The same also applies to north-west Turkey.

Fig. 18 also shows the latest Precambrian–early Cambrian evaporite basins within and near the Tethysides (Zharkov 1984). These basins formed synchronously with, or shortly after, the terminal Pan-African collisions in the present central Tethyside region. The Hormuz evaporite basin is delimited to the north-east by the Zagros suture, but correlative evaporites reappear on the 'Iranian' side of the suture across the Sanandaj–Sirjan zone, near Nayband and north of Kerman (Fig. 18). When the Central Iranian Microcontinent is rotated back to its pre-late Triassic orientation, as mentioned previously (Davoudzadeh *et al.* 1981), the north Kerman Cambrian evaporites become juxtaposed against the Sanandaj–Sirjan zone, which thus becomes the only obstacle separating them from the main Hormuz basin—a further point underlining the previously discussed anomalous position of the Sanandaj–Sirjan zone, and supporting Central Iran's Gondwanian origin.

In Fig. 18 we also show areas that were deformed during the late Precambrian in the North China Foldbelt region (*r*) and in the Qinghai Nanshan (*o*), but no other evidence exists to show that these events, although coeval, may also have been parts of the Pan-African evolution.

Although none of the palaeobiogeographical or geological data we discuss are relevant to the original provenance of the North China block, Lin *et al.* (1985*b*) present both Cambrian stratigraphic and palaeomagnetic data to show that it too belonged to Gondwana Land, as originally speculated by Şengör (1984) on the basis of a broad similarity of early Palaeozoic stratigraphy.

Fig. 14 summarizes the results of this section. The evolutionary scheme outlined in the next section is based on the data reviewed in this and the previous sections. As a detailed analysis of the Phanerozoic facies distribution in the Tethysides and a similarly detailed structural compilation are not yet available, the following evolutionary scheme, summarized in Figs 19 and 20, is only a very generalized and oversimplified picture. Numerous complexities implicit in the evolution of the Tethyside collage could not be dealt with, both because of the chosen scale and lack of detailed information. Only as an illustration of what these complications are, and how they might affect our reconstructions, we will briefly review the late Palaeozoic to late Cretaceous palaeotectonic evolution of Iran and surrounding regions.

Evolution of the Tethyside collage

Figures 19a to 19f display the essential steps in the assembly of the Tethyside orogenic collage. The Tethyan domain is defined with reference to the Permo-Triassic Pangaea, and therefore no 'Tethys' and consequently no 'Tethyan evolution' could have existed before the Pangaea largely formed during the middle Carboniferous. Therefore the history of dispersal of many of the primary components of the Tethyside collage lies outside the Tethyside history *sensu stricto*, because their separation from Gondwana Land predated the middle Carboniferous. If we remove this temporal limitation for a moment, we may view the entire evolution of the Tethyside collage as a diminution in the size of Gondwana Land by continuous calving of continental blocks and their northerly migration to unite with Laurasia.

Fig. 19a shows that, during the late Permian, the terminal Uralide collision was about to occur, and thus the Permo-Triassic Pangaea was in the process of being born. As yet, the entire Cimmerian continent constituted a part of the northern margin of Gondwana Land, with the exception of the eastern part of the Qangtang block (*s″*). Here we place that fragment next to the Yangtze block (*w*), because not only their late Permian biogeographies are similar, but also both contain rift volcanism of basaltic (alkali to transitional types) and silicic types (see Chang *et al.* 1986 for Qangtang, and Fan 1978 for the Yangtze block), known as the Emei Shan basalts in the Yangtze block (Fig. 19a). Fan (1978) interpreted the Emei Shan basalts as products of a marginal-basin opening event. We follow this interpretation and correlate the Emei Shan basalts with the Permian basalts and felsic volcanics of the eastern Qangtang, and interpret them as products of the late Palaeozoic subduction along the future sutures *XXI* and *XXIII*.

This subduction zone may have joined southwards with the one active in the future Iran and Turkey or with the one in the Central Pamirs. The particulars of this possible geometry are shown, for a slightly later time, in the more detailed reconstruction in Fig. 20a, showing the early Triassic palaeogeography. In this reconstruction, the Sanandaj Sirjan zone is interpreted as a north-east-facing magmatic arc that had been active between the late Devonian and the late Permian as a continental-margin

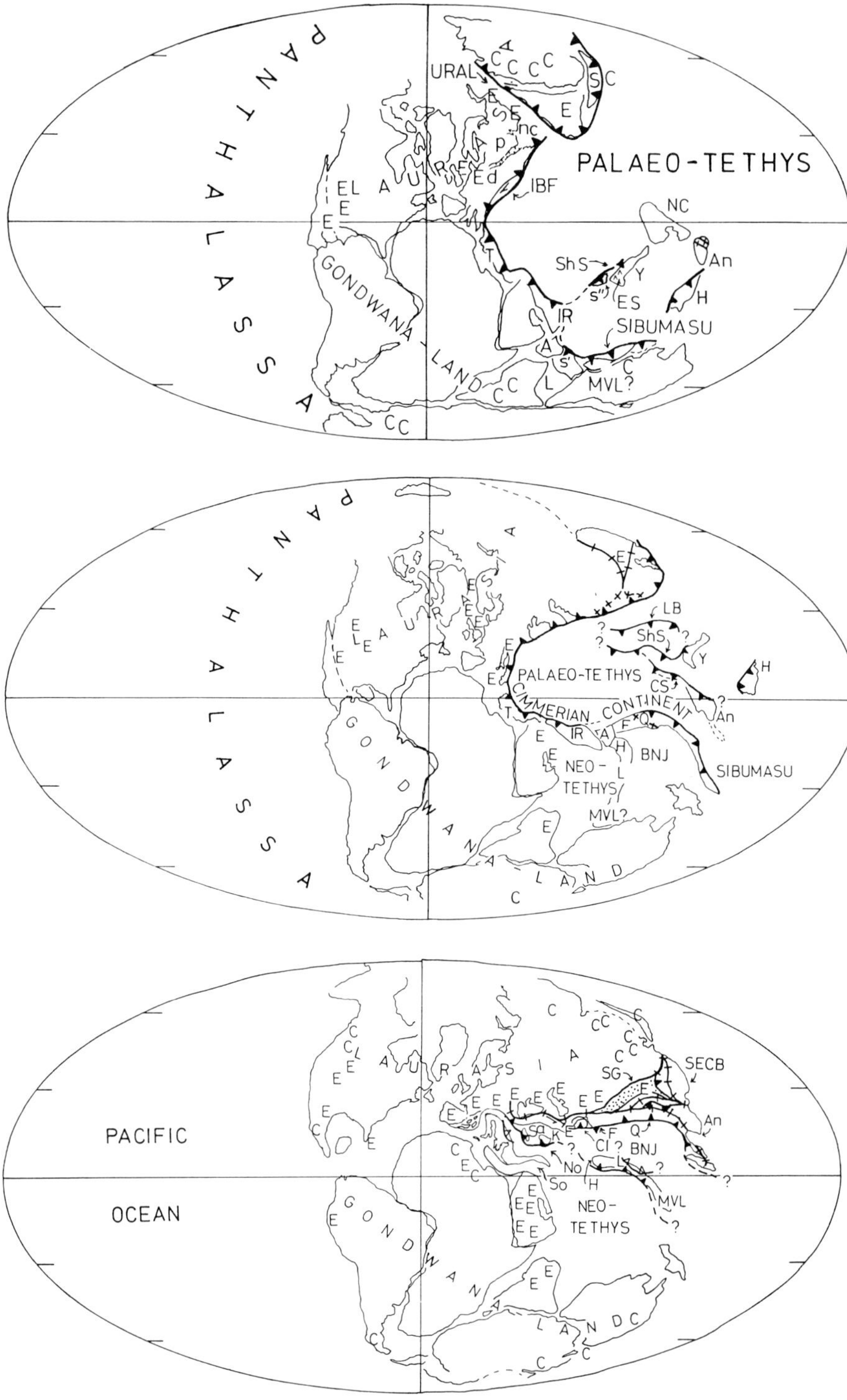

FIG. 19. Highly schematized reconstructions, displaying our present interpretation of the evolution of the Tethyan domain. Abbreviations for all reconstructions in this figure: *A*—Afghan blocks, *An*—Annamia, *B*—Bitlis–Pötürge fragment, *BNJ*—Waser/Rushan–Pshart/Banggong Co–Nu Jiang/Mandalay ocean, *CI*—Central Iranian microcontinent, *CS*—Chola Shan, *d*—Dnyepr–Donetz aulacogen, *ES*—Emei Shan, *F*—Farah block, *H*—Helmand block (*sensu* Şengör 1984), *Hu*—Huanan block

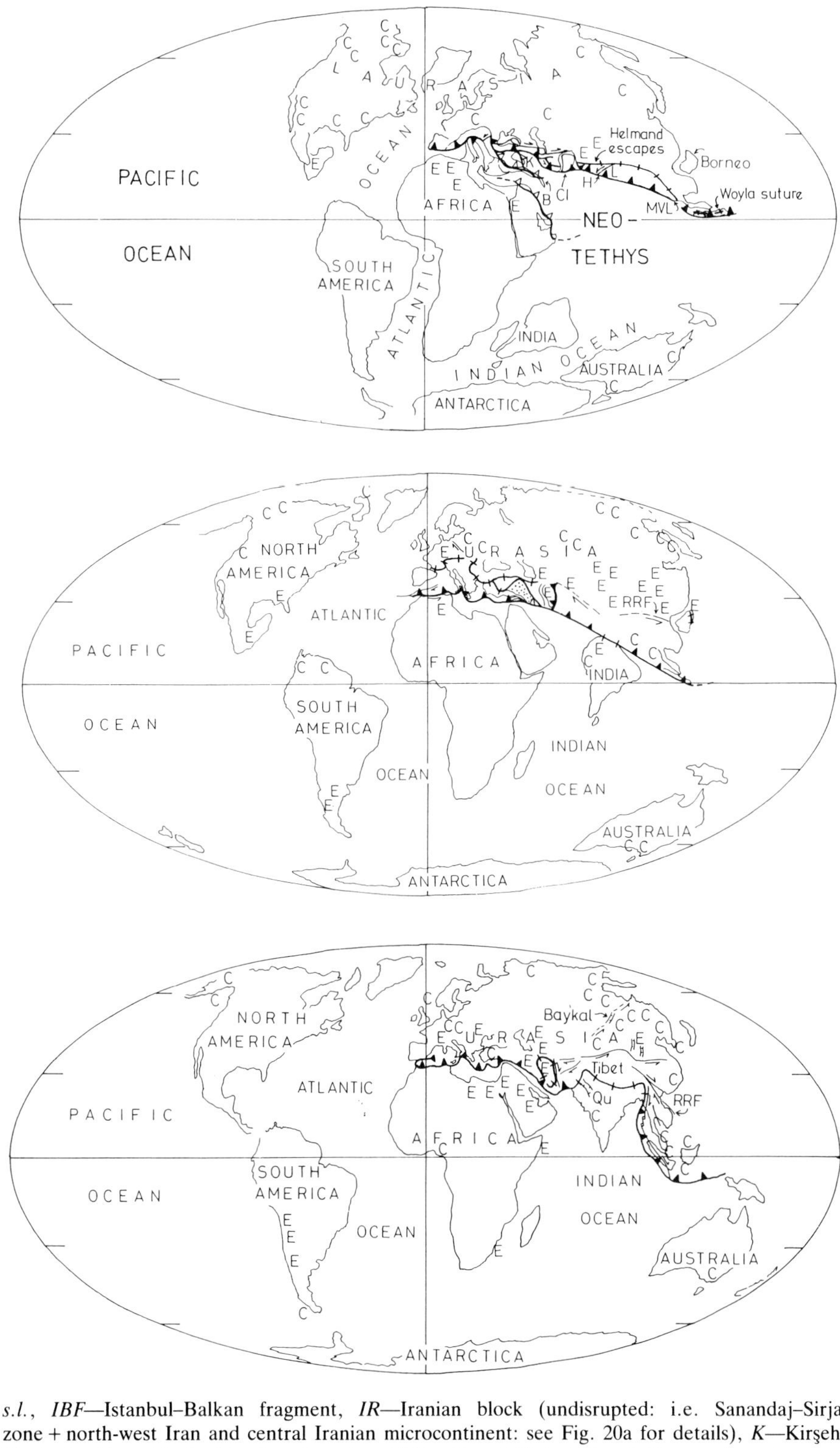

s.l., *IBF*—Istanbul–Balkan fragment, *IR*—Iranian block (undisrupted: i.e. Sanandaj–Sirjan zone + north-west Iran and central Iranian microcontinent: see Fig. 20a for details), *K*—Kirşehir block, *L*—Lhasa block, *LB*—Luochou Arc, *MVL*—Mount Victoria Land, *No*—northern branch of Neo-Tethys, *p*—Pachelma aulacogen, *Qu*—Quetta-Sibi graben, *RRF*—Red River fault, *S*—Tarim, *S′*—Pamir–west Qangtang block, *S″*—east Qangtang block, *Sa*—Sakarya Continent, *SG*—Songpan–

continued over

arc, perched on the northern margin of Gondwanaland. Its western continuation is believed to be represented by the pre-Jurassic basement of the Dzirula, Adzharia–Trialeti and the Artvin–Karabagh zones of the Trans-Caucasus and the Eastern Pontides, and the Sakarya Continent of northern Turkey. Within the Adzharia–Trialeti/Artvin–Karabagh zones, the pre-Jurassic basement crops out mainly in the Khrami and the Loku Massifs. This continental-margin arc is here believed to have been responsible also for the late Palaeozoic compressional deformation in Oman, reported by Michard (1982, 1983).

How far this subduction zone extended eastwards is difficult to decide, but the increased unrest in the Djulfa and Svanetia regions, the late Palaeozoic magmatism in the Gondbad-ı Qabus area of the Alborz Ranges (Stämpfli 1978), in the Anarek region of present-day Central Iran (Roman'ko and Morozov 1983), and in the central and south-east Pamirs (Shvolman 1977) suggests to us that it probably was active at least as far east as the Pamirs. Farther east in north-west Tibet data are sparse, and the question remains open.

Farther west, the same subduction zone probably delimited a southerly-moving Istanbul–Balkan fragment along the north-west margin of Palaeo-Tethys (IBF in Fig. 19a). This fragment now consists of the Istanbul Nappe in north-west Turkey and equivalent late Palaeozoic units in southern Bulgaria (see Şengör *et al.* 1984*a*) and contains a portion of a probably north-east-facing Hercynian foreland fold–thrust belt, plus a molasse basin with a late Carboniferous Euramerian flora (e.g. Kerey 1984). Şengör (1987*b*) postulated that this fragment of the Hercynian orogen had originally belonged to the southern margin of Eurasia and constituted the easternmost prolongation of the southern flank of the European Hercynides (Bard *et al.* 1980). It was later detached along a right-lateral strike-slip fault, and moved southwards into its present place in north-west Turkey. This southerly migration of the Istanbul–Balkan fragment must have occurred between the late Carboniferous and the Lias, although the actual time of transport may have been shorter.

In the extreme east of the Tethyan domain, subduction had already started during the Tournaisian, along a south-dipping subduction zone north of Sibumasu (Fig. 19a), and was also probably active during the Permian.

Throughout the Tethysides, normal faulting and rifting-related volcanicity began during the Permian (from mainland Greece, via Crete eastwards: Fig. 4) and may have already achieved considerable separation between Sibumasu and north-west Australia. South of Sibumasu, this extension generated a marine embayment between it and Australia, in which the glacio-marine pebbly mudstones were deposited.

In China, Huanan-ward dipping subduction and associated late Palaeozoic magmatism on the Huanan block were active.

Along the north, Asiatic, margin of Palaeo-Tethys, the blocks *r*, *q*, *p*, *o*, and *n* had already collided with the Altaid nucleus of Asia before the Permian and Asia-ward dipping subduction zones were active along the sutures *VII*, *VIII*, *XIX*, *XX*, *XXXIIII*, and also probably along *XXXIII′* and *XXXV*, although no late Palaeozoic igneous activity has been reported yet from the last two.

In the later early Triassic, the general geometry of the Tethyan domain may have looked something like that shown in Fig. 19b. Notice here that both the Banggong Co–Nu Jiang and the Neo-Tethyan oceans are in the process of active opening, although west of Afghanistan, this opening may still have been in the form of intracontinental stretching, as indicated by the conditions in Oman. In Fig. 19b we have assumed that the Istanbul–Balkan fragment had already reached its final destination in

---

Ganzi system, *ShS*—Shaluli Shan Arc, *Sibumasu*–China–Burma–Malaya–Sumatra portion of the Cimmerian continent (*sensu* Metcalfe, this volume), *So*—Southern branch of Neo-Tethys, *T*—Turkish blocks.

(a) Late Permian palaeotectonics of the Tethyan domain. Base map is from Scotese (1984). Distribution of coals (*C*) and evaporites (*E*) was taken from Ziegler *et al.* (1979). Map was modified according to data summarized in this paper.

(b) Early Triassic palaeotectonics of the Tethyan domain. Base map and the distribution of coals (*C*), and evaporites (*E*), for this and all subsequent reconstructions, were taken from Parrish *et al.* (1982). This and the subsequent maps were modified according to data summarized in this paper.

(c) Late Jurassic palaeotectonics of the Tethyan domain.

(d) Late Cretaceous palaeotectonics of the Tethyan domain.

(e) Middle Eocene palaeotectonics of the Tethyan domain.

(f) Late Miocene palaeotectonics of the Tethyan domain.

north-west Turkey, but this may be somewhat too early, according to some very preliminary palaeomagnetic data (Sarıbudak *et al.* 1986). The main Song Da subduction zone was active at this time, and may have been gradually propagating westwards to eventually join the active Jinsha Jiang/Nan-Uttaradit/Bentong–Raub subduction zone. North of the Litang subduction zone a new, intra-oceanic arc system, now represented by the Luochou 'arc–trench belt' of Zhang (*XXIII'*: verbal comm. 1985) may have come into being at this time.

Notice in Fig. 19b that we have left the Mount Victoria Land block of Mitchell (in press), still attached to north-west Australia. This fragment may have rifted from this position as late as the late Jurassic (see Şengör 1987; for an alternative and equally viable reconstruction see Hamilton 1983, fig. 1).

Returning to the details of the Iranian case, by the late Triassic, we see that the whole of Iran north-east of the Zagros suture had already been sutured to Laurasia. This suturing event extended between the Greater Caucasus and the Paropamisus, and may have had the geometry shown in Fig. 20b. During this collision, the Farah block probably became severed from (Central) Iran along a strike-slip fault and continued to approach the Paropamisus/Hindu Kush subduction zone.

In the late Jurassic, a considerable portion of the Cimmeride collage had already been assembled (Fig. 19c), and the collisions responsible for this assembly had already disrupted large areas of Eurasia north of the Tethysides. A major segment of the Cimmerian continent north of the Waser/Rushan–Pshart/Banggong Co–Nu Jiang/Mandalay ocean had collided with Laurasia all the way from northern Greece to the Malayan Peninsula. Although the assembly of the extra-Tibetan Cimmeride collage in China had long been complete, its construction mainly by the activity of the peripheral hidden subduction zones on both sides of the Songpan–Ganzi system continued. Palaeomagnetic evidence shows that the tightening of the suture zones in China and Indochina, during the Jurassic, resulted in a 1650 ± 750 km northward motion of Annamia and South China with respect to 'stable' Eurasia (Achache 1984). An inspection of Figs 5 and 6 clearly shows that much of this tightening was also accomplished by intracontinental shortening and crustal thickening. Notice how narrow most magmatic-arc systems still were during the Triassic and the early Jurassic interval, i.e. before, and just after, most of the Cimmeride collisions (Fig. 5). By contrast, magmatic activity is seen to have spread over much wider areas during middle Jurassic–early Cretaceous times, towards the end of a long period of post-collisional tightening of most of the Cimmeride sutures in China and in the Caucasus (Fig. 6).

Jurassic post-collisional intracontinental convergence is also attested to by the evolution of Jurassic compressional structures and persistent topographic highs, along a belt stretching from eastern Iran to China and following the northern margin of the present-day Alpine–Himalayan system. In fact, Hallam (1984) recently argued that the 'anomalous' late Jurassic arid belt of western and central Asia may have developed in the rain shadow of the Cimmeride mountain wall that rimmed the southern periphery of Eurasia at the time, when monsoonal winds possibly had a more dominant role than today (Şengör 1985*b*).

In the late Jurassic the Waser/Rushan–Pshart/Banggong Co–Nu Jiang/Mandalay ocean had already begun to close, both by the obduction of a vast ophiolite sheet on to the Lhasa block (*c.* 170 Ma ago), and by subduction. The Mount Victoria Land end of the Helmand/south-west and south-east Pamir/Lhasa block became finally detached from north-west Australia in late Jurassic times.

In the Mediterranean region, the plate boundary systems of the Neo-Tethys and the Atlantic Ocean had linked up already in the early Jurassic, and ocean-floor spreading began in the middle Jurassic, as testified by numerous ophiolite spreading ages in the Apennines, the Alps, and the Carpathians (Fig. 6). Intra-oceanic subduction in the Vardar ocean had also begun by the end of the Lias (Roddick *et al.* 1979), and the Eo-Hellenic nappes were expelled on to the eastern margin of the African promontory in Yugoslavia and Greece in the late Jurassic, possibly across former fracture zones (Şengör *et al.* 1984*a*). Consumption of the floor of Neo-Tethys was also under way in the late Jurassic, from northern Greece, possibly through Turkey and Iran, to the Himalaya.

In the Middle East, the Sanandaj–Sirjan zone and its equivalents in the Trans-Caucasus and northern Turkey had begun slipping left-laterally. This continental-margin-parallel motion had been going on since the early Jurassic, as the opening of the Slate-Diabase mini-ocean in the Greater Caucasus testifies (Beridze 1984). The Sanandaj–Sirjan zone strike-slip system had been bent into a compressional geometry in Iran, and began tearing Central Iran into pieces (Fig. 20c). This led to the opening of the Sevan–Akera–Qaradagh suture in Transcaucasia and northern Iran (Khain 1975, Ber-

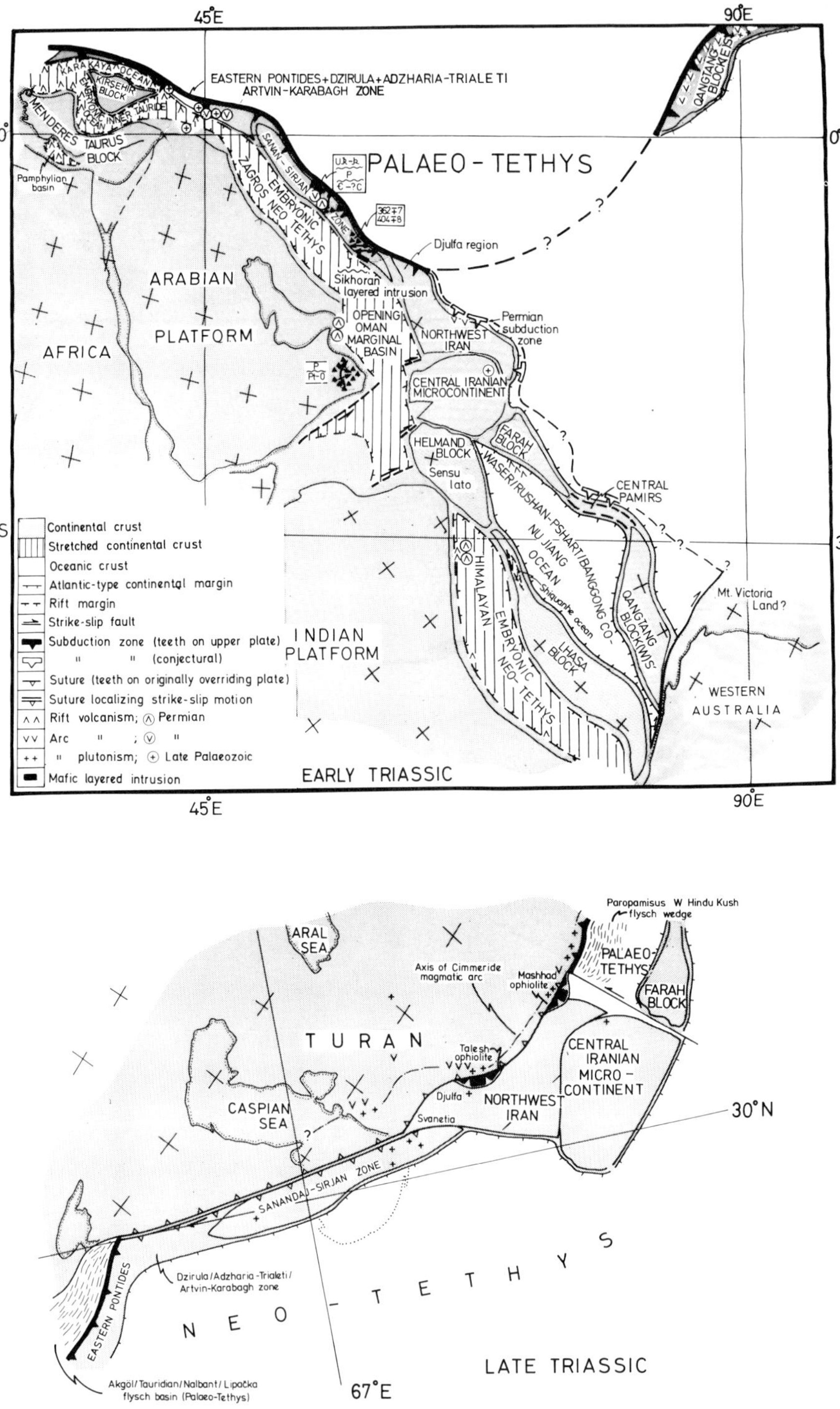

FIG. 20. Tectonic evolution of Iran and surrounding regions, illustrating the complexities introduced by major episodes of strike-slip faulting in the development of a segment of the Tethysides. Comparable complications also exist in other segments (see Fig. 22), where the documentation of the geological evolution is less easy.

(a) Early Triassic (±Griesbachian) palaeotectonics. For sources of data, see caption to Fig. 8a.

(b) Late Triassic (Rhaetian) palaeotectonics of Iran and surrounding regions. For symbols see Fig. 20a.

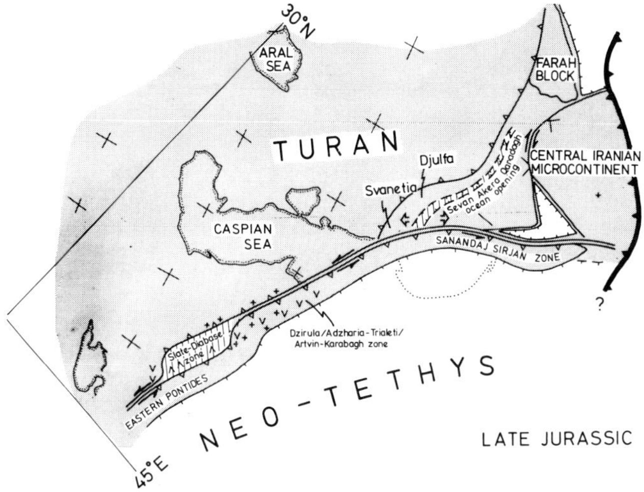

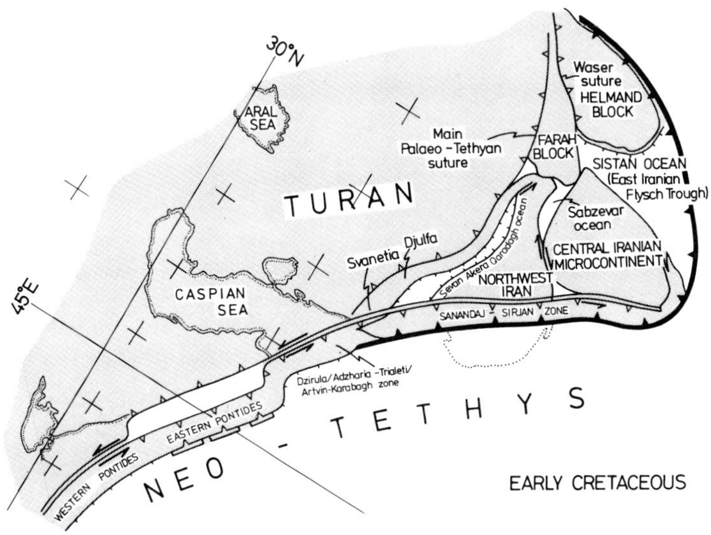

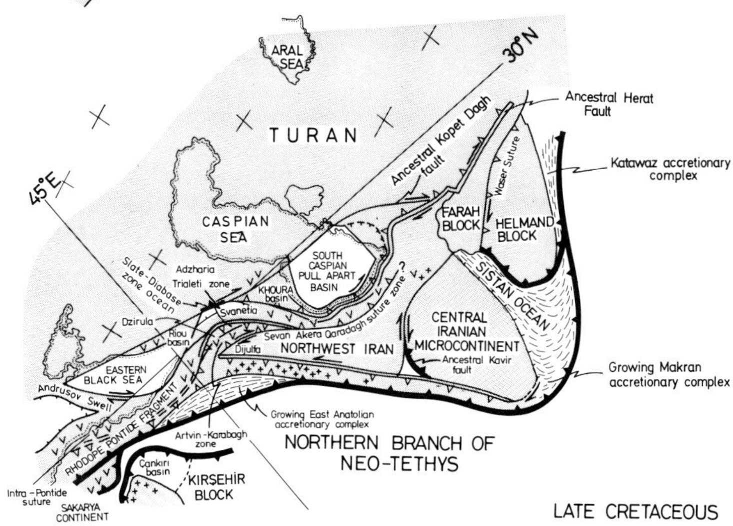

(c) Late Jurassic palaeotectonics of Iran and surrounding regions. For symbols see Fig. 20a.
(d) Early Cretaceous palaeotectonics of Iran and surrounding regions. For symbols see Fig. 20a.
(e) Late Cretaceous palaeotectonics of Iran and surrounding regions. For symbols see Fig. 20a.

berian *et al.* 1981) and the western part of the circum-Central Iranian Microcontinent ocean. This last event was accompanied by the beginning of rotation of the Central Iranian Microcontinent. Therefore, in this geometry, both the Sevan–Akera–Qaratagh and the western half of the circum-Central Iranian Microcontinent oceans seem to be unrelated to the opening of Neo-Tethys. They are independent, small oceans that formed as a result of large-scale strike-slip tectonics along the southern margin of Laurasia.

Fig. 20d shows a further step in the Mesozoic palaeotectonic evolution of Iran and surrounding regions. Notice here that continuing strike-slip motion between the Sanandaj–Sirjan zone and the rest of Iran further rotated the Central Iranian Microcontinent. One result of this ongoing rotation was the opening of the Sabzevar mini-ocean, as evidenced from early Cretaceous ages obtained from the oceanic sediments in this zone (Lindenberg *et al.* 1984).

In the late Cretaceous, the Lhasa block had been welded to Eurasia already, and the northward consumption of the floor of Neo-Tethys was under way all along the southern margin of Eurasia, from Spain to Indonesia, with the singular exception of the northern margin of the African Promontory, where subduction was southward under the Apulian margin (Fig. 19d). Other north-dipping subduction zones also formed within the Neo-Tethys in Turkey, Iran, and Oman, but they were short-lived, because they originated too close to the north-facing passive continental margins. They collided with these margins and resulted in widespread ophiolite obduction from western Turkey to Oman, along a number of fronts, namely along the sutures *14*, *17*, *18*, *19*, *20*, *21*, *23*, *25*, and *27*. These obduction events mostly took place between the Turonian and Maastrichtian, but at least one example is known from the Kırşehir Massif (*K* in Fig. 19; *h* in Fig. 3), where southerly-vergent (present geographical orientation) ophiolite obduction occurred in pre-Cenomanian times.

The formation of these tremendously long Neo-Tethyan subduction zones, which mostly had originated in the early to middle Cretaceous interval was probably one important cause of the late Cretaceous acceleration of global plate motion rates and the consequent late Cretaceous transgression (cf. Hays and Pitman 1973; Pitman 1978). Widespread ophiolite obduction and a few continental collisions which took place in the late Cretaceous resulted in a loss of at least one-third of the Cretaceous subduction zones in the Neo-Tethyan domain. This may have slowed down the global plate mechanism and have caused the latest Cretaceous–early Cainozoic regression.

The generally compressive tectonics of the late Cretaceous in the entire Tethyan belt, also made itself felt in Iran and most of the small, intra-Laurasian oceanic basins such as the Slate-Diabase Zone in the Greater Caucasus, the Sevan–Akera–Qaradagh ocean in northern Iran, the Sabzevar and related oceans all began closing (Fig. 20e). Possibly as a result of this compressive interference, a sliver of the Sanandaj–Sirjan zone became detached and was inserted sideways into the western end of the Sevan–Akera–Qaradagh oceanic space. This sliver now forms the Adzharia–Trialeti–Artvin–Karabagh (*ATAK*) zone of Transcaucasia and separates the Svanetia sliver of 'northwest Iran' from Djulfa (Fig. 20e). The Dzirula Massif is a small fragment of the *ATAK* zone, possibly somewhat rotated during the insertion of the *ATAK* into its present location. The Chorchana–Utslevi zone in Dzirula, and the Chochiani River ophiolites in the Khrami, are remnants of the Palaeo-Tethyan suture in Transcaucasia disrupted by this episode of strike-slip faulting.

The Sistan ocean also began closing at this time, probably as a consequence of the sideways motion of the Helmand block along the former Waser suture. Another possible consequence of the increased rate of convergence in the Tethysides, during the late Cretaceous, was probably the major right-lateral strike-slip system; the southern Caspian and the eastern Black Sea depressions were opened along its pull-apart segments, or at least their opening was helped along. This pull-apart tectonism also disrupted the former Cimmeride Arc in Turan, and redistributed its pieces over such now widely separated regions as Krasnovodsk, the Gondbad-ı Qabus area in the eastern Alborz Mountains, and Aghdarband.

Fig. 19e shows the middle Eocene (Lutetian) geometry of the Tethyan domain. Notice that along a considerable stretch between the Pyrenees and eastern Turkey, Neo-Tethys and the Alpine branch of the Atlantic Ocean sutured at this time, and the north-west African, Calabrian, Hellenic, and Cyprus subduction zones—connected with the main Neo-Tethyan subduction zone farther east via Zagros—overtook the main convergence between Europe and Africa. In the east, India's north-west corner touched Asia, and the subcontinent began rotating counterclockwise around this point of pinning.

By late Miocene (Vindobonian) time, the Tethysides began acquiring their present-day appearance (Fig. 19f). At about this time both the western Mediterranean (Fig. 19f) and the peri-Arabian Alpide oceans vanished; very shortly thereafter, Timor (*T* in Fig. 19f) began climbing on to the Sahul shelf. Only in the eastern Mediterranean, the Arabian Sea and the Somali Basin, the Bay of Bengal, and in the immediate foreland of the Sumatran subduction systems, do the last remnants of Neo-Tethys still survive and are being actively subducted. Soon, no remnant of the maternal Tethys will be left. Wiens (1985/86) has shown that south of the Indian subcontinent, between the Chagos Bank and the Ninetyeast Ridge, the Indian Ocean is already feeling the birth-pangs of yet another mountain range, which, however, will be no longer of Tethyan descendence and may initiate a new mountain dynasty.

Discussion and conclusions

In the preceding sections we reviewed the available evidence to establish the timing of ocean closing and opening along the sutures of the Tethyside superorogenic complex, and the original provenance and subsequent path of migration of primary components of its vast orogenic collage—forming about one-third of the bulk of Eurasia. The Palaeo-Tethys was born when the Permo-Triassic Pangaea was assembled and therefore it is meaningless to speak of the 'opening of Palaeo-Tethys'. It was born as an embayment, and was essentially a giant gulf of Panthalassa.

We have noticed that not only do the Cimmeride sutures not continue into those of the Hercynides and the Altaids, except in the case of two sutures (southern Ghissar and Mongolo-Okhotsk), but also the Alpides, are entirely independent of the Cimmerides, with the singular exception of the continuity between the Waser and the Sistan sutures. Similarly, when a new mountain range arises from the future obliteration of the Indian Ocean, it will have direct connection with the Alpides along only four sutures out of thirty-five—namely the Zagros, Waziristan, Indo-Burman, and Timor sutures. This simple observation supports Şengör's (1979*a*, 1984) classification of the Tethysides as a distinct orogenic system south of the Hercynides and the Altaids, and his recognition within them of two independent orogenic systems, the Cimmerides and the Alpides.

This recognition has also received independent support (e.g. Belov *et al.* 1985, 1986; Chang *et al.* in press). In the tectonic evolution of the Tethysides, we saw that the elimination of Palaeo-Tethys formed the huge, multi-branched orogenic zone of the Cimmerides, as a consequence of the collision with Eurasia, not only of the Cimmerian continent, but also of an immense orogenic collage that had accreted mostly around its eastern end. By contrast, the closure of Neo-Tethys involved a much smaller number of continental blocks, and the resulting Alpides consequently include a much smaller orogenic collage. In fact, much of the Alpide deformation actually took place on a Cimmeride foundation. This overprinting is the main reason why the recognition of the Cimmerides as an independent Tethyan orogenic system has been so difficult.

We have seen further that Belov's (1981) suggestion of treating the Waser/Rushan–Pshart/Banggong Co–Nu Jiang/Mandalay suture as a 'separate' Meso-Tethys (in the sense of a 'middle Tethys'), independent of both Palaeo- and Neo-Tethys, cannot be supported, despite this ocean's considerable size, because it did not extend any farther west than eastern Iran, but also, in Tibet, it was directly connected with Palaeo-Tethys via the mid-Qangtang suture and actually formed a part of it. Correlations by Belov (1981), of suture segments farther west than eastern Iran, with his Meso-Tethys (e.g. in the Caucasus) similarly do not agree with the available observations, as the previous discussions show.

We found subduction-related magmatism to be a generally reliable guide for the detection and tracing of suture zones. Where the clear linear/arcuate magmatic arcs seem to be interrupted for reasons other than just younger cover—as in the most conspicuous example of the late Palaeozoic to Triassic Turco-Iranian magmatic gap or the late Palaeozoic West Thailand magmatic gap (Figs 4 and 5), we realized that it was commonly a result of post-collisional strike-slip-related disruption of a formerly more regular pattern (Figs 20a–e).

There are also primary magmatic gaps in our maps, such as the North China magmatic gap of Triassic and early Jurassic age (Fig. 5). Such primary gaps in the continuity of subduction- *and* collision-related magmatism, if major (i.e. more than 1000 km), indicate original absence of sutures. In the case of the North China magmatic gap, the interruption corresponds with continental regions separating the Altaid system from the Tethysides.

The character of the *distribution* of convergent magmatism is also diagnostic of tectonic phenomena. Subduction-controlled magmatism generally follows a well-defined, continuous, and narrow band, never wider than a few hundred kilometres, and only occurs on one side of the associated suture. Best examples of this are seen in our maps showing late Palaeozoic and late Cretaceous–early Tertiary magmatic activity, i.e. at times when a minimum of continental collisions occurred in the Tethyside system. By contrast, times of major collisions and intracontinental convergence are also times of generally patchy magmatism occurring in short but wide bands, or even in equant areas commonly extending on both sides of sutures. Examples of such cases are abundantly present in Fig. 6, mainly in East Asia, but also in the Caucasus. In a few cases, subduction-related magmatism, originally confined to only one side of a suture, may appear to occur later on both sides, following pervasive conjugate wrench faulting during continuing suture-perpendicular shortening (Fig. 21). Examples of such cases are present in the Kuen-Lun and in the Qin-Ling Range, and also in eastern Burma and West Thailand.

In sharp contrast to convergent-boundary magmatism, rift-related magmatism appears to be a poor guide for delineating former axes of rifting. If rift magmatism was the only guide, we would have hardly recognized the continuity—in places, even the presence—of the Neo-Tethyan rifted margins. This is hardly surprising when compared with regions of active rifting. Sparsity of rift magmatism is also seen, both along the margins of the Atlantic Ocean (Burke 1976) and along the East African Rift Valleys (Burke and Whiteman 1973).

A further difficulty in the recognition of 'opening'-related magmatism in the Tethysides was encountered in the central and eastern Mediterranean region. As seen in Fig. 4, abundant mafic volcanism characterized much of post-Hercynian Europe and was apparently related to post-collisional phenomena, and/or to a broad shear movement that affected much of Europe in post-Hercynian times (Arthaud and Matte 1977). There seems, however, to have been no natural spatial break between this 'convergent' basaltic volcanism, and the 'rift-related' volcanism in mainland Greece and

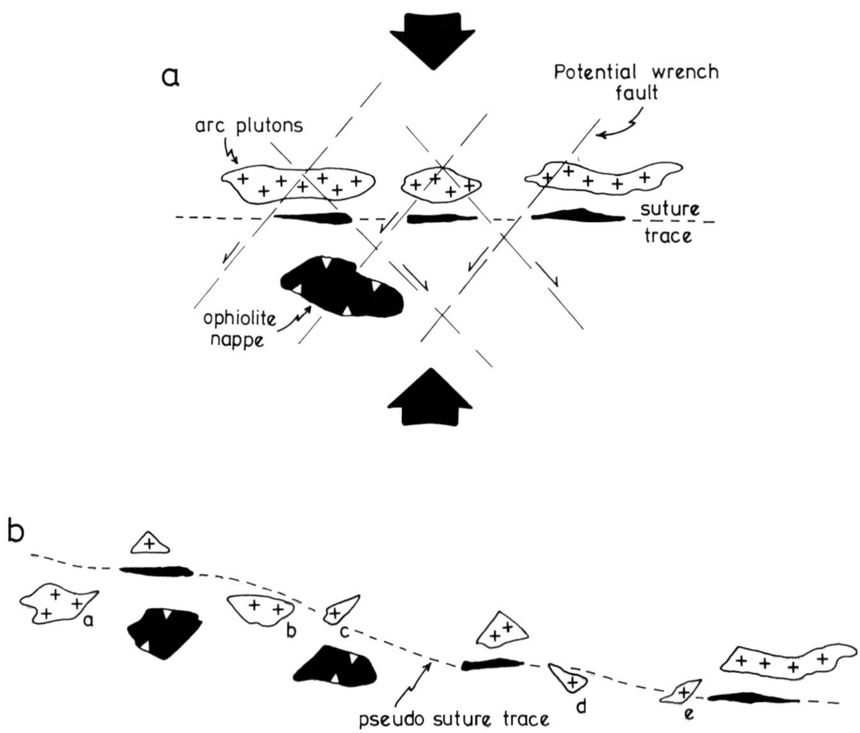

FIG. 21. A mechanism to generate 'pseudo-suture-crossing granites' as seen in northern Tibet. This mechanism also serves to disrupt, and make difficult, the recognition of old sutures.

(a) Before conjugate wrench-fault-dominated post-collisional shortening.

(b) After conjugate wrench-fault-dominated post-collisional shortening. Here the arc-related plutonic rocks *a*, *b*, *c*, *d*, and *e* appear as if they were post-collisional, suture-crossing granitic rocks.

Crete. It is extremely difficult to decide now whether the Cretan late Palaeozoic basaltic volcanism was somehow related to Hercynian events, or to Cimmeride 'back-arc', i.e. rifting events. We prefer the latter interpretation, because the Hercynian orogenic belt did not extend to Crete.

An inspection of Figs 4 through 7, and 19a through 19f, shows how collisions seem to have triggered the generation of new subduction zones. The early to late Carboniferous Altaid collisions in Central Asia probably triggered the coeval onset of the destruction of Palaeo-Tethys; the collision of the Farah–Central Pamir–Western Qangtang fragment (*l,s*) with Laurasia in the early Jurassic seems to have led to the beginning of consumption of the Waser/Rushan–Pshart/Banggong Co–Nu Jiang ocean, probably in the middle Jurassic. The early Cretaceous elimination of the latter greatly accelerated the subduction of Neo-Tethys that had probably begun earlier as a result of ophiolite obduction on to the Lhasa block. Similarly, the Eocene elimination of Neo-Tethys along the Himalayan segment, and the continuing intracontinental convergence since then, appear to be causing the inception of yet another consuming margin south of India at the present day: very much like the earlier cases since the late Palaeozoic. In other segments of the Tethysides, similar relationships between collisions and inception of subduction zones are also seen.

Not only subduction zones may be initiated by collisions, but also spreading centres may be similarly triggered by subduction zones. McKenzie and Weiss (1975) suggested that major spreading centres may be born initially in back-arc basins behind subduction zones, and Şengör (1979) argued that in some places Neo-Tethys may have opened as a series of back-arc basins above Palaeo-Tethyan subduction zones. Later, Şengör *et al.* (1980) have shown that, at least in northern Turkey, the northern branch of Neo-Tethys opened by splitting the Palaeo-Tethyan continental-margin arc. The data reviewed previously now suggest that a large proportion of the Waser/Rushan–Pshart/Banggong Co–Nu Jiang/Mandalay ocean and the Emei Shan marginal basin may also have opened as back-arc basins above south-dipping subduction zones in the Central Pamirs, easternmost Qangtang, and western Thailand. Not only in Turkey, but also in Iran and in Oman, Neo-Tethys appears to have begun as a marginal basin behind the south-west-dipping Palaeo-Tethyan subduction zone. As documented above, there is now increasing evidence to suggest that, in fact, much of the Gondwanan north and north-east margin facing Palaeo-Tethys may have been active, and that the initial disintegration of this margin to open both the Waser/Rushan–Pshart/Banggong Co–Nu Jiang/Mandalay and Neo/Tethyan oceans may have been due entirely to back-arc basin activity.

Tethyside sutures stitch together the Tethyside blocks that form the primary components of the Tethyside orogenic collage. In search of clues concerning the original provenance of these blocks, we have looked at their late Carboniferous to early Cretaceous palaeobiogeographical and late Carboniferous to early Permian palaeoclimatological record. The latter record clearly indicates that all of the Cimmerian continent in Asia east of Iran (*i*, *j*, *k*, *l*, *m*, *s sensu lato*, except *s″*, *t sensu lato*) with the exception of the eastern part of the Qangtang fragment (*s″*) formed the northern margin of Gondwana Land until at least the middle Permian. In the late Permian the *Neoschwagerina* and *Verbeekina* provinces clearly delineate the Cimmerian continent *and* the Chinese blocks, with the exception of *n*, *o*, *p*, *r*, and some places on the northern shelf of Gondwana Land, suggesting that the whole of Turkey and Iran, including Transcaucasia, belonged to the Cimmerian continent, which itself constituted the northern margin of Gondwana Land, a conclusion fully supported by geological observations (e.g. Şengör *et al.* 1984*a*; Belov *et al.* 1986).

The late Carboniferous to early Permian floral record suggests that the North China (*q*), Yangtze (*w*), Huanan (*y*), eastern Qangtang (*s″*), eastern Qaidam (*o*), Annamia (*x*), and with it the eastern Malayan Peninsula (*x‴*) and south-east Sumatra? (*x⁗*) were all near the equator, a conclusion largely corroborated by palaeomagnetic evidence (Lin *et al.* 1985*a*; McElhinny 1985; Opdyke *et al.* 1986), and had a common Cathaysian flora. This picture remained probably similar in the late Permian, as evidenced by the coincidence between the Cathaysian floral and the *Lepidolina* foraminiferal provinces, as shown in Figs 14 and 15. In the middle Permian, this conclusion is further strengthened by the *Monticulifera–Urushtenoidea* brachiopod subprovince, within what Nakamura *et al.* (1985) call the 'Cathaysia–Tethyan subprovince'. Although the geological record shows that these blocks were terminally sutured only by the late-middle to late Triassic, direct 'isthmian' links between most of them probably existed via island-arc systems, as shown in Fig. 19b, although geological evidence is not now sufficient to elaborate this hypothesis. Separation until the Triassic, between the North China and Yangtze blocks is suggested, in addition to palaeomagnetic (Lin *et al.* 1985*a*: McElhinny 1985, Opdyke *et al.* 1986) and geological (Şengör 1985*c*) evidence, by the sharp boundary between the northern and southern subprovinces of the Cathaysian floral realm along the Qin-Ling/Dabie Shan suture zone. In this connection, the northern Cathaysia-type flora of the Qinghai Nan Shan (Fig. 14) is interesting, and may suggest a strike-slip emplacement of this fragment westward into its present location west of the Alxa block from a position farther east, perhaps originally north of the present Qin-Ling, in post-Permian times. A similar strike-slip emplacement of the Istanbul–Balkan fragment in north-west Turkey may have brought the westernmost Turkish occurrence of a late Carboniferous Euramerian flora to its present 'Gondwanan' position (i.e. south of the main Palaeo-Tethyan suture), although similar floras in north-west Turkey, and in Transcaucasia, cannot be explained by the same mechanism. Another possible left-lateral strike-slip displacement is also suggested by the S-shaped *Neoschwagerina* province boundary in western Turkey and Cyprus (Fig. 15).

Such 'anomalous' palaeobiogeographical observations, and also consideration of facies, structure, and even palaeomagnetic evidence suggest that major strike-slip motion in fact did take place along most of the Tethyside sutures before, during, and after collision. Fig. 22 and Table 1 catalogue the best known examples of these. It seems that over 60 per cent (by length) of the Tethyside sutures had considerable strike-slip motion on them. More than 25 per cent had displacements exceeding 1000 km, and over 50 per cent had more than a few hundred kilometres. At least 15 per cent of them experienced more than one major episode of sideways motion. Table 1 shows that, within the

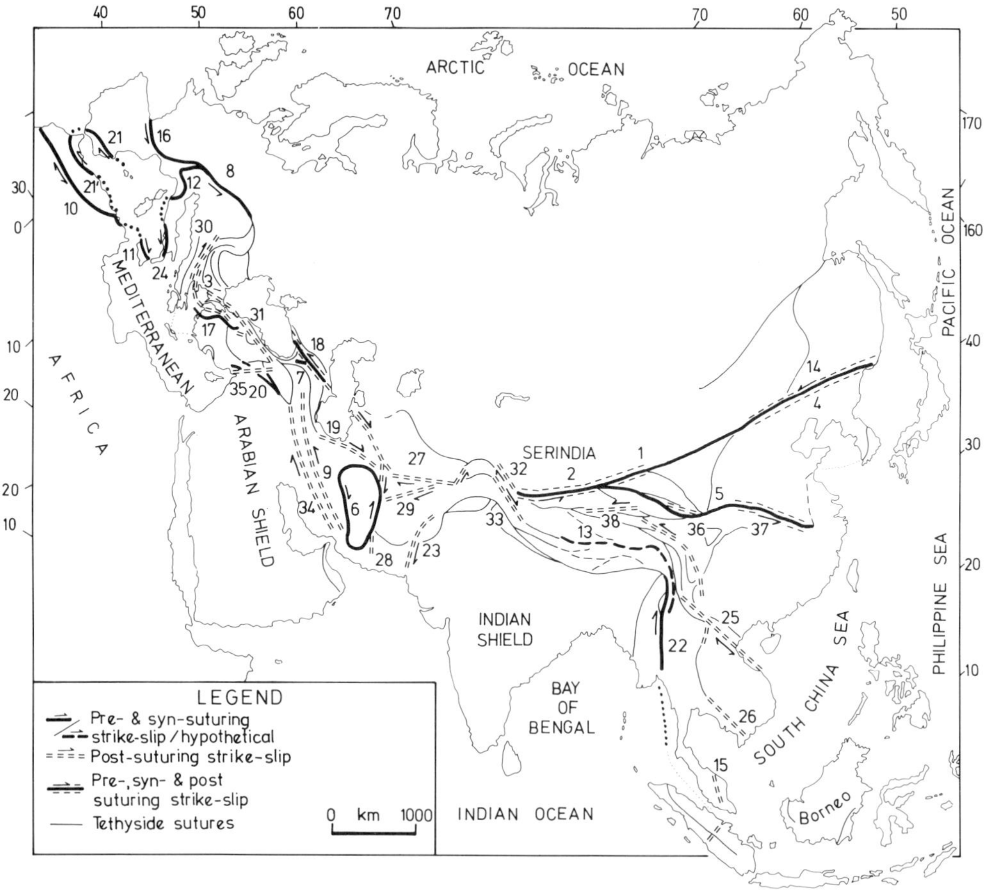

FIG. 22. Map showing pre-, syn-, and post-suturing strike-slip motion along the Tethyside sutures. For numbering and sources see Table 1 (modified after Şengör 1987*a*).

TABLE 1. *Sense, amount, and timing of strike-slip motion on some Tethyside sutures**

Suture†	Sense	Amount	Time	Reference
1 (*XIII*)	Left	?	ePale	Yang *et al.* (1986)
2 (*XI+XIII*)	Left	?	?C–P	Bally *et al.* (1980)
3 (*II*?)	Right	>1000 km	C–?eTr	Şengör (in press, *a*)
4 (*XVI*)	Left	>4000 km	P	Lin *et al.* (1985*a*)
5 (*XX–XXXIII*)	Left	4300 ± 1200 km	P–eJ	Lin *et al.* (1985*a*)
6 (*26*)	Right	~1000 km	lTr–lK	Davoudzadeh and Schmidt (1984)
7 (*VI*)	?	?	P–lTr	Adamia and Belov (1984)
8 (*58*)	Left	~1500 km	lTr–eK	Weissert and Bernoulli (1985)
9 (*IV′*)	Left	~2000 km	lTr–eK	This chapter
10 (*4–5*)	Left Right	10–20 km 10–20 km	e–mJ lJ–eK	Mattauer *et al.* (1977)
11 (*7* south)	?	?	eJ	Catalano and D'Argenio (1982)
12 (*7* north)	?	?	m–lJ	Gianelli and Principi (1977)
13 (*XXIV*)	Left	>500 km	lJ–eK	This chapter
14 (*XIV*)	Left	?	lJ–eK	Klimetz (1983)
15 (*XXXI*)	?	?	Tr–K	Şengör (1986*b*)
16 (*1*)	Left	150–200 km	e–lK	Trümpy (1976)
17 (*14*)	?	?	lK	Şengör *et al.* (1984*a*)
18 (*24*)	Right Left	~2000 km 350–700 km	lTr–lK lK–Pal	This chapter Şengör *et al.* (1984*b*)
19 (*VI + VII*)	Right	~500 km	lK–Pal	Şengör *et al.* (1984*b*) and this chapter
20 (*18,22*)	Right	~1000 km	lK–eE	Y. Yılmaz and O. Sungurlu (pers. comm., 1984)
21 (*2*)	Right	~300 km	lE–lM	Wildi (1983)
21′ (*3*)	Left			
22 (*XL*)	Right	~460 × km	lOl–present	Curray *et al.* (1982)
23 (*28*)	Left	>200 km	e–mT	Lawrence and Yeats (1979)
24 (*7* south)	(north) Left (south) Right	~50 km	T	B. D'Argenio (pers. comm., 1980)
25 (*XXIX*)	Left Right	~500 km A few tens of km	O–M Q	Tapponnier *et al.* (1986) Allen *et al.* (1984)
26 (*XXVII*)	Left	~300 km	O–M	Tapponnier *et al.* (1986)
27 (*VIII*)	Right	?	end E–?M	Tapponnier *et al.* (1981)
28 (*26*)	Right	?	post eM	Tirrul *et al.* (1983)
29 (*IX*)	Right	?	end E	Tapponnier *et al.* (1981)
30 (*12*)	Right	>100 km	M–Pli	Burchfiel (1980)
31 (*13,17*)	Right	>100 km	lM–present	Şengör (1979*b*), A. M. C. Şengör and N. Görür (unpublished data)
32 (*XI,XIII*)	Left	500–600 km	M–present	Tapponnier *et al.* (1981)
33 (*31*)	Right	~200 km	lT	Molnar and Tapponnier (1978)
34 (*25*)	Right	60 km	Pli–present	Tchalenko and Braud (1974)
35 (*21,22*)	Left	~20 km	Pli–present	Şengör *et al.* (1985)
36 (*XXIIII*)	Left	?	lT	Molnar and Tapponnier (1975)
37 (*XXXI*)	Left	several hundred km	post–E	Peltzer *et al.* (1985)
38 (*XIX*)	Left	?	lT	Molnar and Tapponnier (1975)

*Abbreviations: Pale—Palaeozoic, C—Carboniferous, P—Permian, Tr—Triassic, J—Jurassic, K–Cretaceous, T—Tertiary, Pal—Palaeocene, E—Eocene, Ol—Oligocene, M—Miocene, Pli—Pliocene, e—early, m—middle, l—late.

†In the enumeration of sutures, left-hand numerals refer to those in Fig. 23 and right-hand numerals (in parentheses) refer to those in Fig. 4.

Tethysides, strike-slip motion along the sutures took place at all times, with no recognizable temporal preference.

Complications introduced by strike-slip movement are not confined to cases along sutures. Collision-induced escape tectonics (Burke and Şengör 1986) also generates a very complicated strike-slip regime (cf. Şengör *et al.* 1985) on all scales. Major escape-related strike-slip faults that formed or moved during the Tethyside evolution are shown in Fig. 2. Some, but not all, of these are identical with some of the structures depicted in Fig. 22. If one now superimposes Figs 2, 3, and 22, one may begin to appreciate the degree of disruption experienced by not only the original Gondwanan northern margin, which was transported in separate parts northwards and reassembled in a different arrangement to form the Tethyside collage, but also the disruption of the Tethyside collage itself during and after its reassembly.

The imaginary map that the reader is invited to draw in his or her mind's eye, by superimposing Figs 2, 3, and 22 would be a 'suspect terrane map' of the Tethysides. If we now took the further step of considering the innumerable terranes in our imaginary map as having been originally independent entities populating the floor of 'the Tethys', as has been done in other orogenic belts (e.g. Coney *et al.* 1980; Jones *et al.* 1983; Schermer *et al.* 1984; Howell *et al.* 1985), the chaos and confusion this would precipitate would be easy to visualize. That is why in this paper we avoided the recommended 'terrane analysis' and instead began by identifying suture-bound blocks by carefully searching for past evidence of *subduction* and *continental collision*. Suturing by strike-slip apposition of blocks is also possible, as Dewey *et al.* (1986) have shown, but it is not common.

In the present highly disrupted state of the Tethyside collage, one good guide to reconstruction of the pre-dispersal picture of the Tethyside blocks, was seen to be the structure and history of the pre-Tethyside basement. In this regard we found the Pan-African structures to be particularly promising and already helpful in the Middle East, where they are reasonably well-known, both on the Gondwanian foreland (North Africa and Arabian Peninsula), and on the individual Tethyside blocks around it (Şengör *et al.* 1984*b*). Elsewhere in Eurasia, knowledge of the pre-Tethyside Pan-African basement is unfortunately very fragmentary.

Finally, we have seen once more, as frequently emphasized by Şengör (1984, 1985*a*, 1986*a*,*b*) the extremely misleading role of the tripartite subdivision of all Phanerozoic orogenic phenomena into Caledonian, Hercynian, and Alpine 'cycles'. This regrettable terminology of the old fixist model of episodic and synchronous world-wide orogeny unfortunately still pervades much of the local and regional Tethyan geological literature, and leads to serious and widespread misunderstanding and confusion, as documented by Şengör (1985*a*,*b*, 1986*a*,*b*).

We presented six palaeogeographical maps of the entire Tethyan domain, to sketch its evolution between the Permian and the late Miocene. A comparison of these maps with a generalized Mesozoic eustatic sea-level change curve (e.g. see fig. 9 in Şengör 1985*b*) reveals a remarkable correlation of times of major Cimmeride collisions with those of major falls in sea-level. Dewey and Windley (1981) argued that because collision gathers the continental crust into a smaller area, it simultaneously enlarges the capacity of the ocean basins. Moreover, because subduction zones, which facilitate plate motion, are lost during collisions, a general slowing down of global plate motions is observed at times of major collisions. Thus, major regressions coincide temporally with important collision events. It appears that the Cimmeride collisions may have provided one possible factor affecting the early and middle Mesozoic sea-level changes.

Figs 19a–f also disclose a persistent trend in the evolution of the Tethysides since at least the Permian, namely the progressive disintegration of the southern supercontinent, and the northerly flight of its dispersed pieces to unite with Eurasia. This tendency to go north is also seen in the blocks that have been accreted to Korea, Japan (e.g. Lee 1984), and to the North American Cordillera (Howell *et al.* 1985), also since the Permian. This large-scale and persistent migration northwards of continental blocks appears significant, and must somehow be related to the kinematics of the first-order convective circulation, at least in the upper mantle. The history of the Tethysides may thus provide an important constraint for models of convective circulation in the mantle.

Another important point that emerges from our survey of Tethyside geology is the abundance of long and thin strips of continental material, that were successively detached from Gondwana Land during the early stages of the evolution of the Tethysides. Similar long and thin 'ribbon continents' are also abundant in the structure of the Altaid Asia (Şengör 1987*b*). The role of such ribbon continents in the evolution of orogenic collages may be more important than hitherto realized. Vink *et al.* (1984) and

Steckler and ten Brink (1986) suggested a mechanism to show why strip-continents may be preferentially produced during continental rifting.

The Mesozoic evolution of Iran suggests that uncertainties, in the order of 500 to 1500 km, are thus implicit in many places in our Figs 19a–f, with respect to initial reconstructions, motion paths of the collage components, palinspastic restoration of block shapes, and post-collisional distortion of entire collages and the host continents around which they accrete. These are significant dimensions implying, at one extreme, that our present reconstructions may be grossly in error, which could only be avoided if we knew the regional geological details uniformly throughout the Tethysides at more than a reconnaissance level. What is, therefore, most needed in the Tethysides now is more, and more detailed, geological fieldwork to map their suture zones and establish better geological criteria for the place of origin and paths of migration of the blocks they bound. Ages, senses, and amounts of major strike-slip displacement along the Tethyside sutures are among the most outstanding problems in the study of the Tethyside collage. Before all the suture zones, and the major strike-slip boundaries within the Tethysides, are mapped in detail, it will not be possible to interpret satisfactorily the available palaeobiogeographical, palaeoclimatological, and palaeomagnetic data to increase the precision of our reconstructions. A corollary to this statement is that: *all palaeobiogeographical, palaeoclimatological, and palaeomagnetic data must be carefully plotted on suture maps or in reference to sutures*, for them to be useful for work on continental reconstructions.

Most importantly, we hope that this study has underlined sufficiently the necessity of a multidisciplinary, iterative approach to regional tectonic analysis, including regional stratigraphy and structure, petrology, palaeobiogeography, palaeoclimatology, and palaeomagnetism, *in the order given*. This multidisciplinary approach to historical geology, with field data as its fundamental basis, remains as relevant today as when it was first formally introduced into geology by Sir Charles Lyell.

ACKNOWLEDGEMENTS: This paper is a progress report of the Tethyan project, carried out partly in the Istanbul Technical University, in collaboration with Lamont-Doherty Geological Observatory, the Department of Earth Sciences of the University of Oxford, the Geologisches Institut of the Eidgenössische Technische Hochschule in Zürich, and the Geology Department of the Middle East Technical University in Ankara. It also incorporates results obtained as a part of a joint project between the Institute of Geology of the Academia Sinica in Beijing, ETH and ITU, on the tectonics of South China. We are grateful to the following for providing us with information and ideas on various aspects of the Tethyside evolution, without which this paper could not have been written: A. A. Belov, M. Berberian, K. Burke, V. S. Burtman, Chang Cheng-fa, J. F. Dewey, N. Görür, V. E. Khain, Li Jiliang, Sun Shu, O. Sungurlu, Xu Shutong, and Y. Yılmaz. We are also indebted to the following for fruitful discussions: B. C. Burchfiel, Huang Chiqing, İ. Ketin, A. Okay, and M. Sandulescu. Professors M. G. Audley-Charles and A. Hallam invited this contribution to be read during the First Lyell Meeting. Mr İzzet Türkkal drew most of the figures, and the typescript was prepared by the Sürat Daktilo Company. Oya Şengör helped during all stages of the preparation of this paper.

References

Achache, J. (1984). *Paléomagnetisme des zones actives. Croissance, destruction et formation des marges continentales*. Thèse de Doctorat d'Etat, Université Paris VII. (unpaginated).

Adamia, S. A. and Belov, A. A. (1984). Excursion 008A + C. Pre-Mesozoic complexes of the Caucasus. In *Georgian Soviet Socialist Republic, Excursions: 001, 007, 008, 012, 014, 017 Guidebook*. Int. Geol. Congr. 27th Session, Moscow. pp. 108–53, Khelovneba, Tbilisi.

Aiello, E., Bartolini, C., Boccaletti, M., Gocev, P., Karagjuleva, J., Kostadinov, V., and Manetti, P. (1977). Sedimentary features of the Srednogorie zone (Bulgaria): an Upper Cretaceous intra-arc basin. *Sedim. Geol.* **19**, 36–68.

Alavi, M. (1979). The Virani ophiolite complex and surrounding rocks. *Geol. Rdsch.* **68**, 25–52.

Allègre, C. J. *et al.* (1984). Structure and evolution of the Himalaya-Tibet orogenic belt. *Nature* **307**, 17–22.

Allen, C. R., Gillespie, A. R., Han Yuan, Sieh, K. E., Zhang Buchum, and Zhu Chengnan. (1984). Red River and associated faults, Yunnan Province, China. Quaternary geology, slip rates, and seismic hazard. *Bull. Geol. Soc. Am.* **95**, 686–700.

Altermann, W. (1986). The Upper Palaeozoic pebbly mudstone facies of peninsular Thailand and Western Malaysia—continental margin deposits of Palaeoeurasia. *Geol. Rdsch.* **75**, 371–81.

ALTINER, D. (1984). Upper Permian foraminiferal biostratigraphy in some localities of the Taurus Belt. In *Int. Symp. on the Geology of the Taurus Belt.* (ed. O. Tekeli and Göncüoğlu), pp. 255–68, Ankara M. T. A.

AMODIO-MORELLI, L. *et al.* (1976). L'arco Calabro-Peloritano nell'orogene Appenninico–Maghrebide. *Mem. Soc. Geol. Ital.* **17**, 1–60.

ANTONIJEVIČ, I., GRUBIČ, A., and DJORDJEVIČ, M. (1974). The Upper Cretaceous paleorift in Eastern Serbia. In *Metallogeny and concepts of the geotectonic development of Yugoslavia.* pp. 315–39, Faculty of Mining and Geology, Belgrade University, Department of Economic Geology.

ARGAND, E. (1924). *La tectonique de l'Asie:* 13ᵉ Congr. Géol. Int., pp. 171–372, Vaillant-Carmanne, Liège.

ARTHAUD, F. and MATTE, P. (1977). Late Paleozoic strike-slip faulting in southern Europe and northern Africa. Result of a right-lateral shear zone between the Appalachians and the Urals. *Bull. Geol. Soc. Am.* **88**, 1305–20.

AUDLEY-CHARLES, M. G. (1978). The Indonesian and Phillipine Archipelagos. In *The Phanerozoic Geology of the world, II, The Mesozoic, A.* (ed. M. Moullade and A. E. M. Nairn), pp. 165–207, Elsevier, Amsterdam.

BALLY, A. W., *et al.* (1980). *Notes on the geology of Tibet and adjacent areas—report of the American plate tectonic delegation to the Peoples' Republic of China.* U.S. Geol. Surv., Open-file report 80–501. (71 pp.)

BARD, J. P., BURG, J. P., MATTE, P., and RIBEIRO, A. (1980). La chaîne hercynienne d'Europe occidentale en terms de tectonique des plaques. *Colloque C6, 26ᵉ Congr. Géol. Int.* pp. 233–46, Paris.

BASSOULET, J.-P. and BERGOUGNAN, H. (1981). Faune et faciès typiques du domaine sud-téthysien: le Lias du Munzur Dağ (Anatolie orientale). *Bull. Soc. Géol. Fr.* sér. (7), **23**, 83–93.

BASSOULET, J.-P., COLCHEN, M., GUEX, J., LYS, M., MARCOUX, J., and MASCLE, G. (1978). Permien terminal néritique, Scythien pélagique et volcanisme sousmarine, indices de processus tectono-sédimentaires distensifs à la limite Permien-Trias dans un bloc exotique de la suture de l'Indus (Himalaya du Ladakh), *C. R. Acad. Sci. Paris*, **287**, 675–8.

BASSOULET, J.-P., FOURCADE, E., and PEYBERNES, B. (1985). Paléobiogéographie des grands foraminiferes benthiques des marges néo-téthysiennes au Jurassique et au Crétacé inférieur. *Bull. Soc. Géol. Fr.* sér. (8), **1**, 699–713.

BAUD, A. (1985). Geological observations on a geotraverse from the Sutlej to the Yarkand River (W Tibet). In *Himalayan Workshop, Abstracts* (unpaginated). Department of Geology, University of Leicester.

BAZHENOV, M. L. and BURTMAN, V. S. (1986). Tectonics and paleomagnetism of structural arcs of the Pamir-Punjab Syntaxis. *J. Geodyn.* **5**, 383–96.

BELOV, A. A. (1981). *Tectonicheskoye Razvitie Alpinsoki Skladchatoi Oblasti v Paleozoe.* Nauka, Moscow. (211 pp.)

BELOV, GATINSKIY, Y. G., and MOSSAKOVSKY, A. A. (1985). Indosinidi evrazii. *Geotektonika* 1985 (6), 21–42.

BELOV, GATINSKY, Y. G., and MOSSAKOVSKY, A. A. (1986). A précis on pre-Alpine tectonic history of Tethyan paleooceans. *Tectonophys.* **127**, 197–211.

BELYAEVSKY, N. A. (1976). Tectonics of the northern part of the Pamirs–Himalayan Syntaxis: In *International Colloqium on Geotectonics of the Kashmir–Himalaya, Karakorum–Hindu Kush–Pamir Orogenic Belts.* pp. 29–42, Accademia Nazionale dei Lincei, Roma.

BEN-AVRAHAM, Z., NUR, A., and JONES, D. (1982). The emplacement of ophiolites by collision. *J. Geophys. Res.* **87**, 3861–7.

BERBERIAN, M. (1983). The Southern Caspian: a compressional depression floored by a trapped, modified oceanic crust. *Can. J. Earth Sci.* **20**, 163–83.

BERBERIAN, F. and BERBERIAN, M. (1981). Tectonoplutonic episodes in Iran. In *Geodynamic Series* 3, pp. 5–33. Geological Society of America, Boulder, Colorado.

BERBERIAN, M. and KING, G. C. P. (1981). Towards a palaeogeography and tectonic evolution of Iran. *Can. J. Earth Sci.* **18**, 210–65.

BERBERIAN, M., AMIDI, S. M., and BABAKHANI, A. (1981). Discovery of the Qaradagh ophiolite belt, the southern continuation of the Sevan-Akera (Little Caucasus) ophiolite belt in northwestern Iran (Ahar quadrangle); a preliminary field note. *Geol. Surv. Iran* Meshingshar Camp, unedited manuscript, 14 pp.

BERBERIAN, F., MUIR, I. D., PANKHURST, R. J., and BERBERIAN, M. (1982). Late Cretaceous and early Miocene Andean-type plutonic activity in northern Makran and Central Iran. *J. Geol. Soc. London* **139**, 605–14.

BERGOUGNAN, H. and FOURQUIN, C. (1980). Un ensemble d'éléments communs à une marge active alpine des Carpates méridionales à l'Iran central: le domaine irano-balkanique. *Bull. Soc. Géol. Fr.* sér. (7), **22**, 61–83.

BERIDZE, M. A. (1984). *Geosynklinalny Vulkanogeno-osadochny Litogenez.* Akademii Nauk, Gruzinsky SSR, Tbilisi. (182 pp.)

BERNOULLI, D. and JENKYNS, H. C. (1974). Alpine, Mediterranean and Central Atlantic Mesozoic facies in relation to the early evolution of the Tethys. *SEPM, Spec. Pub.* **19**, 129–60.

BERTRAND, J. M. and KIENAST, J. R. (1987). Structural and metamorphic evolution of the Karakorum gneisses in the Braldu Valley (Northern Pakistan). In *Himalaya-Karakorum Workshop, Abstracts*, pp. 3–4, Nancy, France.

BEZNOSOV, N. V., GORBATCHIK, T. N., MIKHAILOVA, I. A., and PERGAMENT, M. A. (1978). Soviet Union. In *The Phanerozoic geology of the world II, the*

Mesozoic, A. (ed. M. Moullade and A. E. M. Nairn), pp. 5–53, Elsevier, Amsterdam.

Bigazzi, G., Ferrara, G., and Innocenti, F. (1972). Fission track ages of gabbro from northern Apennines ophiolite. *Earth Planet. Sci. Lett.* **14**, 242–4.

Blaise, J., Bordet, P., Carbonnel, J.-P., and Montenat, C. (1978). Flysch et ophiolites dans le région de Panjaw: une suture néocimmerienne en Afghanistan central. *Bull. Soc. Géol. France* sér. (7), **20**, 795–8.

Boillot, G. and Capdevila, R. (1977). The Pyrenees: subduction and collision? *Earth Planet. Sci. Lett.* **35**, 151–60.

Boulin, J. and Bouyx, E. (1977). Introduction à la géologie de l'Hindou Kouch occidental. In *Livre à la Mémoire de Albert F. de Lapparent*. pp. 87–105. Mém. h. sér. 8, Soc. Géol. France.

Boztuğ, D., Debon, F., Lefort, P., and Yilmaz, O. (1984). Geochemical characteristics of some plutons from the Kastamonu granitoid belt (Northern Anatolia, Turkey). *Schweiz. Mineral. Petrogr. Mitt.* **64**, 389–403.

Brunnschweiler, R. O. (1966). On the geology of the Indoburman Ranges. *J. Geol. Soc. Austral.* **13**, 137–94.

Buffetaut, E. (1981). Eléments pour une histoire paléobiogéographique du Sud-Est asiatique: l'apport des vertébrés fossiles paléobiogéographique du Sud-Est asiatique: l'apport des vertébrés fossiles continentaux. *Bull. Soc. Géol. France.* **7** (23), 587–93.

Buffetaut, E., Jaeger, J.-J., and Rage, J.-C. (1984). *Paléogéographie de l'Inde, du Tibet et du Sud-Est asiatique: confrontation des données paléontologiques avec les modèles géodynamiques*. Mém. Soc. Géol. France, N.S. 147. (194 pp.)

Bunopas, S. (1985). Review on tectonic events of northeastern Thailand. In *Proceedings of the Conference on Geology and Mineral Resources Development of the Northeast-Thailand*. (ed. P. Thanvarachorn, S. Hokjaroen, and W. Youngme). Khon Kaen Univ., 26–9 November 1985, p. 22.

Burchfiel, B. C. (1980). Eastern European Alpine System and the Carpathian orocline as an example of collision tectonics. *Tectonophys.* **63**, 31–61.

Burchfiel, B. C. and Royden, L. (1982). Carpathian Foreland fold and thrust belt and its relation to Pannonian and other basins. *Bull. Am. Assoc. Petrol. Geol.* **66**, 1179–95.

Burke, K. (1976). Development of graben associated with the initial ruptures of the Atlantic Ocean. *Tectonophys.* **36**, 93–112.

Burke, K. and Sengör, A. M. C. (1986). Tectonic escape in the evolution of the continental crust. In *Geodynamics Series, the Continental Crust* **14**, 41–53.

Burke, K. and Whiteman, A. J. (1973). Uplift, rifting and the break-up of Africa. In *Implications of continental drift to the Earth sciences*. Vol. 2. (ed. D. Tarling and S. K. Runcorn), pp. 735–55, Academic Press, London.

Burret, C. F. (1974). Plate tectonics and fusion of Asia. *Earth Planet. Sci. Lett.* **21**, 181–9.

Cameron, N. R., Clarke, M. C. G., Aldiss, D. T., Aspden, J. A., and Djunuddin, A. (1980). The geological evolution of northern Sumatra, *Proc. Indonesian Petrol. Assoc. 9th Ann. Con.* pp. 149–87.

Cao, R. (1986). Discovery of late Carboniferous glacio-marine deposits in western Yunnan. *Geol. Rev.* **32**, 236–42 (in Chinese with English summary).

Caire, A. (1977). The Central Mediterranean mountain chains in the Alpine orogenic environment. In *The ocean basins and margins, 4B, The western Mediterranean*. (ed. A. E. M. Nairn, W. H. Kanes, and F. G. Stehli), pp. 201–56, Plenum, New York.

Casnedi, R. (1980). Stratigraphical outline of the area between Karakorum and Hindu Kush with probable occurrence of a Hercynian geosynclinal stage. *Mem. Soc. Geol. Ital.* **20**, 277–87.

Catalano, R. and d'Argenio, B. (1982). Infraliassic strike-slip tectonics in Sicily and Southern Apennines. *Rend. Soc. Geol. Ital.* **5**, 4–10.

Chang, C. F., *et al.* (1986). Preliminary conclusions of the Royal Society and Academia Sinica 1985 geotraverse of Tibet. *Nature* **323**, 501–7.

Chang, C. F., Pan, Y. S., and Sun, Y. Y. in press. The tectonic evolution of the Qinghai–Tibet plateau: A review. In *Tectonic evolution of the Tethyan regions*. (ed. A. M. C. Şengör), D. Riedel, Amsterdam.

Chatalov, G. (1979). Two facies types of Triassic in Strandza Mountain, SE Bulgaria. *Riv. Ital. Paleont. Stratigr.* **85**, 1029–46.

Chen, B. W. (1981). The discovery of Middle and Upper Carboniferous tilloids in Baxoi, Xizang (Tibet) and their significance. *Geol. Rev.* **28**, 148–51 (in Chinese with English summary).

Chen, G. (ed.) (1977). *Explanation of the tectonic map of China 1 4,000,000*. Seismology Press, Beijing (in Chinese).

Cherchi, A. and Schroeder, R. (1973). Sur la biogéographie de l'association à Valserina du Barremien et la rotation de la Sardaigne. *C. R. Acad. Sci., Paris* **277**, 829–956.

Chinese Academy of Geological Sciences. (1973). *Geological atlas of the People's Republic of China*. Beijing, Printers of Chinese Geological Maps, 149 pp. (in Chinese).

Choukroune, P., Le Pichon, X., Seguret, M. and Sibuet, J. C. (1973). Bay of Biscay and Pyrenees. *Earth Planet. Sci. Lett.* **18**, 109–18.

Clark, G. C., Davies, R. G., Hamzepour, B., and Jones, C. R. (1975). *Explanatory text of the Bander-e-Pahlavi quadrangle map, 1:250 000*. Geological Survey of Iran, D3, 198 pp.

Cohen, C. (1980). Plate tectonic model for the Oligo-

Miocene evolution of the eastern Mediterranean. *Tectonophys.* **68**, 283–311.

Coney, P. J., Jones, D. L. and Monger, J. W. H. (1980). Cordilleran suspect terranes. *Nature* **288**, 329–32.

Crowell, J. C. (1983). Ice ages recorded on Gondwanan Continents. *Trans. Geol. Soc. South Africa* **86**, 238, 261.

Curray, J. R., Emmel, F. J., Moore, D. G., and Raitt, R. W. (1982). Structure, tectonics and geological history of the northeastern Indian Ocean. In *The ocean basins and margins, 6, the Indian Ocean.* (ed. A. E. M. Nairn and F. G. Stehli), pp. 399–450, Plenum, New York.

Daignieres, M., Gallart, J., Banda, E., and Hirn, A. (1982). Implications of the seismic structure for the orogenic evolution of the Pyrenean range. *Earth Planet. Sci. Lett.* **57**, 88–100.

Davies, R. G., Jones, C. R., Hamzepour, B., and Clark, G. C. (1972). *Geology of the Masuleh Sheet, 1:100 000, Northwest Iran.* Geol. Surv. Iran, Report No. 24. (110 pp.)

Davoudzadeh, M. and Schmidt, K. (1984). A review of the Mesozoic paleogeography and paleotoectonic evolution of Iran. *Neues Jb. Geol. Paläont. Abh.* **168**, 182–207.

Davoudzadeh, M., Soffel, H. and Schmidt, K. (1981). On the rotation of the Central-East-Iran microplate. *Neues Jb. Geol. Paläont., Mh.* 1981(3), 180–92.

Davoudzadeh, M., Lensch, G., and Weber-Diefenbach, K. (1986). Contribution to the paleogeography, stratigraphy and tectonics of the Infracambrian and Lower Paleozoic of Iran. *Neues Jb. Geol. Paläont. Abh.* **172**, 245–69.

Debon, F., Afzali, H., Le Fort, P., and Sonet, J. (1986). Plutonic belts in Afghanistan: typology, age and geodynamic setting. In *Evolution des domaines orogéniques d'Asie méridionale (de la Turquie à l'Indonesie).* (ed. P. LeFort, M. Colchen, and C. Montenat), *Mém. Sci. de la Terre* **47**, 129–53, Nancy.

Dercourt, J., *et al.* (1986). Geological evolution of the Tethys belt from the Atlantic to the Pamirs since the Lias. *Tectonophys.* **123**, 241–315.

Desio, A. (1977). Corrélation entre les structures des chaînes du Nord-Est de l'Afghanistan et du Nord-Quest du Pakistan. In *Livre à la Memoire de Albert f. de Lapparent.* Mém. h. sér. 8, Soc. Géol. France, pp. 179–88.

Dewey, J. F. (1976). Ophiolite obduction. *Tectonophys.* **31**, 93–120.

Dewey, J. F and Windley, B. F. (1981). Growth and differentiation of the continental crust. *Phil. Trans. Roy. Soc. London* **A301**, 189–206.

Dewey, J. F., Hempton, M. R., Kidd, W. S. F., Saroğlu, F., and Şengör, A. M. C. (1986). Shortening of continental lithosphere: the neotectonics of Eastern Anatolia—a young collision zone. In *Geol. Soc. Spec. Pub.* 19, 3–36.

Fan, P. F. (1978). Outline of the tectonic evolution of southwestern China. *Tectonophys.* **45**, 261–7.

Fontaine, H. and Workman, D. R. (1978). Review of the geology and mineral resources of Kampuchea, Laos and Vietnam. In *Proc. 3rd Reg. Conf. on Geol. and Min. Resour. Southeast Asia.* (ed. P. Nutalaya), pp. 539–606, Asian Inst. Technology, Bangkok.

Frisch, W., Neubauer, F. and Satir, M. (1984). Concepts of the evolution of the Austroalpine basement complex (Eastern Alps) during the Caledonian–Variscan cycle. *Geol. Rdsch.* **73.**

Fromaget, J. (1927). Etudes géologiques sur le Nord de l'Indochine centrale. *Bull. Serv. Géol. Indochine* **XVI**. (181 pp.)

Fromaget, J. (1934). Observations et reflexions sur la géologie stratigraphique et structurale de l'Indochine. *Bull. Soc. Géol. France*, sér. (5), **4**, 101–64.

Fuchs, G. (1984). Note on the geology of the Markha–Nimaling area in Ladakh (India). *Jb. Geol. B.-A.* **127**, 5–12.

Gaetani, M., Micora, A., Premoli-Silva, I., Fois, E., Garzanti, E., and Tintori, A. (1983). Upper Cretaceous and Paleocene in Zanskar Range (NW Himalaya). *Riv. Ital. Paleont. Strat.* **89**, 81–118.

Gatinskiy, Y. G. (1986). *Lateralni Strukturno-Formachionniy Analiz.* Nedra, Moscow. (195 pp.)

Gealey, W. R. (1977). Ophiolite obduction and geologic evolution of the Oman mountains and adjacent areas. *Bull. Geol. Soc. America* **88**, 1183–91.

Geological Map of China. (1976). Chinese Academy of Geological Sciences, Beijing.

Gianelli, G. and Principi, G. (1977). Northern Apennine ophiolite: an ancient transcurrent fault zone. *Bull. Soc. Geol. Ital.* **96**, 53–8.

Girardeau, J., Marcoux, J., Allègre, C. J., Bassuollet, J. P., Tang, Y. K., Zao, Y. G., and Wang, X. B. (1984*a*). Tectonic environment and geodynamic significance of the Neo-Cimmerian Donqiao ophiolite, Bangong–Nu Jiang suture Zone, Tibet. *Nature* **307**, 27–31.

Girardeau, J., Marcoux, J., and Zao, Y. G. (1984*b*). Lithologic and tectonic environment of the Xigaze ophiolite (Yarlung Zangbo suture zone, Southern Tibet, China), and kinematics of its emplacement. *Eclogae Geol. Helv.* **77**, 153–70.

Girardeau, J., Marcoux, J., Fourcade, E., Bassoullet, J. P., and Tang, Y. (1985). Xainxa ultramafic rocks, central Tibet, China: tectonic environment and geodynamic significance. *Geology* **13**, 330–33.

Görür, N. in press. Timing of opening of the Black Sea basin. *Tectonophysics.*

Görür, N., Şengör, A. M. C., Yilmaz, Y., and Akkök, R. (1983). Pontidlerde Neo-Tetis' in Kuzey Kolunun açılmasına ilişkin sedimentolojik veriler. *Türk Jeol. Kur. Bülteni* **26**, 11–20.

Görür, N., Oktay, F. Y., Seymen, İ. and Şengör, A. M. C. (1984). Palaeotectonic evolution of the

Tuzgölü basin complex, central Turkey: sedimentary record of a Neo-Tethyan closure. *Geol. Soc. London Spec. Pub.* **17**, 467–82.

Grubiç, A. (1980). Yugoslavie. In *Géologie des Pays Européen (Espagne, Grèce, Italie, Portugal, Yugoslavie)*. pp. 287, 342, Dunod, Paris.

Gruppo di Lavoro Sulle Ofioliti Mediterranee (1977). I complessi ofiolitici e le unita cristalline della Corsica Alpine. *Ofioliti* **2**, 265–324.

Haile, N. S. (1976). Borneo. In *Geol. Soc. London, Spec. Pub.* **4**, 333–47.

Hallam, A. (1984). Continental humid and arid zones during the Jurassic and Cretaceous. *Palaeogeogr., Palaeoclimatol. Palaeocol.* **47**, 195–223.

Hamilton, W. B. (1979). Tectonics of the Indonesian Region. *US. Geol. Surv. Prof. Paper* **1078**. (345 pp.)

Hamilton, W. B. (1983). Configuration of eastern Gondwanaland. *US. Geol. Surv. Circular* **911**, 2–3.

Hays, J. D. and Pitman, W. C., III. (1973). Lithosphere plate motion, sea-level changes, and climatic and ecological consequences. *Nature* **246**, 1-8–22.

Hennig, A. (1915). Zur Petrographie und Geologie von Südwest Tibet. In *Southern Tibet*, **5**, (ed. S. Hedin), Stockholm. (220 pp.)

Herrmann, A. (1923). Serinda. In *Paulys Realenzyklopadie der Altertumswissenschaft*, 2. Reihe (R–Z), 2, J. Metzler'sch Verlagsbuchandlung, Stuttgart.

Howell, D. G., Jones, D. L., and Schermer, E. R. (1985). Tectonostratigraphic terranes of the Circum-Pacific region. In *Tectonostratigraphic terranes of the Circum-Pacific region*. (ed. D. G. Howell), pp. 1–30, Circum-Pacific Council for Energy and Mineral Resources, Earth Science Series 1, Houston.

Hsü, K. J. (1977). Tectonic evolution of the Mediterranean basins. In *The Ocean basins and margins, 4A, the eastern Mediterranean*. (ed. A. E. M. Nairn, W. H. Kanes, and F. G. Stehli), pp. 29–75, Plenum, New York.

Hsü, K. J. (1981). Thin-skinned plate-tectonic model for collision type orogenesis. *Sci. Sinica* **24**, 100–10.

Hsü, K. J. in press. Relic back-arc basins of China and their petroleum potential. In *New perspectives in basin analysis*. (ed. K. Kleinspehn), Springer, Berlin.

Hsü, K. J. and Bernoulli, D. (1978). Genesis of the Tethys and the Mediterranean. In *Init. Rep. DSDP* **42(1)**, pp. 143–949. (K. J. Hsü *et al.*), US Government Printing Office, Washington, DC.

Hsü, K. J. and Schlanger, S. O. (1971). Ultrahelvetic flysch sedimentation and deformation related to plate tectonics. *Bull. Geol. Soc. Am.* **82**, 1207–18.

Hsü, K. J., Nachev, I. K., and Vuchev, V. T. (1977). Geologic evolution of Bulgaria in light of plate tectonics. *Tectonophys.* **40**, 245–56.

Hsü, K. J., Sun, S., and Li, J. L. in press. Mesozoic suturing in the Huanan Alps and the tectonic assembly of South China. In *Tectonic evolution of the Tethyan region*. (ed. A. M. C. Şengör), D. Reidel, Amsterdam.

Huang, T. K. (1945). On major tectonic forms of China. *Mem. Geol. Surv. China Ser. A* **20**. (165 pp).

Huang, T. K. (Huang Chi-Ching). (1978). An outline of the tectonic characteristics of China. *Eclogae Geo. Helvet.* **71**, 611–35.

Huang, K. N., Opdyke, N. D., Kent, D. V., Xu, G. Z., and Tang, R. L. (1986). Further paleomagnetic results from the Permian Emei Shan basalt in SW China. *Kexue Tongbao* **31**, 1195–201.

Hudson, T. and Plafker, G. (1982). Paleogene metamorphism of an accretionary flysch terrane, eastern Gulf of Alaska. *Geol. Soc. Am. Bull.* **93**, 1280–90.

Hudson, T. in press. Southeast Asian segment. In *Tectonic evolution of the Tethyan regions*. (ed. A. M. C. Şengör), D. Reidel, Amsterdam.

Institute of Geochemistry (1980). *Distribution of granites and of fracture zones in South China.* 1:4 000 000 map, Institute of Geochemistry, Academia Sinica, Guiyang, China.

Ishii, K., Okimura, Y., and Ichikawa, K. (1985). Notes on Tethys biogeography with reference to Middle Permian Fusulinaceans. In *The Tethys. Her paleogeography and paleobiogeography from Paleozoic to Mesozoic*. (ed. K. Nakazawa and J. M. Dickins), pp. 139–55, Tokai University Press, Tokyo.

Isler, A. and Pantiç, N. (1980). 'Schistes-lustrés'—Ablagerungen der Tethys. *Eclog. Geol. Helvet.* **73**, 799–822.

Jin, H. H. and Fu, L. W. (1986). The evolution of volcanic rocks in Hunan province and their implications in plate tectonics. *Geol. Rev.* **32**, 225–35 (in Chinese with English summary).

Jones, D. L., Howell, D. G., Coney, P. J., and Monger, J. W. H. (1983). Recognition, character, and analysis of tectonostratigraphic terranes in western North America. In *Accretion tectonics in the Circum-Pacific Regions*. (ed. M. Hashimato and S. Uyeda), pp. 21–35, Terrapub, Tokyo.

Katili, J. A. (1976). Sumara. *Geol. Soc. London Spec. Publ.* **4**, 317–31.

Kerey, I. E. (1984). Facies and tectonic setting of the Upper Carboniferous rocks of Northwestern Turkey. *Geol. Soc. Spec. Publ.* **17**, 123–8.

Khain, V. E. (1975). Structure and main stages in tectomagmatic development of the Caucasus: an attempt at geodynamic interpretation. *Am. J. Sci.* **275–A**, 131–56.

Kligfield, R. (1979). The Northern Apennines as a collisional orogen. *Am. J. Sci.* **279**, 676–91.

Klimetz, M. (1982). The pre-Tertiary geology and Mesozoic tectonics of eastern China and southeast Asia and adjacent regions. Unpublished M.Sc. Thesis, State University of New York at Albany. (159 pp.)

KLIMETZ, M. (1983). Speculations on the Mesozoic plate tectonic evolution of eastern China. *Tectonics* **2**, 139–66.

KLOOTWIJK, C. T. (1981). The India–Asia collision: a summary of paleomagnetic constraints. In *Geological and ecological studies of the Qinghai–Xizang Plateau*, 1. pp. 951–66, Science Press, Beijing, China.

KOSYGIN, Y. A. and PARFENOV, L. M. (1981). Tectonics of the Soviet Far East. In *The ocean basins and margins, 5, the Arctic Ocean*. (ed. A. E. M. Nairn, M. Churkin Jr., and F. G. Stehli), pp. 377–412, Plenum, New York.

KOTANSKY, Z. (1978). The Caucasus, Crimea and their foreland (Scythian Platform), the Black Sea and Caspian Sea. In *Atlas of Alpine Europe and adjoining Alpine Areas*. (ed. M. Lemoine), pp. 545–55, Elsevier, Amsterdam.

KRAVCHENKO, K. N. (1979). Tectonic evolution of the Tien-Shan, Pamir and Karakorum. In *Geodynamics of Pakistan*. (ed. A. Farah and K. A. De Jong), pp. 351–61, Quetta, Geological Survey of Pakistan.

KRÖNER, A. (1984). Late Precambrian plate tectonics and orogeny: a need to redefine the term Pan-African. In *African geology*. (ed. J. Klerkx and J. Michot, pp. 23–8, Musée royal de l'Afrique centrale, Tervuren.

LAWRENCE, R. D. and YEATS, R. S. (1979). Geological reconnaissance of the Chaman Fault in Pakistan. In *Geodynamics of Pakistan* (ed. A. Farah and K. A. De Jong), pp. 351–7, Geological Survey of Pakistan, Quetta.

LEE, G. D. (1984). *Paléomagnetisme des formations crétacées de la Corée du Sud*. Thèse doctorat de 3ème cycle. Institut de Physique du Globe de Paris, 120 pp.

LEE, S. M. (1974). The tectonic setting of Korea with relation to plate tectonics. *United Nations ESCAP, CCOP Tech. Bull.* **8**, 39–53.

LEMOINE, M. and TRÜMPY, R. (1987). Pre-oceanic rifting in the Alps. *Tectonophys.* **133**, 305–20.

LEXIQUE STRATIGRAPHIQUE INTERNATIONAL (1963). VIII, fscl. I: *République Populaire Chinoise*, **I**. CNRS, Paris. (743 pp.)

LEVEN, E. Y. (1981). Permian-Tethys stage scale and correlation of sections of the Mediterranean–Alpine Folded Belt. In IGCP no 5, Newsl. 3 (ed. S. Karamata and F. P. Sassi), pp. 100–12.

LI, C., WANG, Q., LIU, X., and TANG, Y. (1982). *Explanatory notes to the tectonic map of Asia*. Chinese Academy of Geological Sciences, Ministry of Geology, Beijing, China. (49 pp.)

LI, S. G., HART, S. R., ZHENG, S. G., GUO, A. L., LIU, D. L., and ZHANG, G. W. The timing of collision between the North China Block (Sino-Korean Craton) and the South China Block (Yangtze Craton). *Nature* (in press).

LI, T. D. (1980). The development of geological structures in China. *Geo-journal* **4**, 487–97.

LIN, B. Y. (1983). Paleozoic stratigraphy in Xainza Country, Xizang (Tibet). *Bull. Inst. Geol. Chinese Acad. Geol. Sci.* **8**, 1–13.

LIN, J., FULLER, M., and ZHANG, W. (1985*a*). Preliminary Phanerozoic polar wander paths for the North and South China blocks. *Nature* **312**, 444–9.

LIN, J., FULLER, M., and ZHANG, W. (1985*b*). Paleogeography of the North and South China Blocks during the Cambrian. *J. Geodyn.* **2**, 91–114.

LINDENBERG, H. G, GÖRLER, H., JACOBSHAGEN, V., and IBREKEN, H. (1984). Postpaleozoic stratigraphy, structure and orogenetic evolution of the Southern Sabzewar Zone and the Taknar Block (Khorassan, NE Iran). *Neues Jb. Geol. Paläont. Abh.* **168**, 287–326.

LIU, B. P. and CUI, X. I. (1983). Discovery of Eurydesma-fauna from Rutog, Northwest Xizang (Tibet) and its biogeographic significance. *Earth Science—J. Wuhan Coll. Geology* **19**, 79–92.

LIU, D. and LI, X. (1984). Deep structural analysis in Central Anhui. *Sci. Geol. Sinica* 1984(1) 42–50.

LUNNOV, N. P. (ed.) (1971). *Geologiya SSR* **22** (Turkmenia SSR), pp. 444–64. Nedra, Moscow.

MATTAUER, M., TAPPONNIER, P., and PROUST, F. (1977). Sur les mécanismes de formation des chaînes intracontinentales. L'exemple des chaînes atlasiques du Maroc. *Bull. Soc. Géol. Fr.* sér. (7), **19**, 521–6.

MATTAUER, M., MATTE, P., MALAVIEILLE, J., TAPPONNIER, P., MALUSKI, H., XU, Z. Q., LU, Y. L., and TANG, Y. Q. (1985). Tectonics of the Qinling Belt: build-up and evolution of eastern Asia. *Nature* **317**, 496–500.

MCELHINNY, M. W. (1985). Permian paleomagnetism of the Western Yangtze Block, China: a Reinterpretation. *J. Geodynamics* **2**, 115–17.

MCKENZIE, D. P. (1972). Active tectonics of the Mediterranean region. *Geophys. J. Roy. Astron. Soc.* **18**, 1–32.

MCKENZIE, D. P. (1978). Active tectonics of the Alpine–Himalayan belt. The Aegean Sea and surrounding regions. *Geophys. J. Roy. Astron. Soc.* **55**, 217–54.

MCKENZIE, D. P. and WEISS, N. O. (1975). Speculations on the thermal and tectonic history of the Earth. *Geophys. J. Roy. Astr. Soc.* **42**, p. 131.

MEI, H. J., LIN, X. N., CHI, J. X., ZHANG, G. Y., and WU, M. T. (1981). On ophiolite system on Qinghai–Xizang plateau with particular reference to its genesis in West Xizang. In *Geological and ecological studies of Xinghai–Xizang Plateau*. **1**, pp. 545–56, Science Press, Beijing, China.

MICHARD, A. (1982). Contribution à la connaissance de la marge nord du Gondwana: une chaîne plissée paléozoique, vraisemblablement hercynienne, en Oman. *C. R. Acad. Sci. Paris* **295**, 1031–6.

MICHARD, A. (1983). Les Nappes de Mascate (Oman),

rampe épicontinentale d'obduction a faciés schiste bleu, et la dualité apparente des ophiolites Omanaises. *Sci. Géol., Bull.* **36**(1), 3–16.

Milnes, A. G. (1978). Structural zones and continental collision, Central Alps. *Tectonophys.* **47**, 369–92.

Molnar, P. and Tapponnier, P. (1975). Cenozoic tectonics of Asia: effects of a continental collision. *Science* **189**, 419–26.

Molnar, P. and Tapponnier, P. (1978). Active tectonics of Tibet. *J. Geophys. Res.* **83**, 5361–75.

Montadert, L., Winnock, E., Delteil, J.-R., and Reau, G. (1974). Continental margins of Galicia, Portugal and Bay of Biscay. In *The geology of continental margins*. (ed. C. A. Burk and C. L. Drake), pp. 323–42, Springer-Verlag, New York.

Montenat, C., Blaise, J., Bordet, P., Debon, F., Deutsch, S., and Le Fort, P. (1981). Métamorphisme et plutonisme au Paléozoique ancien en domaine gondwan sur la marge nord-ouest des Montagnes Centrales d'Afghanistan. *Bull. Soc. Géol. France.* sér. (7), **23**, 101–10.

Montenat, C., Girardeau, J., and Marcoux, J. (1986). La ceinture ophiolitique néo-cimmérienne au Tibet, dans les Pamirs et en Afghanistan, évolution géodynamique comparative. In *Evolution des domaines orogéniques d'Asie méridionale (de la Turquie à l'Indonesie)*. (ed. P. le Fort, M. Colchen, and C. Montenat), pp. 229, 252, Mém. Sci. de la Terre, **47**, Nancy.

Morgunov, Y. G. (1970). Basic features of tectonics of North Viet Nam. *Int. Geol. Rev.* **12**, 1333–45.

Moullade, M., Peyberbes, B., Rey, J., and Saint-Marc, P. (1980). Interêt stratigraphique et répartition paléobiogéographique des Orbitolinidés mésogéens (Crétacé inférieur et moyen). *Ann. Museum Hist. Natur. Nice* **VI**, 22–41.

Moullade, M., Peyberbes, B., Rey, J., and Saint-Marc, P. (1985). Biostratigraphic interest and paleobiogeographic distribution of Early and Mid-Cretaceous Mesogean Orbitolinids (Foraminiferida). *J. Foram. Res.* **15**, 149–58.

Mountrakis, D. (1986). The Pelagonian zone in Greece: a polyphase-deformed fragment of the Cimmerian Continent and its role in the geotectonic evolution of the Eastern Mediterranean. *J. Geol.* **94**, 335–47.

Murphy, R. W. under review. Southeast Asia: a tectonic triptych. Submitted to: *Transactions, Fourth Circum-Pacific Energy and Mineral Resources Conference.*

Nakamura, K., Shimizu, D., and Zhuo, Ting-L. (1985). Permian palaeobiogeography of brachiopods based on the faunal provinces. In *The Tethys—Her palaeogeography and palaeobiogeography from Paleozoic to Mesozoic.* (ed. K. Nakazawa and J. M. Dickins), pp. 185–98, Tokai University Press, Tokyo.

Nakazawa, K. and Dickins, J. M. (ed.) (1985). *The Tethys—her palaeogeography and palaeobiogeography from Palaeozoic to Mesozoic.* Tokai Univ. Press, Tokyo. (317 pp.)

Norin, E. (1946). *Geological Explorations in Western Tibet.* Publ. 29 of the Sino-Swedish Expedition, Stockholm. (214 pp.)

Norin, E. (1976). The 'Black Slates' Formation in the Pamirs, Karakorum, and Western Tibet. In *International Colloqium on Geotectonics of the Kashmir–Himalaya Karakorum–Hindu Kush–Pamir orogenic belts*. pp. 254–64, Accademia Nazionale dei Lincei Roma, Rome.

Norin, E. (1979). The relationships between the Tibetan platform and the Tarim basin. *Bull. Geol. Inst. Univ. Uppsala* **8**, 17–34.

Okay, A. (1984). Distribution and characteristics of the northwest Turkish blueschists. In *Geol. Soc. London. Spec. Publ.* **17**, 455–66.

Okay, A., Arman, M. B., and Göncüoğlu, M. C. (1985). Petrology and phase relations of the kyanite-eclogites from eastern Turkey. *Contrib. Mineral. Petrol.* **91**, 196–204.

Okimura, Y., Ishii, K., and Ross, C. A. (1985). Biostratigraphical significance and faunal provinces of Tethyan late Permian smaller foraminifera. In *The Tethys—her paleogeography and paleobiogeography from Paleozoic to Mesozoic.* pp. 115–38, Tokai University Press, Tokyo.

Opdyke, N. D., Huang, K., Xu, G., Zhang, W. Y., and Kent, D. V. (1986). Paleomagnetic results from the Triassic of the Yangtze Platform. *J. Geophys. Res.* **91**, 9553–68.

Özgül, N. (1984). Stratigraphy and tectonic evolution of the Central Taurides. In *Geology of the Taurus Belt.* (ed. O. Tekeli and C. Göncüoğlu), pp. 77–90, MTA, Ankara.

Özgül, N. (1985). Alanya tektonik penceresi ve batı kesiminin jeolojisi. In *Ketin Simpozyumu Kitabı*, pp. 97–120, Türk. Jeol. Kur., Ankara.

Parrish, J. T and Curtis, R. L. (1982). Atmospheric circulation, upwelling, and organic-rich rocks in the Mesozoic and Cenozoic eras. *Palaeogeogr., Palaeoclimatol., Palaeoecol.* **40**, 31–66.

Parrish, J. T., Ziegler, A. M., and Scotese, C. R. (1982). Rainfall patterns and the distribution of coals and evaporites in the Mesozoic and Cenozoic. *Palaeogeogr., Palaeoeclimatol., Palaeoecol.* **40**, 67–101.

Pearce, J. A., Alabaster, T., Shelton, A. W., and Searle, M. P. (1981). The Oman ophiolite as a Cretaceous back arc-basin complex: evidence and implications. *Phil. Trans. Roy. Soc. London* **A300**, 299–317.

Peive, A. V., *et al.* (1980). *Tektonika Severnoi Evrazii (Obyasnitelnaya Zapiska k Tektonicheski Karte Severnoi Evrazii Masshtaba 1:5 000 000).* Nauka, Akademii. Nauk SSSR, Moscow. (220 pp.)

Pélissié, T., Peybernes, B., and Rey, J. (1982). Tectonique des plaques et paléobiogéographie des grands Foraminifères benthiques et des Algues calcaires du Dogger à l'Albien sur le pourtour de la Mésogée. *Bull. Soc. Géol. Fr.* sér. (7), **24**, 1069–76.

PELTZER, G., TAPONNIER, P., ZHITAO, Z., and QIN, X. Z. (1985). Neogene and Quaternary faulting in and along the Qingling Shan. *Nature* **317**, 500–5.

PETTERSON, M. G. and WINDLEY, B. F. (1985). Rb–Sr dating of the Kohiston arc-batholith in the Trans-Himalaya of north Pakistan, and tectonic implications. *Earth Planet. Sci. Lett.* **74**, 45–57.

PITMAN, W. C., III. (1978). Relationship between eustasy and stratigraphic sequences of passive margins. *Bull. Geol. Soc. Am.* **89**, 1389–1403.

REUBER, I., MICHARD, A., CHALOUAN, A., JUTEAU, T., and JERMOUMI, B. (1982). Structure and emplacement of the Alpine-type peridotites from Beni Bousera, Rif, Morocco: A polyphase tectonic interpretation. *Tectonophysics* **82**, 231–51.

RICHTER, D., MARIOLAKOS, I., and RISCH, H. (1978). The main flysch stages of the Hellenides. In *Alps, Apennines, Hellenides*, (ed. H. Closs, D. Roeder, and K. Schmidt), pp. 434–8. E. Schweizerbart'sche Verlagsbuchhandlung, Stuttgart.

VON RICHTHOFEN, F. (1877). *China. Ergebnisse Eigener Reisen und Darauf Gegründeter Studien, 1.* Dietrich Reimer, Berlin. (758 pp.)

RICOU, L.-E. (1973). Relations entre les stades paléogéographiques et phases tectoniques sur l'exemple des Zagrides. *Bull. Soc. Géol. Fr.* sér. (7), **14**, 612–23.

RICOU, L.-E. (1976). Evolution structurale des Zagrides: la Région clef de Neyriz (Zagros Iranien). *Mém. Soc. Géol. Fr. N. S.* **125**. (140 pp.)

ROBERTSON, A. H. F. and DIXON, J. E. (1984). Introduction: aspects of the geological evolution of the Eastern Mediterranean. *Geol. Soc. London Spec. Publ.* **17**, 1–74.

RODDICK, J. C., CAMERON, W. E., and SMITH, A. G. (1979). Permo-Triassic and Jurassic ^{40}Ar–^{39}Ar ages from Greek ophiolites and associated rocks. *Nature* **279**, 788–90.

ROMAN'KO, YE. F. and MOROZOV, L. N. (1983). The Anarek–Khvor Massif in Central Iran: structure and history of development. *Geotectonics* **17**, 70–5.

ROSS, C. A. (1982*a*). Paleobiology of Fusulinaceans. *Third North American Paleontological Convention, Proceedings*, **2**, 441–5.

ROSS, C. A. (1982*b*). Paleozoic foraminifera-Fusulinids. In *1982—Foraminifera notes for a short course organized by M. A. Buzas and B. K. Sen Gupta.* (ed. T. W. Broadhead), pp. 163–76. Univ. Tenn. Dept. Geol. Sci., Studies in Geology **6**.

ROSS, C. A. and ROSS, J. R. P. (1981). Late Paleozoic faunas around the Paleo-Pacific margin. In *Evolution Today*. (ed. G. G. E. Scudder and J. L. Reveal), pp. 425–40, Hunt Institute Biological Documentation, Carnegie-Mellon University of Pittsburgh.

ROWLEY, D. B., LOTTES, A. L., NIE, S. Y., YAO, J. P., and ZIEGLER, A. M. (1986). Mesozoic–Cenozoic tectonic evolution of China and adjacent Asia. *1986 Geodynamics Symposium, Mesozoic and Cenozoic Plate Reconstructions, Texas A&M University, Abstracts* (unpaginated).

ROWLEY, D. B., RAYMOND, A., PARRISH, J. T., LOTTES, A. L., SCOTESE, C., and ZIEGLER, A. M. (1985). Carboniferous paleogeographic, phytogeographic, and paleoclimatic reconstructions. *Int. J. Coal. Geol.* **5**, 7–42.

RUTTNER, A. W. (1984). The pre-Liassic basement of the Eastern Kopet Dagh Range. *Neues Jb. Geol. Paläont. Abh.* **168**, 256–68.

SALUN, S. A., GATINSKIY, Y. G., and STRONGANOVA, S. A. (1975). Experiment in characterization of synclinal suture zones, as in certain structures of East Asia. *Int. Geol. Rev.* **17**, 1266–74.

SANDULESCU, M. (1975). Essai de synthése structurale des Carpates. *Bull. Soc. Géol. France*, sér. (7), **17**, 299–358.

SANDULESCU, M. (1978). The Moesic platform and the North Dorogean Orogene. In *Geological Atlas of Alpine Europe and Adjoining Alpine Areas.* (ed. M. Lemoine), pp. 427–42, Elsevier, Amsterdam.

SANDULESCU, M. (1980). Analyse géotectonique des chaînes alpines situées au tour de la Mer Noire occidentale. *Ann. Inst. Geol. Geof.* **56**, 5–54.

SANG, B. L., CHEN, Y. E., and SHAO, G. Q. (1986). The Rb–Sr age and its tectonic significance from metamorphic rocks of Susong Group in the southeastern of Dabie Mountain, Anhui Province (unpublished abstract).

SARIBUDAK, M., SANVER, M., and PONAT, E. (1986). Batı Pontidlerdeki Triyas, Kretase ve Eosen yaşlı kayaçların paleomanyetik sonuçları. *Türkiye Birinci Jeotravers Projesi. 1st Workshop, Abstracts*, Tübitak, Kocaeli, 59–60.

SCHERMER, E. R., HOWELL, D. G., and JONES, D. L. (1984). The origin of allochthonous terranes: perspectives on the growth and shaping of continents. *Ann. Rev. Earth Planet. Sci.* **12**, 107–31.

SCHROEDER, R., CHERCHI, A., GUELLAL, S., and VILA, J. M. (1978). Biozonation par les grands Foraminifères du Jurassique supérieur et du Crétacé infeérieur et moyen des séries néritiques en Algérie N.E. Considérations paléobiogéographiques. *Ann. Mines Géol., Tunis* **28** (II), 243–53.

SCOTESE, C. R. (1984). An introduction to this volume: Paleozoic paleomagnetism and the assembly of Pangea. *Geodyn. Ser.* **12**, 1–10.

SEARLE, M. P. (1983). Stratigraphy, structure and evolution of the Tibetan–Tethys Zone in Zanskar and the Indus suture zone in the Ladakh Himalaya. *Trans. Royal Soc. Edinburgh: Earth Sci.* **73**, 205–19.

SEARLE, M. P. and GRAHAM, G. M. (1982). 'Oman Exotics' oceanic carbonate build-ups associated with the early stages of continental rifting. *Geology* **10**, 43–9.

SEARLE, M. P., LIPPARD, S. J., SMEWING, J. D., and REX, D. C. (1980). Volcanic rocks beneath the

Semail Ophiolite nappe in the northern Oman mountains and their significance in the Mesozoic evolution of Tethys. *J. Geol. Soc. London* **137**, 589–604.

Searle, M. P., Tirrul, R., Rex, A. J., Barnicoat, A., Parrish, R., and Rex, D. C. (1987). Structural, metamorphic and magmatic evolution of the Central Karakoram in the Biafo–Baltoro–Hushe regions, N. Pakistan. In *Himalaya-Karakorum Workshop, Abstracts*, 62, Nancy.

Şengör, A. M. C. (1979*a*). Mid-Mesozoic closure of Permo-Triassic Tethys and its implications. *Nature* **279**, 590–93.

Şengör, A. M. C. (1979*b*). North Anatolian Fault: its age, offset and significance. *J. Geol. Soc. London* **136**, 269–82.

Şengör, A. M. C. (1981). The evolution of Palaeo-Tethys in the Tibetan segment of the Alpides. In *Proc. Symp. Qinghai-Xizang (Tibet) Plateau* **1**, 51–6, Science Press, Beijing, China.

Şengör, A. M. C. (1982). The geology of the Albula Pass area, eastern Switzerland in its Tethyan setting: Paleo-Tethyan factor in Neo-Tethyan opening. Unpublished Ph.D. Thesis, State University of New York at Albany. (405 pp.)

Şengör, A. M. C. (1984). The Cimmeride orogenic system and the tectonics of Eurasia. *Geol. Soc. Am., Spec. Pap.* **195**. (82 pp.)

Şengör, A. M. C. (1985*a*). The story of Tethys: how many wives did Okeanos have? *Episodes* **8**, 3–12.

Şengör, A. M. C. (1985*b*). Die Alpiden und die Kimmeriden: die Verdoppelte Geschichte der Tethys. *Geol. Rndsch.* **74**, 181–213.

Şengör, A. M. C. (1985*c*). East Asian tectonic collage. *Nature* **318**, 16–17.

Şengör, A. M. C. (1986*a*). The dual nature of the Alpine–Himalayan System: progress, problems and prospects. *Tectonophys.* **127**, 177–95.

Şengör, A. M. C. (1986*b*). Die Alpiden und die Kimmeriden: die verdoppelte Geschichte der Tethys: Reply. *Geol. Rndsch.* **75**, 203–12.

Şengör, A. M. C. (1987*a*). Tectonics of the Tethysides: orogenic collage development in a collisional setting. *Ann. Rev. Earth Planet Sci.* **15**, 213–44.

Şengör, A. M. C. (1987*b*). Tectonic subdivisions and evolution of Asia. *Bull. Istanbul Tech. Univ.* **40**, 355–435.

Şengör, A. M. C. and Hsü, K. J. (1984). The Cimmerides of Eastern Asia: history of the eastern and of Paleo-Tethys. *Mém. Soc. Géol. France N. S.* **147**, 139–67.

Şengör, A. M. C. and Kidd, W. S. F. (1979). Post-collisional tectonics of the Turkish–Iranian plateau and a comparison with Tibet. *Tectonophys.* **55**, 361–76.

Şengör, A. M. C. and Yilmaz, Y. (1981). Tethyan evolution of Turkey: a plate tectonic approach. *Tectonophys.* **75**, 181–241.

Şengör, A. M. C., Yilmaz, Y., and Ketin, İ. (1980). Remnants of a pre-late Jurassic ocean in northern Turkey: fragments of Permian–Triassic Paleo-Tethys? *Bull. Geol. Soc. America* **91(1)**, 599–699.

Şengör, A. M. C., Yilmaz, Y., and Sungurlu, O. (1984*a*). Tectonics of the Mediterranean Cimmerides: nature and evolution of the western termination of Paleo-Tethys. *Geol. Soc. London Spec. Pub.* **17**, 77–112.

Şengör, A. M. C., Satir, M., and Akkök, R. (1984*b*). Timing of tectonic evolution and evidence for Pan-African basement in Turkey. *Tectonics* **3**, 693–707.

Şengör, A. M. C., Görür, N., and Şaroğlu, F. (1985). Strike-slip faulting and related basin formation in zones of tectonic escape: Turkey as a case study. In *Strike-slip deformation, basin formation and sedimentation*. (ed. K. Biddle and N. Blick Christie), pp. 227–64, SEPM, Spec. Pub., 37.

Shareq, A. (1981). Geological observations and geophysical investigations carried out in Afghanistan over the period of 1972–1979. *Geodynam. Ser.* **3**, 75–86.

Shvolman, V. A. (1977). *Tektonicheskoe Razvitie Pamira v Melovam i Paleogenovam Periodah*. Nauka, Moscow. (160 pp.)

Shvolman, V. A. (1978). Relicts of the Mesotethys in the Pamirs. *Himalayan Geol.* **8**, 369–78.

Shui, T., Xu, B. T., Lian, L. H., and Qu, Y. S. (1986). Old continental collision zone of Shaoxing–Changshan. *Kexue Tongbao* **6**, 444–8 (in Chinese).

Smith, A. G. (1973), The so-called Tethyan ophiolites. In *Implications of continental drift to the Earth sciences*, 2. (ed. D. H. Tarling and S. K. Runcorn), pp. 977–86, Academic Press, London.

Sobolev, V. S., Lepezin, G. G., and Dobretsov, N. L. (ed.) (1982). *Metamorphic complexes of Asia.* Moscow, Nauka. (350 pp.)

Soffel, H. C. and Förster, H. G. (1984). Polar wander path of the central-East-Iran microplate including new results. *Neues. Jb. Geol. Paläont. Abh.* **168**, 165–72.

Spray, J. G., Bebien, J., Rex, D. C., and Roddick, J. C. (1984). Age constraints on the igneous and metamorphic evolution of the Hellenic–Dinaric ophiolites. *Geol. Soc. London Spec. Publ.* **17**, 619–27.

Srimal, N. (1986). India–Asia collision: implications from the geology of the eastern Karakoram. *Geology* **14**, 523–7.

Stämpfli, G. M. (1978). *Etude géologique générale de l'Erburz oriental au S de Gonad-e, Qabus, Iran N.E.* Thèse de Docteur des Sciences, no. 1868, Université de Genève. (315 pp.)

Stauffer, P. and Mantajit, N. (1981). Late Palaeozoic tilloids of Malaya, Thailand and Burma. In *Earth's Pre-Pleistocene Glacial Record*. (ed. M. J. Hambrey and W. B. Harland), pp. 331–7, Cambridge.

Stauffer, P. and Lee, C. P. in press. Late Paleozoic glacial marine facies in Southeast Asia and its implications. *Geol. Soc. Malaysia. Bull.* **20**.

Steckler, M. S. and ten Brink, U. S. (1986). Lithospheric strength variations as a control on new plate boundaries: examples from the northern Red Sea region. *Earth Planet Sci. Lett.* **79**, 120–32.

Stöcklin, J. (1974). Possible ancient continental margins in Iran. In *The Geology of Continental Margins*. (ed. C. A. Burk and C. L. Drake), pp. 873–87, Springer-Verlag, Berlin.

Stöcklin, J. (1977). Structural correlation of the Alpine ranges between Iran and Central Asia. In *Livre à la Mémoire de Albert F. de Lapparent*. pp. 333–53, Mém. h. sér. **8**, Soc. Géol. Fr.

Suess, E. (1909). *The face of the Earth*. (English translation published in 1909; trans. H. B. C. Sollas). IV, Clarendon Press, Oxford. (673 pp.)

Suess, E. (1911). *Über die Donau*. Vortrag gehalten in der a.o. Festversammlung der k. Akad. Wiss. am 9. Marz 1911. Alfred Hölder, Vienna. (27 pp.)

Szabo, F. and Kheradpir, A. (1978). Permian and Triassic stratigraphy, Zagros basin, south-west Iran. *J. Petrol. Geol.* **1**, 57–82.

Tang, Y. Q. and Wang, F. G. (1986). Plate-tectonic significance of the ophiolites in Northern Xizang (Tibet) Lake district. *Bull. Inst. Geol. Chinese Acad. Geol. Sci.* **13**, 50–65.

Tapponnier, P. (1977). Evolution tectonique du Système Alpin en Méditerranée: poinçonnement et écrasement rigide-plastique. *Bull. Soc. Géol. Fr.* sér. (7), **19**, 437–60.

Tapponnier, P., Mattauer, M., Proust, F., and Cassaigneau, C. (1981). Mesozoic ophiolites, sutures, and large-scale tectonic movements in Afghanistan. *Earth Planet. Sci. Lett.* **52**, 355–71.

Tapponnier, P., Peltzer, G., and Armijo, R. (1986). On the mechanics of the collision between India and Asia. *Geol. Soc. London Spec. Publ.* **19**, 115–57.

Tchalenko, J. S and Braud, J. (1974). Seismicity and structure of the Zagros (Iran): the main Recent fault between 33 and 35°N. *Phil. Trans. Roy. Soc. London* **277**, 1–25.

Teilhard de Chardin, P. and Licent, E. (1924). Observations géologiques sur la bordure occidentale et méridionale de l'Ordos. *Bull. Soc. Géol. Fr.* sér. (4), **24**, 49–91.

Tekeli, O. (1981). Subduction complex of pre-Jurassic age, northern Anatolia, Turkey. *Geology* **9**, 68–72.

Thiele, O., Alavi, M., Assefi, R., Husmandzadeh, A., Seyed Emami, K., and Zahedi, M. (1968). *Explanatory text of the Golpaygan guadrangle map 1/2 500 000*. Geological Survey of Iran. E7, 24 pp.

Tirrul, R., Bell, I. R., Griffis, R. J., and Camp, V. E. (1983). The Sistan suture zone of eastern Iran. *Bull. Geol. Soc. Am.* **94**, 134–50.

Tokay, M. (1973). Kuzey Anadolu Fay Zonunun Gerede ile Ilgaz arasındaki kısmında jeolojik gözlemler. In *Kuzey Anadolu Fayı ve Deprem Kuşağı Simpozyumu* M.T.A., Ankara, 12–29.

TPAO Arama Grubu (1986). Orta Pontidlerin Jeolojisi. *Türkiye Birinci Jeotraverse Projesi, 1st Workshop, Abstracts*, pp. 14–17, Tübitak, Kocaeli.

Trümpy, R. (1971). Stratigraphy in mountain belts. *Quart. J. Geol. Soc. London* **126**, 239–318.

Trümpy, R. (1975). Penninic–Austroalpine boundary in the Swiss Alps: a presumed former continental margin and its problems. *Am. Jour. Sci.* **275-A**, 209–38.

Trümpy, R. (1976). Du Pèlerin aux Pyrénées. *Eclogae Geol. Helv.* **69**, 249–64.

Trümpy, R. (1980). *Geology of Switzerland, a Guide-Book, Part A: an outline of the geology of Switzerland*, Wepf and Co. Publishers, Basel. (104 pp.)

Trümpy, R. and Haccard, D. (1969). Reunion extraordinaire de la Société géologique de France. *C. R. Somm. Seanc., Soc. Géol. France*, Fasc. **9**, 329–96.

Tüysüz, O. (1985). *Kargı Masifi ve Dolayındaki Tektonik Birliklerin Ayırdı ve Araştırılması (Petrolojik İnceleme)*, Doktora Tezi, Istanbul Üniversitesi Fen Bilimleri Enstitüsü. (431 pp.)

Tüysüz, O. (1986). Kuzey Anadolu'da iki farklı ofiyolit topluluğu: eski ve Yeni Tetis'in artıkları. *DOĞA, Tr. Müh. ve Çev.* **10**, 172–9.

Valéry; P., Delteil, J. R., Cottençon, A., Montadert, L., Damotte, B., and Fail, J. P. (1971). La marge continentale d'Aquitaine. In *Histoire structurale du Golfe de Gascogne, 1*. (ed. J. Debysher, X. Le Pichon, and L. Montadert), Editions Technip, Paris, IV. 8–1–N.8–23.

Vereschagin, V. N. and Romov, A. B. (ed.) (1968). *Atlas of the Lithological-Palaeogeographical Maps of the USSR, III*, Triassic, *Jurassic and Cretaceous*. Ministry of Geology and Acad. Sci. USSR, Moscow.

Vergely, P. (1984). *Tectonique des ophiolites dans les Hellenides internes (déformations, métamorphismes et phénoménes sédimentaires). Consequences sur l'évolution des régions Téthysiennes occidentales.* **1** *and* **2**. Thèse de Docteur d'Etat, Université de Paris Sud, Centre d'Orsay. (250 pp., 411 pp.)

Vialon, P., Sabzehi, M., and Hochumandzadeh, A. (1972). Propositions d'un modélé de l'evolution pétrostructurale de quelques montagnes iraniennes comme une conséquence de la tectonique des plaques. *Int. Geol. Congr., 24th. Sess. Sect.* **3**, 196–208.

Vink, G. E., Morgan, W. J., and Zhao, W.-L. (1984). Preferential rifting of Continents: a source of displaced terranes. *J. Geophys. Res.* **89**, 10 072–10 076.

Wang, Q. and Liu, X. (1986). Paleoplate tectonics between Cathaysia and Angaraland in Inner Mongolia of China. *Tectonics* **5**, 1073–88.

Weddige, K. (1984). Zur Stratigraphie und Palaogeographie des Devons und Karbons von NE-Iran. *Senckenbergiana Lethaea* **65**, 179–223.

Weissert, H. and Bernoulli, D. (1985). A transform

margin in the Mesozoic Tethys: evidence from the Swiss Alps. *Geol. Rdsch.* **74**, 665–79.

Wiens, D. A. (1985/86). Historical seismicity near Chagos: a complex deformation zone in the equatorial Indian Ocean. *Earth Planet. Sci. Lett.* **76**, 350–60.

Wildi, W. (1983). La chaîne tello-rifaine (Algérie, Maroc, Tunisie): Structure, stratigraphie et évolution du Trias au Miocène. *Rev. Géol. Dyn. Géogr. Phys.* **24**, 201–97.

Wilson, J. T. (1963). Hypothesis of Earth's behaviour. *Nature* **198**, 925–9.

Winkler, W. and Bernoulli, D. (1986). Detrital high-pressure/low-temperature minerals in a late Turonian flysch sequence of the eastern Alps (western Austria): implications for early Alpine tectonics. *Geology* **14**, 598–601.

Wolfart, R. and Wittekindt, H.-P. (1980). *Geologie von Afghanistan.* Gebrüder Borntraeger, Berlin. (500 pp.)

Wong, W. H. (1929). The Mesozoic orogenic movement of eastern China. *Bull. Geol. Soc. China* **8**, 1–12.

Workman, D. R. (1977). Geology of Laos, Cambodia, South Vietnam and the eastern part of Thailand. *Overseas Geol. and Miner. Resour.* **50**. (33 pp.)

Xu, R. H., Scharer, U., and Allègre, C. J. (1985). Magmatism and metamorphism in the Lhasa block (Tibet): a geochronological study. *J. Geol.* **93**, 41–57.

Xu, S. T. and Lu, J. in press. Tectonic evolution of Tan-Lu fault zone.

Xu, S.T., Dong, S. W., Zhou, H. Y., and Chen, G. B. (1984). Fault rocks and southward thrusting of Dabie Complex in East Dabie Mountains, Anhui, China. *Kexue Tongbao* **29**, 1527–30.

Xu, S. T., Zhou, H. Y., Dong, S. W., Zhang, W. M., and Chen, G. in press. *Deformation and tectonic evolution of the predominate structural elements in Anhui Province, China.* Hefei Geological Bureau.

Yang, Z. Y., Cheng, Y. G., and Wang, H. Z. (1986). *The Geology of China.* Clarendon Press, Oxford. (303 pp.)

Yang, Z. Y., Wu, S. B., and Yang, Q. (1981). Permo-Triassic boundary of South China. *J. Wuhan Coll. Geol.*, **1**, 4–15 (in Chinese).

Yilmaz, O. and Boztuğ, Y. (1986). Kastamonu granitoid belt of northern Turkey: first arc plutonism product related to the subduction of the Paleo-Tethys. *Geology* **14**, 179–83.

Yilmaz, Y. and Şengör, A. M. C. (1985). Paleo-Tethyan ophiolites in Northern Turkey: petrology and tectonic setting. *Ofioliti* **10**, 485–504.

Zhang, Q. W. (1981). The sedimentary features of the flysch formation of the Xikang Group in the Indosinian Songpan–Ganze geosyncline and its tectonic setting. *Chinese J. Geol.* **27**, 405–12.

Zharkov, M. A. (1984). *Paleozoic salt bearing formations of the world.* Springer-Verlag, Berlin. (427 pp.)

Zhu, X., Chen, H., Sun, Z. and Zhang, Y. (1983). The Mesozoic–Cenozoic tectonics and petroliferous basins of China. *Acta Geol. Sinka* **183**(3), 235–42.

Ziegler, A. M., Scotese, C. R., McKerrow, W. S., Johnson, M. E., and Bambach, R. K. (1979). Paleozoic paleogeography. *Ann. Rev. Earth Planet. Sci.* **7**, 473–502.

Ziegler, P. A. (1982). Triassic rifts and facies patterns in Western and Central Europe. *Geol. Rdsch.* **71**, 747–72.

A. M. C. Şengör, İTÜ Maden Fakültesi, Jeoloji Bölümü, Teşvikiye, Istanbul, Turkey. Also at Tübitak-Tbae, Yerbilimleri Bölümü, Gebze, Kocaeli, Turkey.

Altan Cin, Timur Ustaömer, İTÜ Maden Fakültesi, Jeoloji Bölümü, Teşvikiye, Istanbul, Turkey.

Demir Altiner, Ortadoğu Teknik Üniversitesi, Mühendislik Fakültesi, Jeoloji Bölümü, Ankara, Turkey.

K. J. Hsü, Geologiches Institut der ETH-Zentrum, 8092 Zürich, Switzerland.

Lower Palaeozoic facies and faunas around Gondwana

L. R. M. Cocks and R. A. Fortey

ABSTRACT: A new base map for the Lower Palaeozoic of the Gondwanan continental plate incorporates various modifications to older reconstructions. An analysis is attempted of faunas at various times through the Lower Palaeozoic, in particular the Lower Ordovician (Arenig–Llanvirn), the middle and late Ordovician (Caradoc–Ashgill) and the early Silurian (Llandovery). Regressions and transgressions are identified and 'western' and 'eastern' Gondwanan platform faunas identified in contrast to the deeper-water peri-Gondwanan marginal faunas. The effects of climatic variation, both with time and between contemporary areas, are reviewed, and are seen to be the chief controlling factor on platform faunas, with a cline from Boreal to tropical faunas alongside the Gondwanan palaeocontinent. The biogeographical relationships of deeper-water faunas were more complex: certain distinctive taxa may be found distributed widely around the perimeters of the continent, and other faunas have taxa in common with marginal faunas around other contemporary palaeocontinents. Some older-described faunas, particularly brachiopods, trilobites and graptolites, are freshly reviewed, and new faunal data from Libya, Turkey, and Iran are presented and figured.

Introduction

During the past quarter century, not only have the principles of plate tectonics been properly appreciated and developed, but an enormous number of new faunas have been described from what have previously been considered more distant and less accessible parts of the world. Only now are we able to construct a coherent picture of Lower Palaeozoic macrofaunas around and on the vast Gondwanan palaeocontinent. Some pioneering discussions (e.g. Dean 1967) have focused on a so-called 'proto-Tethys', but we prefer to regard this as a single margin relating to the larger picture of Gondwana as a whole. However, compared with relatively reliable reconstructions of the Gondwanan continent in the later Palaeozoic and early Mesozoic, the Lower Palaeozoic still poses many problems. The stable core of the continent—South America, Africa, Peninsular India, Madagascar, Antarctica, and Australia—is not in question, but the status of peripheral areas, which may have been microcontinents separate from the main part of Gondwana or perhaps accreted to it at one time or another, is still under active debate. Even the southern half of Britain itself has been considered separate from Ordovician Gondwana in earlier reconstructions (e.g. Whittington and Hughes 1972; Ziegler *et al.* 1979), whereas recent maps (e.g. Cocks and Fortey 1982) show it as contiguous with the main continent in the early part of the Ordovician, and its detailed geographical position is still undecided.

In this paper we review some of the evidence from faunas on the constitution of Lower Palaeozoic Gondwana and their relationships to palaeolatitude. Although we have many old and new data on the Cambrian (for which also see Palmer 1972), we have been forced through pressures of time and space to concentrate this review on the Ordovician and Silurian, on which we also have the most personal experience; nevertheless we can now present an evolving story which we feel is internally consistent with the considerable data base. Lack of space precludes a complete list of references here, but many of our quoted works themselves contain substantial further documentation. In some areas, for example the Middle and Far East and some parts of South America, the evidence is sketchy, and perhaps we have been forced to make it carry more weight than it should bear.

We have generated a new base map for this paper (Fig. 1). The core of Gondwana is based on a combination of two maps for the Triassic, kindly provided by H. G. Owen using a zenithal equidistant projection with poles in southern Africa and western Australia respectively; to this combination we have added southern Europe and Iran–Afghanistan on the same scale. South-east Asia is split up following the tectonic lines of Mitchell (1981), showing South China, a conventionally cigar-shaped northern Himalayan continent to the north-east of Peninsular India, and a 'Sibumasu' (Siam (Thailand), Burma, Malaya, and Sumatra) continent near

From AUDLEY-CHARLES, M. G. & HALLAM, A. (eds) *Gondwana and Tethys* Geological Society Special Publication No. 37, pp. 183–200.

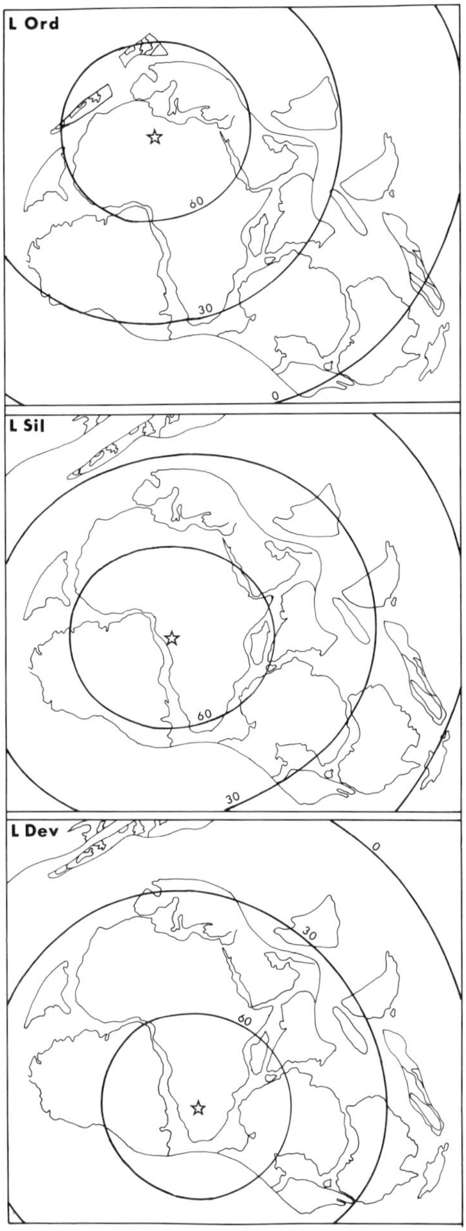

FIG. 1. Reconstructions of the Gondwanan area in Lower Ordovician, Lower Silurian, and Lower Devonian times, showing its movement across the contemporary South Pole and palaeolatitudes. For more details of the map reconstruction, see the text.

Australia, following Burrett and Stait (1985). The New Zealand area is added as a lozenge shape and includes the reconstructed shelf area nearby, as it might have been in Mesozoic and earlier times (R. A. Cooper pers. comm.), and the lesser part of Antarctica is displaced along the eastern fault line. The southern part of South America and a part of Antarctica are omitted, since they represent accretion after Lower Palaeozoic times. The continents of Laurentia and Baltica are not shown on the Arenig–Llanvirn maps since they are outside the map area, but partly included in the end Ordovician to early Devonian maps, as the Iapetus Ocean and Tornquist's Sea closed during the Ordovician and Silurian (Cocks and Fortey 1982). Siberia, Kazakhstan, North China, and the Indo-China continental masses are omitted, since their contemporary positions with respect to Gondwana are unknown: their Ordovician and Silurian faunas probably indicate tropical or north temperate situations. The positions of successive South Poles within Africa (Fig. 1) are based on palaeomagnetic evidence, which has demonstrated a wandering from northern Africa in the early Ordovician to southern Africa at the end of the Silurian, and the 60° and 30° latitude lines and the equators on these reconstructions have been drawn to scale by using modern geographical distances. The coincidence between these polar and equatorial positions and the other diagrams, which we have constructed from our faunal and facies data alone, show a heartening level of agreement.

Biofacies, continents, and palaeolatitude

The factors which determine any particular biofacies, at a particular time, on a particular continent, are both physical and biological. The chief physical factors are (1) plate movement, both with respect to the poles (and hence their palaeolatitude) and also with respect to the closure or separation of faunas on adjacent plates; (2) the depth of water, which affects the actual position of land margins and their adjacent cratonic faunal belts within a plate and which is itself affected by eustatic variations or by changes due to local tectonic activity; and (3) by climatic variation: at some periods in the Lower Palaeozoic the climate seems to have been far more universally equable or extreme than at other times. The chief biological factors are (1) the availability of suitable species at any one time to colonize any particular habitat; (2) the ecological preferences for various life styles, water depths, and substrates; (3) the local availability of food; and (4) the tolerance ranges of individual genera and species.

On the maps (Figs 2 and 8) we have attempted to plot the distribution of faunas belonging to three major categories, in terms of biofacies.

1. Inner-shelf, or cratonic faunas, these being faunas adapted to living over the wide areas of relatively shallow seas that flooded cratonic areas, especially during transgressive phases. Such cratonic faunas are most influenced by local climatic and substrate conditions; these faunas are also regarded as the best indicators of continental siting of any area in question, because they include a suite (or suites) of Gondwanan endemics adapted to climatic and other conditions that pertained on the palaeocontinent. Such faunas can be expected to change in harmony with palaeolatitude, but will intergrade across a palaeolatitude cline.

2. Deeper-shelf faunas. In this category we have included faunas found in sites relatively exterior to the above. Typically this is on a continent to open-ocean transect, but such deeper-water faunas also follow the course of intracratonic basins (e.g. aulacogens). In general, the more or less linear distribution of deeper-shelf faunas can be expected to correspond with the periphery of the continent, and contribute to the recognition of its boundary. On the mapping scale we have to use, we are obliged to 'lump' in this category more than one biofacies; at a regional level these finer biofacies divisions may be very useful biogeographically. So, for example, in the early Ordovician, we include together in the deeper-shelf category the raphiophorid 'community type' of Fortey and Owens (1978), as well as the cyclopygid biofacies (Fortey and Owens 1987), and the nileid–raphiophorid–shumardiid ('Nileid') biofacies found in many peripheral sites. Taxa within the deeper-shelf faunas may be widespread, corresponding with more uniform conditions at greater depth, and hence are not to be regarded in any way as necessarily 'bound' to Gondwana—indeed some of the genera found in these deeper-water faunas may be found on the fringes of several of the Ordovician continents, and are certainly capable of being found all around Gondwana.

3. Oceanic faunas. This category includes faunas of the most exterior kind in the shelf-to-ocean transect, especially oceanic graptolite faunas of the isograptid biofacies, the olenid trilobite biofacies (earlier Ordovician), or the deepest graptolite biofacies in the Silurian. We regard the presence of these biofacies as good evidence of the contiguity of an ocean, and hence they are of particular service in defining Gondwanan margins. We (Fortey and Cocks 1986) have recently plotted the world distribution of the isograptid biofacies, and found it to be unrelated to palaeolatitude or continent plate distribution in general—it includes genuine pandemics even at a time of 'provinciality' in other faunas.

Identification of the biofacies context is a necessary prerequisite before drawing any biogeographical conclusions, because the distributional properties of the faunas differ radically from one biofacies to the next. The inner-shelf faunas are those we regard as the best indicators of palaeogeographical contiguity, and of palaeolatitude; the deeper-water faunas are the best indicators of proximity of an oceanic margin. By combining the evidence from the distribution of more than one biofacies we can arrive at the most plausible palaeogeographical position. Even so, where sparse information only is available, we have sometimes been obliged to decide on biofacies on the basis of small samples, which is not ideal. The placing of microcontinents depends on the comparison of their more inshore faunas with those of recognized inner-shelf faunas elsewhere. However, this still leaves much freedom with regard to actual siting (for example considerable strike-slip movement may be possible, which may not be detectable faunally), and such questions have to be resolved by recourse to other kinds of geological criteria. Even using faunas alone, the data base is very large, and is summarized in the selected references on the figures; this information is not generally repeated in the text.

Dean (1967) emphasized 'migration routes' for faunas along Tethyan lines in the Lower Palaeozoic. So far as the palaeontological calibration allows, migration in deeper-shelf sites was virtually instantaneous: this is shown, for example, by the occurrence of a very similar trilobite fauna at the base of the Tremadoc (with *Niobe*, *Shumardia*, *Macropyge*, *Remizites*, olenids of similar or identical species) in North Wales, China, and South America at the extremes of our Gondwana reconstruction. As far as the inner-shelf faunas are concerned, there may be better evidence of movement of endemic taxa from one area to another, but the control on this is regarded palaeoclimatic or environmental, rather than as the shifting of barriers or the opening up of seaways. We might cite the spread of the *Hirnantia* fauna (and the *Mucronaspis/Dalmanitina* trilobite fauna) from its 'ancestral' position in the temperate or subpolar western Gondwana to a widespread occurrence late in the Ordovician, accompanying the climatic deterioration of the Ashgill glaciation. Again, the trilobite *Neseuretus* is known from the early Arenig only from western Gondwana (Wales, Newfoundland, southern France) but, by the Llanvirn–Llandeilo, occurs over much of Gondwana outside the equatorial belt. It might be tempting to assume a 'migration' in this case, but it seems equally likely to us that the distribution reflects lack of occurrence or lack of collection of the right biofacies in the Arenig of other parts of Gondwana. Similarly, the cyclopygid biofacies is best known from the Arenig of the western ege of Gondwana, but we do *not* regard its subsequent appearance elsewhere as the result of migration—only as the result of a more widespread recovery of the appropriate exterior biofacies (only the eastern edge of Gondwana has the full shelf-to-ocean profile preserved as early as the Arenig). In short, migration of faunas should be viewed with caution, except where it is the product of climatic zonation in shallow-water faunas.

Early Ordovician

The early Ordovician is understood to include the Arenig–Llanvirn on the maps (Figs 2 and 5), a time when 'provincial' differentiation was at a maximum (Whittington and Hughes 1972; Fortey 1975), as a result of strong climatic differentiation and dispersed continental areas. This is shown by the endemic character of the Gondwanan inner-shelf faunas, when compared with inner-shelf faunas of the Baltoscandian craton, and those of Laurentia. There are almost no benthic genera in common between shelf faunas of boreal Gondwana (northern Africa, Armorica, Iberia, Welsh Borderlands), and the Baltoscandian platform centred on Sweden and Estonia, a fact which led Cocks and Fortey (1982) to propose the existence of Tornquist's Sea between them at this time; the separation of Gondwana from Laurentia by an Iapetus Ocean has been familiar for some years. However, even at this time the deeper-shelf faunas may include widespread genera (this fact, with ostracod examples, has misled Schallreuter and Siveter (1985) to suggest that Iapetus may have been narrower than supposed in the middle Ordovician).

Shallower shelf faunas

The relatively shallow shelf faunas can be divided into two types on the basis of endemic trilobites and brachiopods. These two faunas are climatically controlled.

Dalmanitid–calymenacean shelf fauna

This fauna is western Gondwanan, and inhabited high palaeolatitudes. It is found in clastic sediments, associated with sandstones of Grès Armoricain type, and non-calcareous mudstones and ironstones. The sandstones almost invariably contain *Cruziana* trails of a few species (Bergström 1976). We term this fauna the dalmanitid–calymenacean fauna, because of the diversity of these two taxa within the appropriate biofacies.

The dalmanitid–calymenacean fauna includes numerous endemic species, which are found over a wide area comprising hundreds of individual localities in Armorica, Spain and Portugal, Sardinia, North Africa (Morocco, Tunisia, Libya—Fig. 4); excellent monographs of this fauna have recently been published by Henry (1980) and Hammann (1983). Typical elements include: Calymenidae such as *Colpocoryphe*, *Salterocoryphe*, *Calymenella*; Homalonotidae such as *Platycoryphe*, *Iberocoryphe*, *Eohomalonotus*, *Kerfornella*, and *Plaesiacomia*; Dalmanitidae including the various genera of Zeliszkellinae, *Ormathops*, *Phacopidina*, *Eodalmanitina*, and *Crozonaspis*; certain asaphids, such as *Ogyginus*; the pliomerid *Placoparia*; the cheirurid *Eccoptochile*; the odontopleurid *Selenopeltis* and the trinucleid *Myttonia*. Also very typical of the western Gondwanan shallow-water cratonic biofacies is a fauna of characteristic large inarticulate brachiopods (Fig. 3), including the genera *Lingulobolus*, *Pseudobolus*, *Ectenoglossa*, *Lingulepis*, and *Tomasina*: these are known from the whole region from Bohemia and Budleigh Salterton, Devon (UK) in the north to Algeria and Morocco in the south—sometimes they co-occur with one or more members of the dalmanitid–calymenacean trilobite fauna, but more often they appear to have been deposited under even shallower water conditions. In turn, some of the trilobite genera have ranges which extend into more marginal sites where they co-occur with deeper shelf faunas: *Ormathops*, *Colpocoryphe*, *Plaesiacomia*, and *Placoparia*. These are found in Bohemia and central South Wales, for example. Some deeper-water species have reduced visual surfaces.

According to Fortey and Morris (1982), the

calymenid trilobite *Neseuretus* is the most inshore of these western Gondwanan genera and hence of particular use in recognizing a Gondwanan platform. It is the first trilobite to appear in upward-deepening successions, often in units immediately above *Cruziana*-bearing transgressive sandstones (it is the only known genus from the Stiperstones Formation, for example); it is confined to Gondwana, and a detailed list of occurrences was given by Fortey and Morris. The fauna associated with *Neseuretus* is often a sparse one, but it does regularly intermingle with

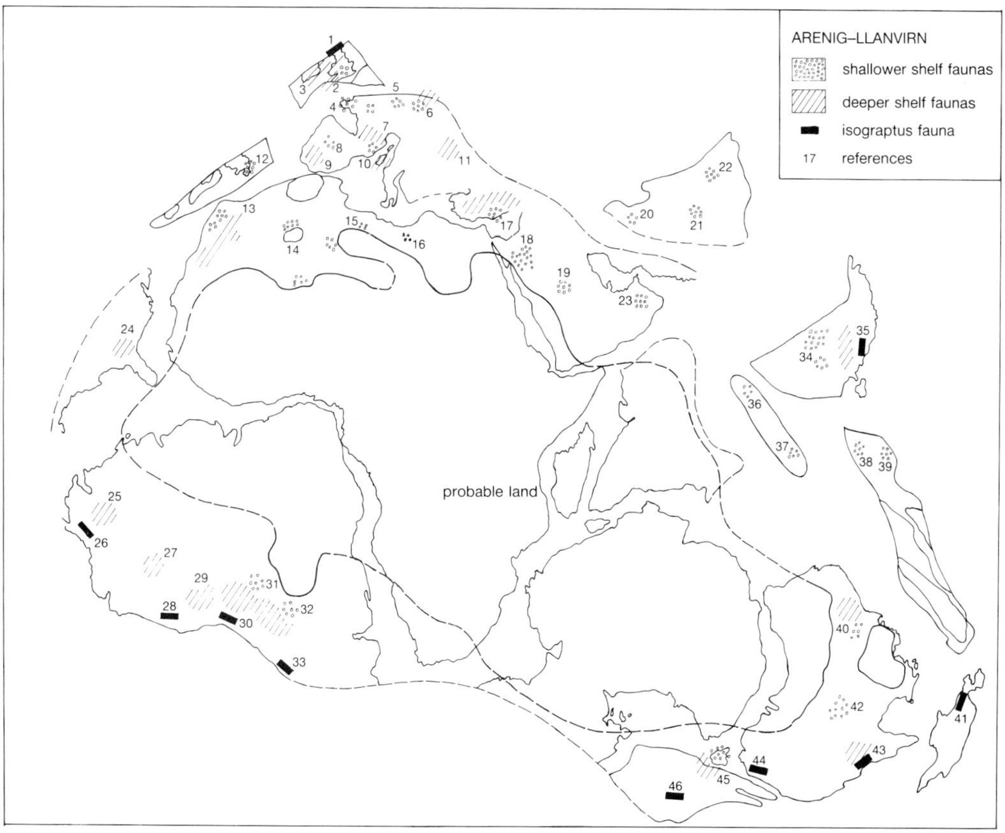

FIG. 2. Distribution of Lower Ordovician (Arenig–Llanvirn) biofacies in the Gondwanan area. References (with further local references therein) are as follows: *1* English Lake District (Jackson 1962), *2* South Wales (Fortey and Owens 1987), *3* Southern Ireland (Brenchley *et al.* 1967), *4* Armorican Peninsula, France (Henry 1980), *5* Thuringia, Germany (Jentsch and Stein 1961; Struve 1975), *6* Bohemia (Havlíček and Vanek 1966; Havlíček 1977), *7* southern France (Courtessole *et al.* 1985; Dean, 1966), *8*, *9* Spain (Hammann 1983; Gutierrez Marco 1982), *10* Sardinia (Serpagli and Gnoli 1984), *11* Bulgaria (Spassow 1958), *12* Newfoundland (Dean and Martin 1978), *13* Morocco (Destombes, in Holland 1985), *14* Algeria (Legrand in Holland 1985), *15* Libya (new data in this chapter), *16* Tunisia and Egypt (Klitzsch, in Holland 1981), *17* Turkey (Dean 1975), *18* Jordan (Wolfart, in Holland 1981), *19* Saudi Arabia (El-Khayal and Romano 1985), *20*, *21* Iran (new data, and Pillet 1973), *22*, *23* Afghanistan (Blaise and Bouyx 1980, Wolfart 1970; in Holland 1981), *24* Florida, USA (Whittington 1953), *25*, *26* Colombia (Harrington and Kay 1951; Turner 1960), *27*, *28* Peru (Turner 1960; Hughes *et al.* 1980), *29–31* Bolivia (Bulman 1931; Harrington and Leanza 1957; Pribyl and Vanek 1980), *32*, *33* Argentina (Harrington and Leanza 1957; Turner 1960), *34*, *35* China (Lu *et al.* 1976; Cooper and Ni 1986), *36* Niti, India (Salter and Blanford 1865, revised herein), *37* Everest area, China (Liu 1976), *38* Shan States, Burma (Fortey and Morris 1982), *39* south-west China (Sheng 1974), *40* Canning Basin, Australia (Legg 1976), *41* New Guinea (Fortey and Cocks 1986), *42* Georgina Basin, Australia (Fortey and Shergold 1984), *43* Queensland (Henderson 1983), *44* Victoria (Thomas 1960), *45* Tasmania (Burrett *et al.* 1983), *46* New Zealand (Cooper 1979).

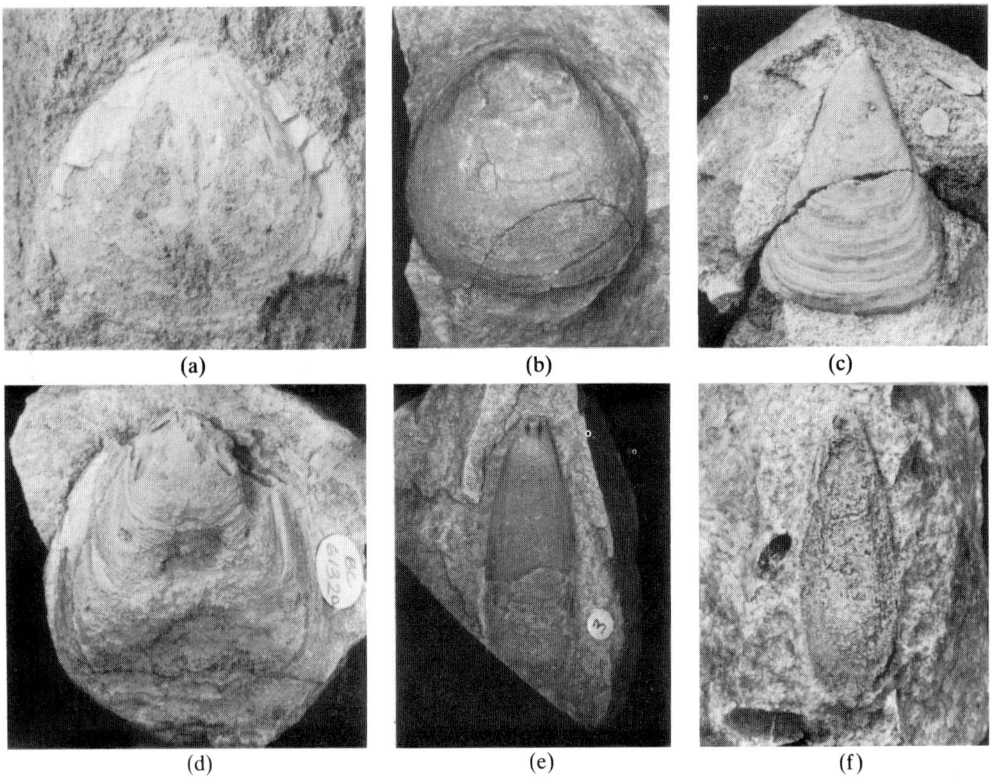

FIG. 3. Characteristic large inarticulate brachiopods of the Lower Ordovician (Arenig) western Gondwanan platform fauna, all ×1. (a) *Lingulobolus hawkei* (Rouault) B 80085, Krusna Hora, Bohemia; (b) *Pseudobolus salteri* (Davidson) B 21504, Budleigh Salterton, Devon (UK); (c) *Lingulepis crassipyxis* Havlíček B 5932, St Remy de Sillé, France; (d) *Lingulobolus brimonti* (Rouault) BC 6132, north of Hoggar, Algeria; (e–f) *Ectenoglossa leseueuri* (Rouault) B 21516, Budleigh Salterton and B 5928, Bagnoles, France.

the dalmanitid–calymenacean fauna in France and Spain. However, its range is much more extensive than that of the dalmanitid–calymenacean fauna as a whole; indeed, *Neseuretus* is found in inshore environments almost all the way around Gondwana, except in limy facies in equatorial regions. It occurs in Bolivia, South China, and the Tibetan block, as well as Saudi Arabia, for example. We think it likely that *Neseuretus* had an exceptional latitudinal tolerance, and was a eurytopic genus adapted to inshore clastic environments. It is interesting to note that Jablonski and Valentine (1981) showed that the most inshore of bivalves tended to be the most eurytopic and to cross more than one latitudinal zone.

The dalmanitid–calymenacean fauna is equivalent to what Whittington and Hughes (1972) termed the *Selenopeltis* province, after a characteristic odontopleurid trilobite that was apparently confined to Ordovician high palaeolatitudes. *Selenopeltis* itself is a genus which is found in more than one biofacies: in the Armorican–Iberian region it occurs with the dalmanitid–cheiruracean fauna, while in Wales it may even occur with deep-water assemblages such as the cyclopygid biofacies. With this spread of occurrences it may well have been epipelagic (Fortey and Owens 1987). The emphasis here is on benthic faunas, especially those of restricted occurrence; hence we prefer to define the shelf faunas in terms of endemic taxa, but the term '*Selenopeltis* province' may be usefully retained as a general description of western Gondwana, to which the eponymous genus is confined.

The Asaphopsis *shelf fauna*

This fauna is mostly eastern Gondwanan, and inhabited relatively low palaeolatitudes. We have adopted the terminology of Whittington

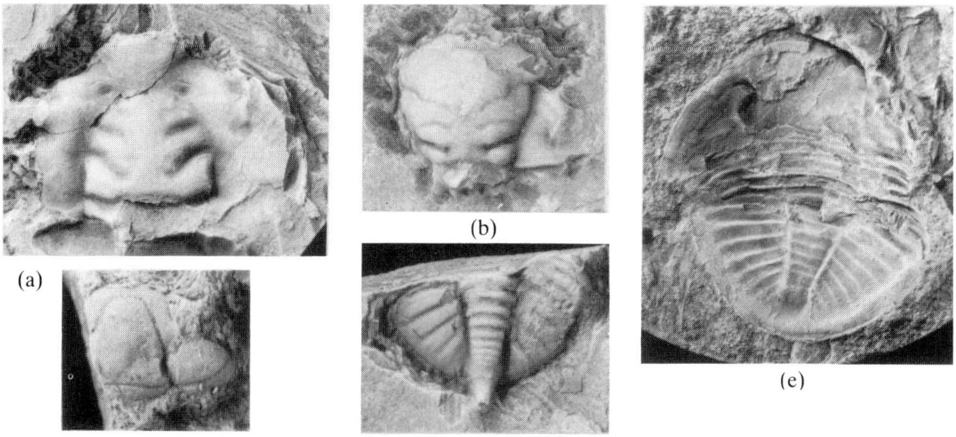

FIG. 4. Undescribed, probably late Llanvirn trilobite fauna from Libya, of typical western Gondwana platform type; Tripolitania, well D1–34, core 14 (BP). (a) *Neseuretus tristani* (Desmarest), cranidium, $\times 1\frac{1}{2}$, It 19924; (b) (c) cranidium and pygidium of *Crozonaspis morenensis* cf. *mayensis* Clarkson and Henry, ×2, It 19922–3; (d) *Plaesiacomia* cf. *vacuvertis* Thomas, cranidium, ×6, It 20467. Bir Tlacsin, Well A1–70, Core 40 (Shell), (e) imperfect exoskeleton of *Ogyginus corndensis* Whittard, $\times 1\frac{1}{2}$, It 19920.

and Hughes here, because it is appropriate for a description of low-latitude Gondwanan faunas. Typical platform faunas are described from China (Lu 1975) and Australia (Fortey and Shergold 1984); although clastic lithologies again seem characteristic, some of the Australian fauna are associated with limestones. The *Asaphopsis* shelf faunas extend into Indo-China and South Korea, on the one hand, and Argentina and Bolivia on the other. Typical elements include: Dikelokephalinidae—a variety of this family is typical including *Asaphopsis* itself and such genera as *Dikelokephalina* and *Dactylocephalus*; Leiostegiidae; *Annamitella* (which extends outside the *Asaphopsis* shelf on volcanic islands) and *Pseudocalymene* (= *Eucalymene*); Trinucleidae: *Ichangolithus*, *Ningkianolithus*, *Ceratolithus*, *Nankinolithus*; Prosopiscidae: *Prosopiscus*; Hammatocnemidae; *Hammatocnemis* (all the earlier history of this genus is eastern Gondwanan): endemic asaphids such as *Birmanites*. The *Neseuretus* fauna extends into China, but not Australia.

There are some differences between the Australian platform and the South Chinese platform, for example the trinucleids are only present in the latter, while the former has its own endemic asaphids. Nonetheless, both Cambrian and Ordovician trilobite faunas of Australia and China resemble one another much more closely than they resemble the eastern Gondwanan faunas. Australia was at tropical palaeolatitudes, and this is reflected by similarities to Laurentia in elements of the fauna which were not confined to Gondwana, and capable of crossing oceanic barriers, such as nautiloids and conodonts, and a few pelagic trilobites (*Carolinites*, *Opipeuter*).

Geographical cline

The relationships between the eastern and western Gondwanan shelf faunas are best understood as a geographical and climatic cline, extending from the extremes of equatorial Australia to high-latitude North Africa, and without sharp breaks (Fig. 5). Distribution of individual genera will be determined by their tolerance ranges along this cline, reflected in overlapping ranges (contrast the sharp truncation of ranges at oceanic barriers, as shown by Tornquist's Sea). This is shown by the fact that 'mixed' eastern and western Gondwanan faunas occur where they should, on palaeolatitudinal grounds, in the Middle East. For example, El-Khayal and Romano (1985) have described a shelf trilobite fauna from the Hanadir Shale, Saudi Arabia, which includes both dalmanitid–calymenacean genera (*Kerfornella*, *Kloucekia*, *Plaesiacomia*) and the Chinese trinucleid *Ningkianolithus*. The eastern Gondwanan genus *Pseudocalymene* extends towards Iran, while Dean, in his various papers on Turkish trilobite faunas, has emphasized their 'mixed' affinities, including even some genera usually regarded as Scandinavian endemics (e.g. *Panderia*), which is

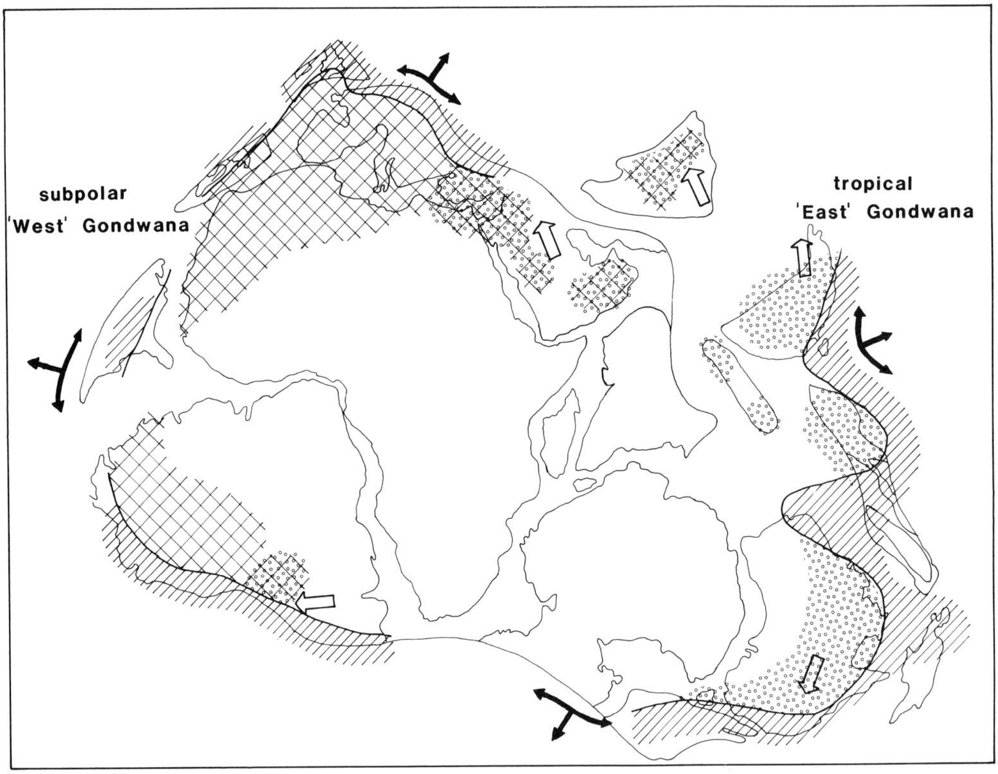

FIG. 5. Lower Ordovician biofacies distribution in Gondwana; the diagonal hatching is the marginal biofacies, whose faunal interchange routes (solid black arrows) extend both round the Gondwanan margins and to neighbouring continents; the cross-hatching is the endemic high-latitude western Gondwana biofacies; the small open circles the low-latitude eastern Gondwana biofacies. The broad open arrows show the faunal interchange routes between the low-latitude eastern Gondwana faunas and those of neighbouring low-latitude cratons (such as North America), and the faunal interchange routes between the mixed central Gondwana biofacies and neighbouring temperate palaeocontinents (such as Baltica).

perhaps not surprising in view of the temperate latitudes which should pertain both in Baltica and this part of Gondwana on our interpretation. Indeed, it should be predicted that this part of Gondwana will prove to be the 'melting pot' for Ordovician faunas, because here the eastern and western Gondwanan faunas will meet, while temperate latitude faunas may be recruited from outside Gondwana. The brachiopods demonstrate this very well, for example the bizarre porambonitacean *Yangzteella* was until now thought to be endemic to South China in the Upper Arenig (Fig. 6a), but recent collecting by W. T. Dean has shown that it is also present in contemporary rocks in southern Turkey (Fig. 6b), and more general studies of the fauna through Ordovician time (Williams 1973) show comparable 'provincial' groupings and gradations.

Not surprisingly, in view of the climatic transect involved, truly pan-Gondwanan genera in shelf faunas are most unusual. One is the dikelokephalinid trilobite *Hungioides* (see Rabano 1983), which is found all around Gondwana regardless of palaeolatitude—but even this genus is much rarer in the western than the eastern sector. Nonetheless, it must have been tolerant of very different conditions; another eurythermal form was the bellerophontoid *Peelerophon*, which is recorded from Argentina, Australia, and Thailand (Jell *et al.* 1984).

Deeper shelf and oceanic faunas

The simple relationship between palaeolatitude and faunal geography that we applied to the shallower-shelf faunas cannot be recognized in the outer shelf and ocean-facing faunas.

The deep-water graptolitic faunas of the

isograptid biofacies are recognized from several sites around the margin of Gondwana, and from different inferred palaeolatitudes. Yet certain graptolite *species* have been identified from these widely separated localities. For example the *Pseudisograptus manubriatus* plexus known from Australia, New Zealand, and the exterior facies belt of China (Cooper and Ni 1986) in eastern Gondwana at low palaeolatitudes, occurs also in Britain (and probably also South America) at high palaeolatitudes, similarity extending to the subspecific level; one might also cite species of *Glossograptus*, *Sigmagraptus* and *Glyptograptus*. Obviously, different biogeographical 'rules' apply to such species by comparison with the inner-shelf faunas. It may have been that such forms occurred below the thermocline, where environmental conditions were more uniform regardless of palaeolatitude. The more inshore graptolite faunas mapped by Fortey and Cocks (1986) apparently include some Gondwanan endemics at this time—for example species of *Azygograptus* and *Corymbograptus*, although there are taxonomic problems in this case which serve to obscure relationships. For example, the Arenig outer-shelf graptolite faunas (*not* isograptid biofacies) from south-west China, described by Mu *et al.* (1979) include a host of new taxa, as well as some described from Britain, but without population studies it is difficult to confirm these identities, or the real distinction of the Chinese forms.

Exterior trilobite biofacies also include much more widespread taxa (here considered at generic level) than was the case for the shallow-shelf faunas, and the distinction between eastern and western Gondwana is less clear-cut.

1. Fortey and Owens (1978) described an olenid biofacies (including *Bienvillia* and *Hypermecaspis*) fauna from the Welsh Basin from eastern Gondwana; the same genera occur in Argentina (Harrington and Leanza 1957) in the appropriate biofacies there, and in Queensland (Henderson 1983)—they also occur at the edge of the Laurentian Plate. The olenid biofacies was evidently independent of palaeolatitudinal constraints.

2. The cyclopygid biofacies is another distinctive, deeper assemblage of probably mesopelagic trilobites; so far as we know its early Ordovician occurrence is confined to deeper-water sites around western Gondwana and Kazakhstan; very similar species appear somewhat later in south-east China, which again implies some independence from the constraints operating on the shallower-water trilobites.

3. Deeper-water trilobite faunas of Arenig

FIG. 6. New Gondwanan brachiopod faunas. (a)(b) *Yangzteella poloi* (Martelli), (a) Upper Dawan Formation (Upper Arenig), Huanghuachang, China, ×2, BB 68364, (b) sandstone equivalent to Sobova Limestone (Upper Arenig), Seydişehir, Turkey (Coll. W. T. Dean), ×2, BC 6829. (c) and (d) *Dalmanella* aff. *multiplicata* (Bancroft), Caradoc beds, Kuh-e-Surmeh, Zagros Mountains, Iran, ×2, BB 56875. (e) *Aegiromena descendens* (Havlíček), Melez Chograne Formation (Caradoc), 35 km north of Mourizede, Libya, ×4, BB 47505.

age, other than olenid and cyclopygid biofacies, have been described from the Welsh Basin by Fortey and Owens (1987) and from the Montagne Noire in western Gondwana (Dean 1966). The distribution of genera in these faunas is quite complex. The faunas include some genera (*Ormathops*, *Colpocoryphe* and *Selenopeltis* are examples) which are related to the western Gondwanan faunas of the adjacent shelf; in the case of the Welsh faunas these include peculiar, blind species adapted to benthic life in the deep-water habitat. A majority of genera ('*Geragnostus*', *Shumardia*, *Taihungshania*, *Hangchungolithus*, *Ampyx*, *Illaenopsis*, *Dionide*) are widespread in deeper-water sites around Gondwana, including China, and may prove to be widespread even at specific level. Added to these, there are occasional genera (*Leioshumardia* from Wales, *Ceraurinella*, *Symphysurus* from the Montagne Noire) which are otherwise known from the faunas of palaeocontinents other than Gondwana. On eastern Gondwana, early Ordovician deeper-water faunas, such as that described from north-west Australia by Legg (1976) are also mixed in palaeogeographical affinities. A few genera (*Prosopiscus*, *Gogoella*, *Fitzroyaspis* from Legg's fauna) belong to shallow-shelf assemblages of eastern Gondwanan, low-latitude type, and there are no western Gondwanan shelf forms; the same fauna includes widespread genera such as *Shumardia*, *Triarthrus*, and *Endymionia*, and some forms (*Plesiomegalaspis*, *Ogygiocaris*) which are also known from Baltica. Pelagic trilobites in Australia (Legg 1976; Henderson 1983; Jell and Stait 1985) are of palaeo-equatorial genera (*Carolinites*, *Opipeuter*) rather than cyclopygid biofacies. Similar comments may be applied to the deeper-water Arenig faunas from Argentina listed in Harrington and Leanza (1957). Until more is known about detailed biofacies relationships and relative abundances of different taxa in deeper-shelf faunas, it would be unwise to generalize further than to state that deeper-shelf faunas are *not* simply differentiated by palaeolatitude, and that they include both peri-Gondwanan genera and some genera known from other continents in marginal facies. Shallow-shelf taxa mixed in with such faunas do relate to those of the adjacent shelves. The general observation that the endemicity of faunas decreases with depth is supported.

Aulacogens

Although the general picture of inner-shelf–outer-shelf–open-ocean profile shown on our maps may be broadly applicable, it is clear that deeper-shelf faunas encroached on to the Gondwanan continent in peripheral areas, in what are presumed to have been intracratonic basins or aulacogens. In the European sector of western Gondwana, for example, the faunas of the Montagne Noire include deeper-shelf faunas, including diverse cyclopygids in the Arenig, which are distinct from the Armorican platform and Iberian platform faunas of inner-shelf type. Hence, we regard it as likely that there was an intracratonic basin, possibly running between the Iberian and Armorican cratons, as far as Sardinia. Some indication of its northern limit is provided by what Henry (1980) termed the 'Synclinal d'Ancenis' in Armorica. This may have been the reason for a 'proto-Tethys' being marked on the reconstruction of Whittington and Hughes (1972). On the other hand, the similarities of the shallow-shelf faunas and associated lithofacies to the north of this structure in Armorica, to those to the south in Iberia and North Africa, indicate that the aulacogen is not likely to have involved much, if any, true oceanic separation, at least by comparison with Tornquist's Sea or Iapetus. Conversely, we can find no evidence of an aulacogen running along the course of the English Channel in the Ordovician, because there is no sign of the expected deeper-shelf biofacies belt between South-west England and the Cotentin Peninsula. Biofacies belts are likely to prove of service in distinguishing other aulacogens elsewhere.

Also shown on Fig. 5 are solid arrows, which indicate that the deeper-water and ocean-facing marginal faunas of Gondwana could be found both around the shores of Gondwana, and also across the surrounding oceans to other continents—if they were not too far away in the case of the deeper-water biofacies, and if they were at any distance in the case of the oceanic biofacies. This happened regardless of palaeolatitude. The open arrows, on the other hand, show the faunal interchange routes of the shallower-water cratonic faunas, and it can be seen that the colder-water subpolar western Gondwanan faunas could find nowhere suitable to go, in contrast with the tropical faunas of eastern Gondwana, which could inhabit (if the adjacent islands and continents were not too far away) other tropical areas such as Laurentia and Kazakhstan. In the central temperate areas of Gondwana, such as Turkey, Saudi Arabia, and central South America, some faunal interchange occurred, both between the more tolerant components of the fauna in both western and eastern Gondwana, and also between adjacent temperate continents such as Baltica. Fig. 5 also demonstrates that the limited evidence from eastern Gondwana indicates (although it is not yet strong enough to prove) that the Arabian, northern Himalayan, South Chinese and Sibumasu blocks may actually have been attached to the main Gondwana continent in Ordovician times; a hypothesis reinforced by the relative contemporary distribution of the local cratonic, deeper-water and oceanic biofacies (Fig. 2).

Middle to end Ordovician

A characteristic high-latitude fauna (usually thought of as 'Bohemian' or 'Mediterranean') persisted during the Llandeilo and most of the Caradoc over the cratons of northern Gondwana. This fauna has been well documented, chiefly by the outstanding work of V. Havlíček, from Bohemia (e.g. Havlíček 1977), Spain (Villas 1985), North Africa (e.g. Havlíček 1971), and from the Armorican area, including the Budleigh Salterton Pebble Bed of Devonshire (e.g. Melou 1976; Cocks and Lockley 1981). It consists mainly of orthides, and is dominated by genera such as *Tissintia*, *Drabovia*, *Destombesium*, and *Tafilaltia*, and early species of *Hirnantia*: the specific diversity is often very low at individual localities. As one might expect, this fauna has also been found at a comparable palaeolatitude in South America, in Bolivia (Havlíček and Branisa 1980). The trilobites and other faunas marginal to these brachiopod-dominated beds show more links with other areas, e.g. the Caradoc fauna described by Hughes *et al.* (1980) from Peru.

However, the growing proximity of the Baltic continent to Gondwana, with the gradual disappearance and subduction of Tornquist's Sea (Cocks and Fortey 1982), coupled with the shift of the European and North African 'northern' sector of Gondwana northwards away from the South Pole, led to an interesting faunal interchange during the Caradoc. From mid-Caradoc (Actonian) times onwards, faunas with their origins in the temperate Baltic latitudes were able to colonize the shelf regions of what is now south Britain and other parts of the Gondwana shelf. For example, a new fauna from Iran (Fig. 6) consists of a virtually monospecific assemblage of *Dalmanella* aff. *multiplicata*, a species common in the Caradoc of Shropshire (UK), but with its earlier generic origins in the Baltic: the fauna occurs in the poorly sorted siltstones typical of a temperate environment. Faunas from Spain (Villas 1983, 1985) are entirely Bohemian in character during the Lower Ordovician, but by Caradoc times they show many species in common with the type Caradoc area of Shropshire. However, the late Ordovician faunas of the high-latitude areas of North Africa remained of very low diversity and endemic to western Gondwana, for example the new record of the Caradoc Bohemian plectambonitacean *Aegiromena descendens* from Libya (Fig. 6e), and the Caradoc faunas in Niger (Gatinsky *et al.* 1966) which also have 'Bohemian' genera.

In the more tropical regions of Gondwana, diverse and in many cases endemic faunas flourished on the craton, for example in the Himalayas and Burma (Mitchell *et al.* 1977). In New South Wales, Australia, trimerellid reefs and other diverse faunas are known (e.g. Percival 1976) and Percival has demonstrated that one-third of the 43 Caradoc brachiopod genera in New South Wales are endemic, and that nearly all of the 49 species have a restricted distribution. The fauna has provincial affinities with both North America (15 genera in common) and Kazakhstan (8 genera in common), demonstrating both the problems and the triumphs of tropical communications at that time, and Webby (e.g. 1985) has documented a comparable story for the contemporary coral faunas. However, the world-wide transgressions and rise in sea-level in the latest Llandeilo and early Caradoc *gracilis* Zone drowned many cratons, and in many cases left a relatively poor shelly-fauna record over much of Gondwana, in contrast to the rich and diverse tropical faunas of, for example, North America (Laurentia).

Because so much has been written elsewhere, the late Ordovician–early Silurian glacial episode will not be considered in depth in this paper, although we have produced a new map (Fig. 7) which shows the distribution of the known *Hirnantia* and *Mucronaspis* faunas, as well as the ice-flow directions in some of the glacial deposits, particularly those of Beuf *et al.* (1971) from the Sahara and others described by McClure (1978) and reviewed by Bigarella (1973) and Spjeldnaes (in Holland 1981). Glacial traces are known from Caradoc times onwards in North Africa, and occur below the latest Ashgill in South Africa (Cocks and Fortey 1986), but there was an undoubted cooling in latest Ashgill times which affected all of the continents. Each of the individual genera of the distinctive and widespread *Hirnantia* fauna had different origins, with *Hirnantia* itself first known from the Llandeilo of Bohemia (Havlíček 1977), but the key combination of genera, *Hirnantia*, *Eostropheodonta*, *Plectothyrella*, *Dalmanella*, *Paromalomena*, *Kinnella*,

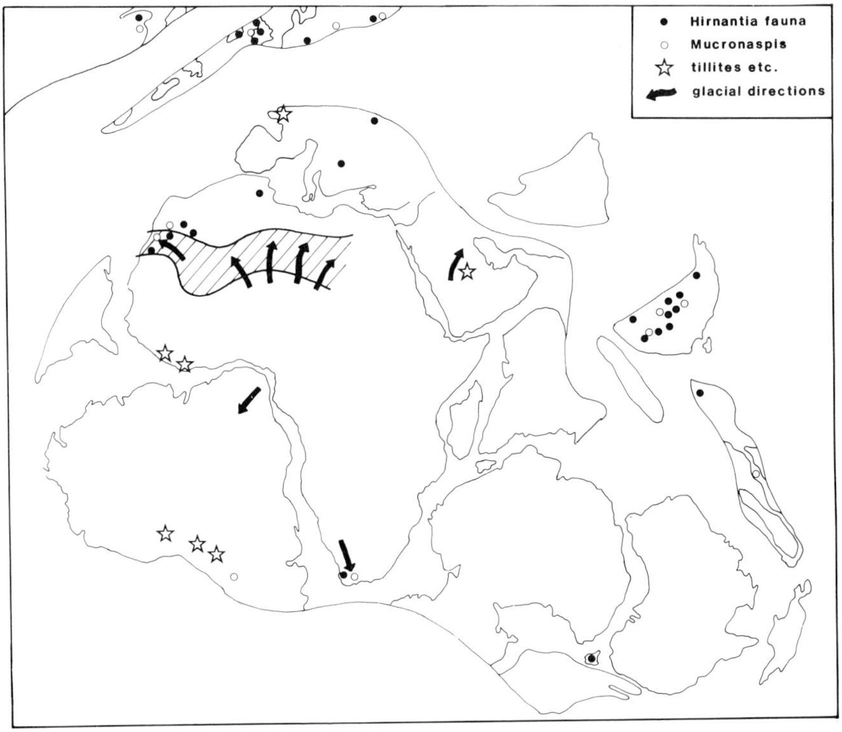

FIG. 7. The Gondwanan palaeocontinent in latest Ordovician (Upper Ashgill) times, showing the glaciogenic deposits (the shaded area in North Africa is from Beuf *et al.* 1971) including ice-flow directions, and the circum-Gondwanan distribution of the *Hirnantia* and *Mucronaspis* faunas.

and others, is not known together before mid-Ashgill times, and clearly represents a colder-water fauna which extended round Gondwana and even as far as North America, as well as being abundant in temperate areas such as China (Rong 1984).

The Silurian

The regression caused by the build-up of ice in the Ordovician–Silurian boundary period, and which may also have been enhanced by ocean-floor configuration changes, led to the wide-spread non-sequences seen in many places between Lower Silurian and earlier rocks. Subsequently, in the early Silurian (Llandovery), the climate appears to have gradually become more equable and rocks were laid down in the consequent transgressions, and by late Llandovery times, faunas are known from many places around the main Gondwanan landmass (Fig. 8). In tropical and temperate areas, the major continents were close enough to permit substantial faunal interchange, and the typical late Llandovery shelly faunas were nearly cosmopolitan. These consisted of well-differentiated depth-related communities on the shelves (Ziegler *et al.* 1968): these were dominated by brachiopods (Cocks and McKerrow 1984), with common genera such as *Lingula*, *Eocoelia*, *Pentamerus*, *Stricklandia*, and *Clorinda*, giving their names to animal communities. Individual transgressions and regressions during the Llandovery have now been documented in many areas, and the consequent lateral shifts in the community distribution bands can be used to differentiate between local tectonic activity and global eustatic events, for example in Wales (Fortey and Cocks 1986) and China (Johnson *et al.* 1985).

These cosmopolitan faunas, which extended to Laurentia, Siberia, Kazakhstan, and Avalonia–Baltica, as well as some of the Gondwanan margins, persisted into middle and late Silurian times. However, during the middle and late Silurian there was some faunal differentiation of the Gondwanan regions bordering the South Pole in southern Africa. This can best be demonstrated by the distribution of the *Clarkeia* fauna, compared with forms such as *Salopina* from the tropical and temperate areas (Fig. 9). The fauna is a complex of low-diversity faunas

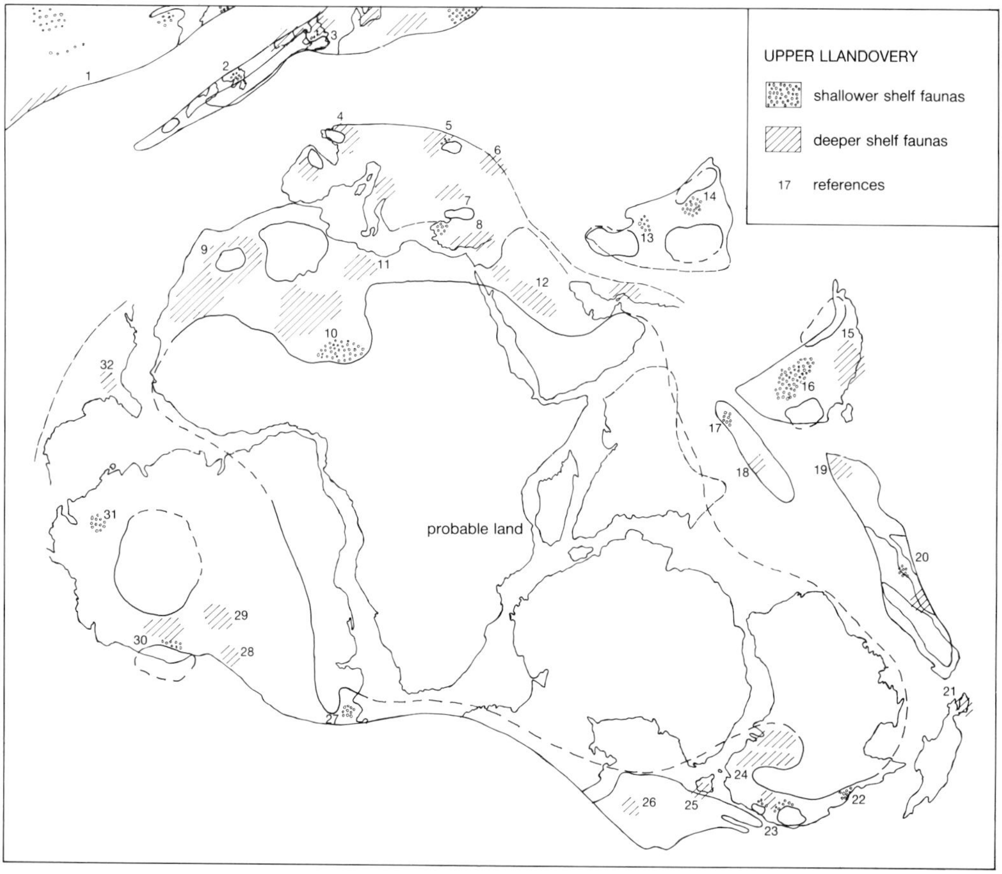

FIG. 8. Biofacies distribution of the Gondwanan region in the early Silurian (Upper Llandovery). References (many with further local references therein) are: *1* Eastern North America (Johnson *et al.* 1985; Berry and Boucot 1970), *2* Newfoundland (McKerrow and Cocks 1986), *3–5* southern Britain, Armorica, and Bohemia (Cocks and Fortey 1982), *6* Western Ukraine (Tsegelnyk 1976), *7* Romania Iordan 1981), *8* Turkey (Berry and Boucot 1972*b*), *9* Morocco (Destombes *et al.*, in Holland 1985), *10* Algeria (Legrand, in Holland 1985), *11* Libya (Massa and Jaeger 1971), *12* Saudi Arabia (Wolfart, in Holland 1981), *13* Iran (Cocks 1979), *14* Afghanistan (Wolfart, in Holland 1981), *15,16* South China (Chen 1984; Johnson *et al.* 1985), *17* Spiti (Reed 1912), *18* Nepal (Strachan *et al.* 1964), *19* Burma (Mitchell *et al.* 1977), *20* Malaya (Berry and Boucot 1972*b*), *21* New Guinea (Visser and Hermes 1962), *22* Queensland (Talent *et al.* 1975), *23* New South Wales (Pickett 1982), *24* Victoria (Garratt 1983), *25* Tasmania (Baillie *et al.* 1978), *26* New Zealand (Cooper and Wright 1972), *27, 28* Argentina (Amos in Berry and Boucot 1972*a*; Amos and Noirat 1971, Isaacson *et al.* 1976), *29* Bolivia (Ahlfeld and Branisa 1960), *30* Bolivia and southern Peru (Laubacher *et al.* 1982), *31* Venezuela (Boucot *et al.* 1972), *32* Florida (Berry and Boucot 1970).

dominated by brachiopods, with the rhynchonellide *Clarkeia* the most common, with scarcer *Heterorthella*, *Eostropheodonta*, *Harringtonina*, *Australina*, *Orbiculoidea*, and sometimes other genera also present sporadically. Accurate dating of each of the various occurrences of the *Clarkeia* fauna is often very difficult (Cocks 1972), but some are known, from graptolite control, to be of Llandovery to Ludlow age. The faunas are precursors, both geographically, and in some cases taxonomically, of the Malvinokaffric faunas so well documented from the early Devonian by A. J. Boucot and his colleagues. To also call these Silurian *Clarkeia* faunas 'Malvinokaffric' may be inappropriate, in that neither the Falkland Islands (the 'Malvinas') or South Africa (Kaffir country) has fossiliferous rocks of Silurian age, but the term is sufficiently entrenched in the literature to make change undesirable.

Over much of western Gondwana, in Armorica and the Iberian Peninsula, through

Romania, Bohemia, Yugoslavia, and Italy, to the whole of North Africa, the Silurian biofacies is dominated by what Berry and Boucot (1967) termed a 'pelycypod–graptolite association', the diverse and numerous bivalve faunas which have been well described by Kříž (e.g. 1979). Associated graptolites give the age of this biofacies as ranging from Llandovery to Ludlow, with its most widespread peak occurring in Wenlock times. The oddest thing about this facies is that, although it seems to represent deposition under rather deeper water than the normal brachiopod-dominated shelly faunas in the northern part of its distribution, for example in Britain and Czechoslovakia, in the southern part, in North Africa, it occurs right up on to old land areas in apparently very shallow-water facies. Presumably, the temperature in North Africa was generally inimical to shelly faunas at that time; the few shells present (Fig. 9) are far outweighed by the bivalve and graptolite-bearing deposits, and thus the bivalve fauna was temperature-adapted to cooler water and able to colonize shallower environments nearer the pole, and progressively deeper sites away from it. This is comparable with the distribution of the North Atlantic conodont 'province' in the earlier part of the Ordovician, and also more directly comparable with the distribution of bivalves in modern seas (J. D. Taylor pers. comm.).

The changing movement of Gondwana with respect to the South Pole is also reflected in the lithofacies. In Bohemia, for example, the late Cambrian and early Ordovician faunas are of low diversity and the lithologies consist entirely of clastic rocks. As the Ordovician and Silurian progressed, and that margin of Gondwana moved gradually northwards into more temperate palaeolatitudes, the diversity of the shelly faunas, particularly the shallower-water brachiopods and trilobites, increased, but deposition of the clastic rocks continued, and it was not until late Silurian (Ludlow) times that the first carbonate deposits (the Kopanina Limestones) were laid down as calcareous shales

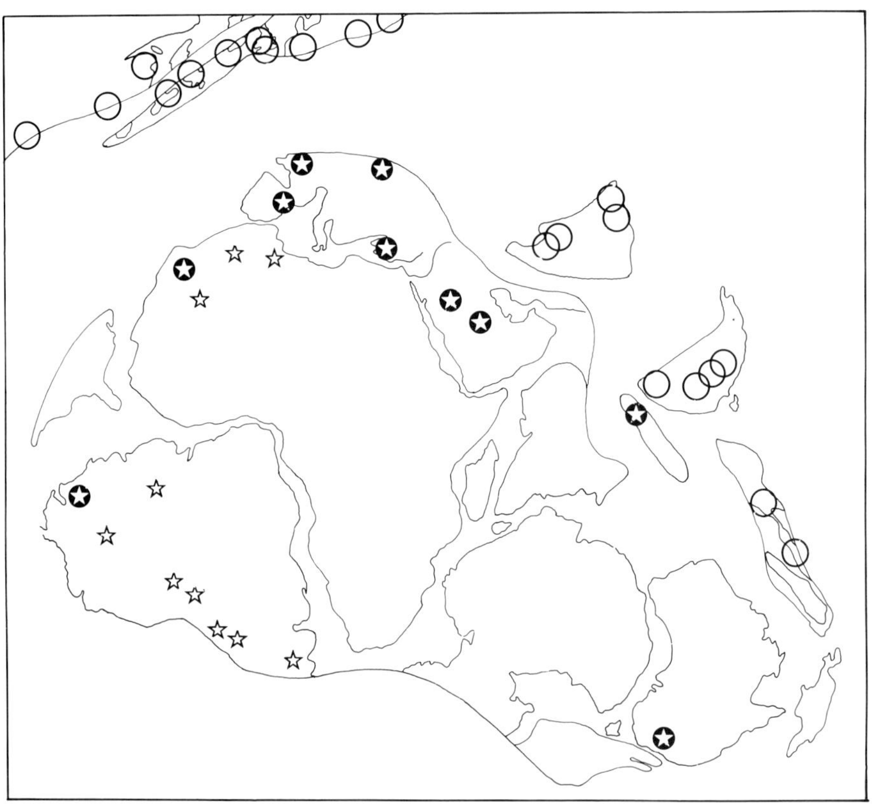

FIG. 9. Contrasts in middle and late Silurian Gondwanan faunas: open stars are *Clarkeia* faunas of South America and comparable low-diversity faunas in Africa. Circles are diverse low-latitude faunas (distribution of *Salopina* is shown). Black circles with stars are lower-diversity temperate faunas.

interbedded with occasional biogenic limestones. By the latest Silurian (Pridoli) there were the first substantial limestone deposits, which continued up into the Devonian. Similarly, in southern Britain, which lay at a lower latitude than Bohemia while the Rheic Ocean opened, there were also only cooler-water clastic rocks deposited on the shallow shelves during the Ordovician; but the first carbonate deposits occur much earlier, with occasional bioherms (such as at Kildare and Keisley) developed in the warmer period of the earlier Ashgill, to cease again during the latest Ordovician–early Silurian glacial episode, and subsequently reappear—first as occasional limy bands during the latest Llandovery, for example in the Welsh Borderland, and then as occasional carbonate build-ups in the early Wenlock (for example at Old Radnor and Woolhope) and finally as substantial carbonate developments, including coral reefs in the late Wenlock (e.g. the Much Wenlock Limestone). However, the distribution of, for example, the coral faunas in the late Silurian (e.g. Oliver and Pedder 1979) confirms the differentiation between the equatorial areas of North America and adjacent Avalonia–Baltica in contrast to the north-western parts of Gondwana.

As the Rheic and other oceans—for example that between Gondwana and Kazakhstan—widened during the Silurian, the cosmopolitan early Silurian tropical and temperate fauna began to lose its coherence. For example, in the late Llandovery, the same species of the strophomenides *Eoplectodonta* and *Protomegastrophia* occurred in both Britain and Bohemia, while by the Wenlock there were different species of the same genera on opposite continental margins (*E. duvalii* and *P. semiglobosa* in Britain, and *E. bidecorata* and *P. miranda* in Bohemia), and by the end of the Silurian and the early Devonian their descendants had evolved further into distinctive genera such as *Plectodonta* and *Plicostropheodonta* on the north-western Gondwanan margin of Bohemia—which are not known from contemporary biofacies in Britain, Belgium, and North Germany. By early Devonian times the degree of global provincialization had become much greater (Boucot *et al.* 1969), and this process must also have been assisted by the larger emergent land areas constructed by the Caledonian and contemporary orogenies, for example the Old Red Sandstone palaeocontinent.

ACKNOWLEDGEMENTS: We thank Dr J. L. Henry for advice on trilobite determinations, Dr H. G. Owen for providing Triassic base maps, and many colleagues for discussion.

References

AHLFELD, F. and BRANISA, L. (1960). *Geologia de Bolivia*. Instituto Boliviano del Petroleo. (245 pp.)

AMOS, A. J. and NOIRAT, S. (1971). A new species of *Ancillotoechia* from the Zapla Formation, northern Argentina. *Smithsonian Contrib. Paleobiol.* **3**, 139–42.

BAILLIE, P. W., BANKS, M. R., and RICKARDS, R. B. (1978). Early Silurian graptolites from Tasmania and their significance. *Search* **9**, 1–2.

BERGSTRÖM, J. (1976). Lower Palaeozoic trace fossils from eastern Newfoundland. *Can. J. Earth Sci.* **13**, 1613–33.

BERRY, W. B. N. and BOUCOT, A. J. (1967). Pelycypod–graptolite association in the Old World Silurian. *Geol. Soc. Am. Bull.* **78**, 1515–22.

BERRY, W. B. N. and BOUCOT, A. J. (ed.). (1970). Correlation of the North American Silurian rocks. *Geol. Soc. Am. Spec. Pap.* **102**, 1–289.

BERRY, W. B. N. and BOUCOT, A. J. (ed.). (1972*a*). Correlation of the South American Silurian rocks. *Geol. Soc. Am. Spec. Pap.* **133**, 1–59.

BERRY, W. B. N. and BOUCOT, A. J. (1972*b*). Correlation of the Southeast Asian and Near Eastern Silurian rocks. *Geol. Soc. Am. Spec. Pap.* **137**, 1–65.

BEUF, S., BIJU-DUVAL, V., DE CHARPAL, O., ROGNON, P., GARIEL, O., and BENNACEF, A. (1971). Les Grès du Paléozoïque inférieur au Sahara. *Publ. Inst. Franç. Pétrole* **18**, 1–464.

BIGARELLA, J. J. (1973). Paleocurrents and the problem of continental drift. *Geol. Rdsch.* **62**, 447–77.

BLAISE, J. and BOUYX, E. (1980). Remarques sur le Paléozoïque inférieur de quelques secteurs de l'Afghanistan central et des domaines ibéro-armoricains: stratigraphie comparée et implications paleogeographiques. *Bull. Soc. Géol. Minéral. Bretagne* **12**, 63–76.

BOUCOT, A. J., JOHNSON, J. G., and SHAGAM, R. (1972). Braquiopodos Siluricos de Los Andes meridenos de Venezuela. *Bol. Geol. Publ. Espec., Caracas* **5**, 585–727, pls 1–34.

BOUCOT, A. J., JOHNSON, J. G., and TALENT, J. A. (1969). Early Devonian brachiopod zoogeography. *Geol. Soc. Am. Spec. Pap.* **119**, 1–111, pls 1–19.

BRENCHLEY, P. J., HARPER, J. C., and SKEVINGTON, D. (1967). Lower Ordovician shelly and graptolitic faunas from south-eastern Ireland. *Proc. Roy. Irish Acad.* **65(B)**, 385–90.

BULMAN, O. M. B. (1931). South American graptolites, with special reference to the Nordenskiöld collection. *Arkiv for Zoologi* **22A**, 1–111, pls 1–11.

BURRETT, C. F. and STAIT, B. (1985). South East Asia as part of an Ordovician Gondwanaland—a palaeobiogeographic test of a tectonic hypothesis. *Earth Planet. Sci. Lett.* **75**, 184–90.

BURRETT, C. F., STAIT, B., and LAURIE, J. (1983). Trilobites and microfossils from the Middle Ordovician of Surprise Bay, southern Tasmania, Australia. *Mem. Assoc. Austral. Palaeontol.* **1**, 177–93.

CHEN, X. (1984). Influence of the late Ordovician glaciation on basin configuration of the Yangzte Platform in China. *Lethaia* **17**, 51–9.

COCKS, L. R. M. (1972). The origin of the Silurian *Clarkeia* shelly fauna of South America, and its extension to West Africa. *Palaeont.* **15**, 623–30, pl. 121.

COCKS, L. R. M. (1979). A silicified brachiopod fauna from the Silurian of Iran. *Bull. Brit. Museum (Natur. Hist.) (Geol.)* **32**, 25–42.

COCKS, L. R. M. and FORTEY, R. A. (1982). Faunal evidence for oceanic separations in the Palaeozoic of Britain. *J. Geol. Soc., London* **139**, 465–78.

COCKS, L. R. M. and FORTEY, R. A. (1986). New evidence on the South African Lower Palaeozoic: age and fossils reviewed. *Geol. Mag.* **123**, 437–44.

COCKS, L. R. M. and LOCKLEY, M. G. (1981). Reassessment of the Ordovician brachiopods from the Budleigh Salterton Pebble Bed, Devon. *Bull. Brit. Museum (Natur. Hist.) (Geol.)* **35**, 111–24.

COCKS, L. R. M and MCKERROW, W. S. (1984). Review of the distribution of the commoner animals in Lower Silurian marine benthic communities. *Palaeont.* **27**, 663–70.

COOPER, R. A. (1979). Ordovician geology and graptolite faunas of the Aorangi Mine area, north-west Nelson, New Zealand. *Palaeont. Bull. N.Z. Geol. Surv.* **47**, 1–127, pls 1–19.

COOPER, R. A. and NI, Y. (1986). Taxonomy, phylogeny and variability of *Pseudisograptus* Beavis. *Palaeont.* **29**, 313–64.

COOPER, R. A. and WRIGHT, A. J. (1972). Silurian rocks and fossils at Hailes Knob, north-west Nelson, New Zealand. *N.Z. J. Geol. Geophys.* **15**, 318–85.

COURTESSOLE, R., PILLET, J., VIZCAINO, D., and ESCHARD, R. (1985). Étude biostratigraphique et sedimentologique des Formations arenacées de l'Arenigien du Saint Chinianais oriental (Hérault) versant sud de la Montagne Noire (France Meridionale). *Mém. Soc. Étud. Scient. Aude* 1–99.

DEAN, W. T. (1966). The Lower Ordovician stratigraphy and trilobites of the Landeyran valley and neighbouring district of the Montagne Noire, southwestern France. *Bull. Brit. Museum (Natur. Hist.) (Geol.)* **12**, 245–353, pls 1–21.

DEAN, W. T. (1967). The distribution of Ordovician shelly faunas in the Tethyan region. *Systematics Assoc. Publ.* **7**, 11–44.

DEAN, W. T. (1975). Cambrian and Ordovician correlation and trilobite distribution in Turkey. *Fossils & Strata* **4**, 353–73.

DEAN, W. T. and MARTIN, F. (1978). Lower Ordovician acritarchs and trilobites from Bell Island, eastern Newfoundland. *Bull. Geol. Surv. Canada* **284**, 1–35, pls 4–7.

EL-KHAYAL, A. A. and ROMANO, M. (1985). Lower Ordovician trilobites from the Hanadir Shale of Saudi Arabia. *Palaeont.* **28**, 401–12.

FORTEY, R. A. (1975). Early Ordovician trilobite communities. *Fossils & Strata* **4**, 331–52.

FORTEY, R. A. and COCKS, L. R. M. (1986). Marginal faunal belts and their structural implications, with examples from the Lower Palaeozoic. *J. Geol. Soc., London* **143**, 151–60.

FORTEY, R. A. and MORRIS, S. F. (1982). The Ordovician trilobite *Neseuretus* from Saudi Arabia, and the palaeogeography of the *Neseuretus* fauna related to Gondwanaland in the earlier Ordovician. *Bull. Brit. Museum (Natur. Hist. (Geol.)* **36**, 63–75.

FORTEY, R. A. and OWENS, R. M. (1978). Early Ordovician (Arenig) stratigraphy and faunas of the Carmarthen district, south-west Wales. *Bull. Brit. Museum (Natur. Hist.) (Geol.)* **30**, 225–94.

FORTEY, R. A. and OWENS, R. M. (1987). The Arenig Series in South Wales. *Bull. Brit. Museum (Natur. Hist.) (Geol.)* **41**, 69–303.

FORTEY, R. A. and SHERGOLD, J. H. (1984). Early Ordovician trilobites, Nora Formation, central Australia. *Palaeont.* **27**, 315–66, pls 38–46.

GARRATT, M. J. (1983). Silurian and Devonian biostratigraphy of the Melbourne Trough, Victoria. *Proc. Roy. Soc. Victoria* **95**, 77–98.

GATINSKY, Y. G., KLOCHKO, V. P., ROZMAN, K. S., and TROFIMOV, D. M. (1966). More facts about the Paleozoic stratigraphy of the Southern Sahara. *Doklady Akademii Nauk SSSR* **170**, 1154–57.

GUTIERREZ MARCO, J. C. (1982). Descubrimiento de nuevos niveles con Graptolitos ordovícicos en la unidad 'Pizarras con Didymograptus'—Schneider 1939—(Prov. Huelva, SW de España). *Comm. Serv. Geol. Portugal* **68**, 241–6.

HAMMANN, W. (1983). Calymenacea (Trilobita) aus dem Ordovizium von Spanien: ihre Biostratigraphie, Ökologie und Systematik. *Abhdl. Senckenbergischen Naturforsch. Gesell.* **542**, 1–177.

HARRINGTON, H. J. and KAY, M. (1951). Cambrian and Ordovician faunas of Eastern Colombia. *J. Paleont.* **25**, 655–68, pls 96–7.

HARRINGTON, H. J. and LEANZA, A. F. (1957). Ordovician trilobites of Argentina. *Spec. Publ. Dept. Geol., Univ. Kansas* **1**, 1–276.

HAVLÍČEK, V. (1971). Brachiopodes de l'Ordovicien du Maroc. *Notes Mém. Serv. Géol. Maroc* **230**, 1–135, pls 1–26.

HAVLÍČEK, V. (1977). Brachiopods of the order Orthida in Czechoslovakia. *Rozpravy Ústředního ústavu geologického* **44**, 1–327, pls 1–66.

HAVLÍČEK, V. and BRANISA, L. (1980). Ordovician brachiopods of Bolivia. *Rozpravy Československé Akademie Věd Řada Matematických a Přírodních Věd.* **90**, 1–54, pls 1–6.

HAVLÍČEK, V. and VANEK, J. (1966). The biostratigraphy of the Ordovician of Bohemia. *Sbornik Geologickych Věd paleontologie* **8**, 7–69, pls 1–16.

HENDERSON, R. A. (1983). Early Ordovician faunas from the Mount Windsor Subprovince, northeastern Queensland. *Mem. Assoc. Austral. Palaeontol.* **1**, 145–75.

HENRY, J.-L. (1980). Trilobites ordoviciens du Massif Armoricain. *Mem. Soc. Géol. Minéral. Bretagne* **22**, 1–250.

HOLLAND, C. H. (ed.) (1981). *Lower Palaeozoic of the Middle East, Eastern and Southern Africa and Antarctica.* John Wiley, Chichester. (331 pp.)

HOLLAND, C. H. (ed.) (1985). *Lower Palaeozoic of north-western and west-central Africa.* John Wiley, Chichester. (512 pp.)

HUGHES, C. P., RICKARDS, R. B., and WILLIAMS, A. (1980). The Ordovician fauna from the Contaya Formation of eastern Peru. *Geol. Mag.* **117**, 1–21, pls 1–4.

IORDAN, M. (1981). Study of Silurian and Devonian faunas from the eastern part of the Moesian Platform. *Mém. Inst. Géol. Geophys., Bucharest* **30**, 115–222, pls 1–42.

ISAACSON, P. E., ANTELO, B., and BOUCOT, A. J. (1976). Implications of a Llandovery (Early Silurian) brachiopod fauna from Salta Province, Argentina. *J. Paleont.* **50**, 1103–12, pl. 1.

JABLONSKI, D. and VALENTINE, J. W. (1981). Onshore–offshore gradients in Recent eastern Pacific shelf faunas and their paleobiogeographic significance. In *Evolution today.* (ed. G. G. E. Scudder and J. L. Reveal), pp. 441–53, Hunt Institute, Pittsburgh.

JACKSON, D. E. (1962). Graptolite zones in the Skiddaw Group in Cumberland, England. *J. Paleont.* **36**, 300–13.

JELL, P. A., BURRETT, C. F., STAIT, B., and YOCHELSON, E. L. (1984). The early Ordovician bellerophontoid *Peelerophon oehlerti* (Bergeron) from Argentina, Australia and Thailand. *Alcheringa* **8**, 169–76.

JELL, P. A. and STAIT, B. (1985). Revision of an early Arenig trilobite faunule from the Caroline Creek Sandstone, near Latrobe, Tasmania. *Mem. Museum Victoria* **46**, 35–51.

JENTSCH, S. VON and STEIN, V. (1961). Neue Fossilfunde im Ordovizium des Ebbe-Satters. *Paläont. Zeits.* **35**, 200–8.

JOHNSON, M. E., RONG, J.-Y., and YANG, X.-C. (1985). Intercontinental correlation by sea-level events in the Early Silurian of North America and China (Yangtze Platform). *Geol. Soc. Amer. Bull.* **96**, 1384–97.

KŘÍŽ, J. (1979). Silurian Cardiolidae (Bivalvia). *Sbornik Geologickych Věd Paleontologie* **22**, 1–157, pls 1–40.

LAUBACHER, G., BOUCOT, A. J., and GRAY, J. (1982). Additions to Silurian stratigraphy, lithofacies, biogeography and palaeontology of Bolivia and southern Peru. *J. Paleontol.* **56**, 1138–70, pls 1–5.

LEGG, D. P. (1976). Ordovician trilobites and graptolites from the Canning Basin, western Australia. *Geologica et Palaeontologica* **10**, 1–37, 10 pls.

LIU, D.-Y. (1976). Ordovician brachiopods from the Mount Jolmo Lungma Region. In *A report of scientific expedition in the Mount Jolmo Lungma Region.* pp. 139–58, pls 1, 2, Fascicule II, Palaeontology, Science Press, Peking.

LU, Y.-H. (1975). Ordovician trilobite faunas of central and southwest China. *Palaeontologia Sinica*, New series B, **11**, 1–463, 50 pls.

LU, Y.-H., *et al.* (1976). Ordovician biostratigraphy and palaeozoogeography of China. *Mem. Nanjing Inst. Geol. Palaeont.* **7**, 1–83, pls 1–14.

MASSA, D. and JAEGER, H. (1971). Données stratigraphiques sur le Silurien de l'Ouest de la Libye. *Mém. Bur. Rech. Géol. Min.* **73**, 313–21.

MCCLURE, H. A. (1978). Early Paleozoic glaciation in Arabia. *Palaeogeogr., Palaeoclimatol., Palaeoecol.* **25**, 315–26.

MCKERROW, W. S. and COCKS, L. R. M. (1986). Oceans, island arcs and olistostromes: the use of fossils in distinguishing sutures, terranes and environments around the Iapetus Ocean. *J. Geol. Soc., London* **143**, 185–91.

MELOU, M. (1976). Orthida de la Formation de Postolonnec (Ordov.), Finistère, France. *Geobios* **9**, 693–717.

MITCHELL, A. H. G. (1981). Phanerozoic plate boundaries in mainland SE Asia, the Himalayas and Tibet. *J. Geol. Soc. London* **138**, 109–122.

MITCHELL, A. H. G., MARSHALL, T. R., SKINNER, A. C., BAKER, M. D., AMOS, B. J., and BATESON, J. H. (1977). Geology and exploration geochemistry of the Yadanatheingi and Kyaukme–Longtawkno areas, Northern Shan States, Burma. *Overseas Geol. & Min. Resourc.* **51**, 1–35.

MU, A. T., GE, M. Y., CHEN, X., NI, Y. N., and LIN, Y. K. (1979). Lower Ordovician graptolites of southwest China. *Acta Palaeont. Sinica* **156(B)**, 1–192, pls 1–48.

OLIVER, W. A. and PEDDER, A. E. H. (1979). Biogeography of late Silurian and Devonian rugose corals in North America, pp. 131–45. In *Historical biogeography, plate tectonics and the changing environment.* (ed. J. R. Gray and A. J. Boucot), Oregon State University Press.

PALMER, A. R. (1972). Problems in Cambrian biogeography. *24th International Geological Congress* **7**, 310–15.

Percival, I. G. (1976). The geology of the Licking Hole Creek area, near Walli, central western New South Wales. *J. & Proc., Roy. Soc. New South Wales* **109**, 7–23.

Pickett, J. W. (1982). The Silurian System in New South Wales. *Geol. Surv. New South Wales Bull.* **29**, 1–264, pls 1–5.

Pillet, J. (1973). Sur quelques trilobites ordoviciens d'Iran Oriental. *Ann. Soc. Géol. du Nord* **93**, 33–8, pls 6–9.

Pribyl, A. and Vanek, J. (1980). Ordovician trilobites of Bolivia. *Rozpřavy Ceskoslovenske Akademie Ved, Rada Matematickych a Prirodnich Ved.* **90**, 1–90, pls 1–26.

Rabano, I. (1983). The Ordovician trilobite *Hungioides* Kobayashi, 1936 (Asaphina, Dikelokephalinidae) from Spain. *Geobios* **16**, 431–41, pl. 1.

Reed, F. R. C. (1912). Ordovician and Silurian fossils from the central Himalayas. *Mem. Geol. Surv. India* (Series 15) **7**, (2), 1–168.

Rong, J.-Y. (1984). Distribution of the *Hirnantia* fauna and its meaning. In *Aspects of the Ordovician System*. (ed. D. L. Bruton), pp. 101–12, Universitetsforlaget, Oslo.

Salter, J. W. and Blanford, H. F. (1865). *Palaeontology of Niti in the Northern Himalaya*. (112 pp., 23 pls.) Calcutta.

Schallreuter, R. E. L. and Siveter, D. J. (1985). Ostracodes across the Iapetus Ocean. *Palaeontology* **28**, 577–98, pls 68–70.

Serpagli, E. and Gnoli, M. (1984). Palaeozoic palaeontology in Sardinia: a review (1857–1983). *Bol. Museo Regionale Scienze Naturale Torino* **2**, 163–80.

Sheng, S. F. (1974). Ordovician trilobites from western Yunnan and their stratigraphic significance. In *Subdivision and correlation of the Ordovician System in China*. (ed. S. F. Sheng), pp. 96–115. Science Press, Beijing. (in Chinese).

Spassow, H. (1958). *Les fossiles de Bulgarie, 1ère Paléozoique*. Bulgarska Akademiya na Naukite, Sofia. (90 pp., 16 pls.)

Strachan, I., Bodenhausen, J. W. A., Booy, T. de, and Egeler, C. G. (1964). Graptolites in the 'Tibetan Zone' of the Nepalese Himalayas (central West Nepal). *Wetenschappelijke Mededel.* 1964, 380–2.

Struve, W. (1975). Die altesten Fossilien Hessens. *Natur und Museum* **105**, 268–82.

Talent, J. A., Berry, W. B. N., and Boucot, A. J. (1975). Correlation of the Silurian rocks of Australia, New Zealand, and New Guinea. *Geol. Soc. Amer. Spec. Pap.* **150**, 1–108.

Thomas, D. E. (1960). The zonal distribution of Australian graptolites. *J. & Proc. Roy. Soc. New South Wales* **94**, 1–58.

Tsegelnyk, P. D. (1976). *Brachiopods and stratigraphy of the Lower Palaeozoic of Volynia–Podolia*. Ukrainian Academy of Sciences, Kiev. (156 pp., 12 pls.)

Turner, J. C. M. (1960). Faunas graptoliticas de America del Sud. *Rev. Asoc. Geol. Argentina* **14**, 5–180, pls 1–9.

Villas, E. (1983). La formaciones de Ordovico medio y superior de la Cadenas Ibericas y su fauna de braquiopodos. *Estud. Geol.* **39**, 359–77.

Villas, E. (1985). Braquiopodos del Ordovicico medio y superior de las Cadenas Ibericas Orientales. *Mem. Museo Paleont. Univ. Zaragoza* **1**, 1–226, pls 1–34.

Visser, N. A. and Hermes, J. J. (1962). *Geological results of the exploration for oil in New Guinea*. Nederlandsche Niew Guinea Petroleum Maatschappij, The Hague. (266 pp.)

Webby, B. D. (1985). Biogeographical significance of some East Australian Ordovician faunas. *Abstr. Geol. Soc. Austral.* **14**, 244–6.

Whittington, H. B. (1953). A new Ordovician trilobite from Florida. *Breviora Harvard Univ.* **17**, 1–6.

Whittington, H. B. and Hughes, C. P. (1972). Ordovician geography and faunal provinces deduced from trilobite distribution. *Phil. Trans. Roy. Soc. London* **B263**, 235–78.

Williams, A. (1973). Distribution of brachiopod assemblages in relation to Ordovician palaeogeography. *Spec. Pap. Palaeontol.* **12**, 241–69.

Wolfart, R. (1970). Fauna, Stratigraphie und Paläogeographie des Ordoviziums in Afghanistan. *Beih. Geol. Jb.* **89**, 1–125, pls 1–21.

Ziegler, A. M., Cocks, L. R. M., and Bambach, R. K. (1968). The composition and structure of Lower Silurian marine communities. *Lethaia* **1**, 1–27.

Ziegler, A. M., Scotese, C. R., McKerrow, W. S., Johnson, M. E., and Bambach, R. K. (1979). Paleozoic palaeogeography. *Ann. Rev. Earth Planet. Sci.* **7**, 473–502.

L. R. M. Cocks and R. A. Fortey, Department of Palaeontology, British Museum (Natural History), Cromwell Road, London SW7 5BD, UK.

Fossil plants as indicators of late Palaeozoic plate positions†

W. G. Chaloner and G. T. Creber

ABSTRACT: Fossil plant records from a late Carboniferous to early Permian time interval are reviewed for the data that they give on connections and migration routes between floristic provinces, and their palaeolatitudinal positions. Caution is urged in using fragmentary leaf remains as a basis for recognizing affinity between floristic provinces, as illustrated by reference to the genera *Phyllotheca* and *Glossopteris*. The distribution of some 38 plant genera from a range of late Palaeozoic areas and localities is made the basis for ordinating them in the form of a dendrogram. An early Permian plate reassembly, based on that of Lin *et al.* 1985, is assessed in terms of its compatibility with the palaeobotanical evidence.

In offering this contribution to the first Charles Lyell Symposium, it is appropriate to note that Lyell himself was well aware of the significance of plant fossils as indicators of palaeoenvironments. In his Manual (Lyell 1855) he illustrates gyrogonites of a freshwater alga *Chara* from the Tertiary of the Isle of Wight, and compares them with a living Italian species of that genus. On this and other evidence, he argues for a non-marine origin for the limestone in which these fossils occur. Although the connection with the theme of the present paper is tenuous, we salute Lyell as a pioneer of the use of plant fossils in reconstructing the past distribution of land and sea. We, like Lyell, are looking at fossil plants as evidence of past environments—although in our case, on the rather different matter of continental plates and their juxtapositions.

Late Palaeozoic floristic provinces

In this review we concentrate our attention on floras of latest Carboniferous and early Permian age. The significance of this interval palaeobotanically is that we see in it for the first time—in the fossil record of land plants—a pronounced differentiation into four major regional floras, with an inescapable palaeolatitudinal element in their distribution. Recently, several authors have suggested that some differentiation is evident as far back as the early Carboniferous (see Vakhrameev *et al.* 1978; Raymond 1985) and even the Devonian (Raymond *et al.* 1985). However, although it is difficult to quantify degrees of floristic differentiation in any consistent way, it appears that the extent of 'regionalization' of Permian floras is more pronounced than at any earlier phase of land-plant evolution.

By about the time of the transition from the Carboniferous into the Permian, four major floristic provinces ('Phytochoria' of some authors) can be recognized. These are (1) the Euramerican flora (also called Euramerian) (2) the Cathaysian (or *Gigantopteris*-) flora, both of palaeoequatorial latitudes, (3) the northern, Angara or Kuznetsk flora, and (4) the southern, Gondwana or *Glossopteris* flora. A fifth province, the North American, represented in the south-western USA, may also be recognized, although of far lesser geographical extent than the others. General reviews of these floristic provinces are given in Chaloner and Meyen (1973), Chaloner and Lacey (1973), and Vakhrameev *et al.* (1978), where earlier references are cited.

In 1937, Halle published a particularly significant map illustrating how, within Asia alone, three of these floral provinces show a juxtaposition having strong implications for continental movement. At a time when 'continental drift' was still heretical to many, he interpreted the *Glossopteris* flora of India, lying close to the Cathaysian flora to the north and east as a result of post-Permian continental movement, and recognized the differentiation of the Cathaysian and Angaran floras and their relationship to the Euramerian flora to the west. The essential features to which Halle drew attention are shown in Fig. 4, which is an updated version of late Palaeozoic plant distribution from the same region. Important added information since Halle's time has included extra knowledge of floras from Sumatra and New Guinea, many new floristic records from China and Russia, and a sharpening of the boundaries between provinces. These are considered further below.

† This paper is a contribution to I.G.C.P. Project No. 237 "Floras of the Gondwanic continents".

From AUDLEY-CHARLES, M. G. & HALLAM, A. (eds) *Gondwana and Tethys* Geological Society Special Publication No. 37, pp. 201–10.

The conceptual limits of these floras, the basis of their definition, and how far they may be subdivided, have been fully explored in the papers cited above and elsewhere. Meyen (in Chaloner and Meyen 1973) recognizes a hierarchy of these somewhat subjective palaeofloristic provinces (phytogeographical units) designated, in descending order as Kingdom (the five units cited above), Area, Province, and District. Within the 'Kingdoms' in Meyen's usage (here called palaeofloristic provinces) he therefore recognized various geographical sub-units, defined in terms of the constituent plants and their changing disposition with time. Chaloner and Lacey (1973) give a simplified version of this, which records only the upper two levels of Meyen's four-tier hierarchy.

In recognizing Palaeozoic palaeofloristic provinces, there is a problem of choice in selecting the level of taxa to be used in defining them, and the length of time interval for which data are considered. The genus is undoubtedly the most convenient, consistently recognized, and appropriate taxonomic level for this purpose, and this case has been argued elsewhere (Chaloner and Lacey 1973) and similar conclusions reached by Raymond (1985) in her analysis of rather older floras.

The time interval for which plant records may reasonably be regarded as more or less synchronous for the purposes of palaeogeographical reconstructions, presents greater problems. The actual age of most late Palaeozoic floras is based largely on the composition of the flora itself (a circular argument with its own perils) although marine faunas may, in some instances, give a more precise age above or below a plant-bearing horizon. If a rather narrow age band is attempted, this then limits the number of plant occurrences to be used in any one map; if too broad a time interval is allowed, then obviously the significance of differences between adjoining provinces may be blurred. We err on the side of a liberal interpretation of our time unit, in including floras of 'late Carboniferous' as well as 'early Permian', and including in some cases plant occurrences when only a broad 'Permian' age can be sustained with confidence. The 'North American province' shown in Fig. 5 ('American') is based on the Permian flora of the south-west of the USA. At one time this flora was believed to show closest affinity to the Cathaysian province, but this is now discounted (see Chaloner and Meyen 1973). The relative isolation of this flora is shown (south-west USA) in Fig. 3. As this bears little on either Gondwanaland or Tethys, it is not considered further here.

Problems with fossil plant taxa

There are many respects in which terrestrial plants differ from, say, invertebrate animal fossils as constituents of biota on which plate reconstructions can be based. Plant fossil assemblages involve elements which may have been transported some distance to the site of fossilization, with resulting separation of constituent organs of one plant, and selectivity in representation. Different preservation states of one organ may be assigned to different taxa, producing spurious repetitions or differences in floristic records. These and related problems are considered in Chaloner and Lacey (1973), and in Raymond (1985), and Raymond *et al.* (1985).

There is obviously a hazard in seeing the distribution of any genus of plant macrofossils as an indication of continuity (or the existence of a migration path) between the regions in which it occurs. The mobility of different plants with different types of reproductive biology is very varied. Broadly, spore-bearing plants have greater mobility but a lesser competitive capacity than seed-bearing plants, particularly those with large seeds (Chaloner and Sheerin 1981). However, an even greater problem probably lies in our inability to consistently recognize taxa which actually represent plants of common, monophyletic stock. We should like to illustrate here just two of the kinds of problems presented by palaeobotanical data which are products of the separation of plant organs, before fossilization, coupled with the occurrence of parallel evolution (homoplasy) in leaf morphology. When palaeobiologists present records of 'common elements' between faunal or floral provinces, all taxa at the same level are often regarded as of equal status in their contribution. We believe it is important to accept that this is rarely so, and that some fossil plant genera may be less satisfactory than others as evidence of a common origin of their constituent species.

The leaf genus *Phyllotheca*, shown in Fig. 1, illustrates the problem of using 'leaf genera' alone as a basis for recognizing affinity between floral provinces. This genus is based on foliage belonging to the class Sphenopsida (Equisetopsida of some authors); it is comparable with the familiar Carboniferous genus *Annularia*, but the leaves, arranged in whorls, are fused at the base (in varying degree, even on the same plant) to form a basal, conical leaf sheath, with distal free leaf tips. Leaves attributable to *Phyllotheca* have been reported widely from the Permian of

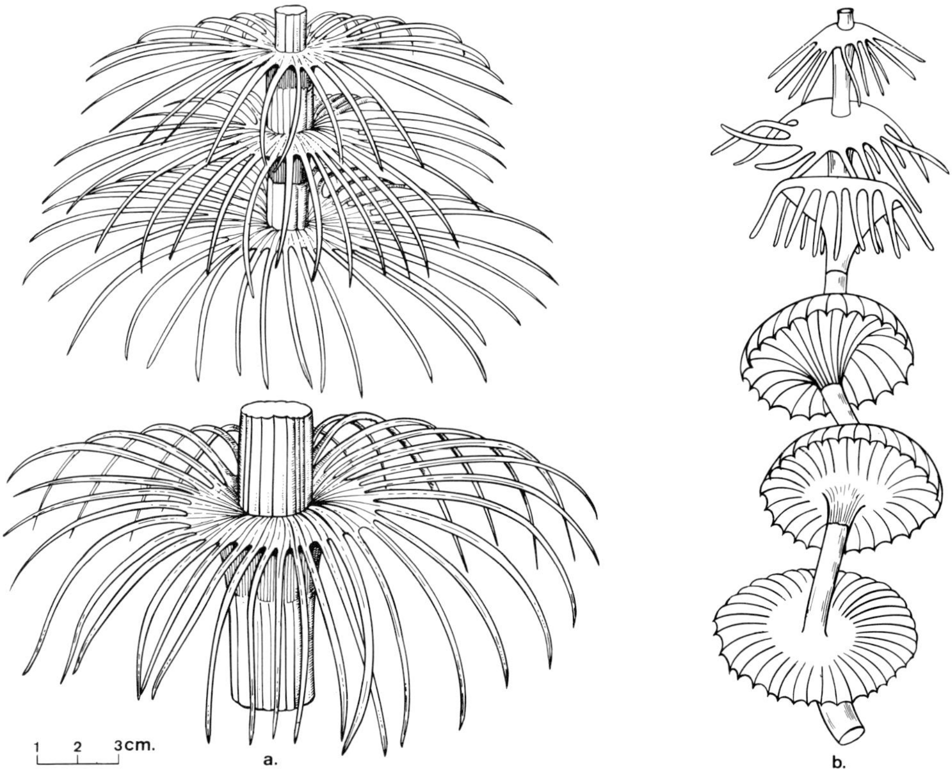

FIG. 1. Reconstructions of leaves of *Phyllotheca* from the Permian of Angara and Gondwana. (a) *P. equisetitoides* from the USSR. (b) *P. etheridgei* from Australia (not to the same scale). Angaran *Phyllotheca* spp. are known in connection with fructifications of *Tschernovia* type, of the family Tchernoviaceae, while Gondwanan *Phyllotheca* spp. had very different fructifications attributed to a different family, the Gondwanostachyaceae. Leaves attributed to the 'form genus' *Phyllotheca* evidently represent distinct Gondwanan and Angaran families, and so cannot be regarded as a significant 'common element' between the Angaran and Gondwanan provinces. (After Boureau 1964.)

Angara (e.g. Fig. 1a, *P. equisetitoides*) but also from Gondwana (e.g. Fig. 1b, *P. etheridgei*). Despite the similarity of these Angara and Gondwana leaves, and their inclusion in the same genus, Meyen has shown that the fructifications (spore-bearing structures) of Phyllothecas from Angara are of a fundamentally different type from those of Gondwana (see Meyen 1969, 1979, and 1987). These two types of fructification have been made the basis of two distinct families within the ubiquitous order of the Equisetales: the Tchernoviaceae, based on the sporangiophores *Tchernovia* from Angara, and the family Gondwanostachyaceae based on the fertile shoots of *Gondwanostachys* Meyen from Australia. The different organization of the reproductive structures suggest that these two families are of quite distinct origin within the order Equisetales, although their leaves are of basically similar architecture. Hitherto the occurrence of *Phyllotheca* in Angara and Gondwana has been seen as a 'common element' linking these two floristic provinces. It is now clear that *Phyllotheca* is in this sense an 'artificial' genus which 'lumps' forms of independent origin from these two floristic provinces. Its value as an indication of floristic affinity is thereby limited, and it is essential that we acknowledge this.

Another genus based on isolated leaves is *Glossopteris*, and its presence in all the constituent parts of Gondwana was one of the reasons for palaeobotanists' affection for 'continental drift' before plate tectonics became accepted dogma (cf. Halle 1937; Chaloner 1959). In the last twenty years our knowledge of 'the *Glossopteris* plant' has expanded enormously, with the linking—in varying degrees of certainty—of the several seed-bearing fructifications, the pollen organs, and the stems and roots of this plant (see e.g. Gould and Delevoryas 1977). But despite the recognition of the real identity of the range of this one genus with the floristic province based upon it, reports of *Glossopteris* well outside Gondwana continue to appear. The problem in this case is rather different from that of the clearly polyphyletic *Phyllotheca*, and amounts to possible mistaken identity. The leaf morphology and venation of *Glossopteris* is seen by some to be idiosyncratic; R. Melville (1969) for example, maintains that a syndrome of venation features characterizes the leaves of that genus, but as Alvin and Chaloner (1970) point out, this venation character is paralleled in the leaves or leaflets of at least five separate classes of Palaeozoic or Mesozoic plants (Fig. 2). W. N. Edwards (1955) has suggested that we should be guarded in accepting any records of *Glossopteris* in implausible locations, when they are not accompanied by their very characteristic fructifications. We believe this applies, for example, to the Russian report (Zimina 1967) of *Glossopteris* from Vladivostok. In this and other Angaraland localities *Glossopteris*-like leaves are not associated with the fructifications known in Gond-

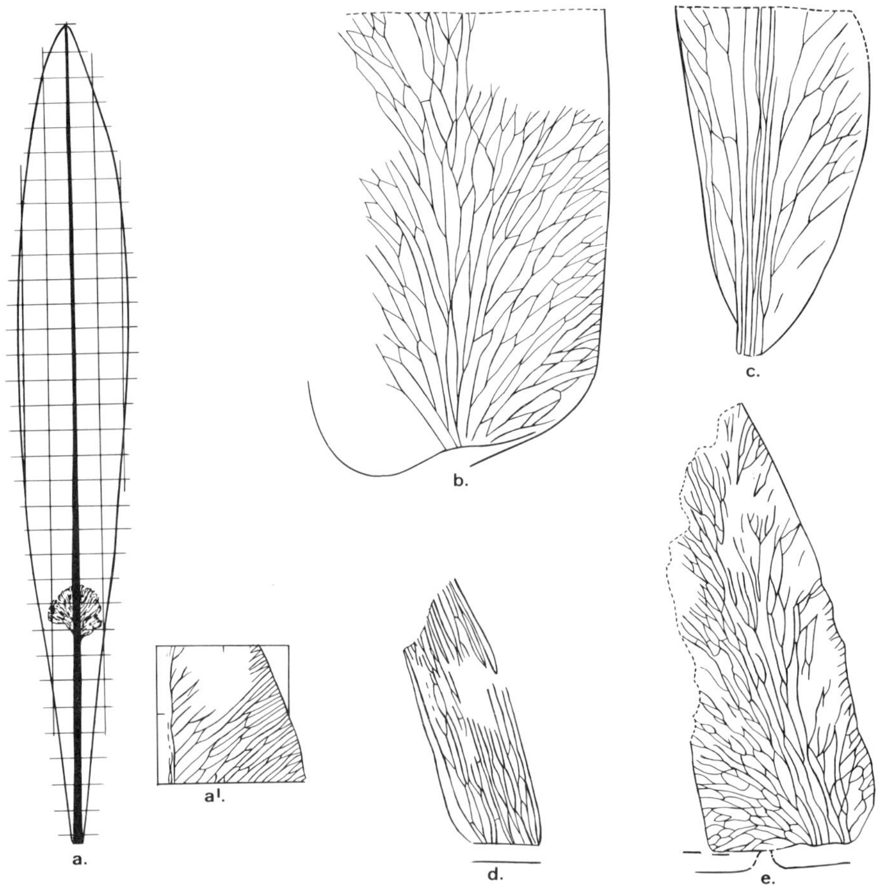

FIG. 2. Leaves of five different orders of gymnosperms, showing convergence in character of venation. (a) one leaf of *Glossopteris pantii* from India. (a—whole leaf, a′—detail of venation; 1 cm graticule gives relative scale); (b)–(e): leaflets (constituent units of compound leaves) of (b) *Linopteris* (Pteridospermales, Carboniferous), (c) *Sagenopteris* (Caytoniales, Jurassic), (d) *Ctenis* (Cycadales, Jurassic), (e) *Dictyozamites* (Bennettitales, Jurassic). (Not to the same scale.) The similarity of leaf form and venation between *Glossopteris* and leaflets of the others underlines the hazard of attributing fragments of leaves or leaflets to *Glossopteris* without supporting evidence of other organs of the plant. ((a) after Chandra and Surange 1979; (b)–(e) after Alvin and Chaloner, 1970.)

wanan *Glossopteris*. Without supporting evidence of fructifications, the relationship of these Angaran leaves to the Gondwanaland *Glossopteris* remains problematical. The same must be said, but to a lesser degree, of the leaves attributed to *Glossopteris* even at Hazro and in New Guinea, although the latter are strengthened by the associated record of the very distinctive *Glossopteris* root casts, *Vertebraria*. Both Hazro and New Guinea may well represent northern extensions of *Glossopteris*, but we are merely recording our view that isolated leaf records of that genus are not all to be regarded as of equal status.

Despite the problems presented by *Phyllotheca* and *Glossopteris* (and indeed many other Palaeozoic plant genera based on leaves) we still regard plant fossils as having unique value in attempts to construct the past positions of continental plates. Land plants have the merit of having grown in the terrestrial subaerial environment, in equilibrium with it and its prevailing climate. Despite the 'noise' induced by taxonomic problems, the recognition of floristic provinces can make a valid contribution to the totality of data bearing on plate positions.

Delimitation of floristic provinces

In their review of late Palaeozoic floras, Chaloner and Lacey (1973) enumerated 33 genera which formed an adequate basis for subjectively characterizing the four main Permian provinces. In order to try to see some objectivity in delimiting these provinces and to quantify the degrees of similarity, we used the Dice Similarity Index to formulate a distance measure of the composition of the floras in the construction of a dendrogram. This is essentially the same procedure as that adopted for Raymond *et al.* (1985) in their cluster analysis of Siegenian–Emsian floras. We have elected to invoke a greater degree of subjectivity in our data, in consciously selecting some 38 macrofossil plant genera which we believe to be adequately defined and consistent in their application. These are the 33 characteristic late Palaeozoic genera tabulated in Chaloner and Lacey (1973, table 1), but with the addition of five further genera (see caption to Fig. 3). Sources of data for the matrix used are cited in that legend. Our aim was to see the order of similarity accorded on this basis to the long-recognized major provinces, and ascertain how the several 'mixed floras' (combining elements typical of more than one province) aligned themselves. We have also elected to include records from quite broad regions (see e.g. 'Europe', Southern and Central Africa) in Fig. 3, while including in the same matrix, records from single or very limited collecting localities (Hazro, Turkey; Unaizah, Saudi Arabia). We hope that our candour in drawing attention to this further subjective procedure will be accepted by our critics as a token of our good intentions, and of our awareness of the limitations of the dendrogram produced. The scaled distance measure of compositional difference between floras, forming the horizontal axis of Fig. 3, ranges from zero units (for floras of generically the same composition) to 25 units, which is the maximum distance accorded to floras with no common elements.

As would be expected, the several Gondwana fragments (Africa south of the Sahara, India, South America, and Australia) cluster at a low distance measure, with Antarctica distanced by a much smaller range of recorded genera. New Guinea is linked with these on the presence of *Glossopteris*, but at a greater distance measure. All the subjectively assigned Cathaysian localities (including Hazro, Ga'ara, and Saudi Arabia, shown as *1*, *2*, and *3* respectively, on Fig. 5) link eventually with the Euramerican province nested within them. Common elements similarly link the Angara flora with the Euramerican. Sumatra, with its rather small number of taxa, is separated from the body of the Cathaysian localities but is, of course, far distanced from Gondwana, lacking any typical plants of that province.

Fig. 3 has an interesting similarity to Meyen's diagram of the time element in floral differentiation (Chaloner and Meyen 1973, fig. 4). Just as the Gondwana flora is seen by him as segregating first in a time sequence, compared with the Cathaysia/Angara/Euramerica differentiation, so is the Gondwana flora distanced at a higher level from the other three major floristic provinces.

It would be naïve to suggest that the horizontal axis (distance measurement of floristic composition) in Fig. 3 is in any direct sense a 'time axis'. The degree of difference in generic composition between more or less synchronous floras is controlled partly by the time elapsed since migration between them was severed (or severely restricted), but also partly by sustained differences in climate. The latter can, of course, in present-day (as presumably in Palaeozoic) floras, produce a gradational relationship of composition, as say between the Canadian Arctic and Florida, even when there is no 'barrier' to migration.

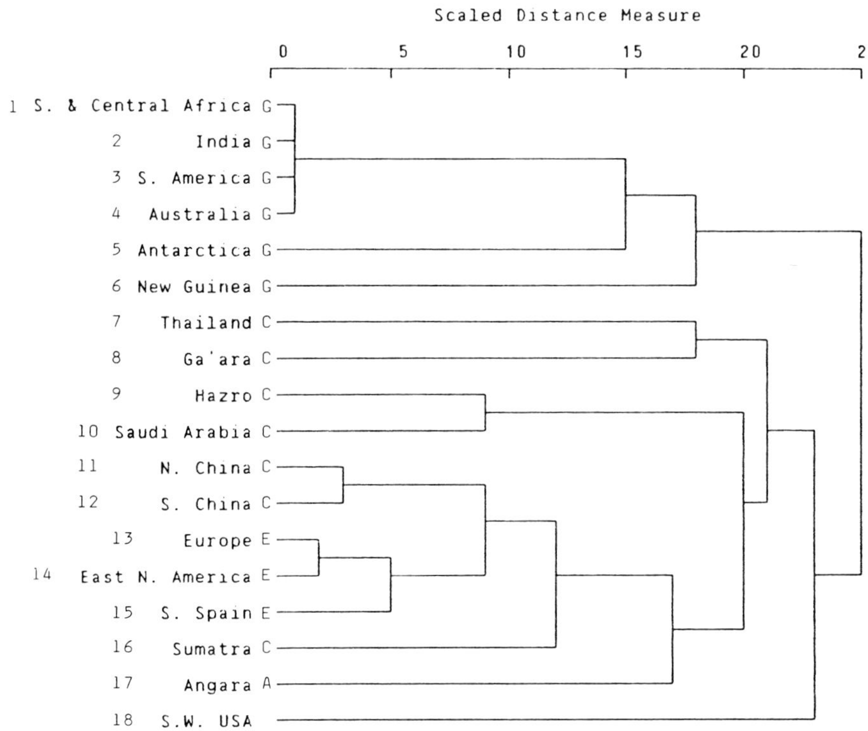

FIG. 3. This dendrogram expresses similarity of floristic composition, at generic level, between eighteen regions or localities. The scaled 'distance' measure, the horizontal axis, is derived from the Dice Index of Similarity (see text, p. 205). The comparison of these floras is based on 38 characteristic and (locally) common plant genera, chosen either for their distinctiveness (reliability of identification) or their constant association with one or more floristic provinces. These genera are the 33 listed in Chaloner and Lacey 1973, table I, with the addition of *Rhipidopsis*, *Gondwanostachys*, *Botrychiopsis*, *Fascipteris*, and *Lepidodendron s. s.* The letters *G*, *C*, *E*, and *A* identify floras as subjectively assigned to the Gondwana, Cathaysia, Euramerica, and Angara provinces. General sources of data are from Chaloner and Meyen (1973), Chaloner and Lacey (1973), Vachrameev *et al.* (1978); individual records were also derived from the following sources, for the localities or regions shown: *1–5*: Plumstead (1973), *6*: Jongmans (1940), Hopping and Wagner (1962), *7*: Kon'no (1966), *8*: Ctyroky (1973), *9*: Archangelsky and Wagner (1983), *10*: El-Khayal and Wagner (1985), *11*,*12*: Asama (1984); Kon'no (1966); Li and Yao (1982); Wang (1985), *15*: Broutin (1982), *16*: Jongmans (1937).

Contribution of palaeofloristic data to plate reconstructions

We have attempted to summarize significant features of late Palaeozoic floristic data in the context of Tethys and Gondwana, in the form of two maps (Figs 4, 5).

Fig. 4 shows the juxtaposition of floral provinces brought together by the closure of the eastern end of Tethys. The main features have frequently been pointed out, and reviewed extensively elsewhere (see e.g. Kon'no 1966, fig. 1; Li and Yao 1979, 1982; Asama 1984, fig. 3; Hallam 1986, fig. 4). One of the most significant features of the distributions recorded in Fig. 4 is the continuity of floras of broadly Cathaysian type extending from southern China through Thailand and Malaya into Sumatra (the Djambi flora). We know of no indisputable records of *Glossopteris* extending into this part of the Cathaysian province. The Djambi flora is in need of re-collecting and revision in the light of recently increased knowledge of the Cathaysian flora generally, but on present evidence, its Cathaysian affinity (Figs 4 and 5) implies migrational continuity from South China into that part of Sumatra in Permian times. This contrasts with the several records of a flora including *Glossopteris* (made more secure by the presence of its very distinctive root, *Vertebraria*) from Western New Guinea (Hopping and Wagner 1962).

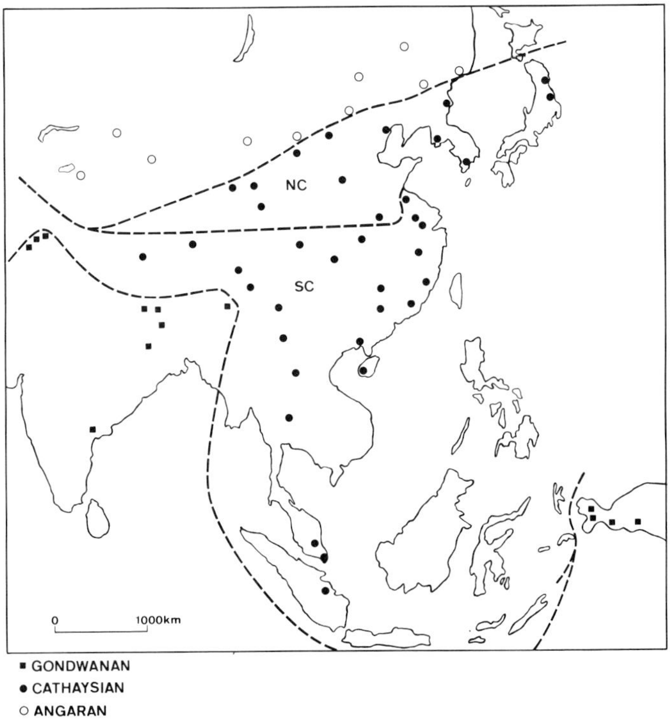

FIG. 4. Selected records of fossil floras in eastern Asia of broadly late Carboniferous and Permian age, attributed to the Angaran, Cathaysian, and Gondwanan floristic provinces. The dashed lines represent approximate boundaries between the provinces. *NC* = North China, *SC* = South China. (After Li and Yao 1979, and Asama, 1984.)

Taken at their face value, these data certainly favour this part of New Guinea, at least, having been in juxtaposition with the Australian plate and presumably on the south side of Tethys in Permian times, with some barrier or gap separating it from the southern part of the Cathaysian province. This is scarcely compatible with the Carboniferous through Triassic reconstructions of Audley-Charles (1983, figs 1 and 2) which show sustained juxtaposition of Sumatra and New Guinea. On these grounds we prefer the Permian reconstruction of Lin *et al.* (1985, fig. 4) assuming that Western New Guinea retained contact with the northern edge of the Australian Plate at a time when the South China block lay well to the north in palaeoequatorial latitudes.

The *Glossopteris* flora—or fragments of parts of Gondwana that bore the *Glossopteris* flora—are now reported in the southern part of Tibet, so that on this basis, the present-day 'junction' of Cathaysia and Gondwana there actually lies within Tibet. Further mapping of floras within the western part of the South China block should clarify this boundary.

Recent Chinese work has emphasized the distinction between the North and South China provinces (or subprovinces) within Cathaysia (Li and Yao 1979; Li and Yao 1982; Wang 1985). However, most, if not all, of the genera of the Palaeozoic floras characteristic of the Cathaysian province occur on both the North and South China blocks. Differences are of relative frequency; *Gigantopteris* for example, was widespread in the southern subprovince but only rarely recorded in the northern. *Taeniopteris* shows the reverse, being commoner in the north (Li and Yao 1979). Despite differences at specific level, the differentiation between the North and South China blocks is not of the order of that between either and the Angara and *Glossopteris* floras to the north and south respectively (Fig. 3). The difference between Permian floras of North and South China is consistent with their being on separate plates (cf. Lin *et al.* 1985, fig. 4, and our Fig. 5), but the floristic evidence alone can hardly be said to require this disjunction.

The boundary and time relationship between the Cathaysian and Angaran provinces is reviewed fully in Li and Yao 1982. The controversy concerning the synchroneity of the two floras in the 'Nanshan section', and their spatial relationship, was commented on by Halle (1937). Li and Yao (1982) now suggest that there is indeed a mixed flora with elements of both Angara and Cathaysia types, in the Nanshan area. They suggest that 'by at least the Late Permian the Tianshan–Khingan sea trough by which the two floras were previously separated, no longer existed, thus enabling the Angara flora to move southward . . . and mingle with the Cathaysia flora'. Meyen however (pers. comm. Jan. 1987) suggests that the Nanshan flora cannot be called truly mixed (i.e. containing both Angaran and Cathaysian plants) because taxa which can be qualified as typically Cathaysian are problematical in Nanshan. Li and Yao's suggestion is attractive in palaeobotanical terms, but palaeomagnetic evidence favours a considerable gap (some 20°) between the North China and Siberian Plates (Lin *et al.* 1985, and our Fig. 5).

The westward extension of floras of Cathaysian type into Asia Minor presents several problems. These concern the floras of Hazro, Turkey (for recent revision see Archangelsky and Wagner 1983), of Ga'ara, Iraq (Ctyroky 1973) and of Unaizah, Saudi Arabia (see somewhat conflicting versions of El-Khayal *et al.* 1980, Lemoigne 1981, and El-Khayal and Wagner 1985). These three floras are shown (as *1*, *2*, and *3*) on the Asia Minor 'block' in Fig. 5, and are treated individually in the dendrogram of Fig. 3. The occurrence of leaves at Hazro, Turkey, apparently of *Glossopteris* type, in considerable isolation from the main body of the *Glossopteris* flora and at a low palaeolatitude, remains one of the most puzzling features of all the Permian floras. Despite the *Glossopteris* record at Hazro, the other Cathaysian elements from that locality (*Lobatannularia*, *Gigantopteris s. l.*) cause it to link, together with Ga'ara and Saudi Arabia, to the main body of the Cathaysian province (North and South China) in Fig. 3. However, these three floras at least fall in plausible juxtaposition to—and at the same palaeolatitude as—the South China block in Fig. 5.

The record by Broutin (1982) of a 'mixed flora' from Seville, southern Spain, with elements suggestive of Angara-like plants (compared with *Koretrophyllites*) and a

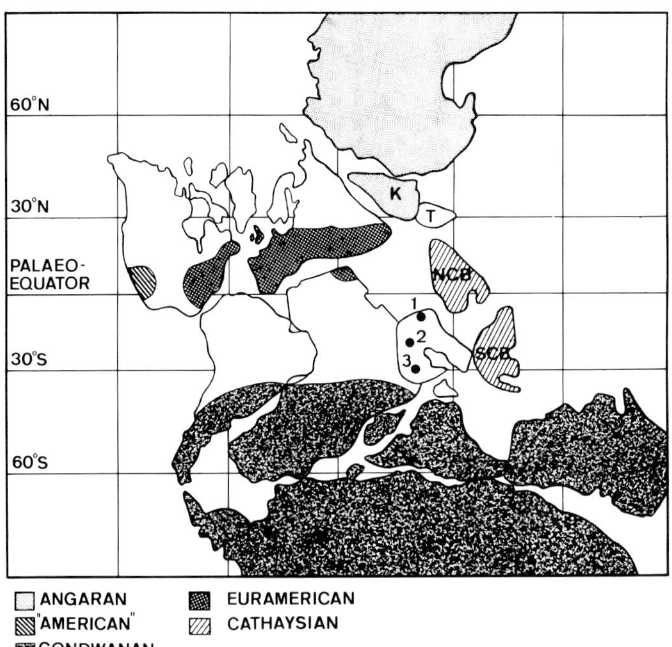

FIG. 5. Permian reconstruction of plate positions, showing approximate areas to be attributed to the floristic provinces of south-western USA ('America'), Euramerica, Gondwana, Angara, and Cathaysia. (*K*: Kazakhstan block; *T*: Tarim block; *NCB*, *SCB*: North and South China blocks. *1*: Hazro, Turkey; *2*: Ga'ara, Iraq; *3*: Unaizah, Saudi Arabia.) The floras *1–3* have some Cathaysian affinity, but their attribution is considered in the text. (After Chaloner and Lacey 1973; Lin *et al.* 1985.)

Ginkgophytopsis (suggestive of Gondwana affinity) presents features quite as puzzling as those of the Hazro flora. However, on broad generic composition (of those genera included in our dendrogram) the southern Spanish flora clusters with the Euramerian province in Fig. 3, but clearly the implications of Broutin's discovery need further exploration.

Of the several plate reassemblies available for the late Permian, we favour that shown in our Fig. 5, based on Lin *et al.* 1985, for reasons already indicated. The most significant difference from the corresponding palaeofloristic map (in Chaloner and Lacey 1973) is the changed relationship of the Siberian Plate and the totality of the Cathaysian province. The Smith, Briden, and Drewry model used (and referenced) in Chaloner and Lacey, placed the Angaran and Cathaysian provinces in comparable latitudes between approximately 25° and 70°N. The separation of these provinces, and the restriction of the Cathaysian flora (North China block (NCB), South China block (SCB) in Fig. 5) to the palaeoequatorial zone, with the Siberian Plate and its Angaran flora in higher latitudes, is far more compatible with the palaeobotanical evidence.

The particular relationship of South America and Africa, *vis-à-vis* North America and Europe, implicit in this map, gains no particular support from the floristic data. The only constraint that fossil floras offer in this aspect of Pangaea is the similarity of north-west African Carboniferous floras to those of Europe. This, as Hallam (1983) has pointed out, is one of several diverse pieces of evidence favouring a 'Pangaea A' type reconstruction (with north-west Africa in juxtaposition with south-west Europe, rather than having been, in Permian time, far to the east of its present longitudinal relationship to Europe, as in Pangaea B and Pangaea C, Hallam 1983). The particular features of the Lin *et al.* reconstruction used here, that are conformable with the floristic data are the palaeoequatorial position of NCB and SCB and the several Asian floras (*1–3*, legend to Fig. 5) and the generally high northern palaeolatitude of the Angara province. We hold no particular brief for other aspects of this reconstruction, noting that Livermore *et al.* (1986), on a reinterpretation of the palaeomagnetic data, favour (with Hallam) a version of Pangaea A which would involve a more westerly position of Gondwana in relation to Europe and North American than that used here.

ACKNOWLEDGEMENTS: We record our warm thanks to Dr A. Wroot, Department of Computer Science, Royal Holloway and Bedford New College, for help with computational work in preparation of the dendrogram of Fig. 3, and to Mrs Angela Christie for her help in preparing the remaining figures.

References

ALVIN, K. and CHALONER, W. G. (1970). Parallel evolution in leaf venation: an alternative view of angiosperm origins. *Nature* **226**, 662–3.

ARCHANGELSKY, S. and WAGNER, R. H. (1983). *Glossopteris anatolica* sp. nov. from uppermost Permian strata in south-east Turkey. *Bull. Brit. Museum Natur. Hist. (Geol.)* **37**, 81–91.

ASAMA, K. (1984). *Gigantopteris* flora in China and Southeast Asia. In *Geology and palaeontology of Southeast Asia*. Vol. 25 (ed. T. Kobayashi, R. Toriyama, and W. Hashimoto), pp. 311–23.

AUDLEY-CHARLES, M. G. (1983). Reconstruction of eastern Gondwanaland. *Nature* **306**, p. 48.

BOUREAU, E. (1964). *Traité de paléobotanique Vol. 3: Sphenophyta, Noeggerathiopsida*. Masson et Cie, Paris.

BROUTIN, J. (1982). Importance paléobiogéographique de la découverte d'une flore continentale mixte dans le Permien inférieur du Sud-Ouest de la Péninsule ibérique. *C. R. Acad. Sci., Paris* **295**, 419–22.

CHALONER, W. G. (1959). Continental drift. *New Biology* **29**, 7–30.

CHALONER, W. G. and LACEY, W. S. (1973). The distribution of Late Palaeozoic floras. In *Organisms and continents through time*. (ed. N. F. Hughes), Spec. Pap. Palaeont. **12**, 271–89.

CHALONER, W. G. and MEYEN, S. V. (1973). Carboniferous and Permian floras of the northern continents. In *Atlas of palaeobiogeography*. (ed. A. Hallam), pp. 169–86, Elsevier, Amsterdam.

CHALONER, W. G. and SHEERIN, A. (1981). The evolution of reproductive strategies in early land plants. In *Evolution today*. (ed. G. C. E. Scudder and G. L. Reveal), pp. 93–100, Proc. Second Intern. Congr. System. Evol. Biol. Hunt Institute of Botanical Documentation, Pittsburgh, PA.

CHANDRA, S. and SURANGE, K. R. (1979). *Revision of the Indian species of Glossopteris*. Monograph No. 2. Birbal Sahni Institute of Palaeobotany, Lucknow, India. (pp. 1–291.)

CTYROKY, P. (1973). Permian flora from the Ga'ara region (Western Iraq). *Neues Jb. Geol. Paläont. Mh.* **7**, 383–8.

EDWARDS, W. N. (1955). The geographical distribution of past floras. *Adv. Sci.* **46**, 1–12.

El-Khayal and Wagner, R. H. (1985). Upper Permian stratigraphy and megafloras of Saudi Arabia: palaeogeographic and climatic implications. *C. R. 10e Congrès l'avancement étud. stratigr. Carbonifère, Madrid 1983* **3**, 17–26.

El-Khayal, A. A., Chaloner, W. G., and Hill, C. R. (1980). Palaeozoic plants from Saudi Arabia. *Nature* **285**, 33–4.

Gould, R. and Delevoryas, T. (1977). The biology of *Glossopteris*: evidence from petrified seed-bearing and pollen-bearing organs. *Alcheringa* **1**, 387–99.

Hallam, A. (1983). Supposed Permo-Triassic megashear between Laurasia and Gondwana. *Nature* **301**, 499–502.

Hallam, A. (1986). Evidence of displaced terranes from Permian to Jurassic faunas around the Pacific margins. *J. Geol. Soc. London* **143**, 209–16.

Halle, T. G. (1937). The relation between the Late Palaeozoic floras of eastern and northern Asia. *C. R. 2e Congrès l'avancement étud. stratigr. Carbonifère, Heerlen 1935, Maastricht* **1**, 237–45.

Hopping, C. A. and Wagner, R. H. (1962). Plant determinations. In *Geological results of the exploration for oil in Netherlands New Guinea.* (ed. W. A. Visser and J. J. Hermes), p. 53 and fig. legends, Verhandelingen van het Koninklijk Nederlands geologisch mijnbouwkundig genootschap, Geologische serie, deel XX, Speciaal nummer.

Jongmans, W. J. (1937). The flora of the Upper Carboniferous of Djambi (Sumatra, Netherl., India) and its possible bearing on the palaeogeography of the Carboniferous. *C. R. 2ème Congrès l'avancement étud. stratigr. Carbonifère, Heerlen 1935, Maastricht* **1**, 354–62.

Jongmans, W. J. (1940). Beiträge zur Kenntnis der Karbonflora von Niederländisch Neu-Guinea. *Mededel. Geol. Stichting 1938–39*, 263–74.

Kon'no, E. (1966). Some connection between the Cathaysian flora and the *Glossopteris* flora in India during the later Permian age. *Palaeobotanist* **14**, 26–35, 97–132.

Lemoigne, Y. (1981). Flore mixte au Permien supérieur en Arabie Saoudite. *Géobios* **14** (5), 611–35.

Li, X. X. and Yao, Z. Q. (1979). Carboniferous and Permian floral provinces in East Asia. *Papers presented at the 9th International Congress of Carboniferous Stratigraphy and Geology, 1979.* Academia Sinica, Nanjing, China.

Li, X. X. and Yao, Z. Q. (1982). A review of recent research on the Cathaysia flora in Asia. *Am. J. Botan.* **69**, 479–86.

Lin, J. L., Fuller, M., and Zhang, W. Y. (1985). Preliminary Phanerozoic polar wander paths for the North and South China Blocks. *Nature* **313**, 444–9.

Livermore, R. A., Smith, A. G., and Vine, F. J. (1986). Late Palaeozoic to early Mesozoic evolution of Pangaea. *Nature* **322**, 162–5.

Lyell, C. (1855). *Manual of elementary geology.* (5th edn). J. Murray, London.

Melville, R. (1969). Leaf venation patterns and the origin of angiosperms. *Nature* **224**, 121–5.

Meyen, S. V. (1969). New data on relationship between Angara and Gondwana Late Palaeozoic floras. In *Gondwana Stratigraphy, IUGS Symposium, Buenos Aires, 1967.* pp. 141–57, Unesco, Paris.

Meyen, S. V. (1979). Relation of Angara and Gondwana floras: a century of controversies. *Fourth International Gondwana Symposium, Calcutta, India, 1977*, pp. 45–50.

Meyen, S. V. (1987). *Fundamentals of palaeobotany.* Chapman & Hall, London.

Plumstead, E. P. (1973). The late Palaeozoic *Glossopteris* flora. In *Atlas of palaeobiogeography.* (ed. A. Hallam), pp. 187–205, Elsevier, Amsterdam.

Raymond, A. (1985). Floral diversity, phytogeography and climatic amelioration during the early Carboniferous (Dinantian). *Palaeobiol.* **11** (3), 293–309.

Raymond, A., Parker, W. C., and Barrett, S. F. (1985). Early Devonian phytogeography. In *Geological factors and the evolution of plants.* (ed. B. H. Tiffney), pp. 129–67, Yale University Press, New Haven.

Vakhrameev, V. A., Dobruskina, I. A., Meyen, S. V., and Zaklinskaya, E. D. (1978). *Paläozoische und mesozoische Floren Eurasiens und die Phytogeographie dieser Zeit.* G. Fischer Verlag, Jena, East Germany.

Wang, Z. Q. (1985). Palaeovegetation and plate tectonics: palaeophytogeography of North China during Permian and Triassic times. *Palaeogeogr., Palaeoclimatol., Palaeoecol.* **49**, 25–45.

Zimina, V. G. (1967). *Glossopteris* and *Gangamopteris* from the Permian deposits of the South Maritime Territory. *Paleont. J.* **2**, 98–106. (English language translation from Russian *Palaeontologicheske Zhurnal* 1967, (2), 113–21.)

W. G. Chaloner and G. T. Creber, Biology Department, Royal Holloway and Bedford New College, University of London, Egham Hill, Egham, Surrey TW20 0EX, UK.

The nature, extent, and subsequent dispersal of the Permian Tethys

J. B. Waterhouse

ABSTRACT: On the basis of the sedimentary and biotic facies characteristics of Permian strata deposited in different climatic zones, scepticism is expressed about some recent palaeogeographical reconstructions that demand a vast wedge-shaped Tethys in Asia, or movement over considerable latitudes of allochthonous terranes.

Definition and extent of Tethys

Suess (1901) extended his original Tethys from the ancient Permian–Mesozoic of the Mediterranean eastwards through Asia as far as Timor, but it would seem unnatural to disrupt the Tethys at the Atlantic Ocean, because rocks and faunas of the Caribbean region in Mexico, north South America and south and west United States are allied. Tethys may also be extended around parts of the Pacific Ocean, through China, Japan, Kamchatka, Sikhote Alin, and southwards beyond Timor into New Guinea and on as far as New Zealand. The belt includes a wide range of facies, ranging from uplands through epicontinental seas, to troughs and volcanic arcs. The Permian of the Mediterranean Tethys is characterized by its tropical faunas, with compound rugose corals and fusulines through much of the sequence, and the Euramerican (or Euramerian) flora. To the east, in comparable rocks, lies the Cathaysian flora, in China, and Japan. By contrast, Gondwana, also proposed by Suess as a vast southern continent, or supercontinent, was characterized during the Permian period by the *Glossopteris* flora, and by the widespread distribution of glacial sediments. Coal measures and terrestrial sediments were also well developed, and faunas showed cold-water attributes, entirely lacking in compound rugose corals, or fusulines. Between typical Tethys and Gondwana is a widespread transition zone, with diamictites and cold-water faunas in the basal Permian, and warm-water faunas higher in the column, sometimes bearing compound rugose corals and fusulines. The zone extends through southern Asia, including the Salt Range, Pamirs, Himalayas, Burma, West Thailand, Malaysia, (Thai-Shan), New Guinea, etc., and is usually referred to Tethys. The northern region of the Permian world, Laurasia, was characterized by fusulines and rugose corals at some intervals, had a characteristic Angara flora, lacked diamictites, and occasionally developed seas characterized by peculiarly high salinity (Zechstein, Phosphoria, Kazanian–Kungurian of Russia, etc.). Some glacial diamictites are found in north-east Siberia in Verchoyan and Kolyma, which lacked fusulines and rugose corals, and were somewhat comparable to Gondwana. Overall, the Permian world showed a fair degree of asymmetry in climate, with southerly glaciation extending much nearer the equator—even overlapping the subtropics in the early Permian (Waterhouse and Bonham-Carter 1976).

Dispersal and deformation of Tethys

The dispersal of glacial sediment over scattered fragments of Gondwana has long provided support for continental displacement, because the Southern Hemisphere diamictites now lie as far north as 38°N, and the Himalayas near 27°N. The Permian equatorial belt passes through latitudes as high as 45°N, so that there has been a general northward dispersion, as well as lateral pulling apart since the Permian. But the concept of a vast wedge-shaped Tethys, or Panthalassa in Asia (Tollmann and Kristan-Tollmann 1985), has never received support from most geologists working in the field in Asia. For example, in that model, the Karakorum and Pamirs were placed on the north side of Panthalassa, some 50° north of the Himalayas. Yet the Permian faunas of the three regions belonged to the same faunal provinces: and rock types persist from the Himalayas into the Karakorum (Waterhouse 1982). The rocks and faunas of central Tibet are related to those of south of the Yarlungzangbu suture, and include diamictites, similar flora and faunas,

From AUDLEY-CHARLES, M. G. & HALLAM, A. (eds) *Gondwana and Tethys* Geological Society Special Publication No. 37, pp. 211–12.

with even a substantial Gondwana fauna 250 km north of the suture at Rutog (Liu and Gui 1983; Waterhouse 1987). In short, the Indus–Yarlungzangbu suture was of negligible significance during the Permian, and does not mark the line of a former vast Permian ocean. Moreover, equatorial rocks and faunas lie north of these southerly Tethyan areas, in Fergana and North Tibet. Under the wedge-shaped Tethys, they should be of high latitude types. Similarly, there is evidence, palaeomagnetic, as well as faunal and stratigraphic, that Thai-Shan lay in the Southern Hemisphere, and that the Indo-China and South China blocks lay in the palaeotropics during the Permian, not far to the north.

More recently, small continental blocks, or terranes, or island arcs and derived sediment, have been postulated as being scattered through the large oceanic Tethys, grading eastwards into a huge Pacific Ocean. Subsequently, according to this model, the blocks have converged over thousands of kilometres, and the ocean crust was subducted. It is, of course, not easy to adequately test a model based substantially on crust that has now disappeared! But few of the current proposals are convincing. The extraordinary proposal by McElhinny *et al.* (1981) that Kolyma occupied a station near 40°N, near south-east Asia is denied by rock types and by faunas: Kolyma lay well north, and close to Verchoyan, with close similarities to Arctic Canada and Novaya Zemlya; Sikhote Alin lay further south; south-east Asia lay substantially in the Southern Hemisphere. The recent proposals for several far-spaced terranes in New Zealand by Bishop *et al.* (1985), are not supported by detailed mapping, which has found rocks of allegedly separate originally far-flung terranes in stratigraphic sequence, and in turn demonstrably tied stratigraphically to a nearby volcanic arc (that allegedly formed another far-off terrane, Waterhouse 1985). A somewhat better model for the Pacific is offered by Ross and Ross (1983), but even here, there is evidence to support Tozer (1982) in suggesting that the accreted terranes originally formed archipelagos not very far offshore from the continents to which they are fused.

References

Bishop, D. G., Bradshaw, J. D., and Landis, C. A. (1985). Provisional terrane map of South Island, New Zealand. In *Tectonostratigraphic terranes of the circum-Pacific region*. (ed. D. G. Howell), Circum Pacific Council Energy & Mineral Resources. pp. 515–21, Houston, Texas.

Liu Benpei and Gui Xinsheng (1983). Discovery of *Eurydesma* fauna from Rutog northwest Xizang (Tibet), and its biogeographic significance. *Earth Sci. J. Wuhan Coll. Geol.* **1**, (19), 79–92.

McElhinny, M. W., Embleton, B. J. J., Ma, X. H., and Zhang, Z. K. (1981). Fragmentation of Gondwana in the Permian. *Nature* **293**, 212–16.

Suess, E. (1901). *Das Antlitz der Erde* Vol. 3, pt 1, Tempsky & Freytag, Prague. (508 pp.)

Ross, C. A. and Ross, J. R. P. (1983). Late Palaeozoic accreted terranes of western North America. In *Pre-Jurassic rocks in western North America suspect terranes*. (ed. C. H. Steven), pp. 7–22, Pacific section, Soc. Econ. Paleont. Mineral.

Tollmann, A. and Kristan-Tollmann, E. (1985). Paleogeography of the European Tethys from Paleozoic to Mesozoic and the Triassic relations of the eastern part of Tethys and Panthalassa. In *The Tethys—her paleogeography and paleobiogeography from Paleozoic to Mesozoic*. (ed. K. Nakazawa and J. M. Dickins), pp. 3–22, Tokai University Press, Tokyo.

Tozer, E. T. (1982). Marine Triassic faunas of North America: their significance in assessing plate and terrane movements. *Geol. Rdsch.* **71**, 1077–104.

Waterhouse, J. B. (1982). How wide was the Tethyan 'Ocean'? *Contrib. Himalayan Geol.* **2**, 245–9.

Waterhouse, J. B. (1985). Preliminary account of geology of Malte Brun Range, New Zealand. *Austral. Geol. Newsl.* **57**, 7–13.

Waterhouse, J. B. (1987). Aspects of the Permian Tethys. Its definition, interface with Gondwana, original disposition and subsequent deformation. In *International Symposium on Shallow Tethys* **2**, 1986. (ed. K. G. McKenzie), pp. 131–48, A. A. Balkena, Rotterdam (544 pp).

Waterhouse, J. B. and Bonham-Carter, G. (1976). Range, proportionate representation and demise of brachiopod families through Permian time. *Geol. Mag.* **113**, 401–28.

J. B. Waterhouse, Department of Geology, University of Queensland, St Lucia, Brisbane, Australia.

Unexpected microfaunal communities within the Triassic Tethys[1]

Edith Kristan-Tollmann

ABSTRACT: Modern research has demonstrated a remarkable uniformity in Triassic faunas throughout the Tethyan region. This is true not only for planktonic and pelagic elements but also for vagile and sessile benthic organisms, and involves a surprisingly large number of species of macro- and microfauna, including numerous zone fossils. In this work, further striking microfossil examples are selected from potentially important groups that have been little utilized previously for biogeographical research.

The extent of faunal similarity

The similarity of Triassic macrofauna and also microfauna in the whole Tethyan region shows itself to be closer than previous knowledge and published faunal lists had led us to assume. This phenomenon concerns the whole range of organisms, planktonic, nektonic, and both vagile and sessile benthonic. The reason for the earlier underestimation of the extent of similarity may, in the first place, result from lack of an opportunity to make personal comparisons of suites of fossils from the Far East, Middle East, and Occident. Secondly, many researchers, working on isolated samples collected during expeditions to regions far from Europe, have not risked giving similar-looking fossils the same names, but on account of the great distances involved, have preferred to create new local names.

In the pioneering days of macrofossil research there was already some tendency to place the free-swimming ammonites and the halobiid lamellibranchs into the same genera and species. This was especially true of researchers on Alpine material such as E. Suess, to whom were assigned for study collections from the eastern part of Tethys. The personal collecting and assessments of these classical faunal elements in widely separate Tethyan localities by the present generation of researchers, including E. T. Tozer, R. Assereto, L. Krystyn, and F. Tatzreiter, has already established that they indeed spread uniformly along the whole extent of Tethys. The author has therefore applied herself to the study of Triassic macro- and microfaunal groups which had hardly, or not at all, been studied in terms of possible Tethys-wide distribution. Indeed, such a distribution had been dismissed as highly improbable for certain organisms with a sessile benthonic mode of life. At an early stage in the research, the surprising fact became apparent that, to a much greater extent than anticipated, not only vagile benthic but also sessile groups such as crinoids and corals had Tethys-wide distributions (Kristan-Tollmann *et al.*, 1979, 1980; Kristan-Tollmann and Tollmann, 1983*a*). Subsequently, it was confirmed that many species of both macro- and microfauna had extended not only across Tethys but also into the circum-tropical belt of Panthalassa (Kristan-Tollmann and Tollmann 1983*b*; Kristan-Tollmann 1986*a*, 1987). In these publications, and also in Kristan-Tollmann and Tollmann (1982), the problem of means of dispersal is discussed. It is concluded that larvae and pseudoplanktonic adults of sessile forms, attached to seaweed and driftwood, were able to drift in an east–west direction across the Triassic Pacific Ocean to Tethys. Since further examples of widely distributed macrofaunal elements have been dealt with very recently (Kristan-Tollmann, 1987), the author will here report on some relatively unfamiliar microfauna, whose Tethyan distributions have hitherto been largely ignored.

If now in the following, I report on elements that extend throughout Tethys at the same stratigraphic level, one should not forget that there have also been numerous species with restricted distributions, confined to narrowly defined faunal provinces. Of course the last word has not been spoken in such instances, but I will turn

[1]Publication in association with IGCP Programme No. 203: 'Permo-Triassic events of the Eastern Tethys and their intercontinental correlations' and Project No. P 5632 supported by the Österr. Fonds zur Förderung d. wissenschaftlichen Forschung.

From AUDLEY-CHARLES, M. G. & HALLAM, A. (eds) *Gondwana and Tethys* Geological Society Special Publication No. 37, pp. 213–23.

now to documentation of the Tethys-wide existence of some notable microfaunal elements.

Examples of Tethys-wide distribution of Triassic microfauna

Algae (Fig. 1)

The alga *Clypeina besici*, first described by Pantić (1965), may serve as a good example for the Tethys-wide distribution of numerous algae, because it is not only, like many other algal species, an excellent zone fossil, but has also proved itself to be facies-independent. It occurs in both limestones and marls of Carnian age, and can therefore be recognized both by its characteristic appearance in thin sections and its calcified, isolated whorls (Fig. 1).

Clypeina besici Pantić is a zone fossil of the lower Upper Triassic (Cordevolian, Julian). It probably continues up into the Tuvalian (Upper Carnian). Pantić (1965) first described this alga from the Lower Carnian of Montenegro. Subsequently, it has been discovered in limestones at many localities in the western Tethyan region, such as the Wetterstein Limestone in the Northern Calcareous Alps, Carnian limestones in the Southern Alps, Megalodont limestones in Western Anatolia and the island of Chios, and also in Karaburun (Ott 1972). Isolated whorls, which clearly disintegrated readily after death, are sometimes even common in marls of different types of Carnian litho- and biofacies.

Such whorls have only been illustrated previously once (Kristan-Tollmann in Tollmann 1976, p. 143, fig. 63), and so examples from different Alpine localities are portrayed here: (1) Numerous whorls from the Reingraben Marl (Middle Carnian–Julian) of Steiglweg, north-west of the Gosausee, Upper Austria (Fig. 1.1,

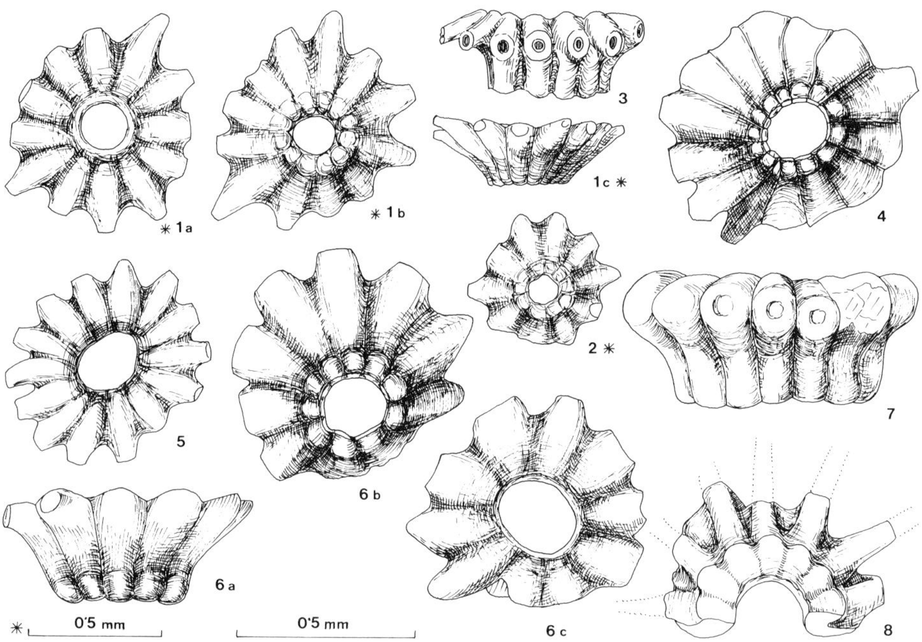

FIG. 1. Isolated whorls of the alga *Clypeina besici* Pantić from the Carnian (lower Upper Triassic) of Tethys.
1 and *2*: Reingraben Marl, Middle Carnian (Julian), Steiglweg, north-west of Gosausee, Upper Austria. *1a*: whorl from above, *1b*: whorl from below, *1c*: lateral view: whorl from below. *3 and 8*: Thigaungdaung Limestone, presumed (Upper Ladinian) Lower Carnian age, *c.* 5 km east of Hsa Hmong Hkam on the Kalaw–Heho road, Southern Shan State, Burma (fossil locality 8 of Gramann *et al.* 1972; pl. 3, fig. 1,2). *3*: Weathered whorl in lateral view (compare *7*), *8*: half whorl from below. *4*: Cassian Marl, Carnian, southern slope of Pralongia east-south-east of Corvara (sample *X*23), South Tirol, Italy. Whorl from below. *5,6,7*: Raibl Beds, Middle Carnian (Julian), left branch of spring of Kämpferbach, 1.2 km west of Raibl, Julian Alps, Italy (sample U968). *5*: Whorl from above. *6*: Whorl in lateral view (*a*), from below (*b*), and from above (*c*). *7*: Flattened whorl in lateral view.

1.2). (2) Several whorls from the Cassian Marl (Carnian) of the southern slope of Pralongia near Corvara, South Tirol, Italy (Fig. 1.4). (3) Numerous isolated whorls from the Raibl Beds (Middle Carnian—Julian) of Kämpferbach near Raibl, Julian Alps, Italy (Fig. 1.5–7).

In addition, my material at hand includes an iron oxide cast of a dissolved-out whorl from the Hallstatt Limestone (Cordevolian) of Saklibeli, near Antalya, Taurus Mountains, Turkey (not illustrated). Finally, there are some isolated whorl fragments from the Thigaungdaung Limestone (?Lower Carnian) of Hsa Hmong Hkam, Southern Shan State, Burma, that Gramann *et al.*, 1972 (plate 3, figs 1–3) portrayed as *Clypeina* sp., from which two examples are once more illustrated (Fig. 1.3, 1.8).

Crinoids (Fig. 2)

Again and again it has turned out that numerous Triassic stalked crinoids (and therefore sessile forms) have a Tethys-wide distribution. Moreover, because of their short stratigraphic range, they can serve as good zone fossils (compare Kristan-Tollmann and Tollmann 1983*a*; Tollmann and Kristan-Tollmann 1985; Kristan-Tollmann, 1986*b*, 1987). This is not, however, the case with the example presented here, *Holocrinus*? *quinqueradiatus* (Bather). With increased knowledge of this diminutive crinoid, whose columnals have been found in the residue of microfossil samples, its stratigraphic range has been extended from the Upper Ladinian–Lower Carnian boundary (to which it originally had been thought to be confined) to the whole Upper Triassic, including the Sevatian.

Holocrinus? *quinqueradiatus* was created by F. A. Bather in 1911 on the basis of a few columnals. For a long time the description of this fossil, discussed in the Cassian Beds of Bakony, Hungary, was forgotten or ignored, so that R. Zardini in 1973 created a new name, *Encrinus varidentatus*, for the similarly rare trochites of this species from the Cassian Beds of the South Tirol Dolomites. It was not until the discovery of huge numbers of columnals from all parts of the stem, in Upper Ladinian/Lower Carnian Hallstatt Limestone at Saklibeli (Taurus, Turkey), that it was possible on the one hand to reconstruct the whole stem (Kristan-Tollmann, p. 278ff., fig. 19, in Kristan-Tollmann and Krystyn 1975), and on the other hand to demonstrate the possibility of a much wider regional distribution than had been presumed originally. Knowledge of the stem structure of this species, found predominantly in microfossil samples, was of decisive help in further identifications in other Triassic samples from the Tethyan realm. It transpired that the stem had a round cross-section only in the lower, distal part, from which the trochites had been originally described. In the middle part of the stem, the section becomes rounded pentagonal, further up angular pentagonal, and in the proximal part, pentastellate. Lacking knowledge of this trochite sequence one could be easily misled, and the highest, pentastellate columnals could be assigned to a quite different genus from the rounded columnals from the stem base (compare Kristan-Tollmann, 1975, fig. 19, A8 and A4, A1).

From further research it can be determined that *Holocrinus*? *quinqueradiatus* has a Tethys-wide distribution in a wide range of facies. Up to now it has been recovered from the following horizons and localities.

1. Cassian Beds, Lower Carnian, Bakony, Hungary (Bather 1911)

2. Cassian Beds, Lower Carnian, South Tirol Dolomites, Italy (Zardini 1973)

3. Hallstatt Limestone, Upper Ladinian/Lower Carnian, Saklibeli, Taurus Mountains, Turkey (Kristan-Tollmann 1975 in Kr.-Tollm. and Krystyn); Julian, Mezarlik, Taurus (fig. 2.3, 2.10)

4. Tuffaceous sediments, Upper Ladinian, Aghdarband, near Mashhad, Iran (Kristan-Tollmann 1986*b*)

To these can be added:

5. Raibl Beds, Middle Carnian, Kämpferbach, near Raibl, Italy (Fig. 2.1, 2.2)

6. Hallstatt Limestone, Upper Ladinian (Langobardian), Sommeraukogel, west of Hallstatt, Upper Austria (Fig. 2.5, 2.6)

7. Marl Bed in Hallstatt Limestone at transition to Zlambach Marl, Upper Sevatian (Rhaetian), Zwischenkögel, west of Hallstatt, Upper Austria (Fig. 2.4, 2.7, 2.8)

8. Marl Bed in Hallstatt Limestone, Sevatian (Rhaetian), Steinbergkogel, near Hallstatt, Upper Austria (Fig. 2.4, 2.7, 2.8)

With the most recent discoveries (Fig. 2), the stratigraphic range of *Holocrinus*? *quinqueradiatus* (Bather) can be extended from the

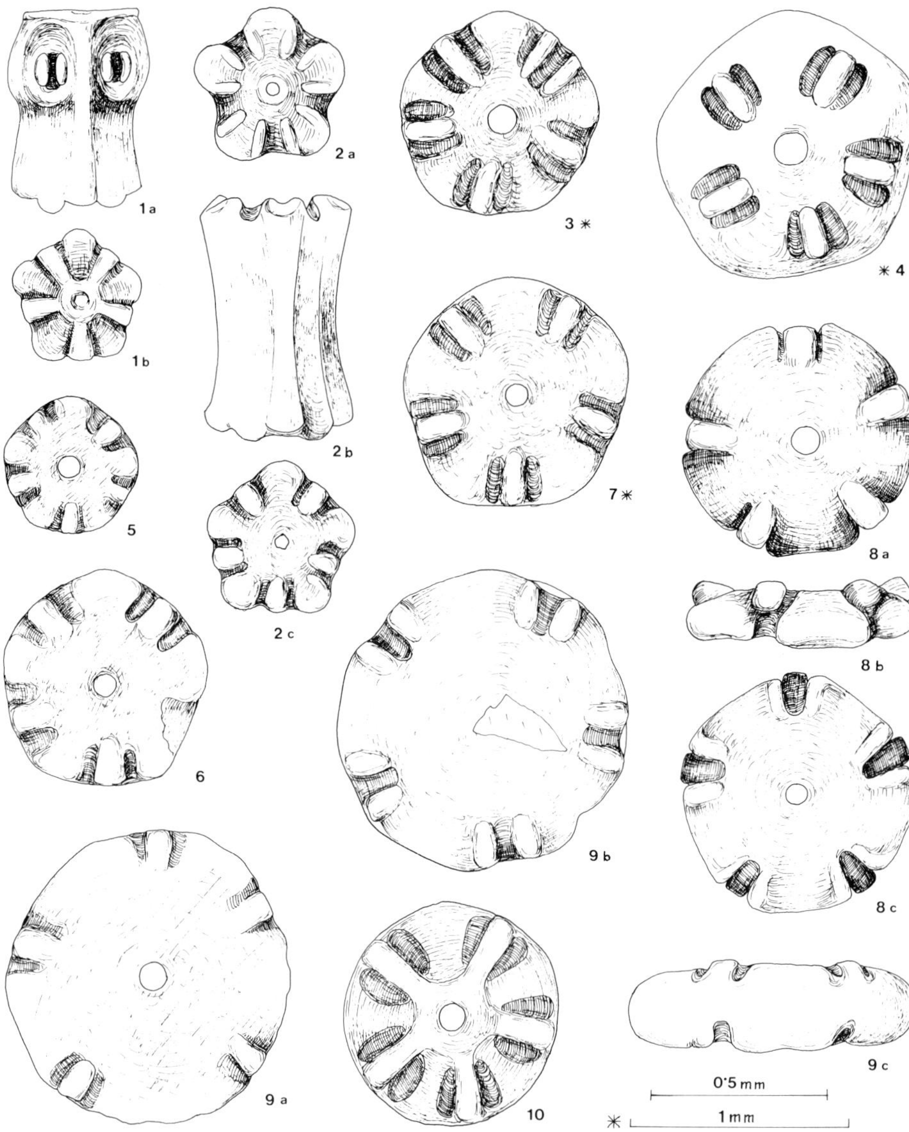

FIG. 2. *Holocrinus*? *quinqueradiatus* (Bather) from the upper Middle Triassic and Upper Triassic of Tethys. *1,2*: Very juvenile but quite elongate columnals from the proximal part of the stem. Raibl Beds *s.l.*, Middle Carnian (Julian). Left branch of Spring of Kämpferbach, 1.2 km west of Raibl, Julian Alps, Italy (Sample U968). *3,10*: Columnals from the proximal (*3*) and distal (*10*) part of the stem. Red Hallstatt Limestone, Middle Carnian (Julian) of Erenkolu Mezarlik a few km south of Saklibeli, south-west of Antalya, Taurus Mountains, Turkey (sample L. Krystyn EM4). *4,7,8*: Columnals from the middle to upper part of the stem. Marl bed in Hallstatt Limestone at the transition to the Zlambach Marl (Rhaetian), Zwischenkögel, west of Hallstatt, Upper Austria (sample R234). *5,6*: Columnals from the upper (proximal) part of the stem. Hallstatt Limestone, Upper Ladinian (Langobardian), Sommeraukogel, west of Hallstatt. *9*: Flat columnal from the middle part of the stem. Marl bed in Hallstatt Limestone (Rhaetian), Steinbergkogel, 2.3 km west-north-west of Hallstatt (sample S358).

Upper Ladinian (Langobardian) to the Lower Rhaetian (Sevatian). The regional distribution within Tethys is known up to now to extend from the Alps, via Hungary and Turkey, to Iran.

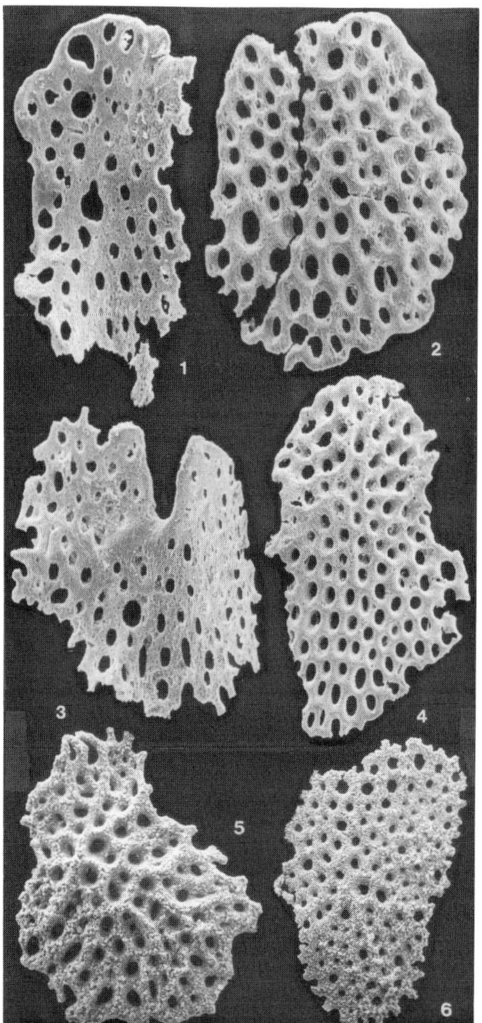

FIG. 3. *Calclamnoidea canalifera* Kristan-Tollmann (Holothuroidea) from the Tethyan Carnian. *1–4*: Sclerites from the Hallstatt Limestone, Lower Carnian (Cordevolian) of Saklibeli 30 km south-west of Antalya, Taurus Mountains, Turkey. *1,3*: From the slightly concave side, 43×. *2*: From the convex, thickened side, 43×. *4*: Like *2*, 32×. *5,6*: From the Hallstatt Limestone, Upper Carnian (Tuvalian), Raschberg, 9 km north-west of Bad Aussee, Salzkammeryut. *5*: From the convex, thickened side, 43×. *6*: Like *5*, somewhat flatter platelet, 34×.

Holothurians (Figs 3, 4, 5)

Triassic holothurians also have many species that combine a wide distribution across Tethys with a narrow stratigraphic range, as will be shown with these new examples.

1 *Calclamnoidea canalifera* Kristan-Tollmann 1963

This characteristic and abundant species was first described from the Cassian Beds (Cor-

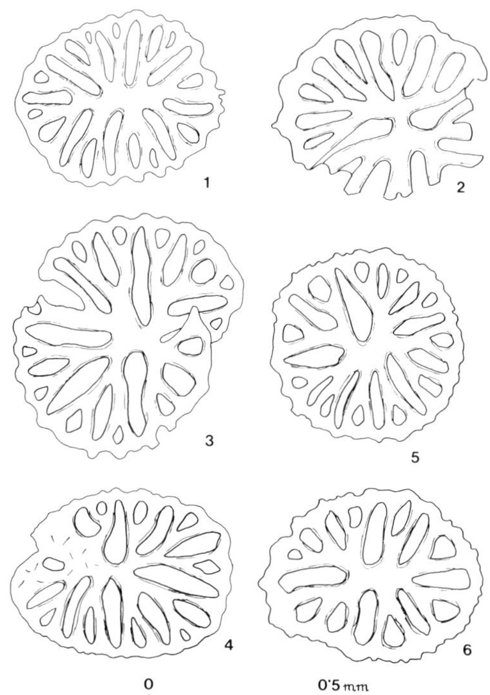

FIG. 4. *Fissobractites subsymmetrica* Kristan-Tollmann (Holothuroidea) from the Tethyan Upper Triassic. *1,2*: Sclerites from marl bed in Sevatian/Rhaetian Pötschen Limestone, Thülerergraben, 3 km north of Lofer, Salzburg (sample S736). *3*: Sclerite from marl bed in Upper Sevatian Hallstatt Limestone at transition to Zlambach Marl, Zwischenkögel, west of Hallstatt (sample R234). *4*: Sclerite from Kuti Marl (Sevatian–Rhaetian) near Kuti in north-eastern Kumaun, Himalaya, India. *5,6*: Sclerites from Thigaungdaung Limestone, of presumed (Upper Ladinian) Lower Carnian age, *c.* 5 km east of Hsa Hmong Hkam on the Kalaw–Heho road, Southern Shan State, Burma (fossil locality 8 of Gramann *et al.*, 1972, pl. 4, fig. 10, 11).

devolian) of Pralongia in South Tirol, Italy. Subsequently, it has been recognized in almost all facies types—reefs excluded—in the western part of Tethys, as the following demonstrate: Hallstatt Limestone of Upper Ladinian (Langobardian) age, Sommeraukogel, near Hallstatt, Upper Austria, and Hallstatt Limestone of Upper Carnian (Tuvalian) age, Raschberg, near Bad Aussee, Upper Austria (Fig. 3.5, 3.6); Reifling Limestone (Ladinian/Carnian) of Gross Reifling, Lower Austria; Mürztal Beds (Lower Carnian) of the Mürz Gorge near Frein, Steiermark (Austria); Halobia Beds (Middle Carnian–Julian) of Steiglweg near the Gosausee, Upper Austria; finally Seeland Beds (Lower Carnian) of the Seeland Alps, Plätzwies, South Tirol Dolomites, Italy. As the species has now also been proved in the Lower Carnian Hallstatt Limestone of Saklibeli, Taurus Mountains, Turkey (Fig. 3.1–3.4), it was evidently distributed also as a wide area in the middle Tethyan region. The stratigraphic range of *Calclamnoidea canalifera* extends from the Upper Ladinian to the Upper Carnian, with the maximum development in the Lower and Middle Carnian.

2 *Fissobractites subsymmetrica* Kristan-Tollmann 1963 (Fig. 4).

Fissobractites subsymmetrica belongs to the most abundant and striking holothurian sclerites of the Rhaetian. As reported here for the first time, it is also the Triassic species with the widest Tethyan distribution proved so far. However, increasing knowledge has shown that it is not confined to the Rhaetian, although this represents the time of its maximum development, but extends down into the Lower Carnian (Seeland Beds—see below), if not an even older horizon.

Since its first description from material collected from the Zlambach Marls of Salzkammergut, Austria, *Fissobractites subsymmetrica* has been repeatedly reported from the Rhaetian of the Northern Calcareous Alps and the Carpathians. Some further cases can be mentioned here: Roschitzbuch north-west of Mitterndorf, Salzkammezut, Styria; Thälerer Valley near Lofer, Salzburg (Fig. 4.1, 4.2); Lanzing near Hohe Wand, Lower Austria; and Zwischenkögel near Hallstatt, Upper Austria (Fig. 4.3). Moreover, the recurrence of numerous sclerites in the Lower Carnian (Cordevolian) Seeland Beds of the Seeland Alps, near Schluderbach, South Tirol (not illustrated) proves the existence of the species, at least to the beginning of the Upper Triassic.

It is noteworthy that the species has also been proved from two distant localities of the eastern Tethys: (1) the Rhaetian Kuti Marl of Kumaun, Himalaya (Fig. 4.4; Kristan-Tollmann and Gupta 1987), (2) Thigaungdaung Limestone (presumed Lower Carnian), Southern Shan State, Burma (Fig. 4.5, 4.6; fig. 10, 11, plate 4 of Gramann *et al.* 1972).

3 *Achistrum triassicum* Frizzell and Exline 1955 (Fig. 5).

The sclerites of *A. triassicum* (Fig. 5.1–5), so typical of, and abundant, in the Cassian Marl, also occur frequently in the Carnian sediments of the Northern Calcareous Alps, as for example in the Halobia Beds of Schneckengraben near Gössl/Grundlsee, Steiermark, Austria.

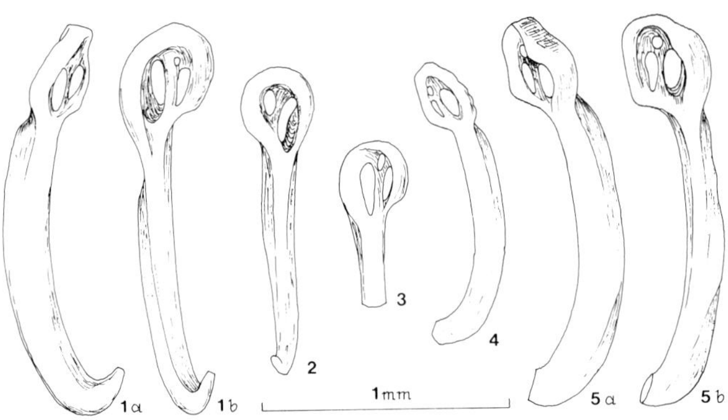

FIG. 5. *Achistrum triassicum* Frizzell & Exline (Holothuroidea) from Cassian Marl (Carnian) of South Tirol, Italy.
1: Landslip west-north-west of St Leonhard-Abtei above Pedraces (sample X82). *2,3*: Ruones Wiesen, west slope of Pralongia, south-east of Corvara (sample X20). *4,5*: South slope of Pralongia, east-south-east of Corvara (sample X26).

Recently, there was also reported from Iran: highest Ladinian (Sutherlandi Zone) of the Aghdarband Formation, Aghdarband, 100 km east of Mashhad (Kristan-Tollmann, 1986*b*, plate 1, fig. 7,8).

Ostracods (Figs 6–8)

With increasing knowledge of Triassic ostracod faunas of the middle and eastern parts of Tethys, these also prove to have been distributed across the whole of the Tethyan realm, both more

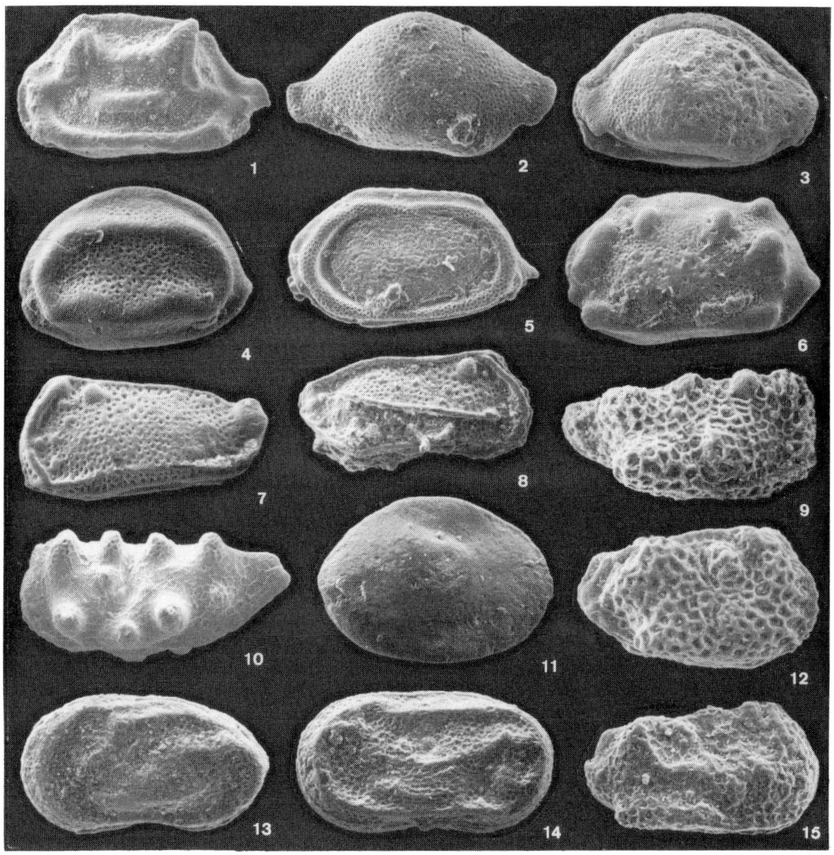

FIG. 6. Ostracods from the Rhaetian marl of the Salt spring profile north-east of Bagrabad, near Isfahan, Iran (*12* from sample R838, all others from R836). All the species illustrated here are also known from the eastern Alps.
1: *Dicerobairdia bicornuta kollmanni* Kristan-Tollmann, 30×
2: *Hiatobairdia subsymmetrica* Kristan-Tollmann, 35×
3: *Hiatobairdia labrifera* Kristan-Tollmann, 26×
4: *Carinobairdia triassica triassica* Kollmann, 35×
5: *Carinobairdia alpina* Kollmann, 26×
6: *Nodobairdia mammilata* Kollmann, 30×
7: *Kerocythere (Kerocythere) hartmanni* (Bolz and Kozur), 35×
8: *Kerocythere (Rekocythere) mostleri* (Bolz and Kozur), 35×
9: *Mostlerella dizluense* (Kristan-Tollmann), 70×
10: *Triceratina fortenodosa* (Urlichs), 53×
11: *Cytherella acuta* Urlichs, 44×
12: *Judahella (Judahella) andrusovi* Kozur and Bolz, 70×
13: *Leviella triassica* (Bolz), 44×
14: *Leviella fraterna valida* (Bolz), 35×
15: *Judahella (Costahella) triassica* (Kozur and Bolz), 53×.

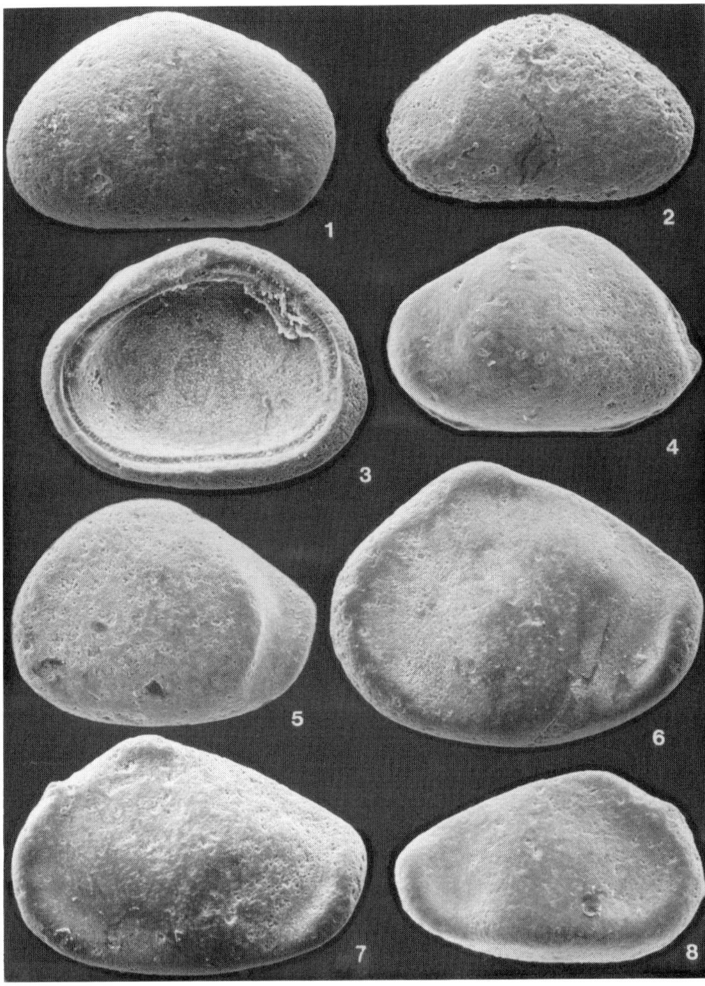

FIG. 7. Hungarelline ostracods from the Sevatian–Rhaetian, Zlambach Marl, Schneckenkogel, north-east of St Agatha, Upper Austria (sample R27).
1,2: *Triadohealdia trigonia* Kristan-Tollmann, 52×; *1*: left valve; *2*: right valve; both adult. *3,4*: *Signohealdia robusta* Kristan-Tollmann; *3*: interior of left instar valve; 52×; *4*: right valve, adult 48×. *5*: *Torohealdia opisthocosta* Kristan-Tollmann, 48×; left valve, adult. *6–8*: *Torohealdia tuberosa* Kristan-Tollmann; *6*: left valve, female adult, 52×; *7*: left valve, male adult, 52×; *8*: right valve, male adult, 48×.

extensive and numerous than ever anticipated. An example has been sketched in outline by (Kristan-Tollmann 1986*a*, p. 207ff.): A borehole drilled on the north-west edge of the Australian shelf south of Timor, 'Sahul Shoals No. 1', encountered marine Upper Triassic at 1880–1890 metres in depth, which contained an ostracod fauna with a high percentage of species known from the western end of Tethys. The dominant species is *Cytherella acuta* Urlichs, which was first described from the Kössen Beds (Rhaetian) of Kössen in Tyrol, Austria, and can be found as a common element in the micro-fauna of the Rhaetian marl of the Salt spring profile near Bagerabad in central Iran (Kristan-Tollmann *et al.* 1980). Further typical Tethyan ostracods from this borehole include *Nodobairdia mammilata* Kollmann and *Cytheropteron koessenense* Urlichs, a younger synonym of *Grammella austriaca* Kozur & Bolz.

Analysis of the ostracod fauna of the Bagerabad deposits mentioned above has shown that, of 25 species determined, 21 were already known from the Eastern Alps, and only four were new (Kristan-Tollmann *et al.* 1980). The composition of this fauna signifies, in ecological terms, a peripheral habitat, with marl layers intercalated between small coral-like layers and mounds. This inference is supported by the other macro- and microfauna, and by comparison, for example, with the fauna of Plackles, Hohe Wand, Lower Austria. Fig. 6 shows most of the characteristic Rhaetian ostracods of the Saltspring profile, that occur also in the eastern Alps.

From another facies realm, namely the open marine basin, a suite of small ostracods is found in the Rhaetian Kuti Marl of Kumaun, Himalaya, which had its facies equivalent in the Zlambach Marl of Salzkammergut. This fauna is strongly dominated by representatives of the Hungarellinae, such as *Triadohealdia trigonia*, *Torohealdia opisthocostata*, *Torohealdia tuberosa*, and *Signohealdia robusta* (Kristan-Tollmann and Gupta, 1987). All these species were first described from the western end of Tethys; Fig. 7 portrays examples of the above-named species from the Zlambach Marl, for which they are characteristic.

As a further example, the ostracod fauna from Upper Anisian marls, of the Leidapo Subformation at Guiyang in South China (Kristan-Tollmann, 1983) can be cited. This fauna obtained from test samples, contains 30 specifically determinable taxa, of which 24 are new, the remainder, however, being already known from the Middle and Upper Triassic of the western Tethys. It must be added that this Leidapo fauna signifies the first large suite of Middle Triassic Tethyan ostracods to have been made known. One may therefore assume that, with increasing knowledge of Middle Triassic marine ostracods, significantly more species with Tethys-wide distributions will be recognized.

Although, as yet, few complete suites of Triassic ostracods have been described, it is apparent that the individual species have a very wide regional distribution and can exhibit marked facies independence and long stratigraphic range. An impressive example is *Judahella (Judahella) tuberculifera* (Gümbel). This species occurs in both the Germanic and Alpine facies realms, from the highest Scythian to the Rhaetian, from Europe to China (Kristan-Tollmann, 1986*a*, p. 212, fig. 5). A further, less spectacular example is portrayed in Fig. 8: *Gruendelicythere ampelsbachensis* Kozur & Bolz has only been proved in the Tethyan realm, but from the Upper Anisian to the Rhaetian, from Western Europe to South China.

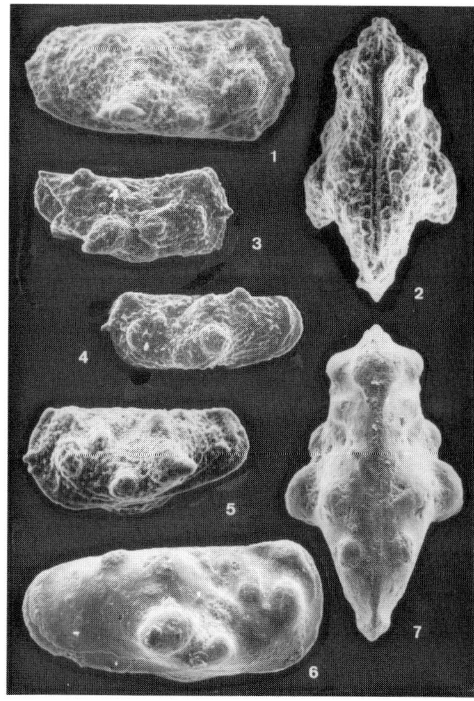

FIG. 8. *Gruendelicythere ampelsbachensis* Kozur & Bolz (Ostracoda) from the Tethyan middle and late Triassic. *1,2*: Shell from right and above; Seeland Beds, lowest Carnian, Seelandbach, exposure south-west of Plätzwies Fortress (sample U844), 3 km north-west of Schluderbach, South Tirol Dolomites, Italy, 100×. *3*: Right valve, rear part damaged, marl in Leidapo Subformation, Upper Anisian, Leidapo, 30 km south of Guiyang, South China (sample CH40), 70×. *4*: Left valve. Same locality as *3*. Sample CH48, 60×. *5*: Shell from left. Zlambach Marl, Lower Rhaetian, south of Rossmoosalm, western Stambach–Quellgraben, (3 km north-east of Gaisern, Sulzkammerzut (sample R342)), 100×. *6,7*: Steinkern from right and above. Raibl Beds *s.l.*, Carnian, left branch of Kämpferbach spring (sample U968), 1.2 km west of Raibl (Cave de Predil), Julian Alps, Italy, 100×.

Anomuran coprolites (Fig. 9)

Anomuran coprolites serve as an example of a microfossil group that is especially striking, and sometimes relatively abundant in Triassic sediments. Representatives are known that extend from the western end of Tethys to the other side of the Pacific, on the western margins of North America.

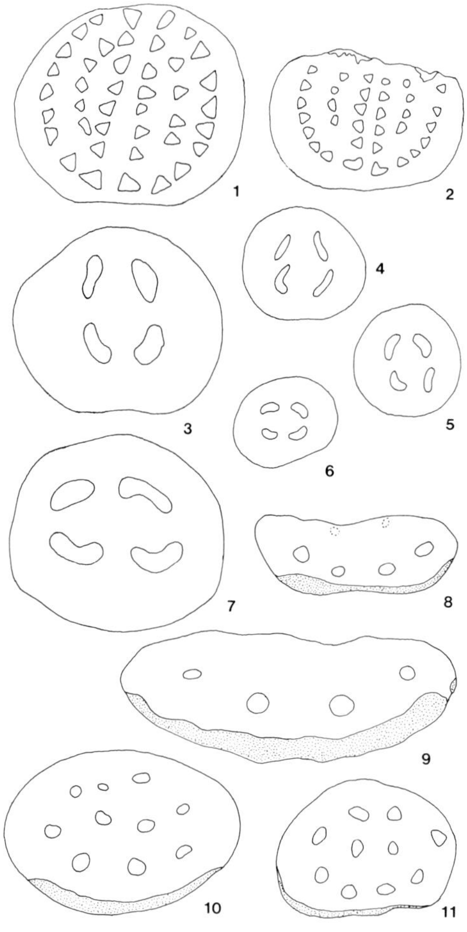

FIG. 9. Anomuran coprolites from the Rhaetian of Europe and North America (magnifications not correlated, coprolite diameter 0.3–1.0 mm).
1,2: *Parafavreina thoronetensis* Bronnimann *et al.*
1: Holotype; Thoronet, Provence, France.
2: Devils Canyon, Shasta District, California.
3,7: *Palaxius salataensis* Bronnimann *et al.*
3,7: Val Salata near Cortina d'Ampezzo, Italy.
4–6: Round Mountains, Shasta District, California.
8,9: *Thoronetia alata* Bronnimann *et al.*
8: Thoronet, Provence, France.
9: Round Mountains, California.
10,11: *Thoronetia quinaria* Bronnimann *et al.*
10: Round Mountains, California.
11: Thoronet, Provence.

Parafavreina thoronetensis Bronnimann *et al.* 1973*b* (Fig. 9.1, 9.2) is known from numerous sections in the western and middle Tethys, with a stratigraphic range from Norian to Middle Liassic (Kristan-Tollmann and Tollmann 1983*b*, p. 236; Tollmann and Kristan-Tollmann 1985, fig. 4c). Recently, this species was also found in Rhaetian limestones of the Modin Formation in Devil's Canyon near Big Bend, Shasta District, California (Kristan-Tollmann and Tollmann, 1983*b*, pl. 15; fig. 12, 13), which proves its Tethys-wide/peri-Pacific distribution.

Still more impressive, an assemblage of three coprolite species, *Palaxius salataensis*, *Thoronetia quinaria*, and *Thoronetia alata*, originally described by Bronnimann *et al.* (1973*a*, *c*) from the Rhaetian of France and Italy, has also been found in the West Round Mountains of the same district of California (Kristan-Tollmann and Tollmann, 1983*b*, p. 232 *et seq.*, pl. 12, 14; figs. 9, 11).

ACKNOWLEDGEMENTS: I want to express my gratitude to Prof. Hallam for translating this paper, and to both him and the Geological Society for inviting me to this first Lyell symposium.

References

BATHER, F. A. (1911). Triassic Echinoderms of Bakony. *Res. wiss. Erforsch. Balatonsee*, 1. Bd., 1. Tl., *Anhang: Paläont.1.* Bd., VI, Hölzel, Wien, 1–288.

BRONNIMANN, P., CARON, J.-P., and ZANINETTI, L. (1973*a*). New galatheid anomuran (Crustacea, Decapoda) coprolites from the Rhaetian of Provence, Southern France. *Mitt. Ges. Geol. Bergbaustud.* **21**, 905–20.

BRONNIMANN, P., CARON, J. P. and ZANINETTI, L. (1973*b*). *Parafavreina*, n. gen., a new thalassinid anomuran (Crustacea, Decapoda) coprolite form—genus from the Triassic and Liassic of Europe and North Africa. *Mitt. Ges. Geol. Bergbaustud.* **21**, 951–6.

BRONNIMANN, P., CROS, P., and ZANINETTI, L. (1973*c*). New thalassinid anomuran (Crustacea, Decapoda) coprolites from infraliasic limestones of the Dolomites, Italy. *Mitt. Ges. Geol. Bergbaustud.* **21**, 921–8.

GRAMANN, F., LAIN, F., and STOPPEL, D. (1972). Paleontological evidence of Triassic age for limestones from the Southern Shan and Kayah States of Burma. *Geol. Jb. B1*, 1–33.

KOZUR, H. and MOCK, R. (1972). Neue Holothurien-Sklerite aus der Trias der Slowakei. *Geol. Paläont. Mitt. Innsbruck* **2**, (12), 1–47.

KRISTAN-TOLLMANN, E. (1963). Holothurien-Sklerite aus der Trias der Ostalpen. *Sitzber. österr. Akad. Wiss. m.–nat. Kl.* **172**, 351–80.

KRISTAN-TOLLMANN, E. (1983). Ostracoden aus dem Oberanis von Leidapo bei Guiyang in Südchina. *Schriftenr. erdwiss. komm. österr. Ak. Wiss.* **5**, 121–76. Wien.

KRISTAN-TOLLMANN, E. (1986*a*). Observations on the Triassic of the south-eastern margin of the Tethys—Papua/New Guinea, Australia and New Zealand. *Neues Jb. Geol. Paläont. Mh.* 1986, 201–22.

KRISTAN-TOLLMANN, E. (1986*b*). Echinoderms from the Triassic Aghdarband Formation (NE Iran). *Jb. Geol. Bundesanst. Wien* (in press).

KRISTAN-TOLLMANN, E. (1987). Triassic of the Tethys and its relations with the Triassic of the Pacific Realm. In *Shallow Tethys*, Proc. Int. Symp. Shallow Tethys, Vol. 2. (ed. K. G. McKenzie), 169–86, A. A. Balkema, Rotterdam.

KRISTAN-TOLLMANN, E. and GUPTA, V. J. (1987*a*). Remarks on the microfauna of the Rhaetian Kuti-Marls from Kumaun, Himalaya. *N. Jb. Geol. Paläont. Mh.* 467–92, Stuttgart.

KRISTAN-TOLLMANN, E. and GUPTA, V. J. (1987*b*). Die Übereinstimmung der unterrhätischen Mikrofauna der Zlambach- und Kutimergel (Ostalpen-Himalaya). *Proceed. Int. Zechstein Sympos. 1987*, Hannover (in press).

KRISTAN-TOLLMANN, E. and KRYSTYN, L. (1975). Die Mikrofauna der ladinisch-karnischen Hallstätter Kalke von Saklibeli (Taurus-Gebirge, Türkei). *Sitzber. österr. Ak. Wiss. m.-nat. Kl.* **184**, 259–340, Wien.

KRISTAN-TOLLMANN, E. and TOLLMANN, A. (1982). Die Entwicklung der Tethys-Trias und Herkunft ihrer Fauna. *Geol. Rdsch.* **71**, 987–1019.

KRISTAN-TOLLMANN, E. and TOLLMANN, A. (1983*a*). Überregionale Züge der Tethys in Schichtfolge und Fauna am Beispiel der Trias zwischen Europa und Fernost, speziell China. *Schriftenr. Erdwiss. Komm. österr. Akad. Wiss.* **5**, 177–230.

KRISTAN-TOLLMANN, E. and TOLLMANN, A. (1983*b*). Tethys-Faunenelemente in der Trias der USA. *Mitt. österr. geol. Ges.* **76**, 213–72.

KRISTAN-TOLLMANN, E., TOLLMANN, A., and HAMEDANI, A. (1979). Beiträge zur Kenntnis der Trias von Persien. I. Revision der Triasgliederung, Rhätfazies im Raum von Isfahan und Kössener Fazieseinschlag bei Waliabad SE Abadeh. *Mitt österr. geol. Ges.* **70**, 119–86, 17 Textfig., 1 Tab., 5 Plates, Wien. II.

KRISTAN-TOLLMANN, E., TOLLMANN, A., and HAMEDANI, A. (1980). Beiträge zur Kenntnis der Trias von Persien. II Zur Rhätfauna von Bagerabad bei Isfahan (Korallen, Ostracoden). *Mitt. österr. geol. Ges.* **73**, 163–235.

OTT, E. (1972). Die Kalkalgen-Chronologie der alpinen Mitteltrias in Angleichung an die Ammoniten—Chronologie. *N. Jb. Geol. Paläont. Abh.* **141**, 81–115.

PANTIĆ, S. (1965). *Clypeina besici* sp. nov. des sediments Triassiques des Dinarides externes. *Bull. Geol.* **4**, 133–41.

TOLLMANN, A. (1976). *Analyse des klassischen nordalpinen Mesozoikums. Monographie der Nördlichen Kalkalpen*, 2, Deuticke, Wien. (580 pp.)

TOLLMANN, A. and KRISTAN-TOLLMANN, E. (1985). Paleogeography of the European Tethys from Paleozoic to Mesozoic, and the Triassic relations of the eastern part of Tethys and Panthalassa. In *The Tethys*. (ed. K. Nakazawa and J. M. Dickins), pp. 3–22, Tokai University Press, Tokyo.

ZARDINI, R. (1973). *Fossili di Cortina*. Atlante degli echinodermi Cassiani (Trias medio-superiore) etc. Foto Ghedina, Cortina d'Ampezzo. (29 pp.)

EDITH KRISTAN-TOLLMANN, Institut für Geologie, Universität Wien, A-1010 Wien, Austria.

Structure and palaeogeography of the western Tethys during the Jurassic: tests based on ammonite palaeobiogeography

Jacques Thierry

ABSTRACT: Comparisons between the recently proposed palaeogeographical reconstructions for the Tethys and the palaeobiogeography of Jurassic ammonites are important in the study of the structural evolution and the palaeogeography of the Tethys and its margins. The geographical distribution of Jurassic ammonites is considered to be due to a combination of historical and ecological factors, with the existence of biota and bioma (*sensu* Croizat 1958). 'Boreal' and 'Tethyan biota' are used and preferred to the 'Boreal' and 'Tethyan realms' (or provinces), as previously used to designate ammonite communities with similar geological and evolutionary history; several 'Bioma' are defined for faunal communities with similar morphofunctionary adaptations (for instance on very shallow platforms or in deep oceanic regions). The overlap of biota and bioma better explains the compound ammonite faunas. The recently proposed new maps of the Tethys show that in the west the oceanic areas are much more restricted in area than in previous reconstructions; a large oceanic region is constantly extended throughout the Jurassic, only in its eastern part. Subsequently, the total area west of the so-called Apulian Plate, and in Western Europe, takes a prominent part in the faunal ammonite distribution. The Arctic Basin sometimes acted as an evolutionary trap, where Boreal faunas were totally isolated from the Tethyan ones; sometimes an important eustatic transgression allowed faunal exchanges to be made between Boreal and Tethyan stocks. Several examples, taken from the Pliensbachian to the Callovian, serve to demonstrate the influence on ammonite distributions of either the possible deep-oceanic areas or very shallow platforms and emergent areas.

Introduction

The geographical distribution of fossils has been studied by palaeontologists for more than a century. Neumayr (1872, 1883), and Uhlig (1911) first proposed faunal provinces partly based on ammonites. The boundaries between these realms were mostly qualitative; they also had a certain climatic meaning based upon presence–absence criteria: 'Boreal realm' for northern Europe and 'Tethyan realm' (or 'Mediterranean') for southern regions.

During the last two decades, plate tectonics, and better knowledge of the systematics and the evolution of Jurassic ammonites, have stimulated palaeobiogeographical researches, which have reached a dynamic and interpretive phase after a static and descriptive one. Exhaustive lists of references of Jurassic ammonite palaeobiogeography can be found in Enay (1980), Enay and Mangold (1982), Thierry (1976, 1982), and Cariou *et al.* (1985). However, in all these contributions, the reasons put forward for the geographical distribution of ammonites remain traditional and qualitative. An attempt to split the provinces into the so-called 'Sub-Boreal', 'Sub-Tethyan' or 'Sub-Mediterranean' provinces has not served to clarify the problem, but has instead introduced more confusion in the classification of biogeographical realms. Other authors considered that the geographical distribution of ammonites may be the result of ecological restrictions. This concept was first stated by Kilian (1895, 1903) and Haug (1900, 1903), subsequently forgotten, and rediscovered in the middle of the century (Scott 1940; Lucas 1942).

In fact, we have to accommodate the two approaches, as did biogeographers such as Croizat (1958), Blondel (1979, 1986), Dommergues and Marchand (1987), Thierry (1987), Marchand and Thierry (1986). Beyond a descriptive and a qualitative palaeobiogeography, which remains the first and indispensable step for every study, we must proceed more and more towards an interpretive and quantitative palaeobiogeography, based on a wide variety of data. All these requirements are met in Jurassic ammonites: many fundamental palaeontological investigations provide good information for most of the major taxa; new concepts in ammonite palaeoecology can be taken into account, and ammonite distributions can be related to sedimentary facies distribution and inferred masses and oceanic areas.

From AUDLEY-CHARLES, M. G. & HALLAM, A. (eds) *Gondwana and Tethys* Geological Society Special Publication No. 37, pp. 225–34.

In the examples discussed below, I will try to show that the biogeography of Jurassic ammonites can be a test of the structural framework of the evolution of Tethys. I do not expect that my paper will make invalid previous work on ammonite distribution and palaeogeography, but I wish to give an interpretation specifically linked with a new structural model (Dercourt *et al.* 1985). I have chosen to illustrate some particular times where biological and geological events seem to be closely related; other Jurassic stages should be investigated with the same concepts.

In this paper, the terms 'bioma' and 'biota' are used in the meaning defined by Croizat (1958) and Blondel (1979, 1986). A bioma is a community of organisms which has evolved in adaptation to environmental factors; it has an ecological meaning. A biota is a community of organisms which has evolved in the same geographical area, and consequently has the same evolutionary history; it has a historical meaning.

Faunal exchanges and oceanic areas in the western Tethys

The palinspastic reconstructions given for the western Tethys and the epicontinental seas extending beyond, show an important reduction of the oceanic areas. The '*Mesogée*' (*sensu* Biju-Duval *et al.* 1977—) has completely disappeared; the Liguro-piedmontan area, with its extension as far as Gibraltar, is highly reduced compared with previous maps (Dercourt 1970; Smith 1971). Large shallow platforms extend from the Iberian Peninsula to the Apulian Plate, between Western Europe and North Africa, divided by narrow and deep oceanic straits. A large and deep oceanic area (the Tethys *s. s.*) exists on the west of the Apulian Plate, which completely separates the north and south Tethyan margins (Dercourt *et al.* 1985).

The successive steps of the construction of this oceanic area may be discussed in relation to the distribution of Lower and Middle Jurassic ammonite faunas.

Pliensbachian: ecological endemism and geographical barrier; possibly no oceanic furrow

After the generally cosmopolitan faunas of the Lower Lias, the Pliensbachian stage shows evidence of strong endemism, with the individualization of two faunal realms. On the one hand, the so-called Euro-Caucasian province (Sapunov 1974; Dommergues 1984, 1986), as a part of the Boreal realm, extends over the whole of Middle Europe and on the northern Tethyan margin, on the north side of emergent or very shallow-marine areas: Briançonnais, Corso-Sardinian block, Balearic and Sub-Betic platforms, Lusitanian Basin (Dommergues 1982). On the other hand, the so-called Mediterranean province, belonging to the Tethyan realm, coincides with the southern margin of the Tethys, and with countries extending south of the emergent or shallow-marine areas previously recorded. In these reconstructions the physical barriers which separated the Boreal and Tethyan faunas are approximately latitudinal and parallel with the ecological limits; Boreal and Tethyan biota and bioma therefore overlap and are mostly merged (Dommergues and Marchand 1987). *Platypleuroceras*, *Uptonia*, *Acanthopleuroceras*, *Androgynoceras*, *Beaniceras*, etc., are typical of the Euro-Caucasian realm, while *Protogrammoceras*, *Reynesocoeloceras*, *Bettoniceras*, and *Prodactylioceras* are index taxa for the Mediterranean realm. Frequent faunal exchanges took place between the two provinces, at both ends of the barrier. The first, on the eastern part, corresponds with the western end of the oceanic Tethys (Northern Austrian Alps, Bakony Mountains, Slovenian karst); the second took place along the Iberian–Moorish strait (Cariou *et al.* 1985).

On the whole, biogeographical and palaeogeographical data seem to match quite well. The Briançonnais, the Corso-Sardinian and Provence platforms acted like swells in the Alpine region—from Austria to southern Spain on the palinspastic map (Fig. 1A). The eastern exchange route corresponds with a region where an oceanic strait existed. The European margin may not have been a real impediment; the Boreal faunas, in spite of doubtless different ecological constraints compared with the Tethyan ones, spread widely over the Apulian Plate. The western migration route may equally correspond with a region where the reconstructions suggest that no important geographical barrier existed to prevent faunal migration (Fig. 1A). However, the pronounced endemism of the Lusitanian faunas, at any time during the Carixian, will probably necessitate some palaeogeographical modifications between Portugal and Morocco.

During the late Pliensbachian, the well-known spreading of the Boreal amaltheids to North Africa (Donovan 1967; Elmi *et al.* 1974;

Mouterde and Dommergues 1980) may be easy to explain by a global deepening of the epicontinental seas, if we agree with the idea that these ammonites might have been more abundant in moderately deep to deeper-marine regimes (Tintant *et al.* 1982). In the Toarcian, Boreal influence lessened and finally disappeared in all areas of the western Tethys, while Mediterranean faunas spread to the north (Mouterde and Dommergues 1980). The occurrence of bouleiceratids, from the Arabo-Malagasian realm to North Africa, the Iberian Peninsula and the Apulian Plate, is also a well-known and significant Tethyan migration (Faugères and Mouterde 1979). These ammonites are considered as nektobenthic and adapted to life on very shallow platforms (Tintant *et al.* 1982), and they may have lived only in epicontinental seas (Thierry, 1982; Enay, 1985); if a deep oceanic strait had existed at this period, they would never have spread so easily.

Thus, it appears that the definition of 'province' in the Pliensbachian corroborates the combined action of two restrictive factors: on one hand, the presence of a positive structure acting as a swell and a very reduced oceanic strait spreading in the north; on the other hand, the presence of different ecological conditions on each side of this barrier. This last argument may be supported by the distribution of the phylloceratids adaptated to deep water according to Tintant *et al.* (1982). These ammonites existed on the eastern part of the Apulian Plate, and on the margins of the Tethyan ocean as far as North Africa. They never spread over Western Europe except during very short periods: the Pliensbachian endemic genus *Tragophylloceras* which, interpreted as morphologically convergent with the Euro-Boreal ammonites, seems to have been adapted to shallower seas (Cariou *et al.* 1985). The same arguments, as well as an increase in the ongoing Jurassic eustatic transgression, may explain the spread of the oxycone phylloceratid *Zetoceras* as far as Britain. One can also point out the proportion of phylloceratids within the total ammonite faunas, which decreases from east to west (Geczy, 1984), while at the same time their bioma is restricted.

Bajocian–Bathonian: isolation and endemism of Boreal faunas, Arabo-Malagasian faunal migrations; possible occurrence of the ocean hiatus

Endemic faunas in the Boreal realm arose from a total isolation, resulting from a world-wide regression and a widespread development of non-marine areas which extended between it and the Tethyan realm: emergent areas with brackish water and deltaic sediments extended between the North American landmass and the British Isles and as far as the Scandinavian peninsula. However, on the south side of this barrier, there are some differences between the southern and the northern margins of the Tethys (Enay 1980; Enay and Mangold 1982). The Arabo-Malagasian ammonite invaders, with a simplified suture line, are restricted to Moghrebian areas, e.g. *Ermoceras* in the late Bajocian. This suggests that deep-water regions may have existed between North Africa and Europe, which were insuperable obstacles for these ammonites interpreted as nektobenthonic and adapted to very shallow sea (Tintant *et al.* 1982). Such conditions may have been discontinuous in time, because in the middle to late Bathonian, *Micromphalites* and *Clydoniceras*, living in the same environment, were able to spread as far as France and England (Arkell 1951–9, 1952; Thierry 1982; Enay 1985). Other groups, like the tulitids which are interpreted as nektopelagic (Tintant *et al.* 1982), extended widely over all Europe (Thierry 1976). In the opposite direction, migrations from the north to the south involved only the so-called nektopelagic taxa, such as parkinsonids and morphoceratids, which reached Morocco and Algeria (Cariou *et al.* 1985).

Thus, the construction of a Betico-Moghrebian oceanic strait in Bajocian–Bathonian times may have taken place in several steps, inferred by irregular and selective faunal exchanges. On the contrary, the Liguro-Piedmontan oceanic furrow seems to have been more widely opened, with important migrations of taxa from the northern and southern margins. In the present-day Bakony Mountains (Geczy 1973, 1984; Cariou *et al.* 1985), this oceanic feature can also be inferred from a higher proportion of phylloceratids in the Middle than in the Lower Jurassic.

Callovian and Oxfordian: possible wide opening of the oceanic strait

The global transgression which had begun in the early Jurassic reached a maximum in the Callovian, and led to a restriction of the shallow carbonate platforms and emergent areas still well developed in the late Bathonian. The so-called 'North Sea strait' (Callomon 1985) opened again, and Boreal faunas spread widely over Western Europe. In this context, valuable information has been obtained from kosmoceratid and cardioceratid distributions, which

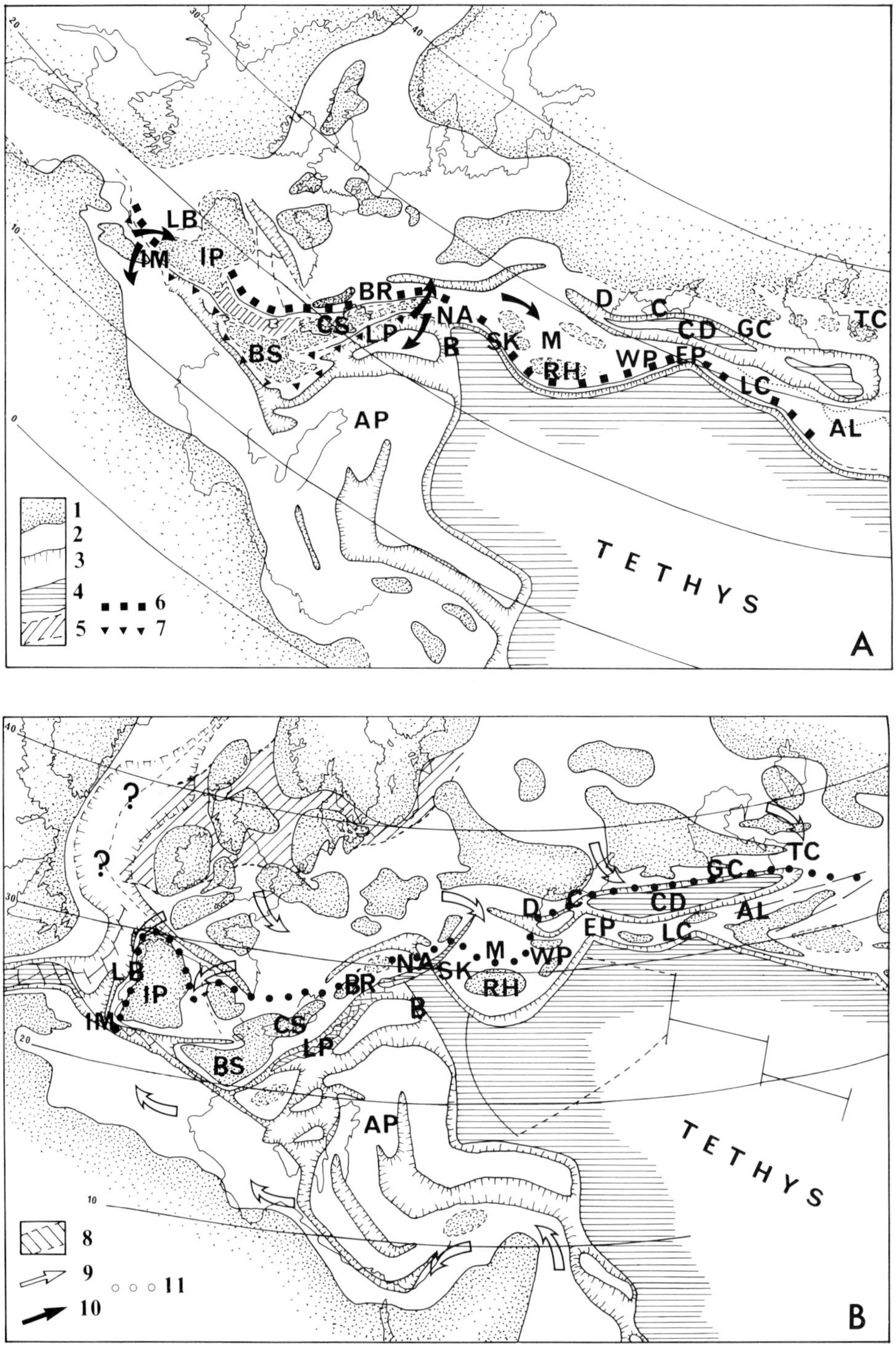

FIG. 1. Palaeogeography of the western Tethys and its surroundings in Pliensbachian (A) and Bathonian–Callovian (B), modified from Dercourt *et al.* (1985) and Cariou *et al.* (1985).
1—probable land areas, *2*—epicontinental seas, *3*—continental margins of oceanic areas and rifting areas, *4*—oceanic areas, *5*—shallow calcareous platform areas in Carixian, *6*—southern limit of Euro-Boreal faunas in Carixian, *7*—northern limit of Mediterranean faunas in Carixian, *8*—continental or

have been discussed many times (Brinkmann 1929; Tintant 1963; Hallam 1969; Thierry 1976, 1987; Enay 1980; Cariou *et al.* 1985; Callomon 1985; Marchand 1986; Dommergues and Marchand 1987).

In the western Tethys, these groups gradually spread over the northern margin, in the Callovian and Oxfordian respectively. The boreal kosmoceratids are known in the Algarve (Portugal) and Iberian Ranges (Spain), while cardioceratids reached Portugal (Montejunto) and the South of France (Causses, Cevennes, Languedoc, Alpes-Maritimes). Even though they are minor elements of the ammonite faunas of these areas, it is important to note that they never reached the southern Tethyan margin which, at that time, considering the new palinspastic reconstructions, was separated from the northern margin by the Liguro-Piedmontan oceanic strait as far as the Sub-Betic Ranges and Gibraltar. This deep and wide furrow acted as an impassable barrier for both kosmoceratids and cardioceratids considered as nektobenthic ammonites (Tintant *et al.* 1982).

Once more, biological and palaeogeographical patterns seem to match well. For instance, the complete lack of kosmoceratids and cardioceratids in the Callovian of Sicily, versus the abundance of perisphinctids, reineckeids and phylloceratids, seems to agree very well with Sicily's location on the south Tethyan margin as a part of the Apulian Plate. The opening of a west Tethyan oceanic furrow, a probable connection with the opening Atlantic, and the extension of oceanic environments over western Europe, may be corroborated by the increasing proportion and widening bioma of phylloceratids. Phylloceratids are totally missing in north-west Portugal, whereas they are well represented in the Algarve. Although there is a very high number of macrocephalitids, perisphinctids, and reineckeids, the depth of this Portuguese Atlantic sea seems to have been too shallow for the phylloceratids. Thus, a barrier may have existed between the present-day Algarve and North Tagus areas. Moreover, at the eastern end of the oceanic furrow, for instance in the Bakony Mountains, the oceanic environment is more marked than in any other place, with the phylloceratid ratio increasing to 60 per cent (Geczy, 1984).

The possible expansion eastwards of reineckeids (Cariou 1984) might also have been closely related to the same causes. As a result, ammonites such as perisphinctids, macrocephalitids, etc., possibly nektopelagic, but certainly less dependent on bathymetric conditions, extended over all the West European and North African regions (Cariou *et al.* 1985; Marchand and Thierry 1986).

Ammonite faunas and palaeogeography of the north-western Tethyan margin

On the previously discussed maps for the Pliensbachian and Callovian (Dercourt *et al.* 1985), the north-western Tethyan margin looks rather complex. One can observe from west to east.

1. A Balkano-Moesian Plate embracing the present-day Moesian area, the Rhodope Massif, and the western Pontides. To the north, it is bordered by an extension of the Apusenian sphenochasm and the western end of the Caucaso-Dobrodgean furrow which, when matched, almost separates it from the rest of the European platform.

2. A west to east elongated Caucaso-Iranian Plate, which is built with the Eastern Pontides, the Lesser Caucasus and the Elborz, Samandaj, and Sirjan chains. It is kept apart from the European platform (Crimea, Greater Caucasus and Trans-Caspian areas) by the Caucaso-Dobrodgean furrow, less widely opened in the Pliensbachian than in the Callovian.

These two plates are immediately bordered on the south by the Tethyan Ocean, which widely separated them from the southern Tethyan margin. However, this oceanic realm is less extended in its western part, for instance along the Apulian Plate. The geographical distribution of all the ammonite faunas of this north-western margin of the Tethys seems to be

brackish sediments in Bajocian–Bathonian, *9*—faunal exchange seaways in Bajocian–Bathonian, *10*—faunal exchange seaways in Callovian, *11*—southern limit of Boreal faunas in Callovian. *IM*: Ibero-Moorish strait, *LB*: Lusitanian Basin, *IP*: Iberic Peninsula, *BS*: Balearic and Sub-Betic platforms; *CS*: Corso-Sardinian platform, *AP*: Apulian Plate, *LP*: Liguro-Piedmontan furrow, *BR*: Briançonnais, *B*: Bakony Mountains, *NA*: Northern Alps, *SK*: Slovenian karst, *M*: Moesian platform, *RH*: Rhodope Mountains, *D*: Dobrodgea, *C*: Crimea, *CD*: Caucaso-Dobrodgean furrow, *WP*: western Pontides, *EP*: eastern Pontides, *GC*: Greater Caucasus, *LC*: Lesser Caucasus, *AL*: Alborz, *TC*: Trans-Caspia.

strongly influenced by such a structural framework.

The relative isolation of the Balkano-Moesian area

In Pliensbachian times, the western Balkan faunas (Stara Planina) clearly belong to the Euro-Boreal realm (Dommergues *et al.* 1986). However, one should note the lack of endemic, but temporarily abundant groups in Western Europe, like *Prodactylioceras* and *Protogrammoceras*. Moreover, the absence of characteristic 'Mediterranean' genera like *Arieticeras* and *Reynesoceras* remains surprising, in view of the southern position of this area. During the middle Jurassic we find again the same reduction of faunal diversity. This is an important fact for this period; the lack or extreme scarcity of parkinsonids, garantianids, sphaeroceratids, and clydoniceratids in the Bajocian–Bathonian, tulitids, pachyceratids, kosmoceratids, cadoceratids, cardioceratids, phlycticeratids, distichoceratids, and reineckeids in Callovian.

Since the same phenomenon recurs in the Oxfordian, the idea of a relative isolation of this region is plausible. According to possible different modes of life of various ammonite groups, only the supposed nektopelagic ones (perisphinctids, macrocephalitids, phylloceratids, etc.) might have been relatively independent of the bottom environment. The others, supposedly nektobenthic, might have been restricted by the deep furrows of the Apusenian sphenochasm and Dobrodgeo-Caucasian oceanic areas. This conclusion agrees with the old concept of the 'Oriental Island of Pompeckj', reconsidered by Arkell (1956) and Sapunov (1974).

In the last two cited areas, the oceanic aspect of the ammonite faunas is indicated by the high proportion of phylloceratids and lytoceratids. The most significant example is the Bathonian–Callovian fauna of eastern Serbia (Mihajlovic 1969) where, in a 'Klaus Schichten' (Greben) facies, these two families represent more than 50 per cent of the fauna whereas, on the proximate Moesian platform, they only attain 20 per cent (Dommergues *et al.* 1986).

The Caucaso-Dobrodgean furrow

According to the previously cited palinspastic maps, this wide oceanic area separates the Dobrodgea, the Crimea, and the Greater Caucasus in the north, from the eastern Pontides and the Lesser Caucasus in the south. This deep basin may have acted as an effective barrier, preventing the Boreal ammonite expansion to the south. On the Russian platform the faunas which extended as far as the Greater Caucasus, never reached the Lesser Caucasus.

This fact is strongly demonstrated in the Callovian, where the faunas are highly differentiated. In the Greater Caucasus and Trans-Caspian areas, there are typical Euro-Boreal ammonites such as kosmoceratids, cadoceratids, and cardioceratids. The faunal diagrams of these countries (Ammanniazov 1962; Lominadze 1982), briefly recorded here (Fig. 2), are very similar to those of northern France, Swabia in Germany, and southern Poland (Matyja and Gizejewska 1979). On the other side, the faunas of the Lesser Caucasus (Abdoulkasoumzade 1963; Lominadze 1982) are lacking in these Boreal taxa; conversely, the phylloceratids, perisphinctids, and reineckeids are omnipresent. The same ratios exist as in the Western Pontides, which were at this time not very far to the west, and in a very similar structural position according to Dercourt's reconstructions.

It is also important to point out the distinctive character of the Greater Caucasus, where kosmoceratids, cadoceratids, and cardioceratids occur with phylloceratids during all of the Callovian stage. One can note the same phenomenon in other European areas like Pro-

FIG. 2. Examples of Callovian ammonite distributions in the western Tethys. Same legend as in Fig. 1 for the palaeogeographical framework. Each faunal diagram represents the composition of the fauna, in per cent per family, in several stratigraphic units of this stage. Data from the author and various publications of Abdoulkasoumzade, Ammanniazov, Elmi, Lominadze, Matyja and Gizejewska, Mihajlovic, Sequeiros.

Lo: Lower Callovian, *Mi*: Middle Callovian, *U*: Upper Callovian, *M*: Macrocephalus zone, *G*: Gracilis zone, *P*: Patina subzone, *J*: Jason zone, *C*: Coronatum zone, *A*: Athleta zone, *L*: Lamberti zone. *1*: Cadoceratinae–Cardioceratinae, *2*: Gowericeratinae–Kosmoceratinae, *3*: Proplanulitinae, *4*: Tulitidae, *5*: Macrocephalitidae, *6*: Pachyceratidae, *7*: Phlycticeratinae, *8*: Oppeliinae–Hecticoceratinae, *9*: Peltoceratinae–Aspidoceratinae, *10*: Reineckeinae, *11*: Perisphinctinae, *12*: Phylloceratidae–Lytoceratidae, *13*: Clydomieratidae.

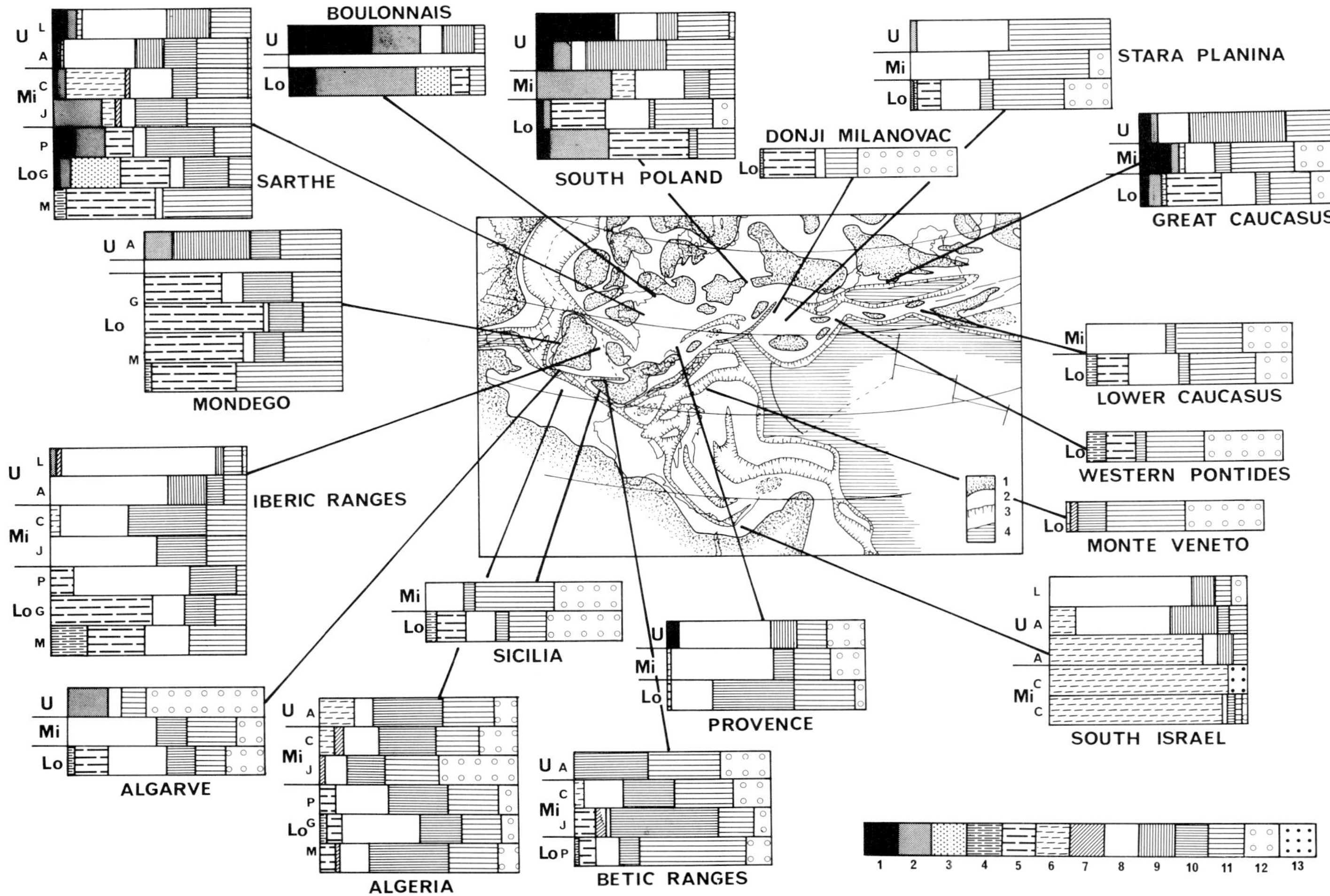

BOULONNAIS
SOUTH POLAND
STARA PLANINA
DONJI MILANOVAC
GREAT CAUCASUS
SARTHE
MONDEGO
LOWER CAUCASUS
WESTERN PONTIDES
MONTE VENETO
IBERIC RANGES
SICILIA
PROVENCE
SOUTH ISRAEL
ALGARVE
ALGERIA
BETIC RANGES

vence in France (Dardeau *et al.* 1979), and southern Poland (Matyja and Gizejzewska 1979). This may be due to the overlapping of the Boreal and Tethyan biotas with the phylloceratid bioma. The composite ammonite faunas are better explained by the action of historical and ecological factors, rather than by the so-called 'Sub-Boreal' and 'Sub-Tethyan' provinces, as previously assumed.

The ammonite fauna and the palaeogeography of this part of the northern Tethyan margin seem to match well. However, Turkish authors (Şengör and Yılmaz 1981) consider that in the middle Jurassic the Caucaso-Dobrodgean furrow ('Palaeo-Tethys') was totally closed; subsequently, the Pontic platforms and Laurasia were in continuity. In this case, Boreal nektobenthic ammonites would have been able to reach the Pontic chains, unless these very shallow platforms were impassable barriers acting like a deep-oceanic furrow. A more exhaustive study of the very poorly known Middle Jurassic ammonite faunas of Turkey should lead to a more precise answer.

Conclusions

The few examples discussed here serve to demonstrate that the composition of the ammonite faunas, and their geographical distribution and variation in space and time, are not the consequence of random causes. However, the combined action of historical and ecological factors upon their distribution is very complicated and difficult to unravel. A quantitative approach is the most promising way of elucidating the respective importance of the two factors.

One must also be careful to note that the respective roles of these factors are not apparent to the same degree for all ammonite taxa. The ecological factor for most ammonite taxa can be made evident from local investigations (one or several outcrops with well-known sedimentary environments), sometimes at a regional level. This is illustrated, for instance, by Tethyan phylloceratids. The historical factors are more evident on a regional scale, such as in the Boreal and Tethyan realms.

Until recently, ammonites were the unrivalled and always unequalled biostratigraphic tool for the Jurassic period, but it is now seen that they also have considerable palaeogeographical value as well, providing a means for testing models of structural evolution of the ancient seas and oceans. Palaeobiogeography and palaeoecology thus appear as two complementary approaches, inseparable from classical systematic palaeontology and biostratigraphy in the study of Earth and life history. But we must remain careful in our conclusions, because many of the geographical distributions of a number of ammonite groups do not yet have a satisfactory explanation. This is particularly evident when we work at a time resolution down to the level of substage or even standard zone, subzone, or horizon.

ACKNOWLEDGEMENTS: I am grateful to Professor J. H. Callomon (University College, London) for all his fruitful comments and constructive criticisms about the manuscript of this paper. I also thank Mrs Sebedio for her help in writing this paper in English.

References

Abdoulkasoumzade, M. R. (1963). Stratigraphie et faune du Jurassique supérieur de la partie Nord-Est du Petit Caucase. *Publ. Acad. Sci. Azerbaidjan, Ser. Sci. Terre, Bakou*. (107 pp.)

Ammanniazov, K. (1962). Stratigraphie et ammonites des depots du Jurassique supérieur du Touar-Kyr. *Inst. Geol. Akad. Nauk-Turkmenskaya* SSR. Edit. Ashkabad. (110 pp.)

Arkell, W. J. (1951–9). Monograph of the English Bathonian ammonites. *Jour. Paleont. Soc.* 102–13, 264, 33 pl.

Arkell, W. J. (1952). Jurassic ammonites from Jebel Tuwaiq, Central Arabia. *Phil. Trans. Roy. Soc. London* **B236**, 241–313, pl. 13–31.

Arkell, W. J. (1956). *Jurassic geology of the world*. Oliver & Boyd, Edinburgh. (306 pp.)

Biju-Duval, B., Dercourt, J., and Le Pichon, X. (1977). From the Tethys ocean to the Mediterranean seas: a plate tectonic model of the evolution of the western alpine system. In *Internat. Symp. struct. hist. Medit. bassin*, Split, Yugoslavia, 25–30 Oct. 1976, pp. 134–64, Ed. Technip., Paris.

Blondel, J. (1979). *Biogéographie et écologie*. Coll. d'Ecologie, 15, Masson Edit., Paris. (173 pp.)

Blondel, J. (1986). *Biogéographie evolutive*. Coll. d'Ecologie, 20, Masson Edit., Paris. (221 pp.)

Brinkmann, R. (1929). Monographie der gattung Kosmoceras. *Abh. Ges. Wiss. Gottingen Math. Phys. Kl. N.F.* **4**, VII. (123 pp., 1 pl.)

Callomon, J. H. (1985). The evolution of the Jurassic ammonite family Cardioceratidae. *Spec. Pap. Palaeont.* **33**, 49–90.

Cariou, E. (1984). Structure, origine et paléobiogeographie de la famille des Reineckeidae, Ammonitina, du Jurassique moyen. *C. R. Acad. Sci. Paris, Sér. II* **298**, 245–8.

Cariou, E., Contini, D., Dommergues, J. L., Enay, R., Geyssant, J., Mangold, Ch., and Thierry, J. (1985). Biogéographie des ammonites et évolution structurale de la Tethys au cours du Jurassique. *Bull. Soc. Géol. Fr.* Sér. (88), **1**, (5), 619–97, 7 fig.

Croizat, L. (1958). Panbiogeography or an introduction synthesis of zoogeography, phytogeography and geology. 1: The New World, 2 (a and b): The Old World. Published by the author, Caracas. 1–1018/1–1731.

Dardeau, G., Marchand, D., and Thierry, J. (1979). Relations entre la répartition des facies et la composition des faunes d'ammonites au cours de l'évolution du bassin callovien des Alpes-Maritimes. *Bull. Soc. Géol. Fr.* Sér. (7), **21**, (6), 753–7.

Dercourt, J. (1970). L'expansion océanique actuelle et fossile; ses implications geotectoniques. *Bull. Soc. Géol. Fr.* Sér. (7) 1,(3), 261–309.

Dercourt, J., *et al.* (1985). Présentation de 9 cartes paléogeographiques au 1/20 000 000 s'étendant de l'Atlantique au Pamir pour la periode du Lias a l'actuel. *Bull. Soc. Géol. Fr.* Sér. 8, 1, 5, 637–52.

Dommergues, J. L. (1982). Le provincialisme des ammonites boreales au Lias moyen; une crise faunique sous controle paleogeographique. *Bull. Soc. Géol. Fr.* Sér. 7, 24, (5–6), 1047–51.

Dommergues, J. L. (1984). L'évolution des Ammonitina au Lias moyen (Carixien, Domerien basal) en Europe occidentale. Thèse Doct. État, Univ. Lyon I, 303 pp.

Dommergues, J. L. (1986). Les Dactylioceratidae du Carixien et du Domerien basal, un groupe monophyletique. Les Reynescocoeloceratinae nov. sub. fam. *Bull. Sci. Bourg.* **39**, (1), 1–26, Tpl.

Dommergues, J. L. and Marchand, D. (1987). Paléobiogeographie historique et ecologique, application aux ammonites du Jurassique. In *2nd International Cephalopod Symposium. Cephalopods: present and past; Tubingen, 1985* (in press).

Dommergues, J. L., Marchand, D., Sapunov, I. G., and Thierry, J. (1986). Les faunes d'ammonites du Jurassique bulgare: une confirmation paléobiogeographique de l'isolement relatif de la region balkano-moesienne. *Bull. Soc. Géol. Fr.* Sér. 7, 3, (4), 107–12.

Donovan D. T. (1967). The geographical distribution of Lower Jurassic ammonites in Europe and adjacent areas. In *Aspects of Tethyan Biogeography*. (ed. C. G. Adams and D. V. Ager), pp. 111–34. The Systematics Association, London, 7.

Elmi, S., Atrops, F., and Mangold, C. (1974). Les zones d'ammonites du Domerien-Callovien de l'Algérie Occidentale—1ère partie: Domerien–Toarcien *Docum. Lab. Geol. Fac. Sci.* Lyon **61**, 1–83.

Elmi, S. (1985). Influences des hauts-fonds sur la composition des peuplements et sur la dispersion des ammonites. In *Géodynamique des seuils et des hauts-fonds. Comité Trav. Hist. Sci., Bull. Sect. Sci.* **9**, 217–28.

Enay, R. (1980). Evolution et relations paléobiogeographiques de la Tethys mésozoique et cenozoique. In *Géologie des chaînes alpines issues de la Tethys*, 26th Int. Geol. Congr., Paris; C5, 276–83.

Enay, R. (1985). Seuils et voies d'échanges: rôle dans la distribution et l'évolution des ammonites jurassiques. In *Geodynamique des seuils et des hauts-fonds*, pp. 203–15, Comité Trav. Hist. Sci. Bull. Sect. Sci. 9.

Enay, R. and Mangold, Ch. (1982). Dynamique biogeographique et évolution des faunes d'ammonites au Jurassique. *Bull. Soc. Géol. Fr.* Sér. (7), **24**, (5–6), 1025–46.

Faugères, T. C. and Mouterde, R. (1979). Paleobiogeographie et paleogeographie aux confins atlantico-mésogéens; données fournies par le Lias sud-rifain (Maroc) 7ème R.A.S.T. Lyon, p. 183, Soc. Géol. Fr. Edit.

Geczy, B. (1973*a*). Plate tectonics and paleogeography in the East-Mediterranean Mesozoic. *Acta Geol. Acad. Sci. Hung.* **17**, 421–8.

Geczy, B. (1973). The origin of the Jurassic faunal provinces and the Mediterranean plate tectonics. *Ann. Univ. Scient. Budapest.* 99–114.

Geczy, B. (1984). Jurassic ammonite provinces of Europe. *Acta Geol. Hungarica* **27** (1–2), 67–71.

Hallam, A. (1969). Faunal realms and facies in the Jurassic. *Paleont.* **12**, 1–18.

Haug, E. (1900). Les géosynclinaux et les aires continentales. Contribution à l'étude des transgressions et des regressions inconnues. *Bull. Soc. Géol. Fr.* Sér. (3), **28**, 617–711.

Haug, E. (1903). Remarque sur les communications entre le bassin anglo-parisien et la 'Mediterranée centrale' au Jurassique. *Bull. Soc. Géol. Fr.* Sér. (4), **3**, p. 653.

Kilian, W. (1895). Note stratigraphique sur les environs de Sisteron et contributions à la connaissance des terrains secondaires du Sud-Est de la France. *Bull. Soc. Géol. Fr.* Sér. (3), **23**, 659–803.

Kilian, W. (1903). La question du genre de vie des ammonites. *Bull. Soc. Géol. Fr.* Sér. (4), **3**, p. 668.

Lominadze, T. A. (1982). Callovian ammonites of the Caucasus. Academy of Sciences USSR. (272 pp.) (in Russian).

Lucas, G. (1942). Description géologique et petrographique des Monts de Ghar Rouban et du Sidi el Abed. *Bull. Serv. Carte Géol. Algerie*, 2ème. Sér., **16**, 536 pp.

Marchand, D. (1986). L'evolution des Cardioceratinae d'Europe occidentale dans leur contexte paleobiogeographique (Callovien supérieur-Oxfordien moyen). Thèse Doctorat d'Etat, Dijon. (601 pp.)

MARCHAND, D. and THIERRY, J. (1986). Relations entre les evenements calloviens et l'evolution des peuplements d'ammonites en Europe occidentale. *Bull. Centre Rech. Explor. Prod. Elf-Aquitaine* **10**, (3), 383–92.

MATYJA, B. A. and GIZEJEWSKA, M. (1979). Distribution of Callovian and Lower Oxfordian ammonite faunas in Poland. *Acta Geol. Polonica*, **29**, (2), 177–85.

MIHAJLOVIC, M. (1969). La faune d'ammonites des couches de Klauss (Greben) des environs de Donji Milanovac (Serbie orientale). *Bull. Museum d'Hist. Natur. Belgrade* Sér. A, **24**, 45–125.

MOUTERDE, R. and DOMMERGUES, J. L. (1980). Cheminements des invasions mesogéennes en domaine subboreal au Domerien inférieur dans le Sud-Ouest de l'Europe. *8ème R.A.S.T.* Marseille, 255. *Soc. Géol. Fr. Edit.*

NEUMAYR, M. (1872). Über Jura-Provinzien. *Verhandl. k.k. Geol. Reichanst, Wien* **3**, 54–7.

NEUMAYR, M. (1883). Über klimatische zonen während der Jura und Kreidezeit. *Densksch. Akad. Wiss. Wien, (Min. Kl)*, **47**, 277 pp.

SAPUNOV, I. (1974). Notes on the geographical differentiation of the Lower Jurassic ammonite faunas. In *Colloque du Jurassique, Luxembourg, 1967. Mem. BRGM* **75**, 263–70.

SCOTT, G. (1940). Paleoecological factors controlling the distribution and mode of life of Cretaceous ammonoids in the Texas area. *J. Palaeont.*, **14**, (4), 299–323.

ŞENGÖR, A. M. C. and YILMAZ, Y. (1981). Tethyan evolution of Turkey: a plate tectonic approach. *Tectonophys.* **75**, 181–241.

SEQUEIROS, L. (1979). El jurasico de las zonas externas de las Cordilleras Beticas. In *II° Coloquio de estratigrafia y paleogeografia del Jurasico de Espana, Granada.* Cuad-Geol. Granada, **10**, 115–34.

SMITH, A. G. (1971). Alpine deformation and the oceanic areas of the Tethys, Mediterranean and Atlantic. *Bull. Geol. Soc. Am.* **82**, 2039–70.

THIERRY, J. (1976). Palaeobiogéographie de quelques Stephanocerataceae (Ammonitina) du Jurassique moyen et supérieur: une confrontation avec la théorie mobiliste. *Geobios* **9**, (3), 291–331.

THIERRY, J. (1982). Tethys, Mesogée et Atlantique au Jurassique: quelques reflexions basées sur les faunes d'ammonites. *Bull. Soc. Géol. Fr.* Sér. (7), **XXIV** (5–6), 1053–67.

THIERRY, J. (1987). La répartition géographique des faunes d'ammonites: provincialisme et/ou ecologie? L'exemple du Callovien inférieur et moyen (Jurassique moyen) en France. In *2nd International Cephalopod Symposium. Cephalopods: present and past*, Tübingen, (in press).

TINTANT, H. (1963). Les Kosmoceratides du Callovien inferieur et moyen d'Europe occidentale; essai de paleontologie quantitative. (500 pp.)

TINTANT, H., MARCHAND, D., and MOUTERDE, R. (1982). Relations entre les milieux marins et l'evolution des ammonoides: les radiations adaptatives du Lias. *Bull. Soc. Geol. Fr.* Sér. (7), **24**, (5–6), 951–61.

UHLIG, V. (1911). Die marinen Reiche der Jura und der Unterkreide. *Mitt. Geol. Ges., Wien* **4**, (3), 329–448.

JACQUES THIERRY, Centre des Sciences de la Terre de l'Université de Bourgogne and U.A. CNRS no. 157 'Géodynamique sédimentaire et Évolution géobiologique'. 6, Bd. Gabriel, 21100 Dijon, France.

Middle Jurassic ammonite biogeography supports ambi-Tethyan origin of Tibet

Gerd E. G. Westermann

ABSTRACT: In contrast to the latest Jurassic–earliest Cretaceous ammonite faunas of the Spiti Shales in the Tethyan Himalaya—which were early recognized to be closely akin to Indo-West Pacific faunas of south-eastern Tethys—the scarce Middle Jurassic subjacent faunas were previously identified with mostly European or late Jurassic taxa. Current revision of all known Middle Jurassic ammonites from Tibet, however, indicates that Bajocian and Callovian ammonites of the Tethyan Himalaya are closely affiliated to, firstly, northern (and western) Australasia, and secondly, to western India, Madagascar and Kenya, i.e. they belonged to the ecotone of Himalayan and Ethiopian provinces (or subprovinces of the Indo-West Pacific province). In contrast, the few species known from the Lhasa block of South Tibet and the Qamdo block of North Tibet appear to be typically western Eurasian. This supports current plate-tectonic theory, which holds that the Jurassic location of the Qamdo and Lhasa terranes was, respectively, at and in proximity to Eurasia, while the Tethyan Himalaya belonged to the Indian Shield, then at the northern margin of Gondwana.

Introduction

Megatectonics

Much of southern and south-eastern Asia is a collage of continental blocks which were assembled sequentially from the latest Palaeozoic to the Palaeogene. A variety of geological histories for Tibet has been offered in recent years (in Lie *et al.* 1981). One of the most lucid and most recent reconstructions is that of Şengör (1984, 1985). Fig. 1 shows a modification of Şengör's palinspastic map for the later part of the Jurassic, with north-eastern Gondwana, according to Hamilton (1979, 1983), and the Somali Basin already open. The more northerly position of the Indian Shield provides room for the proven and, mostly, controversial terranes in the Indian Ocean; it improves the biogeographical 'fit' of Permian to Jurassic Himalayan faunas with those of the northern Australasian margin (eastern Indonesia–New Guinea, etc.). The age of the Jurassic magnetic 'quiet zone', when Madagascar is said to have separated from Africa (Rabinowitz *et al.* 1983; Bosellini, 1986), is now extended downward from Callovian–Oxfordian to Bajocian (Ogg and Steiner 1985, and pers. comm.), and abundant Bajocian oceanic ammonoids at Mombasa indicate pre-Bajocian separation (Westermann 1975).

According to Şengör (1984, 1985), the Cimmerian Continent became separated from northern Gondwana at the end of the Palaeozoic and, while moving northward across Tethys, split lengthwise into two parts. The northern part was assembled with Laurasia in the early Mesozoic (mid-Triassic–earliest Jurassic), thus closing the original Tethyan ocean, i.e. Palaeo-Tethys. Northern Cimmeria is now the Cimmerides orogenic belt, ranging from the Caucasus to Indo-China and includes the North Tibetan Qamdo terrane.

During the middle and late Jurassic, the Tanggula Ocean separated the southern parts of Cimmeria (the later south Tibetan or Lhasa block) from Eurasia. The early Cretaceous subduction of the Tanggula ocean along the margin of the Qamdo terrane closed this small ocean and welded the last part of Cimmeria to Laurasia/Eurasia. Since Şengör's map is for the late Jurassic, the Bajocian–Callovian Tanggula ocean was probably somewhat larger than shown here.

The Indus–Yarlung Zangbo suture (fracture) defines the southern margin of the Lhasa terrane, marking the Eocene collision of the Indian Shield and the closure of Neo-Tethys. The Himalaya belongs to the Alpide orogenic zones stretching from the Mediterranean to Indonesia. Most of our localities, i.e. in the Tethyan Himalaya, are from near the northern shore (and ?slope) of the Indian Shield, now the western end of the Indo-Australian Plate.

Whereas the uppermost Jurassic to basal Cretaceous ammonite faunas of the Tethyan Himalaya, i.e. the Spiti Shales, have long been known to be intimately related to eastern

From AUDLEY-CHARLES, M. G. & HALLAM, A. (eds) *Gondwana and Tethys* Geological Society Special Publication No. 37, pp. 235–9.

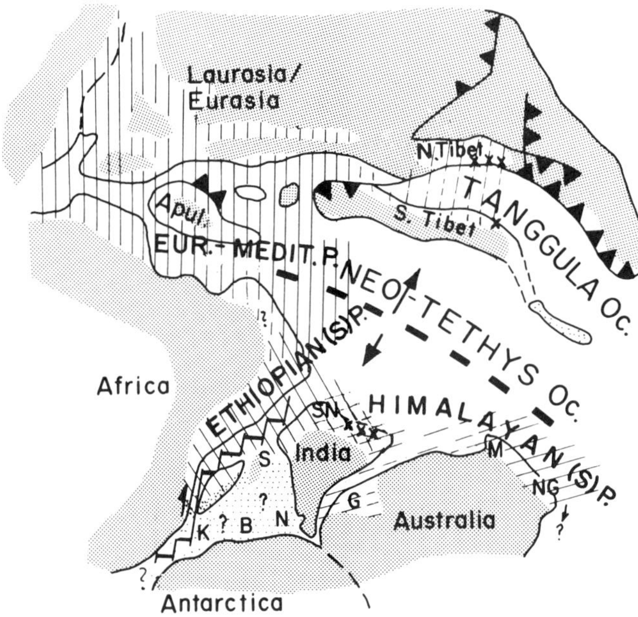

FIG. 1. The middle to late Jurassic position of the continents around Neo-Tethys, according to Şengör (1985) for south-eastern Eurasia, and modified after Hamilton (1979, see text) for the Indo-Madagascar region.

Indonesia and New Guinea (Uhlig 1911), the Middle Jurassic Himalayan ammonites have not previously been recognized as typically south-east Tethyan.

Bajocian

Arkell (1953) attached great importance to the Bajocian ammonite fauna from 'Kampadzong' in the Tethyan Himalaya (Fig. 2, Loc. 1): 'the Tibetan occurrence is the only known link (besides the little known one in the Pamir) between the Bajocian occurrences of East and West Asia'. Arkell identified this supposedly 'condensed' assemblage, with mostly European *Euhoploceras*, *Witchellia*, *Dorsetenia*, and *Frogdenites*. Subsequently, scientific expeditions by the Academia Sinica found some new localities. Bajocian ammonites are now known from the eastern Tethyan Himalaya (Fig. 2, Locs 2, 3, 4) and from North Tibet (Locs 5, 6). Later taxonomic work followed Arkell (Wang and Chong 1974; Wang, in Zhao 1976; Wang *et al.* 1979).

Alleged Bathonian

'*Delecticeras*' (Oppeliidea) recorded from the Tethyan Himalaya by Yang in his 'Atlas of Paleontology of Tibet I' (unpublished) is probably a Triassic *Dittmarites* (Wang Yi-gang, pers. comm.). The supposed Boreal '*Arctocephalites (Cranocephalites)* sp.' identified by Wang from eastern Heilongjiang, is now transferred to early Cretaceous Desmoceratidae (Wang Yi-gang, pers. comm.).

Callovian

Ammonite-bearing marine beds are widespread and thirteen localities (Fig. 2, Locs 3, 7–18) are now known (Zhao 1976; Wang and Chen, in Wang *et al.* 1979; Westermann and Wang in press). The ammonites were previously identified with European and Indian macrocephalitids, Boreal kosmoceratids, Oxfordian mayaitids, as well as Tethyan perisphinctids and reineckeids. Recent fieldwork at Locs. 2 and 13 in 1985 has resulted in significant changes in biostratigraphy, taxonomy, and biogeography (Westermann and Wang in press).

Biogeography

According to the plate-tectonic subdivision of Tibet (Fig. 2) the Middle Jurassic ammonite

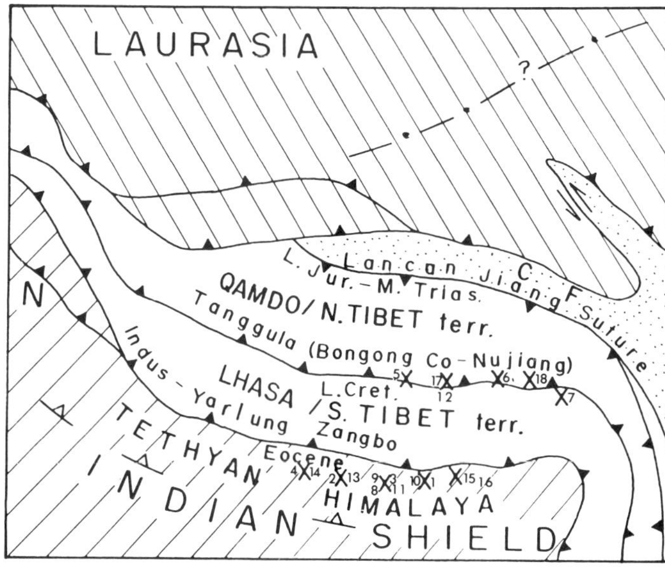

FIG. 2. Middle Jurassic ammonite localities and principal tectonic subdivision of Tibet.

localities belong to: (1) the north Tibetan Qamdo block or terrane (Locs 5, 6, 12, 17, 18) which since early Jurassic was amalgamated with Eurasia, (2) the north-eastern margin of the South Tibetan/Lhasa block or terrane (Loc. 7) which was still separated from Eurasia (including North Tibet) by the narrow Tanggula Ocean, and (3) the Tethyan Himalaya area (Loc. 1–4, 8–11, 13–16) along the northern margin of the Indian Shield, then part of Gondwana at the southern shore of Tethys. The samples from the first two areas are, unfortunately, small and poorly preserved (1), or small and not available to us for examination (2). The affinities of Himalayan ammonite species to other ammonite provinces are listed in Table 1 and illustrated in Fig. 3, including the northern and western Australasian area of extreme south-eastern Tethys (eastern Indonesia and Western Australia).

Bajocian

The poorly known ammonite fauna from North Tibet includes *Dorsetensia* cf. *romani* (Oppel), a species of mainly European, but also Andean and possibly East African distribution (Westermann 1975).

The ammonite fauna from the Tethyan

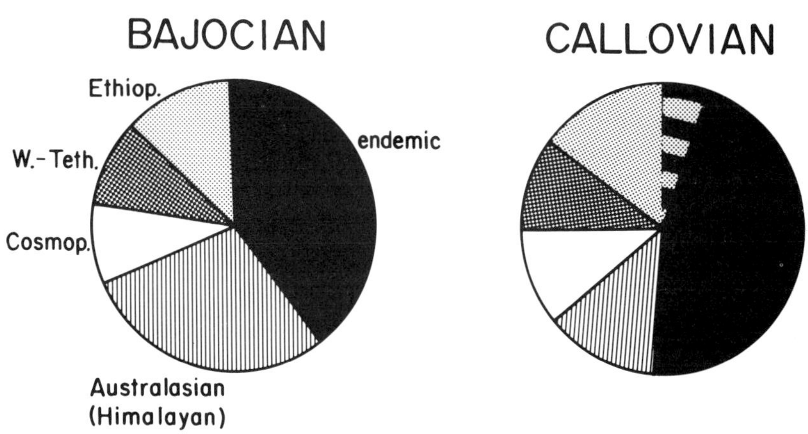

FIG. 3. Pie diagrams for the geographical distribution (other occurrences) of ammonite species from the Tethyan Himalaya (see Table 1).

TABLE 1. *Biogeographical analysis of Himalayan fauna from Tibet**

Species	Pan-Tethyan	West Tethyan	Indo-Madagascar (Ethiopia)	E. Indonesia and W. Australia	Endemic
Bajocian					
Euhoploceras cf. *subdecoratum*	(+)				
Witchellia cf. *australica*				(+)	
W. tibetica					+
W. cf. *sutneri*		(+)	(+)		
Fontannesia haydeni					+
F. kiliana				+	
Subtotal	0.5	0.5	0.5+	1.5	2
Callovian					
Jeanneticeras				+	
cf. *anomalum/adelum*		(+)			
Grayiceras blanfordi					+
G.? waageni					+
G.? n. sp.			?		(+)
Subkossmatia cf. *opis*			+		
Erymnoceras n. sp. aff. *coronatum*					+
Choffatia cf. *madani*	(?)		(+)		
C. cf. *balinensis*	(+)				
C. cf. *funata*	(+)	(+)			
C. (*Grossouvria*)* *propinqua*					+
Subtotal	1	1	1.5+	1	4.5
Total score	1.5	1.5	2.5	2.5	6.5

*Uncertain identifications placed in parentheses and scored half unless genus restricted similarly.

Himalaya of Tibet is mainly endemic (score = 2); it has its closest affinity to eastern Indonesia, i.e. the Sula Islands and Irian Jaya (1.5); whereas the Pan-Tethyan to cosmopolitan, Indo-Madagascan (Ethiopian) and western Tethyan elements are minor (0.5 each). The scarcity of Indo-Madagascan elements, however, is probably misleading, because coeval marine faunas are known from there in only a few specimens from Madagascar (Collignon 1958, figs 28–30) and Kenya (Westermann 1975, pl. 2, figs 1, 3).

Callovian

The ammonite fauna of North Tibet is known only from a large incomplete *Macrocephalites* cf. *macrocephalus* (Schloth.) of typically European appearance, and from the cosmopolitan *Choffatia* spp.

The analysis of the ammonite fauna of the Tethyan Himalaya of South Tibet is also provisional. Endemism appears to be high (score = 4.5), with the affinities of the remaining elements being probably stronger with India-Madagascar than to Pan-Tethys, western Tethys (?) and eastern Indonesia. Single reported Boreal elements were presumably misidentified. As in case of the Bajocian fauna, however, resemblances to the Indo-Madagascan and eastern Indonesian faunas may be underrated, due to differences in biofacies and incomplete stratigraphic records.

Conclusion

The affinities of the Middle Jurassic ammonite species from the Tethyan Himalaya of Tibet support the 'mobilistic' (plate-tectonic) model, i.e. that their habitat was at the south-eastern margin of Tethys, with connections to western India, East Africa, and Madagascar, as well as to north-western Australasia. The limited evidence from North Tibet, on the other hand, is consistent with a Eurasian position. The latest Jurassic (Tithonian) ammonoid affinities of the Tethyan Himalaya become extremely close to north-western Australia, which, therefore, Uhlig (1911) correctly included in the Himalayan province. The Tethyan Himalaya, thus lay at the junction (ecotone) of the north–south extending Indo-Madagascan (Ethiopian) and the west–east-trending Himalayan ammonoid provinces (or subprovinces of Indo-West Pacific province).

References

Arkell, W. J. (1953). Bajocian ammonites collected by Sir Henry Hayden near Kampdazong, Tibet. *Geol. Mag.* **90**, 331–6, pls 13–14.

Bosellini, A. (1986). East African continental margins. *Geology* **14**, 76–8.

Collignon, M. (1958). *Atlas des fossiles caracteristique de Madagascar.* Fasc. **1**, 2, pls 1–33, Tananarive.

Hamilton, W. (1979). Tectonics of the Indonesian region. *Geol. Surv. Prof. Pap.* **1078**. (345 pp.)

Hamilton, W, (1983). Configuration of eastern Gondwanaland. *U.S. Geol. Surv. Circ.* **911**, 2–3.

Lie, Dong-sheng (ed.) (1981). Geological and ecological studies of Qinghai–Xizang Plateau. I. Geology, geological history and origin of Qinghai–Xizang Plateau. *Proc. Symp. on Qinghai–Xizang (Tibet) Plateau*, (Beijing, China) Science Press, Beijing. (967 pp.)

Ogg, J. and Steiner, M. B. (1985). Jurassic magnetic polarity time scale; current status and completion. In *International Symposium on Jurassic Stratigraphy, Erlangen, Sept. 1–8, 1984.* Symp. Vol. 3, pp. 777–94.

Rabinowitz, P. D., Coffin, M. F., and Falvey, D. (1983). The separation of Madagascar from Africa. *Science* **220**, 67–9.

Şengör, A. M. C. (1984). The Cimmeride orogenic system and the tectonics of Eurasia. *Geol. Soc. Am. Spec. Pap.* **195**, 82 pp.

Şengör, A. M. C. (1985). The story of Tethys: how many wives did Okeanos have? *Episodes* **8**, (1), 3–12.

Uhlig, V. (1911). Die marinen Reiche des Jura und der Unterkreide. *Mitt. Geol. Ver. Wien.* Heft **3**, 329–448.

Wang, Yi-gang and Chang, Ming-liang (1974). *The stratigraphy of Mount Jolmo Lungma region: Jurassic. Report on scientific expedition in the Mount Jolmo Lungma region (1966–1968): Geology* pp. 127–46. Beijing, Science Press (in Chinese).

Wang, Zheng, Zhuo-guan, and Chen, Guo-long (1979). *Atlas of fossils in Qinghai: Cephalopoda*, pp. 3–59, 17 pls, Geological Publishing House, Beijing (in Chinese).

Westermann, G. E. G. (1975). Bajocian ammonoid fauna of Tethyan affinities from the Kambe Limestone Series of Kenya and implications to plate tectonics. *Newsl. Stratigr.* **4**, (1), 23–48.

Westermann, G. E. G. and Wang, Yi-gang (in press). Middle Jurassic Ammonitina of Tibet and the lower Spiti Shales. *Palaeontology*.

Zhao, Jin-ke (1976). *Jurassic and Cretaceous ammonoids from the Mount Jolmo Lungma region, and report of scientific expedition in the Mount Jolmo Lungma region (1966–1968): palaeontology* (Fasc. 3). pp. 503–74, Science Press, Beijing (in Chinese).

Gerd E. G. Westermann, Dept. of Geology, McMaster University, Hamilton, Ontario, Canada.

Mesozoic Turkey as part of Europe

Derek V. Ager

ABSTRACT: Evidence from shallow-water, sessile benthos, mainly brachiopods, suggests that Mesozoic Turkey should be placed on the northern side of Tethys in palaeogeographical reconstructions, not on the southern side. Triassic, Jurassic, and Cretaceous faunas, whether from north or south of the supposed Central Anatolian suture, are wholly 'European' in character and include no obvious 'African' elements. The faunas also contrast markedly with those of the Middle East and (to a lesser extent) with those of the Adriatic Microplate.

Introduction

Ten years ago, at the first meeting of European Geological Societies, I suggested that Turkey was part of the Eurasian Plate in Mesozoic times (Ager 1975). My reasons were wholly palaeontological, and I have since been encouraged in my beliefs by further discoveries.

In palaeogeographical reconstructions, Turkey always seems to be placed, as a single entity, on the southern shores of the Mesozoic Tethys (e.g. by Smith *et al.* 1973). Discussions with geophysicists have, however, failed to reveal any real reason why this should be so, and it is well known that a considerable margin of error is possible in palaeomagnetic fixes.

I am relying largely for my contrary suggestion on the distribution of Mesozoic articulate brachiopods. As was pointed out by Valentine and Jablonski (1983), 'Living articulates for which data are available have without exception nonplanktotrophic development'. That is to say, their larvae only last in the free-swimming stage for days or hours. These larvae are produced in far fewer numbers than those of planktotrophic invertebrates, and generally do not feed during this stage of their development. It is obvious therefore that they cannot travel far from their parents. Their migration potential is limited, and it is impossible for them to cross oceans before settling on the sea-floor to begin their normal adult life. This seems to conform with the data available about the distribution of Mesozoic brachiopods generally and probably applies to Palaeozoic forms as well.

Why Turkey was Europe

I was first impressed by the European nature of the Mesozoic rocks of Turkey on a visit there nearly thirty years ago. Two of the field trips I made on that occasion are particularly relevant to this paper. One was with my good friend Professor Kemal Ernuvanli to the Black Sea coast near Sile in the north (all the localities mentioned in this paper are shown in Fig. 1). I had recently been in Italy studying the *argille scagliose* of the Apennines, and felt confident in identifying jumbled deposits of Cretaceous boulders overlying Eocene nummulitic deposits as a similar submarine landslip deposit (Ager 1958). I was even more impressed by a cliff section near here in the uppermost Cretaceous. This was described in the Turkish literature as a white limestone with chert nodules and with a strange-sounding list of fossils. What I saw in fact was typical Upper Chalk with flint nodules and the usual fossils, such as *Micraster*, *Echinocorys* and *Porosphaera* with which I am familiar in southern England. Later, I saw similar deposits in Bulgaria and in Georgia, in the southern part of the USSR.

Brachiopods were not particularly obvious, but Dr Ellis Owen tells me (pers. comm. 1985) that, in his long experience of Cretaceous brachiopod faunas of Turkey, Israel, and North Africa 'the Turkish fauna is more akin to the European Cretaceous faunas of France, Belgium, Holland and Poland. This is particularly apparent in the Upper Chalk where some of the Rhynchonellidae (*Cretirhynchia*, *Orbirhynchia*) are very close to northern European species, including the British Upper Campanian and Lower Maastrichtian forms'.

My second trip back in 1957 was under the guidance of Dr Cahide Kiragli, to the Lower Jurassic section at Yakacik in central Anatolia, north-west of Ankara (Fig. 1). There I collected a large fauna of early Jurassic (Sinemurian–Pliensbachian) brachiopods, which I subsequently described (Ager 1959). These were wholly European in affinities and included such

From AUDLEY-CHARLES, M. G. & HALLAM, A. (eds) *Gondwana and Tethys* Geological Society Special Publication No. 37, pp. 241–5.

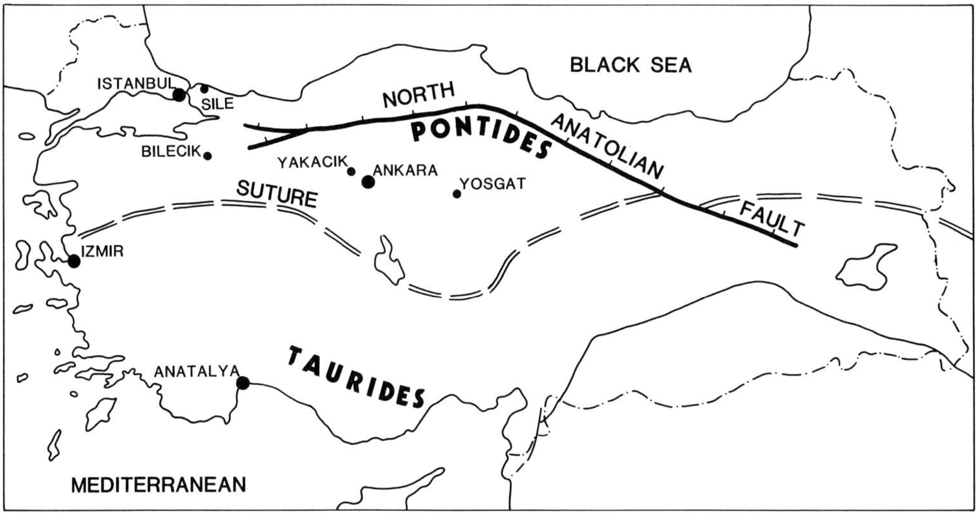

FIG. 1. Sketch-map of Turkey, showing the localities mentioned in the text and the position of the North Anatolian Fault (solid line) and the supposed suture (double line).

familiar north-west European extra-Alpine species as *Homoeorhynchia acuta* (J. Sowerby), *Cincta numismalis* (Lamarck), and *Aulacothyris* cf. *resupinata* (J. Sowerby). The one obviously Alpine type (though rare) was what would now be called *Nucleata aspasia* (Meneghini). In the same paper, I noted other Turkish Lower Jurassic brachiopods on the basis of collections in Istanbul and Ankara, notably those described by Vadasz (1913). These included further familiar European forms, such as *Lobothyris punctata* (J. Sowerby) and *Spiriferina alpina* (Oppel). I saw a precisely similar fauna from Bilecik, farther west.

Several years later, Dr Thierry Juteau visited me at Swansea, bringing further collections made by a French team working in southern Turkey, north of Antalya on the Mediterranean coast. The faunas included Triassic as well as early Jurassic brachiopods, which we subsequently described (Ager *et al.* 1980). Most remarkable perhaps were specimens of the large and distinctive rhynchonellinids *Carapezzia geyeri* (Bittner) and *C. globosa* (Carapezza and Schopen), only previously known from the Rhaetian of the Gailtaler Alps of Austria and from Sicily. They were accompanied by the characteristic Upper Triassic species *Halorella amphitoma* (Bronn) and by the rarer, typically Rhaetian, forms *Fissirhynchia fissicostata* (Suess), *Austrirhynchia cornigera* (Schafhautl), and *Rhaetina* sp. Less spectacular, though of greater interest to me, were the Lower Jurassic brachiopods from this area, notably at Sorkun Yaylasi, the well-known north-west European species *Calcirhynchia calcaria* Buckman of the Hettangian–lowermost Sinemurian at Mentesi near Sandliki, and a Pliensbachian fauna from Sorkun Yaylasi that was virtually identical with those I had previously seen from Yakacik in the north and a collection from Bilecik, much farther west, kindly sent to me by Miss Fusun Alkaya. This is true—not only species for species, but also in colour, preservation, and size. The fauna included a few further north-west European species, such as *Cuneirhynchia oxynoti* (Quenstedt), *C. dalmasi* (Dumortier), *Zeilleria perforata* (Piette), and *Spiriferina* cf. *tumida* (Buch) which may indicate a late Sinemurian rather than early Pliensbachian age.

A further fauna was sent to me by Dr Cazibe Sayar from Yosgat east of Ankara. This proved to be virtually identical with the late Triassic fauna described from the south of the USSR by Dagis (1963), with species such as *Caucasorhynchia baliana* (Bittner), *Laballa suessi* Winkler, *Triadispira caucasica* Dagis, and *Rhaetina pyriformis* (Suess). All of these indicate a Norian or Rhaetian age, and all are species only otherwise known from southern Europe. Bilgutay (1967) also described another species known from the southern USSR—*Oxycolpella oxycolpos* (Emmrich)—from this region.

Subsequently, Professor A. Farinacci of the University of Rome brought to me several collections from the western Pontides in north-west Turkey, notably from sections at Gunoren and Bairkoy Trafo near Bilecik.

These were again of early Jurassic (Pliensbachian) age, and were also incredibly similar in specific composition, preservation, colour, and size to the collections from the other Turkish Liassic localities. It is hoped to describe these faunas (and their accompanying molluscs) at a later date. Of special interest here was the presence of the southern European rhynchonellid *Prionorhynchia* and the unusual biplicate terebratulid *Plesiothyris verneuili* (Eudes-Deslongchamps) only previously known from Spain on the south side of the Pyrenees. The other notable difference here was the abundance of the typically Alpine brachiopod *Nucleata aspasia*, which had been rare at Yakacik and completely absent in the south. This species is commonly thought of as occurring in deeper-water facies, and it is noteworthy that the sediment here is of a *rosso ammonitico* type, on which Professor Farinacci is a considerable authority (Farinacci and Elmi 1981).

It may be deduced therefore that the early Jurassic brachiopod faunas of the Pontide belt in the north are more of an Alpine type, while those in the centre of Anatolia are much closer to those of extra-Alpine Europe. Those with an abundance of *N. aspasia* are particularly reminiscent of the faunas of the Mesozoic hills of the Pannonian Basin of Hungary, as described by Vigh (1943). It is noteworthy that in Bulgaria extra-Alpine forms predominate in the main part of the Balkan Range, but Alpine species appear in the east.

The almost incredible feature about Turkish early Jurassic faunas is the virtually identical nature of their brachiopods whatever the locality from which they come. They belong to the same species, they are similar in size range (i.e. apparently stunted), and they are even similar in preservation and colour. The red, nodular calcareous marl seems to be very widespread, and is highly reminiscent of the *rosso ammonitico*. This is confirmed by Professor Anna Farinacci who collected the material from the western Pontides.

The *rosso ammonitico*, however, is not noted for its brachiopods, but rather for pelagic forms (as its name implies). At only one other locality known to me do stunted brachiopods occur superabundantly in a *rosso ammonitico* facies. This is below the dam at Venta de Contreras in northern Spain.

It may be said that if the Turkish Mesozoic brachiopod faunas do not relate to those of Africa itself, they may relate to those of the area on both sides of the Adriatic, extending up into the Eastern Alps, which many regard as the battering ram of Africa driving into Europe to form the Alpine chains. I have called this the Adriatic Microplate (Ager 1980) for it seems to be only partly independent of Europe. Thus *Halorella*, which is particularly characteristic of the Norian of the Eastern Alps, extends all the way, presumably along the north shore of Tethys, via the Himalayas to south-east Asia (Ager 1967, 1968). On the other hand, *Carapezzia* was only previously recorded from southern Austria and Sicily. However, there are notable Adriatic absentees in Turkey, including the sulcate rhynchonellids such as *Pisirhynchia*, which are particularly characteristic of the Adriatic area.

Also generally absent in northern Turkey are the giant Upper Triassic bivalves, the megalodontids, which are particularly characteristic of the Adriatic microplate. However, these have been recorded in the Taurides of southern Turkey (Vegh Neubrandt *et al.* 1976).

Even if there are affinities with the Adriatic area, this area is still more European than African in its faunas.

Why Turkey was not Africa

Having indicated the close connections in terms of shallow-water sessile benthos between Turkey and Europe, it is therefore relevant to demonstrate why it is not part of Africa or a southern continent of Gondwana. To a certain extent we are hampered by the fact that we just do not know what African faunas 'should' look like for much of the Mesozoic. We know virtually nothing, for example, about African Triassic faunas. We do have the late Triassic faunas described by Hudson and Jefferies (1961) from Oman, which have several elements such as *Hagabirhynchia* which are completely unknown in Turkey and the whole of Europe.

There is the further complication that the Maghreb region of north-west Africa is—in effect—part of Europe (Ager 1980, p. 457). The Rif chain along the north coast is a mirror image of the Betic Cordillera of southern Spain and has similar Alpine faunas.

The Atlas chains farther south have mainly European faunas, but some species identical with those of East Africa. This mixture is to be expected at the narrow end of the Tethys.

So far as Turkey is concerned, the most relevant African area is the so-called Ethiopian province, which I here interpret as extending from Egypt to Malagasy in Mesozoic times. The East African seaway began to open late in early Jurassic times, and forms such as *Spiriferina* reached Malagasy (Madagascar), with the species *S. madagascarensis* Thevenin, which is limited to the basal Toarcian of Madagascar and Saudi Arabia.

The best-known Mesozoic brachiopod faunas in the Ethiopian province are undoubtedly those originally described from the Callovian and Oxfordian rocks of British Somaliland (now Somalia) by Weir (1929), and subsequently by Muir-Wood (1935). Many elements in these faunas have subsequently been recorded from Ethiopia (Jordan 1971) and are known from Kenya and Tanzania (S. Kapilima pers. comm. 1983). They are also found in Kutch in north-west India, on the evidence brought to Swansea by Dr Sushant Kanjilal.

Most distinctive of all the 'Ethiopian' brachiopods is the highly unusual rhynchonellid *Septirhynchia* in the Callovian (subsequently described by Muir-Wood 1935, and Muir-Wood and Cooper 1951) from Somalia and, for example, by Mancenido and Walley (1979) from southern Tunisia. It also extends up into Sinai and possibly also occurs in Kutch in north-west India, but it does not extend into Turkey.

In the succeeding Oxfordian strata, the species *Somalirhynchia africana* (Weir) is extremely common in East Africa and extends round to the Western High Atlas of Morocco, accompanied by distinctive terebratulids which may be attributed to the East African form *Bihenithyris* (Ager 1974). None of these, however, extend into Europe, where forms such as *Loboidothyris zieteni* (de Loriol) were flourishing at this time. They certainly did not reach Turkey, where such large and distinctive forms could hardly have been missed.

General geological considerations

When I originally suggested Turkey as part of Europe in 1975, Professor Achil Tzagarelli of Tbilisi drew my attention to Soviet work on Lower Jurassic conglomerates in the Crimea (Kamyshan and Babanova 1962). These conglomerates were thought to have been derived from the south, and an imaginary country was synthesized, from which they must have been derived. This imaginary country turned out to be remarkably like northern Turkey in geology.

It may be suggested therefore that Turkey was close enough to the Crimea to provide the boulders for the Jurassic conglomerates, and that the Black Sea only opened at a later date. It should also be noted that the boulders were associated with a European extra-Alpine brachiopod fauna, including *Gibbirhynchia curviceps* (Quenstedt), *Labothyris punctata* (Sowerby), *Aulacothyris waterhousei* (Davidson), and *Zeilleria* (*Cincta*) *numismalis* (Quenstedt).

It is noteworthy that the Balkan Mountains of Bulgaria have very similar faunas to those mentioned above, and to those of Turkey (e.g. Tchoumatchenco 1970, 1972). In the latter paper, Platon Tchoumatchenco commented particularly on the close affinities of his brachiopod faunas with those of England. It is noteworthy that the Balkanides do not continue out into the Black Sea, but terminate at the coastline (Ager 1972, p. 264). It was formerly supposed that the Balkanides continued under the Black Sea to the Crimea, but this has now been rejected (Ager 1980, p. 349). It is more likely that they continue into Turkey as the 'Anti-Balkans' and then the Pontides (Ager 1980, p. 414).

Across the centre of Turkey, from west to east, runs the North Anatolian Fault and what is taken to be a major suture, with ophiolitic nappes pushing south the vast olistostromes. It is conventional, therefore, to regard southern Turkey, with the Taurides, as quite different from northern Turkey and the Pontides. All I can say is that my Mesozoic brachiopods did not know about the conventions of plate tectonics, for the faunas to the south of the suture are every bit as European as those to the north, and I am convinced that articulate brachiopods never went far from their European mothers. At the other end of the Black Sea there is a wide graben, which mainly opened in Palaeogene times (Ager 1980, p. 440). This was presumably the time when Turkey separated from Europe.

In the conclusions of the joint paper on the brachiopods from southern Turkey (Ager *et al.* 1980), two hypotheses were suggested to explain the supposed anomaly of the European nature of the brachiopods in the Tauride belt. One concerned the actual position of the limit between the Pontides and the Taurides. The other hypothesis was that the Mesozoic brachiopods evolved independently from a common Palaeozoic stock.

Though I was a joint author of this paper I did not subscribe to this particular theory (my function having been simply to identify the fossils). In my view, the faunas are far too similar between the two chains, being virtually all conspecific with European forms, to have evolved separately. What is more, this applied to other groups, such as the megalodontid bivalves, as well as the brachiopods. There is no evidence of them becoming differentiated with time, as suggested in those conclusions.

Therefore, whatever the palaeomagnetists and tectonicians may say, in my view, southern

Turkey was never far from northern Turkey, and neither area was far from the main Eurasian plate. At least they were never too far apart for brachiopod larvae to swim between them.

ACKNOWLEDGEMENTS: I am grateful to all the French, Turkish, and Italian workers who so generously let me examine their collections. Mr J. H. Edwards drew the map.

References

AGER, D. V. (1958). On some Turkish sediments. *Geol. Mag.* **95**, 83–4.

AGER, D. V. (1959). Lower Jurassic brachiopods from Turkey. *J. Paleontl.* **33**, 1018–28.

AGER, D. V. (1967). Some Mesozoic brachiopods in the Tethys region. In *Aspects of Tethyan biogeography*. (ed. C. G. Adams and D. V. Ager), pp. 135–51. Systematics Association Publication No. 7.

AGER, D. V. (1968). The supposedly ubiquitous Tethyan brachiopod *Halorella* and its relations. *J. Palaeont. Soc. India* (1960–64), 5–9, 54–70.

AGER, D. V. (1972). Summer field meeting in Bulgaria, 1971. *Proc. Geol. Assoc.* **83**, 239–68.

AGER, D. V. (1974). The Western High Atlas of Morocco and their significance in the history of the North Atlantic. *Proc. Geol. Assoc.* **85**, 23–41.

AGER, D. V. (1975). Mesozoic Turkey as part of Europe. *Abstr. First Meeting Europ. Geol. Soc.*, Reading.

AGER, D. V. (1980). *The geology of Europe*. McGraw-Hill (U.K.), Maidenhead, Berks. (535 pp.)

AGER, D. V., GUTNIC, M., JUTEAU, T., and MONOD, O. (1980). New Early Mesozoic brachiopods from southern Turkey. *Bull. Min. Res. Explor. Inst. Turkey* **91**, (1978), 59–75.

BILGUTAY, U. (1967). A study on some Triassic macrofossils of the Hasanoglan–Bayramdere Region, Ankara, Turkey. *Bull. Miner. Res. Explor. Inst., Ankara* **70**, 67–9.

DAGIS, A. S. (1963). Upper Triassic brachiopods of the southern U.S.S.R. *Izdatel Nauk SSSR*, Siberskoe Otdelenie, Moscow. (24 pkp.) (in Russian).

FARINACCI, A. and ELMI, S. (ed.) (1981). *Rosso ammonitico symposium proceedings*. Edizioni Tecnoscienza, Rome. (112 pp.)

HUDSON, R. G. S. and JEFFERIES, R. P. S. (1961). Upper Triassic brachiopods and lamellibranchs from the Oman Peninsula, Arabia. *Palaeont.* **4**, 1–41.

JORDAN, R. (1971). Megafossilien des Jura aus dem Antalo-Kalk von Nord-Athiopien. *Beiheft geol. Jb.* **116**, 141–72.

KAMYSHAN, V. P. and BABANOVA, L. J. (1962). Discovery of boulders of Lower Jurassic limestone in the region of Karadag (Crimea). *Doklady Akad. Nauk SSSR* **145**, 384–5. (in Russian)

MANCENIDO, M. O. and WALLEY, C. D. (1979). Functional morphology and ontogenetic variation in the Callovian brachiopod *Septirhynchia* from Tunisia. *Palaeontol.* **22**, 317–37.

MUIR-WOOD, H. M. (1935). Jurassic Brachiopoda. In *The geology and palaeontology of British Somaliland*. pp. 75–147, Pt. II, **7**. London.

MUIR-WOOD, H. M. and COOPER, G. A. (1951). A new species of the Jurassic brachiopod genus *Septirhynchia*. *Smithsonian Miscell. Collections* **116**, 1–6.

SMITH, A. G., BRIDEN, JU. C., and DREWRY, G. E. (1973). Phanerozoic world maps. In *Organisms and continents through time*. (ed. N. F. Hughes), pp. 1–42, Palaeont. Assoc. Spec. Pap. 12.

TCHOUMATCHENCO, P. (1970). Lower Jurassic brachiopods of the Central and Western Stara Planina and Krayshchto. *Bulgarska Akad. Hauhite, Geol. Inst.* (22 pp.) (in Bulgarian).

TCHOUMATCHENCO, P. (1972). Stratigraphical distribution of Lower Jurassic brachiopods in the central and western Balkans Kraysti. *Bull. Geol. Inst., Stratigr. Lithol. Ser.* **21**, 63–84. (in Bulgarian with French résumé).

VADASZ, M. E. (1913). Liasfossilien aus Kleinasien. *Mitt. Jb. Konigliche Geol. Reichsanst.* **21**, 57–82.

VALENTINE, J. W. and JABLONSKI, D. (1983). Larval adaptations and patterns of brachiopod diversity in space and time. *Evolution* **37**, 1052–67.

VEGH NEUBRANDT, E., *et al.* (1976). Megalodontidae du Trias supérieur dans la chaîne Taurquie (Turquie meridionale). *Geobios* **9**, 199–222.

VIGH, G. (1943). A Gerecse heygeseg eszaknyugati reszenek foldtani es oslenytani viszonyaia. *Foldtani Kozlony* **73**, 301–612. (in Hungarian with German abstract).

WEIR, J. (1929). Jurassic fossils from Jubaland, East Africa, collected by V. G. Glenday, and the Jurassic geology of Somaliland. *Monogr. Geol. Dept. Hunterian Museum, Glasgow* **3**, 1–63.

DEREK V. AGER, Dept. of Geology, University College, Swansea, W. Glamorgan SA2 8PP, UK.

The trans-Pacific spread of equatorial shallow-marine benthos in the Cretaceous

P. W. Skelton

ABSTRACT: From Aptian to Maastrichtian times, the development of broad tracts of anomalously shallow ocean floor in the Pacific, studded with reef-rimmed seamounts and islands, allowed some Caribbean benthos to spread westwards by larval drift between staging posts in Tethyan/Atlantic waters debouching between the Americas. There is as yet no compelling evidence that the Pacific was completely traversed in this manner during the first and major phase of intra-plate volcanothermal activity in the Aptian–Albian: evidently there still existed a sufficiently wide region of Pacific abyss to the west to serve as a barrier to larval dispersal. However, clear evidence for the 'landfall' of Caribbean-derived species on eastern Tethyan shores in Campanian to Maastrichtian times is provided by several examples of disjunct endemics from the two regions. Such taxa include rudist bivalves, gastropods, larger foraminifera, and calcareous algae. This episode coincides with a second, smaller phase of mid-plate volcanothermal activity in the Pacific, which, in conjunction with the relict topography from the earlier phase, evidently furnished a temporarily continuous chain of staging posts for the trans-Pacific spread of the shallow-marine benthos.

Introduction

Schlanger *et al.* (1981) have summarized the evidence for widespread volcanothermal activity within the Pacific and Farallon Plates of the Pacific Ocean during Aptian to Maastrichtian times. From large benthic foraminifera of Campanian to Maastrichtian age, and of Caribbean affinities, that were encountered amidst redeposited reefal limestones recovered in a number of Deep Sea Drilling Project cores from the western Pacific, and an isolated record of the same from Papua, New Guinea, they inferred that the constellation of shallow-water plateaux and seamounts so produced had served as staging posts for the trans-Pacific larval dispersal of the foraminifera, from the Caribbean to at least as far as Papua, and possibly beyond into the eastern Tethyan ocean.

The purpose of this paper is to present evidence for the arrival at several eastern Tethyan sites of shallow-marine benthic taxa derived in this fashion from the Caribbean/ Central American area, and to make the case that this episode of complete crossing of the Pacific by such taxa was limited to Campanian–Maastrichtian times. Despite the abundance of sunken Middle Cretaceous reefs in the present-day western Pacific (see, for example, Winterer and Metzler 1984, and Konishi 1985), there is as yet no compelling evidence for the 'landfall' of Caribbean-derived taxa anywhere in the eastern Tethys prior to the Campanian.

Volcanothermal activity in the Pacific

Since the work of Hamilton (1956), the mountainous topography of the Western Pacific Ocean floor has been known to contain numerous sunken reefs, of mid- to late Cretaceous age, built upon volcanic foundations. Examples have been recorded from the Ogasawara Plateau seamounts (Konishi 1985), around the Mariana Basin (Moberly *et al.* 1983*a*), the Mid-Pacific Mountains (several references reviewed in Winterer and Metzler 1984), and along the Marshall and Line Islands chains (Schlanger and Premoli-Silva 1981, and Schlanger *et al.* 1981). Moreover, Schlanger *et al.* summarized the evidence for similar edifice-building activity on the Farallon Plate, which floored a large part of the Eastern Pacific in Cretaceous times: allochthonous ophiolitic terranes, accreted on to western Central and North America during subduction of the Farallon Plate, incorporate shallow-water limestone facies with faunal components like those of the drowned reefs of the Pacific Plate. Tarduno *et al.* (1985) have further attempted to match the development of one such edifice, now incorporated in the Franciscan Complex of

From AUDLEY-CHARLES, M. G. & HALLAM, A. (eds) *Gondwana and Tethys* Geological Society Special Publication No. 37, pp. 247–53.

northern California, as a Farallon Plate counterpart to the Shatsky Rise on the Pacific Plate, postulating that the pair formed together over a hot-spot sited beneath a part of the Pacific/Farallon mid-oceanic ridge.

The manner of formation of the volcanic foundations of these reefs has been the subject of much debate, summarized by Menard (1984). The present understanding of them, as outlined by Menard for the Western Pacific, involves both the generation of sequential chains of volcanic seamounts as oceanic plates moved over hot-spots, and the subsequent addition of intrusives and volcanics over a vast mid-plate swell (the 'Darwin Rise')—an amalgam of the various different models that had been proposed over the previous few decades for the mountainous physiography of the Western Pacific floor. A synoptic stratigraphy of the volcanic investments and reef developments encountered in DSDP cores from the Nauru Basin (Marshall Islands), the Line Islands, the Central Pacific Basin lying between them, and the Mid-Pacific Mountains to the north is given by Schlanger and Premoli-Silva (1981, figure 3). Although there is evidence for volcanism throughout the Aptian to Maastrichtian interval, peaks of activity have been identified in the Aptian–Albian and the Campanian–Maastrichtian stages, in this and later studies (e.g. Moberly *et al.* 1983*a* and *b*).

Finally, by way of background, it is worth pointing out at this stage that the Darwin Rise would have been situated in the eastern Central Pacific area in early to mid-Cretaceous times, with the Farallon Plate to its north-east (see Schlanger *et al.* 1981, figure 4). The Darwin Rise subsequently drifted broadly north-westwards in the late Cretaceous and Cainozoic to its present Western Pacific position.

Disjunct distributions and trans-Pacific dispersal

Extrapolation of Schlanger *et al.*'s trans-Pacific dispersal theory into the eastern Tethys Ocean has recently been argued by Skelton and Wright (1987), who have described well-preserved specimens of the distinctive hippuritid rudist bivalve *Torreites* from several sites in the United Arab Emirates and Oman, confirming an earlier report of the genus from the region, based on a single worn fragment (Grubić 1979). Prior to these discoveries, *Torreites* was considered endemic to the Caribbean province (Kauffman 1973). The quality of preservation of the material and the distinctiveness of the genus, which can be diagnosed from at least four constructionally independent and marked modifications of 'normal' hippuritid morphology ('autapomorphies' in the language of the cladists), allowed Skelton and Wright to reject homeomorphy as a possible explanation for the Old and New World populations. Faced then with the need to explain the highly disjunct distribution of this late Cretaceous genus, they opted for the connecting dispersal route across the Pacific rather than via the superficially more obvious, shorter Mediterranean Tethys/Atlantic route (see Fig. 1) for the following reasons: rising levels of endemism in the Caribbean province from mid-Aptian times onwards (Coates 1973; Kauffman 1973; Skelton 1982) suggests that the widening Atlantic would have been a significant barrier to rudist larval dispersal by late Cretaceous times (no exclusively Campanian–Maastrichtian rudist genera, for example, are shared by the Caribbean and western Mediterranean regions); moreover, Philip (1982, 1985) has shown that there was a marked faunal divergence between the 'Aquitano–Pyrenean' area and the eastern and central Mediterranean area in Campanian/Maastrichtian times, which he attributed to barriers to dispersal created by the collision of the Adriatic block with Europe. A few eastern Mediterranean taxa did find their way to south-eastern Spain, however, although none of these is known to have crossed the Atlantic. It is unlikely that *Torreites* could have hurdled these barriers that lay between Arabia in the east and the Caribbean in the west, which other rudists found impassable. But, if it was an exception in this respect, its lack of a fossil record in the relatively well-inspected Upper Cretaceous rudist limestones of the Mediterranean would indeed be a curious coincidence, in view of it having been recovered from several Caribbean and Arabian sites. The trans-Pacific dispersal route of Schlanger *et al.* (1981), on the other hand, offers a highly tempting solution to the *Torreites* problem, particularly since the appearance of *Torreites* in the eastern Tethys in Maastrichtian times coincides neatly with the age of the large benthic foraminifera cited by them as evidence for the dispersal route.

The Skelton and Wright model for the distribution of *Torreites* can only be directly tested by a search for further specimens of the genus: if it is correct, then discovery of specimens in DSDP material from the Pacific, or from Upper Cretaceous limestones caught up in the Tethyan suture zone of, say, Tibet, may be expected. In contrast, the recovery of specimens from, for example, Spain or north-western Africa, would militate against the hypothesis.

An alternative approach to the problem is to seek other, similar cases of apparent disjunct endemism between the Caribbean and eastern Tethys. The more cases that can be revealed, the weaker becomes any argument for coincidental preservational loss in the Mediterranean Tethys (or for coeval homeomorphy), and the more likely the trans-Pacific spread theory as an explanation. And if sufficient examples are indeed forthcoming, their age relations on either side of the Pacific can then be used as a test for the east to west pattern of dispersal postulated in the theory (based on the reconstructed oceanic circulation pattern of the Cretaceous oceans, shown in Fig. 1: see, for example Luyendyk *et al.* 1972; Gordon 1973; Berggren and Hollister 1974, 1977; and Lloyd 1982). The east to west pattern of dispersal ought to have left a preponderance of stratigraphically older (pre-dispersal) records in the Caribbean/Central American area. Other records might still be virtually synchronous on a geological time scale, where taxa had arisen after the corridor had been opened up, since the actual process of larval dispersal along the chain of staging posts would presumably have taken place over an 'ecological time scale' of but a few years.

Although still only at a preliminary stage, my quest for other examples of disjunct distributions at either end of the Pacific has already yielded a small number of cases, more or less explicitly described in the literature (Figs 1 and 2). Future primary taxonomic work will, no doubt, bring further examples to light, as synonymies between widely separated taxa, unimagined by their original authors, are established. But such a programme will entail some years of work; in this paper I simply intend to emphasize what seems to be apparent even from my preliminary compilation of data.

Apart from the examples of *Torreites* described by Skelton and Wright (1987) and the pseudorbitoid foraminifera discussed by Beckmann (1976), and Schlanger and Premoli-Silva (1981), the following cases may be considered.

(1) *Thyrastylon* is a radiolitid rudist genus erected by Chubb (1956), and found in Guatemala and the Antilles in association with *Titanosarcolites* (a 'Middle' to Upper Maastrichtian genus according to Sohl and Kollmann 1985, Figure 19), and in Iran, together with *Dictyoptychus* and *Loftusia* (approximately 'mid-Maastrichtian' according to my own field observations in Oman and the United Arab Emirates).

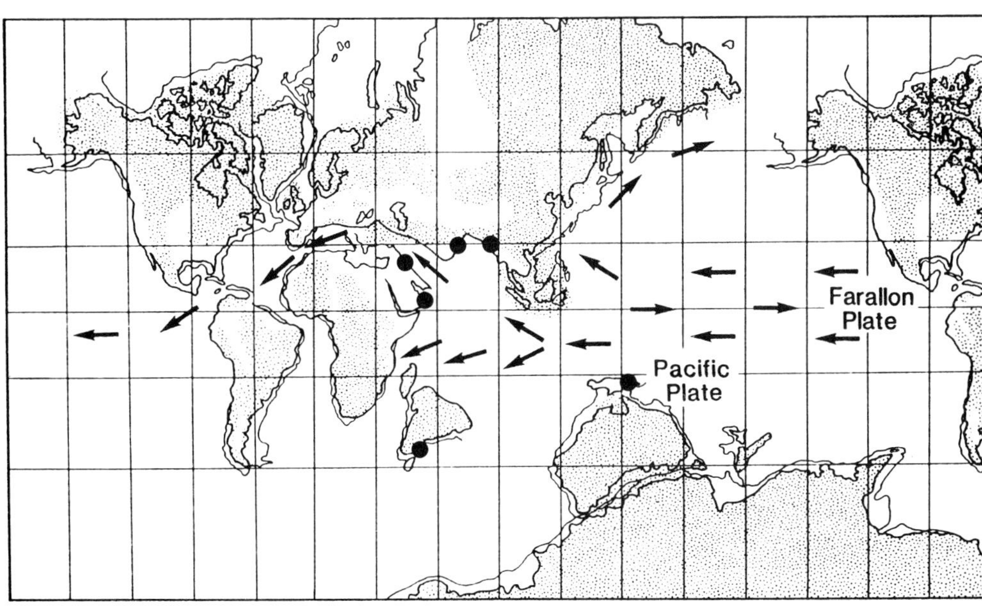

FIG. 1. Inferred Campanian–Maastrichtian palaeogeography, showing sites around the eastern Tethyan ocean embayment from which shallow-marine benthic taxa with Caribbean affinities have been recorded (black dots). Oceanic surface currents (arrows) are based on sources cited in the text. Based on text-fig. 11 in Skelton and Wright (1987), with continental positions derived from Smith and Briden (1977), and land (dotted ornament) and sea (white), distributions from Ziegler *et al.* (1983).

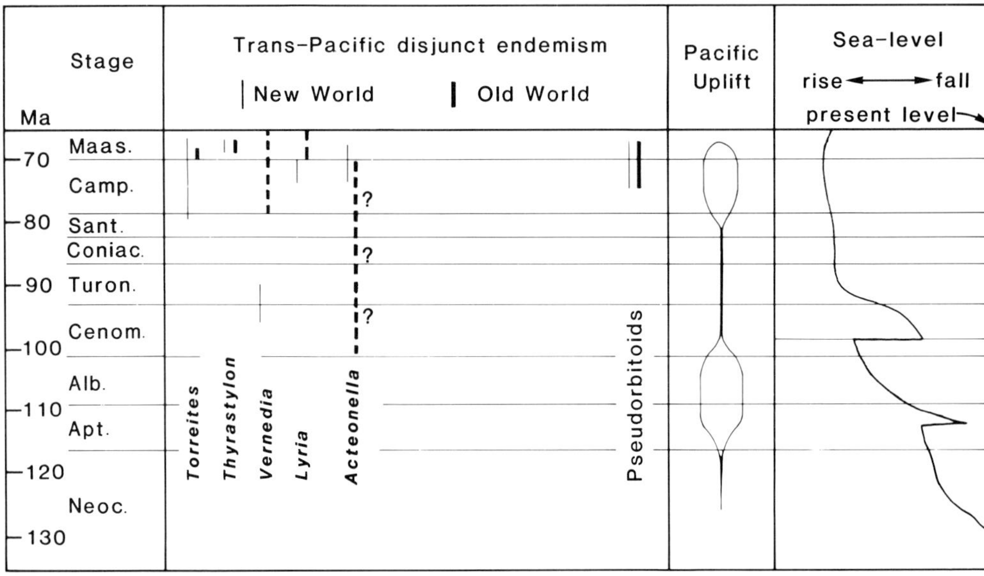

FIG. 2. Stratigraphic ranges, in the New World and the Old World, of taxa showing disjunct endemism in the Caribbean/Central American area and eastern Tethys (see text for details). Uncertain ranges are indicated by broken lines. Main phases of volcanothermal uplift in the Pacific and eustatic sea-level changes, as reconstructed by Vail *et al.* (1980), are shown at the right, for comparison.

(2) *Vernedia* is an itieriid (nerineacean) gastropod genus with several species. One Mexican species, *Vernedia friesi*, Kollmann and Sohl 1980, of late Cenomanian to early Turonian age, was stated by its authors to have much closer affinities with a Campanian to Maastrichtian species from southern India, *V. globoides* (Stoliczka), than with any known European or Trans-Caucasian species.

(3) *Lyria* is a volutid gastropod genus. One species from the Upper Campanian Ripley Formation of Missouri State, USA, *L. keownvillensis* (Sohl) (see Sohl, 1964) shows a striking similarity to *L. crassicostata* Stoliczka 'var. *omanica* Lees' from the Maastrichtian of Oman (see Lees 1928). A further similar specimen from the Maastrichtian of Tibet was illustrated by Douvillé (1916 pl. VIII, fig. 6), labelled simply as '*Lyria* sp.' while other comparable material is known from the ?Campanian of Ecuador (pers. comm. N. J. Morris and R. Cleevely). Further work on these is needed to confirm their suggested affinities.

(4) *Acteonella* is a gastropod genus with many species. In a recent review of those from the New World, Sohl and Kollmann (1985) report the finding of *A. borneensis* Nuttall and Leong from the Campanian–Maastrichtian of Mexico and Cuba. Previously, it had only been recorded from a limestone block of uncertain, but suggested Cenomanian to Campanian age in Sabah, Borneo. Nuttall and Leong (1972) point out that their dating relies solely on the *Acteonella* specimens themselves, which bear broad resemblances to some of the later *Acteonella* species from elsewhere. Thus, its age in the Old World must be considered unresolved.

(5) *Ovulites*, a codiacean alga, shows a disjunct distribution in the Upper Cretaceous between the Caribbean and northern Iraq, Afghanistan and Tibet, according to Elliott (1981). Unfortunately no more precise ages are given for these occurrences.

Though few, and beset by some uncertainties, these data already suggest an interesting stratigraphic pattern (Fig. 2). If the very uncertain age of the Old World *Acteonella borneensis* is ignored, the earliest recorded ages of the New World examples all either predate, or are more or less coeval with, those of their Old World counterparts. This is consistent with the east to west pattern of dispersal, as discussed earlier (p. xx). Also, none of the Old World records (again excepting the problematical *Acteonella borneensis*, as well as unfocused dates for *Ovulites*) predates the Campanian. The 'landfall' of the taxa in question in eastern Tethys appears, thus far, to have been a purely Campanian–Maastrichtian phenomenon.

Conclusions and discussion

The arrival in eastern Tethys of equatorial shallow-marine benthos, derived from the Caribbean/Central American region by larval dispersal along a trans-Pacific chain of shallow-water staging posts (Schlanger *et al.* 1981) seems to have been restricted to Campanian–Maastrichtian times, from a preliminary survey of taxa showing apparent disjunct endemism at either end of the Pacific.

Should further data substantiate this conclusion, it is worth considering the possible reasons for the timing of this episode of trans-Pacific spread. From mid-Aptian times onwards, the shallow-marine benthos of the Caribbean Province of Kauffman (1973) included many endemic genera of bivalves (Kauffman 1973), especially rudists (Coates 1973; Skelton 1982), as well as certain distinctive lituolacean foraminifera (Dilley 1973). Although certain of these endemic elements have been recovered from some of the sunken mid-Cretaceous reefs of the Pacific (e.g. Matthews *et al.* 1974), discussed on p. 247, none has been recorded from eastern Tethys. The somewhat sparse rudist and orbitoline fauna of the late Aptian/Albian of Afghanistan (Douvillé 1926*a*; Montenat *et al.* 1982), northern Pakistan (Douvillé 1926*b*; Pudsey *et al.* 1985), and Japan (see Hayami 1975), for example, bear close resemblances to forms from the Mediterranean Tethyan area, and show no otherwise exclusive affinities with any distinctively Caribbean taxa. Similarly, my own field observations on shelf limestone sequences of this age in Oman have, as yet, revealed nothing of distinctively Caribbean provenance. Thus it would seem that the peak of volcanothermal edifice-building activity at that time (p. 248) only permitted larval dispersal from the Caribbean across the then eastern and central Pacific where this new physiography developed (Fig. 3). Presumably, at that time, the western Pacific must have been floored by older, cold ocean floor, devoid of sufficient prominences to serve as staging posts. Only when the relict topography of this earlier event (on the Pacific Plate) had drifted further to the west, and had been both rejuvenated and added to on younger areas of ocean floor by the second peak of activity (in the Campanian–Maastrichtian see p. 248), was the chain of staging posts sufficiently extended to allow a complete range extension from the Caribbean area to the eastern Tethys (Fig. 3). Subsequently, the communication again failed, in the Tertiary, as revealed by the exclusively Old World character of the large benthic foraminifera in redeposited reefal material of Palaeogene age in the Nauru Basin/Marshall Islands and Line Islands of the Central Pacific (Schlanger and Premoli-Silva 1981). The diminution of volcanic activity, with the growth of relatively smooth, younger ocean floor in the East Pacific at this time (Fig. 3), presumably created an oceanic barrier to westward larval dispersal, somewhat like the barrier that existed in the older West Pacific in early Cretaceous times.

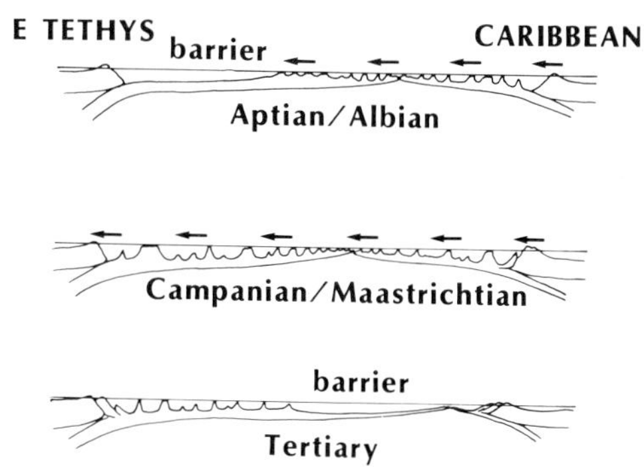

FIG. 3. Cartoon illustrating the plate-tectonic context for the history of the trans-Pacific spread of shallow-marine benthic taxa from the Caribbean/Central American area in Cretaceous times, as postulated in this paper. *Top*: the initial, major phase of volcanothermal activity in the Aptian/Albian creates staging posts for the westward spread, by means of planktonic larval dispersal (arrows), of the New World taxa. However, they do not reach eastern Tethys, possibly because of an oceanic barrier zone in the Eastern Pacific, floored by older, cold oceanic lithosphere. *Centre*: the westward drift of the relict topography from the earlier volcanism and reef formation on the Pacific Plate, with rejuvenation, and further edifice building on younger ocean floor (due to the second episode of volcanothermal activity) in the Campanian–Maastrichtian, creates a complete chain of staging posts across the Pacific, allowing the penetration of eastern Tethys by some New World taxa. *Bottom*: the resumption of smoother, normally subsident oceanic lithosphere growth in the Eastern Pacific, during the Tertiary, reintroduces a barrier to the westward spread of New World taxa.

ACKNOWLEDGEMENTS: Noel Morris, Ron Cleevely and Pat Nuttall (British Museum (Natural History)) were most helpful in discussing the taxonomic contents of this paper, while Paul Wright (Bristol) contributed significantly to the palaeobiogeographical thinking. Responsibility for errors or unwarranted speculation is mine alone.

References

BECKMANN, J. P. (1976). Shallow-water foraminifera and associated microfossils from Sites 315, 316 and 318, DSDP leg 33. In *Init. Rep. DSDP* **33**. pp. 467–90, US Government Printing Office, Washington DC.

BERGGREN, W. A. and HOLLISTER, C. D. (1974). Paleogeography, paleobiogeography and the history of circulation in the Atlantic Ocean, 126–86. In *Studies in paleo-oceanography*. (ed. W. W. Hay), pp. 126–86, Spec. Publ. Soc. Econ. Palaeont. Miner., Tulsa, 20, Tulsa, Oklahoma.

BERGGREN, W. A. and HOLLISTER, C. D. (1977). Plate tectonics and paleocirculation—commotion in the ocean. *Tectonophys.* **38**, 11–48.

CHUBB, L. J. (1956). *Thyrastylon*, a new rudist genus from the Upper Cretaceous of Guatemala, the Antilles, and Persia, with a discussion of the functions of rudist oscules and pillars. *Palaeontogr. Am.* **4**, 31–48.

COATES, A. G. (1973). Cretaceous Tethyan coral-rudist biogeography related to the evolution of the Atlantic Ocean. *Spec. Pap. Palaeont.* **12**, 169–74.

DILLEY, F. C. (1973). Cretaceous larger foraminifera. In *Atlas of palaeobiogeography*. (ed. A. Hallam), pp. 403–19, Elsevier, Amsterdam.

DOUVILLÉ, H. (1916). Le Crétacé et l'Éocène du Tibet central. *Mem. Geol. Surv. India Palaeont. Indica, New Ser.* **5** (Mem. 3).

DOUVILLÉ, H. (1926*a*). Description de quelques fossiles Crétacés de l'Afghanistan. *Rec. Geol. Surv. Ind.* **58**, (4), 345–8.

DOUVILLÉ, H. (1926*b*). Fossiles recueillis par Hayden dans le Kashmir en 1906 et les Pamirs en 1914; leur description. *Rec. Geol. Surv. Ind.* **58**, (4), 349–57.

ELLIOTT, G. F. (1981). The Tethyan dispersal of some chlorophyte algae subsequent to the Palaeozoic. *Palaeogeogr., Palaeoclimatol., Palaeoecol.* **32**, 341–58.

GORDON, W. A. (1973). Marine life and ocean surface currents in the Cretaceous. *J. Geol.* **81**, 269–84.

GRUBIĆ, A. (1979). *Torreites milovanovici* sp. nov. iz Omana, *T. coxi* sp. nov. i *T. chubbi* sp. nov. sa Jamajke, nov prikaz roda *Torreites* Palmer i osvrt na značajnjegovog paleogeografskog rasprostranjenja. *Vesn. zavod Geol. i Geofiz. istraž* **37**, (A), 81–95. (In Serbo-Croat with English summary).

HAMILTON, E. L. (1956). Sunken islands of the Mid-Pacific Mountains. *Mem Geol. Soc. Am.* **64**, 97 pp.

HAYAMI, I. (1975). A systematic survey of the Mesozoic Bivalvia from Japan. *Bull. Nat. Sci. Mus., Tokyo* **10**, 1–228, 10 pls.

KAUFFMAN, E. G. (1973). Cretaceous Bivalvia. In *Atlas of palaeobiogeography*. (ed. A. Hallam), pp. 353–83, Elsevier, Amsterdam.

KOLLMAN, H. A. and SOHL, N. F. (1980). Western Hemisphere Cretaceous Itieriidae gastropods. *Prof. Pap. US Geol. Surv.* **1125-A**, 15 pp.

KONISHI, K. (1985). Cretaceous reefal fossils dredged from two seamounts of the Ogasawara Plateau. In *Preliminary report of the Hakuhō Maru Cruise KH 84–1*. (ed. K. Kobayashi), pp. 169–79, Ocean Research Institute, University of Tokyo.

LEES, G. M. (1928). The geology and tectonics of Oman and parts of southeastern Arabia. *Quart. J. Geol. Soc. London* **84**, (336), 535–70.

LLOYD, C. R. (1982). The mid-Cretaceous Earth: paleogeography; ocean circulation and temperature; atmospheric circulation. *J. Geol.* **90**, 393–413.

LUYENDYK, K. P., FORSYTH, D., and PHILLIPS, J. D. (1972). Experimental approach to the paleocirculation of the oceanic surface waters. *Bull. Geol. Soc. Am.* **83**, 2649–64.

MATTHEWS, J. L., *et al.* (1974). Cretaceous drowning of reefs on Mid-Pacific and Japanese guyots. *Science* **184**, (4135), 462–4.

MENARD, H. W. (1984). Darwin reprise. *J. Geophys. Res.* **89**, (B12), 9960–68.

MOBERLY, R., *et al.* (1983*a*). Geology of Site 585, Deep Sea Drilling Project Leg 89. *Initial Core Descriptions of the Deep Sea Drilling Project*. pp. 39–55, University of California, Scripps Institute of Oceanography.

MOBERLY, R., *et al.* (1983*b*). Deep Sea Drilling Project Leg 89—the Mesozoic superocean. *Nature* **302**, p. 381.

MONTENAT, C., MOULLADE, M., and PHILIP, J. (1982). Le Crétacé inférieur à orbitolines et rudistes d'Afghanistan central. *Géol. Mediterranéenne* **9**, (2), 109–22.

NUTTALL, C. P. and LEONG, K. M. (1972). Occurrence of *Acteonella* (opisthobranch gastropod) in the Cretaceous of Sabah. *Geol. Surv. Malaysia, Geol. Pap.* **1**, 1–9.

PHILIP, J. (1982). Paléobiogéographie des rudistes et géodynamique des marges mésogéennes au Crétacé supérieur. *Bull. Soc. Géol. Fr.* Sér. (7), **24** (5–6), 995–1006.

PHILIP, J. (1985). Sur les relations des marges téthysiennes au Campanien et au Maastrichtien déduites de la distribution des rudistes. *Bull. Soc. Géol. Fr.* Sér. (8), **1** (5), 723–31.

PUDSEY, C. J., SCHROEDER, R., and SKELTON, P. W. (1985). Cretaceous (Aptian/Albian) age for island-arc volcanics, Kohistan, N. Pakistan, pp. 150–68. In *Geology of Western Himalayas*. Contrib. Himalayan Geol. Vol. 3. (ed. V. J. Gupta *et al.*), pp. 150–68, Hindustan Publishing Corporation, India.

SCHLANGER, S. O. and PREMOLI-SILVA, I. (1981). Tectonic, volcanic, and paleogeographic implications of redeposited reef faunas of Late Cretaceous and Tertiary age from the Nauru Basin and the Line Islands. *Init. Rep. DSDP* 61, US Government Printing Office, Washington DC, pp. 817–27.

SCHLANGER, S. O., JENKYNS, H. C., and PREMOLI-SILVA, I. (1981). Volcanism and vertical tectonics in the Pacific Basin related to global Cretaceous transgressions. *Earth Planet. Sci. Lett.* **52**, 435–49.

SKELTON, P. W. (1982). Aptian and Barremian rudist bivalves of the New World: some Old World similarities. *Cretaceous Res.* **3**, 145–53.

SKELTON, P. W. and WRIGHT, V. P. (1987). Island-hopping across the Pacific in the Late Cretaceous: a Caribbean rudist bivalve in Oman. *Palaeont.* **30** (3), 505–29.

SMITH, A. G. and BRIDEN, J. C. (1977). *Mesozoic and Cenozoic paleocontinental maps.* Cambridge University Press. (63 pp.)

SOHL, N. F. (1964). Neogastropoda, Opisthobranchia, and Basommatophora from the Ripley, Owl Creek and Prairie Bluff formations. *Prof. Pap. US Geol. Surv.* **331-B**, 153–344, pls 19–52.

SOHL, N. F. and KOLLMANN, H. A. (1985). Cretaceous actaeonellid gastropods from the Western Hemisphere. *Prof. Pap. US Geol. Surv.* **1304**. (108 pp.)

TARDUNO, J. A., MCWILLIAMS, M., DEBICHE, M. G., SLITER, W. V., and BLAKE, M. C., JR. (1985). Franciscan Complex Calera limestones: accreted remnants of Farallon Plate oceanic plateaus. *Nature* **317**, 345–7.

VAIL, P. R., MITCHUM, R. M., JR., SHIPLEY, T. H., and BUFFLER, R. T. (1980). Unconformities of the North Atlantic. *Phil. Trans. Roy. Soc. Lond.* **A294**, 137–55.

WINTERER, E. L. and METZLER, C. V. (1984). Origin and subsidence of guyots in Mid-Pacific Mountains. *J. Geophys. Res.* **89**, (B12), 9969–79.

ZIEGLER, A. M., SCOTESE, C. R., and BARRETT, S. R. (1983). Mesozoic and Cenozoic paleogeographic maps. In *Tidal friction and the Earth's rotation II.* pp. 240–52. Springer-Verlag, Berlin.

P. W. SKELTON, Department of Earth Sciences, Open University, Walton Hall, Milton Keynes MK7 6AA.

Gondwana, Tethys, and terrestrial vertebrates during the Mesozoic and Cainozoic

Jean-Claude Rage

ABSTRACT: During the Triassic, free dispersal of terrestrial vertebrates was possible in Pangaea; however provinciality appeared in Gondwana. Distribution of Triassic vertebrates may be explained even if the Earth diameter did not expand and if Pacifica did not exist. During the Jurassic, a free land communication persisted between Gondwana and Laurasia, and provinciality is not demonstrated. A Gondwanan pattern emerged during the Cretaceous, but since that time it is not possible to identify a global difference between Gondwanan faunas and Laurasian ones. An African/South American terrestrial community developed until the Aptian. Subsequently, the opening of South Atlantic hindered interchanges between these two areas, but did not end them; exchanges might have taken place then via trans-Atlantic filter, or sweepstakes, routes. For the Cretaceous/Eocene time interval, crossings of the Tethys occurred intermittently, the major event being the interchange that took place between North and South America by latest Cretaceous. By the Cretaceous/Palaeocene transition, a terrestrial route probably linked the still southern Indian Plate to Laurasia. It is uncertain whether this route reached Madagascar. After the Eocene/Oligocene transition, exchanges between Gondwanan continents vanished, whereas exchanges between Gondwanan areas and Laurasia sometimes occurred.

Introduction

Since the beginning of the Mesozoic, land vertebrates have represented an important part of terrestrial biotas. Land vertebrates (including freshwater forms) provide a good test of terrestrial communications. Freshwater animals are even more significant than strictly terrestrial forms; they were probably unable to make use of narrow land bridges (the biotas of which were probably inconsistent with environments needed by these forms) and, obviously, could not island-hop from one land area to another.

One of the main difficulties that faces palaeontologists is represented by the 'absences'. Is absence of particular fossils from a region an accident of fossilization, or does it mean that the taxon concerned never existed here. This is a question that is often put to palaeontologists. Moreover, reports of taxa are sometimes doubtful. When not reappraised, some of them have to be disregarded, which may lead to an open question, the solution of which depends on the usual 'possible re-evaluation to come' (if the fossils are still available). Literature on which palaeobiogeographical studies rely is overstocked with such examples (see below, *Lystrosaurus* in Indo-China, *Laplatasaurus* in India and Madagascar). Studies often come up against these two major problems. In any case, palaeontologists, especially experts in land vertebrates (who quite often work on limited localities), cannot provide original maps. They are able to propose connections, or barriers, between areas, but they cannot reconstruct outlines of these areas and connections, more especially because the location of communications is not always known precisely (e.g. Africa/Eurasia relations during the Cretaceous/Eocene times).

Triassic times and the emergence of faunal provinces

It is nearly always agreed that a homogeneous terrestrial fauna, probably partly developed in earlier times (late Permian), was present on the Triassic Pangaea and that, only later, a distinction appeared between Gondwanan and Laurasian faunas.

For instance, Cox (1973, 1974) and Kalandadze (1974), after comparison of faunas, based on the use of mathematical indices, concluded that the Triassic land masses were not subdivided into faunal provinces. In fact, if several Triassic groups were actually more or less cosmopolitan, the view that Pangaea was a wide continental area, on which a world-wide land fauna radiated without important barriers, is probably oversimplified.

Pangaean faunal evidence

During the early Triassic, two terrestrial families of labyrinthodont amphibians were widely dis-

From AUDLEY-CHARLES, M. G. & HALLAM, A. (eds) *Gondwana and Tethys* Geological Society Special Publication No. 37, pp. 255–73.

persed in Pangaea: the Lydekkerinidae and Brachyopidae. As early as the Induan (early Triassic) the Lydekkerinidae, which are restricted to the early Triassic, were present in South Africa, Antarctica, India, Australia, Greenland, and the European USSR (Cosgriff 1984), and perhaps Madagascar (Swinton 1956; Carroll and Winer 1977). The Brachyopidae, known since the latest Permian in Australia, were widely dispersed during the early Triassic (South Africa, Antarctica, Australia, and Tasmania, European USSR, Spitsbergen, and Greenland); they reached North America during the middle Triassic. Moreover, the terrestrial Indobrachyopidae probably represent the sister group of the (probably) shore-dwelling Rhytidosteidae (Cosgriff and Zawiskie 1979). The Indobrachyopidae/Rhytidosteidae assemblage spread widely over the southern continents during the earliest Triassic (Induan): the Indobrachyopidae were present in Australia, Tasmania, India, and Madagascar, whereas the Rhytidosteidae occurred in Australia, Tasmania, South Africa, and Antarctica; the latter family reached Spitsbergen somewhat later (Olenekian) during the early Triassic (Cosgriff 1984; Cosgriff and Hammer 1984). Another labyrinthodont family, the Capitosauridae, shows an almost world-wide distribution, but its members might have been more or less adapted to an aquatic existence. Thus, palaeobiogeographical inferences drawn from their range may be somewhat less reliable than those provided by true terrestrial forms. Capitosaurids are restricted to the Triassic, and as early as the Scythian (early Triassic) they were known in Southern and East Africa, Antarctica, Australia, Tasmania, European and most western Asian USSR, Greenland, Spitsbergen, and perhaps India (Howie 1970; Lozovskiy and Shishkin 1974; Cosgriff 1984); they spread to North America during the middle Triassic and to South America during the late Triassic (Bonaparte 1978). The example of the Trematosauridae has been often reported. This group, known only from the early Triassic, occurred in Southern Africa, Australia, Madagascar, India, Pakistan, European USSR, Greenland, and Spitsbergen (Carroll and Winer 1977; Cosgriff 1984). But the members of this family that linked Gondwanan and Laurasian areas were aquatic, perhaps marine (Lehman 1979); however, they were probably only littoral forms unable to cross oceans (Cosgriff 1984). Besides amphibians, some reptiles display an almost Pangaean range. The dicynodont, perhaps semi-aquatic, genus *Lystrosaurus* is known in the Lower Triassic beds of South Africa, Antarctica, India, China, European USSR, and perhaps Laos (see below for a discussion about the problem of China) and Colbert (1984*a*) reports that it might be present in South America. Other dicynodonts, the Kannemeyeriidae, were known only in South America, South Africa, and apparently Australia (Thulborn 1983*a*, 1983*b*) during the early Triassic, but they increased their range during the middle Triassic (south-western and East Africa, South America, India, China, and the European USSR) and late Triassic (Morocco, South America, North America). The traversodontid cynodonts, the geographical range of which was also restricted during the early Triassic (South America), dispersed in Pangaea during the middle (into South and East Africa, South America, and perhaps the European USSR) and late Triassic (India, South America, Canada), (Battail 1977, 1984; Bonaparte 1978, 1982; Dutuit 1980; Hopson 1984). Dispersal in both Gondwanan and Laurasian areas was still possible by the end of the Triassic, as exemplified by the prosauropod dinosaurs; this group abruptly spread over Pangaea during the late Triassic: into North and South America, Europe, southern China, Southern and East Africa, India and, according to Molnar (1982), Australia. At that time, the tritylodont cynodonts also radiated widely (South Africa, South America, China, Western Europe, North America) and even primitive mammals, the Morganucodontidae, dispersed in both Gondwana and Laurasia: Western Europe, southern China and southern Africa (Jenkins and Crompton 1979). It should be noted that, because the precise age of some bed regarded here as uppermost Triassic is questioned, some of these faunal dispersals might be of early Jurassic age (Attridge *et al.* 1985).

From these distributions, it appears that no insurmountable intra-Pangaean barriers obstructed the wandering of Triassic land vertebrates. But, in spite of these possibilities of dispersal, a rather clear provincialization became noticeable. For instance, it should be noted that the above-mentioned widely distributed amphibians were poorly represented in South and North America. On the other hand, excepting the case of the kannemeyeriids and prosauropods, the nearly world-wide reptiles did not occur in Australia, which appears as a peculiar province during the early Triassic. According to Thulborn (1984), it is difficult to determine what ecological causes can account for the peculiarities of the Triassic faunas of Australia. However, Battail (1981) notes that the predominance of amphibians, and the fact that most reptiles were semi-aquatic, suggest that the Australian Triassic environments were wet and different from those of the neighbouring continents. Besides this eastern–western 'gradation', a northern–southern one is also discernible. Although differences are not clear, it appears that labyrinthodont amphibians dominated the faunas of northern continents, whereas therapsid reptiles ruled over the faunas of the southern continents (Colbert 1984*a*), except Australia. We are here reaching the problem of the emergence of a possible Gondwanan fauna, and it should also be noted that freshwater lungfishes from the late Triassic of India and Madagascar are different from the Asian forms, whereas freshwater lungfishes from North Africa and Texas were still very closely related at that time (Martin 1981*a*, 1983; Bonis *et al.* 1985). Connections between north-western Africa and south-eastern North America are also illustrated by the freshwater redfieldiid fishes (Martin 1980), and apparently by a dicynodont reptile (Dutuit 1980). This is consonant with the opening of the Tethys Ocean in eastern Pangaea, and the retention of a western connection between Gondwana and Laurasia. Judging by marine (but littoral-dwelling) reptiles, the western bottom of the Tethyan gulf reached, as an epicratonic sea, the German realm during the middle Triassic (Mazin and Martin 1984).

Triassic Gondwana and faunal provinces of the southern continents

Although more or less characteristic Gondwanan and Laurasian faunas began to emerge during the Triassic, it is not possible to distinguish a Gondwanan fauna from a Laurasian one. Moreover, there is no group restricted to Gondwana and distributed on every Gondwanan continent. It is only possible to list, in an untidy way, affinities between faunas of two or three continents.

Strong affinities have been recognized between the earliest Triassic of South Africa and Antarctica; in particular, six reptile species are common to these two continents: *Procolophodon trigoniceps*, *Thrinaxodon liorhinus*, *Ericiolacerta parva*, *Lystrosaurus murrayi*, *L. curvatus*, and *L. maccaigi*. The cynodont *Cynognathus*, from the early Triassic, occurs in South Africa and South America. The Galesauridae were present during the earliest Triassic in South Africa and Antarctica; they maintained a niche in South Africa during the late early Triassic and appeared in South America at that time (Battail 1984). The Stahleckeriidae were present in the middle Triassic of East Africa, India, and South America; the species *Rechnisaurus cristarhynchus* is common to East Africa and India. The traversondontid *Exaeretodon* from the late-middle and early-late Triassic of South America has been found in the late Triassic of India (Chatterjee 1982). The small labyrinthodont family Chigutisauridae is known in the early-late Triassic of South America and the early Triassic of Australia (Bonaparte 1978; Warren 1984). The freshwater lungfishes of the *Ptychoceratodus hislopianus* group occur only in the late Triassic of India and Madagascar (Martin 1983). Such a list could be lengthened, but this would be tedious and useless.

Therefore, there is no global Gondwanan fauna and, even within Gondwana, faunal provinces may be distinguished from one another. Besides Australia, which was somewhat unusual during the early Triassic (see above), two different faunal provinces may be recognized within Gondwana during the late Triassic.

In the northern part of Gondwana, a rather well-demarcated province, that overlapped Laurasia, may be defined by late Triassic times. Its fauna was different from that of southern Gondwana areas; Beltan and Dutuit (1977) noted that this fauna displays a marked 'northern Gondwanan' pattern. This province is characterized by the frequent association of metoposaur amphibians and phytosaur reptiles that are lacking in South America and in southern Africa. It stretched across North Africa, from western North America to Western Europe (where it encroached slightly upon the northern border of the bottom of the Tethys embayment) and, on the southern border of the Tethys Ocean, it reached India and Madagascar. Phytosaurs, without metoposaurs, are known in southern China and Thailand; this might represent an extension of the province on the northern side of the Tethys, where late Triassic beds are rare (Fig. 1). This province was appropriately called 'peritethysian' or 'circumtethysian' by Martin (1981*b*) and named 'Metoposaur/Phytosaur Empire' by Tucker and Benton (1982). The fact that India belonged to this province, which spanned southern Laurasia and northern Gondwana, could account for the presence of Laurasian forms in the Triassic of India. The presence of such fossils in India is awkward to accommodate in the hypothesis of a simple Gondwanan/Laurasian faunal opposition, and Chatterjee (1985) did not hesitate to connect India directly to Asia in his reconstruction of Pangaea, in order to explain this distribution. But, as seen above, Gondwana and

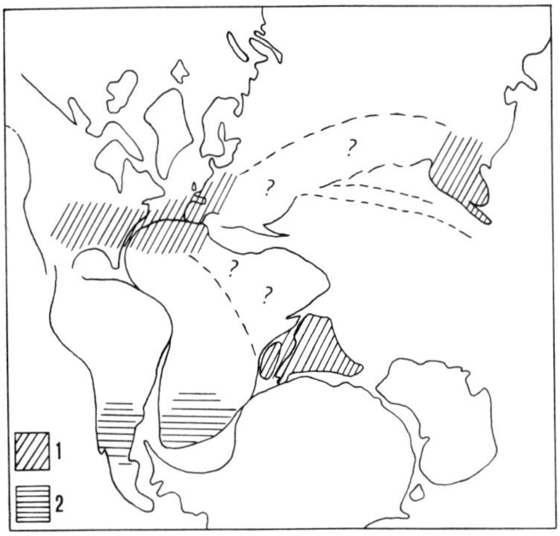

FIG. 1. Late Triassic provinciality. *1:* Peri-Tethyan province (phytosaur/metoposaur empire). *2:* southern Gondwanan province. (Outlines from Şengör 1985 and partly from Buffetaut and Martin 1984; simplified.)

Laurasia faunas were not completely opposite.

The presence of phytosaurs on both sides of the Tethys Ocean may theoretically be the result of either a vicariance process (the split of a previously narrower area by the opening of the Tethys) or a dispersal on one (or the other?) side of the Ocean. But the vicariance hypothesis may be rejected because it implies a very late and rapid opening of the Tethys. Dispersal in marine waters, along shorelines, has been suggested by Dutuit (1983), but Buffetaut and Martin (1984) note that metoposaurs were probably not good swimmers; moreover the capabilities of both groups to disperse on land is demonstrated by their presence in western North America. In spite of this aptitude, metoposaurs and phytosaurs are lacking in southern Gondwana. Late Triassic faunas of South America and southern Africa were dominated by therapsid and non-phytosaur reptiles (late Triassic faunas are unknown in Antarctica and Australia).

The causes of this regionalization remain unknown, but it is difficult to evade palaeoecological questions. Buffetaut and Martin (1984) rule out climatic causes; they suggested that most of these southern Gondwanan faunas could have inhabited upland zones lacking lacustrine and fluviatile biotas that correspond with phytosaur and metoposaur environments.

The analysis of the regionalization problem in Gondwana suffers from the lack of late Triassic faunas between the Peri-Tethyan province and the southern Gondwanan one. For the time being, no convincing explanation can be put forward.

The problem of China

The early Triassic dicynodont *Lystrosaurus* displays a broad, but disjunct, geographical range, which comprises a Gondwanan area (southern Africa, Antarctica, India) on one hand, and a Laurasian one (European USSR, Xinjiang, Shanxi plus a doubtful report from Laos) on the other hand. This distribution has puzzled Colbert (1974, 1982) and Battail (1981) who propose several hypotheses that may account for it. Colbert (1982) is reluctant to admit a dispersal route that would have passed round the western end of the Tethys. He notes that such a route would have been too long and inconsistent with the close relationships shared by Chinese and southern Gondwanan forms. According to Colbert, the presence of *Lystrosaurus* in China is only the result of the Gondwanan origin of China. Battail (1981) is not convinced by this opinion; he proposes two main possibilities: (1) dispersal either from a northern centre of origin (European USSR) or from a southern one; or (2) crossing of the Tethys. The role of the Gondwanan origin of China should apparently be ruled out. During the Permian, the Tarim block (location of the Xinjiang (Sinkiang) *Lystrosaurus*) was already accreted to the northern continents (McElhinny *et al.* 1981) and, according to Lin *et al.* (1985) the North China block (on which was located the Shanxi (Shansi) *Lystrosaurus*) was connected to the Siberian block as early as the late Permian. Both blocks were far from Gondwana by the Permo-Triassic transition; crossing of the Tethys was perhaps possible. The south-eastern Asia and the South China block, plus probable terranes (Hallam 1986), drifted between Gondwana and Laurasia, and made up insular stages. On the other hand, a western dispersal route cannot be rejected. It was long but, as seen above, other tetrapods dispersed very widely during the Triassic. Moreover, earliest Triassic deposits are very rare along this terrestrial route between the southern and northern *Lystrosaurus* ranges. Therefore, the lack of evidence along this potential dispersal way is not surprising, and the northern distribution of *Lystrosaurus* does not require the presence of Chinese blocks along northern Gondwana during the early Triassic. Furthermore, the distribution of land vertebrates in East Asia during the Triassic does not require the alteration of the Earth's diameter, as sometimes suggested (but not published for Triassic times) by palaeontologists (a small diameter, involving a very narrow eastern Tethys, or even its closure, would account for the crossing of this ocean between

Gondwana and East Asia). For instance, Owen's map (1983) for the Triassic with a reduced Earth diameter, shows East Asia very close to Australia, which would account for an easy dispersal of *Lystrosaurus*. However, such a proximity of East Asia and Australia would have caused similarities between the faunas of these two areas, but such affinities do not appear.

Incidentally, a particular point (particular, but of general interest) may be tackled here. Can terranes have played a role, as 'Noah's arks', in dispersal? This idea deserves consideration, especially where colonization of eastern Asia by forms of southern origin is concerned. Terranes might have made up—along with cratonic blocks—a series of islands across eastern Tethys (see above), thus perhaps allowing terrestrial faunas to cross this ocean. But could terranes convey living forms that were able to colonize a broader region after landing? This seems quite unlikely. Biotas of these small areas were probably quite unbalanced, and terranes were probably unable to harbour living continental faunas. Displaced fossil invertebrate marine faunas are well known (Hallam 1986) but, to my knowledge, only one case of displaced fossil vertebrates, euryhaline fishes, is known (Sosson and Martin 1985). Apparently, the role of terranes in colonization of continents may be *a priori* rejected.

The problem of south-eastern Asia

The idea that south-eastern Asia was formerly a part of Gondwana (Ridd 1971, 1980; Tarling 1972) seems now quite widely accepted (but see Helmcke 1985). No fossil vertebrates corroborate this hypothesis but none of them is inconsistent with it. On the other hand, continental vertebrates from the late Triassic of Thailand (more precisely from the Indo-China block) provide information about the date of collision with Asia, if any. A freshwater lungfish is very similar to *Ptychoceratodus szechuanensis* from the late Triassic of Sichuan (Szechwan), and it demonstrates that south-eastern Asia (at least the Indo-China block) and the South China block were connected at that time. Moreover, a labyrinthodont, *Cyclotosaurus* cf. *posthumus*, is a member of an almost entirely Laurasian group that might have, however, slightly overlapped most of northern Gondwana (Morocco; Dutuit 1976). The Thai *Cyclotosaurus* is closely related to a German species (Ingavat and Janvier 1981). Phytosaur reptiles comprise *Mystriosuchus*, also known in Western Europe (Buffetaut 1985*a*,*b*) and a form very similar to the North American

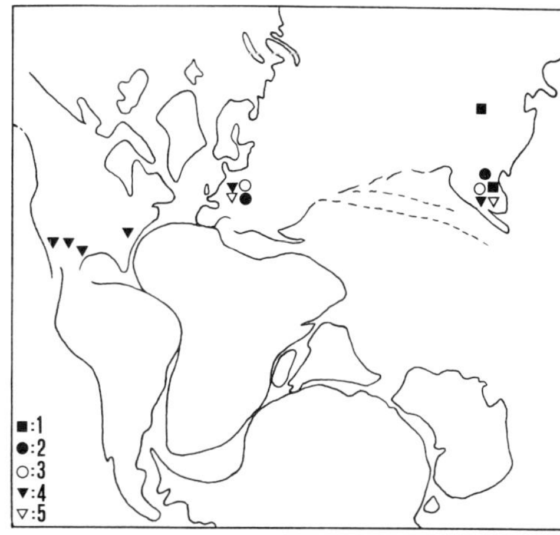

FIG. 2. Laurasian affinities of the Indo-China block by late Triassic; *1: 'Ptychoceratodus' szechuanensis* (freshwater lungfish), *2: Cyclotosaurus posthumus* group (amphibian), *3: Proganochelys* (turtle), *4: Belodon/Rutiodon* group (phytosaur reptiles), *5: Mystriosuchus* (phytosaur reptile). (Outlines from Şengör 1985; simplified.)

Rutiodon and European *Belodon* (Buffetaut and Ingavat 1982). Some remains of turtles might also be reminiscent of Laurasian forms (Broin *et al.* 1982). Therefore, the late Triassic fauna of south-eastern Asia displays a Laurasian pattern (Fig. 2) and this presumed fragment of Gondwana was accreted, along with the intervening Chinese blocks, to mainland Asia—at the latest during the (probably early) late Triassic (Buffetaut 1985*a*,*b*; Buffetaut and Martin 1985). Such a date is the latest one possible, but an earlier collision is quite possible.

Jurassic

The record of terrestrial vertebrates is very scarce for the Jurassic. According to Olsen and Galton (1984), the Triassic provinciality turned to homogeneity in the early Jurassic. It is true that, in spite of the western advance of the Tethys, similarities between Gondwanan and Laurasian faunas still existed during the Jurassic. Apart from the possible early Jurassic interchanges (see above for the problem of the age of some supposed Upper Triassic beds), affinities between Laurasian and Gondwanan forms still occurred during middle and late Jurassic times. The sauropod dinosaur *Cetiosaurus*, reported from the middle Jurassic to early Cretaceous in Western Europe, is also known in the middle Jurassic of North Africa (Monbaron and Taquet 1981). The Middle Jurassic Xaishaximiao Formation of Sichuan has yielded *Shunosaurus*, a sauropod dinosaur that could be similar to the approximately contemporaneous *Rhoetosaurus* from Australia (Molnar 1984), and to the early, or middle, Jurassic *Barapasaurus* from India (Dong *et al.* 1983; Dong 1985*a*). Carnosaurs and sauropods from the middle/late Jurassic of Argentina, are more or less related to middle and late Jurassic forms from Laurasia and perhaps Africa (Bonaparte 1979). Rich dinosaur faunas have been yielded by the late Kimmeridgian Morrison Formation (western United States) and Tendaguru Beds (Tanzania). Charig (1973) points out the affinities shared by these two faunas. According to Galton (1977*a*, 1977*b*), five dinosaur genera are reported to be common to the Morrison and Tendaguru localities. However, he clearly indicates that, owing to the preservation of the material, the presence of only one genus (the ornithopod *Dryosaurus*) on both continents is established. However, even if not congeneric, the four other pairs of species correspond with related forms. Two stegosaur genera, *Chialingosaurus* and *Tuojiangosaurus*, from the late Jurassic Shangshaximiao Formation in Sichuan, might be related to *Kentrosaurus* from the Tendaguru Beds (Dong *et al.* 1983; Dong 1985*a*). *Barapasaurus* from the early, or middle, Jurassic of India could be reminiscent of *Apatosaurus* from North American late Jurassic. Colbert (1984*b*) infers from this Indian occurrence that India was not isolated during early (or middle) Jurassic; although *Barapasaurus* was lacking in Africa, he favours a connection with this latter continent, which is not surprising. Mammals could also give evidence of land connections between Gondwana and Laurasia at that time; the peramurid pantotheres, known in the Portlandian (and perhaps late Bathonian) of England, might be also present in the Tendaguru Beds (Kraus 1979).

About these late Jurassic interchanges, Galton (1977*a*,*b*) considers two possible land routes. An Africa–Europe–North America route would be supported by the presence, in the upper Kimmeridgian of Portugal, of *Brachiosaurus* (also known in North America and perhaps Africa) and *Apatosaurus* (also known in North America). But Galton (1980) reports that the generic allocation of these Portuguese dinosaurs is probably inaccurate. He is inclined to think that these dinosaurs made use of an Africa–South, Central and North America route that was apparently intermittently continuous (Hallam 1981). But the reconstruction of the western Tethyan area proposed by Enay (1972) offers a more direct route (Africa–North America) that should not be neglected. The early Jurassic of Australia and the middle Jurassic of China (Junggar Basin and Sichuan) have yielded the last labyrinthodont amphibians (Warren and Hutchinson 1983; Dong 1985*b*). These fossils have been referred to the Brachyopidae but, in spite of these close relationships, they probably represent relicts, independently descended from the Triassic inhabitants of Australia and China, and they are not indicative of geographical relations between these Gondwanan and Laurasian areas during the Jurassic. Besides, Warren recently (1984) assigned the Australian fossil to the Chigutisauridae, a family closely related to the Brachyopidae.

Owing to the scarcity of Jurassic data, it is practically fruitless to investigate interrelationships within Gondwana. Schaeffer (1972) refers a fish discovered in the lacustrine early Jurassic of Antarctica to the freshwater family Archaemaenidae, elsewhere known only in continental beds of the Australian early Jurassic. Grande and Eastman (1986) corroborate this allocation. However, Molnar (1980) notes that Australian faunas from the Jurassic display no special similarity to other Gondwanan ones. It may be noted that, apart from the Archaemaenidae, none of the known Gondwanan continental forms of the Jurassic is the counterpart of Jurassic fossils from another Gondwanan continent.

Finally, from the poor Jurassic record, we can only infer that interchanges between Gondwana and Laurasia were still possible.

Cretaceous–Eocene: a series of events

The Cretaceous–Eocene period corresponds with a critical phase for terrestrial vertebrates. A series of important palaeobiogeographical events had strong repercussions on the history of these animals.

Apparently, Gondwana and Laurasia were effectively separated by the Tethys barrier, although some isolated interchanges occurred (one of them, of short duration, allowed important exchanges between South and North America; see below). A Gondwanan pattern, partly based on negative evidence, can be recognized.

The Gondwanan fauna

The distribution of most land vertebrates from the early Cretaceous is probably the result of a vicariant process: the geographical ranges were derived from the late Jurassic distribution (Cox 1980). With regard to Gondwana, it is only possible to report that, during the early Cretaceous, the crocodilian genera *Sarcosuchus* and *Araripesuchus*, as well as the freshwater actinistian fish *Mawsonia*, were present in both Africa and South America. The freshwater lungfish *Neoceratodus africanus*, present in the whole African Cretaceous, has been recovered in South American beds that might be of Cenomanian age or slightly older. Moreover, the turtle *Araripemys*, known in the Aptian of Brazil, may also be present in contemporaneous beds of Africa (Broin, 1980).

During the late Cretaceous, a more widespread Gondwanan pattern appeared. The titanosaurid dinosaurs, previously known in the late Jurassic and early Cretaceous of Africa, were present in the late Cretaceous of South America, Africa, India, and Madagascar; they reached a Laurasian continent, i.e. North America, only by latest Cretaceous (Bonaparte 1984*a*). Madtsoiine boid snakes are encountered in Senonian beds of South America (Albino 1986), Africa, and Madagascar (Rage 1984*a*); the trematochampsid crocodiles display a similar stratigraphic and geographical distribution (Buffetaut 1985*c*). Such a geographical range may be the result of vicariance events and/or dispersals (see below). The freshwater lungfishes of the *Ptychoceratodus madagascariensis* monophyletic group occurred in the late Cretaceous of South America and Madagascar, and in the early and late Cretaceous of Australia, where the group persisted in undated levels of the Cainozoic (Martin 1982, 1983); however, it is not absolutely impossible that this range corresponds with a vicariance process, the initial dichotomy having perhaps taken place in the late Jurassic (Martin pers. comm.). The dinosaur *Laplatasaurus* was defined in the late Cretaceous of Argentina; the genus has been reported from coeval beds of India and Madagascar, but these reports are based on rather poor material, and the presence of *Laplatasaurus* in India and Madagascar may be questioned (Fig. 3). According to Molnar (1980), Cretaceous faunas from Australia did not show close relationships to Gondwanan (or Laurasian) ones. It should be noted here that the discovery of probable Eocene polydolopid marsupials in Antarctica (Woodburne and Zinsmeister 1984) is the only tangible proof of the role of Antarctica in southern dispersal of marsupials.

On the other hand, the Laurasian fauna is characteristic. Bearing in mind that prudence is required when taking absences into account, the lack of several Laurasian forms in Gondwanan continents deserves attention. Tyrannosaurid

FIG. 3. Late Cretaceous Gondwanan land vertebrates; *1:* pipid frogs, *2:* madtsoiine snakes, *3:* trematochampsid crocodiles, *4: Laplatasaurus* (dinosaur), *5:* late Cretaceous *'Ptychoceratodus' madagascariensis* group (freshwater lungfishes), *6:* early Cretaceous freshwater lungfishes closely related to *'P.' madagascariensis*.

and ornithomimid dinosaurs, restricted to the Cretaceous of Laurasia, never entered a Gondwanan area. As for hadrosaurid and ceratopsian dinosaurs, that were widely distributed in Laurasia and restricted to the Cretaceous—they only succeeded in reaching South America by the latest Cretaceous (Bonaparte 1984*a*).

After the Cretaceous, the distinction between Gondwanan and Laurasian faunas became blurred. This is the result of the breakup of Gondwana and, mainly, of the scarcity of Palaeogene faunas from Gondwanan territories.

Trans-Tethyan routes

Although the Gondwanan fauna was rather different from that of Laurasia, some interchanges took place locally across the Tethys. Some exchanges might have taken place between Eurasia and Africa during the early Cretaceous. The ornithopod dinosaur *Valdosaurus*, previously reported from the early Cretaceous of Europe, has been discovered in the Aptian of Niger (Galton and Taquet 1982). However, these authors note that *Valdosaurus* is closely related to the late Jurassic *Dryosaurus* from North America and Africa and they do not rule out parallel evolution: the African *Valdosaurus* species could have arisen from the African *Dryosaurus*, whereas the European *Valdosaurus* could be a descendant of the Laurasian *Dryosaurus*. In this case, the genus *Valdosaurus* would be polyphyletic, and its distribution would not require a trans-Tethyan land route in the early Cretaceous. The pachycephalosaurid dinosaurs are present in the early Cretaceous of England, and in the late Cretaceous of North America and Asia, but a representative of this family has been reported from the late Cretaceous of Madagascar by Sues and Taquet (1979). They do not rule out the role of vicariance, but Galton and Taquet (1982) believe that the presence of the family in Madagascar results from a dispersal from Europe during the early Cretaceous. If the form from Madagascar actually belongs to the Pachycephalosauridae (as in other reptiles, pachyostosis, the major characteristic of the family, might be liable to parallel evolution), then a trans-Tethyan land route was probably present in the Eastern Hemisphere sometime in the Cretaceous, but not necessarily during the early Cretaceous (see later, the problem of Madagascar). In conclusion, these early Cretaceous interchanges are not yet really ascertained. By the late Cretaceous, important interchanges took place between North and South America (Rage 1978, 1981). Dinosaurs (hadrosaurids, ceratopsians) and perhaps lizards (teiids), and mammals (condylarths), were involved in a southward dispersal, whereas other dinosaurs (titanosaurids), snakes (aniliids, boines, tropidopheines), and perhaps iguanid lizards and bothremydid turtles passed from South America to North America (Bonaparte 1984*a*, 1984*b*, 1984*c*). The distribution of avisaurs, possible dinosaurs (Brett-Surman and Paul 1985), and didelphid mammals (Sigé 1972; Crochet 1979) is also consistent with this land connection, but the direction of their dispersal remains unknown. Among those vertebrates that support the presence of a land route between North and South America, the Hadrosauridae rank among the more demonstrative, but Milner and Norman (1984) propose a vicariance hypothesis that would annul the palaeogeographical significance of the group for late Cretaceous times. However, this vicariance hypothesis implies the presence of hadrosaurids in East Asia during the middle and late Jurassic, and in the whole of Laurasia plus a part of Gondwana during the early Cretaceous. However, hadrosaurids are known only from the late Cretaceous, and this vicariance hypothesis appears questionable. Anyhow, whatever the history of hadrosaurids may be, the presence of a land route between North and South America by late Cretaceous—a hypothesis rejected by most palaeontologists for a long time—seems to be accepted now. These interchanges apparently occurred during Campanian–Maastrichtian times (Rage 1981; Bonaparte 1984*c*). Geodynamic data corroborate both the presence and the age of this terrestrial connection (Schmidt-Effing 1980; Seyfried and Sprechmann 1986). Estes and Baez (1985), and Rage (1986), recently reviewed this faunal interchange problem.

Gingerich (1985) believes that such a dispersal event again took place in the late Palaeocene; edentates, notoungulates and uintatheres apparently passed from South to North America. If a vicariance event is not the cause of this geographical distribution, the dispersal might have been accidental, because if a true terrestrial route was present, more land vertebrates would probably have been implied in this event.

The comparative proximity of Eurasia and Africa in palaeogeographical reconstructions leads palaeontologists to suppose that interchanges took place during the Cretaceous–Eocene period. But evidence is still poor. Palaeoryctid mammals from the Palaeocene of Morocco (Cappetta *et al.* 1978), lipotyphlan and creodont mammals from the Eocene of Algeria

(Mahboubi *et al.* 1984), and embrithopod mammals from the early Oligocene of Egypt (Sen and Heintz 1979), indicate exchanges between these two continents (Fig. 4). A further piece of evidence relevant to this problem is the presence in Africa and on the other side of the Tethys of groups that are probably related. The theridomyid rodents, present in the European

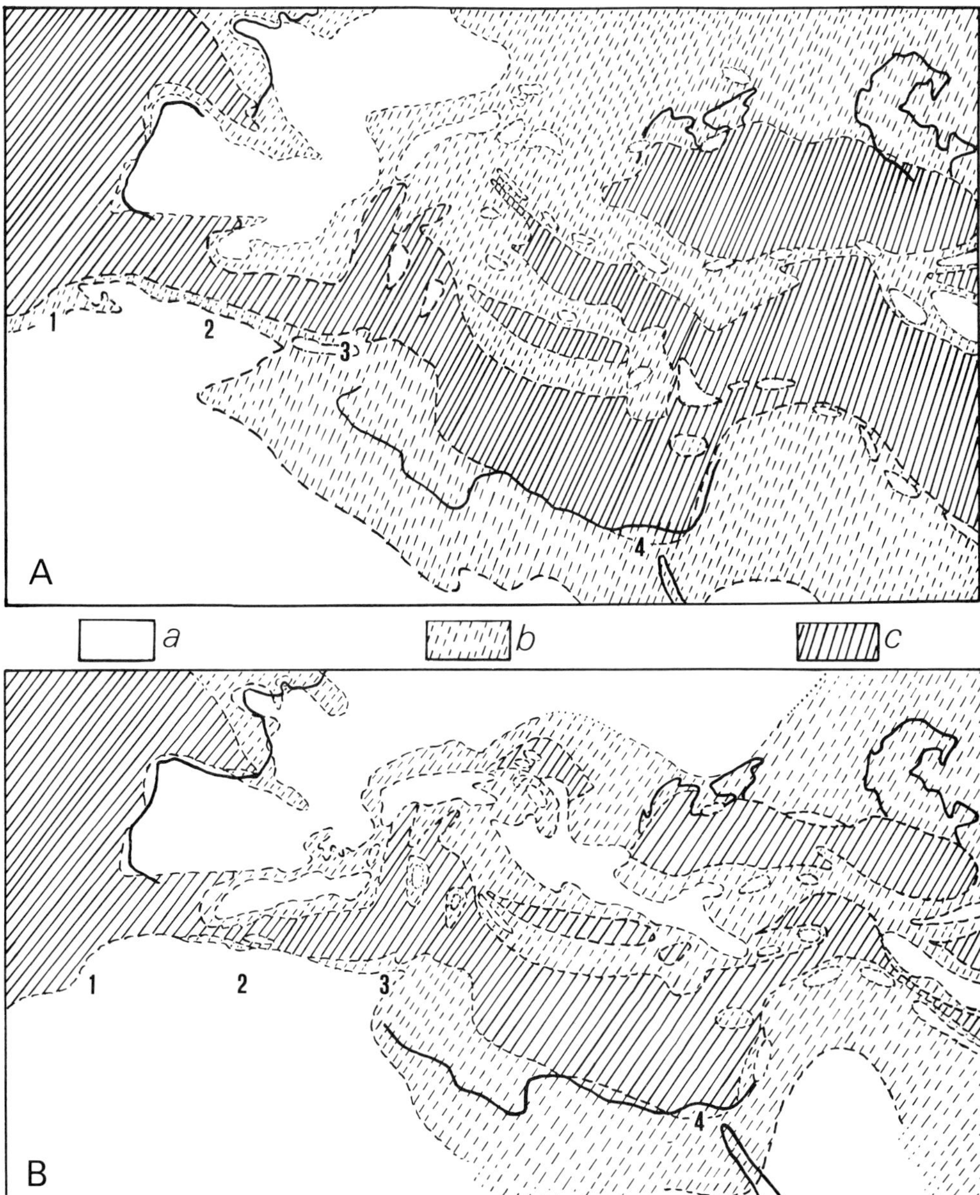

FIG. 4. Palaeogene African fossils with Laurasian affinities—Europe, Africa, and the Apulian Plate. The Apulian Plate theoretically offered possibilities of exchanges between Europe and Africa, but such interchanges are still poorly documented. *A:* Cretaceous/Palaeocene transition, *B:* Eocene/Oligocene transition. *a:* emerged land, *b:* thick continental crust, *c:* oceanic crust and thin continental crust. *1:* Palaeocene, Palaeoryctid mammals, *2:* early Eocene: ziphodont crocodile, lipotyphlan and creodont mammals, *3:* early Eocene, anomalurid rodents, macroscelidea, *4:* early Oligocene, embrithopod mammals. (Outlines after Dercourt *et al.* 1985; very simplified.)

Eocene, probably represent the sister group of the anomalurid rodents known in the African Eocene (Hartenberger *et al.* 1985); the Macroscelidea, apparently related to the Laurasian condylarths, have been discovered in the African Eocene (Hartenberger 1986). Occasional trans-Tethyan connections would have provided routes for these exchanges between Africa and Eurasia. Ranid amphibians (Rage 1984*b*) and cordyloid lizards (Estes 1982) might also support this. Finally, these Cretaceous–Eocene interchanges were rather rare, which is surprising, considering the narrowness of the Tethys between Africa and Eurasia. This is even more astonishing because the Apulian Plate provided a series of emergent lands between these two continents at that time (Dercourt *et al.* 1985). Therefore, the role of a climatic filter (Hartenberger 1986) deserves consideration. On the other hand, if a direct communication linked Europe and Africa, several European groups would have entered Africa, and African forms would have reached Europe. Apparently, such interchanges did not occur, and it is quite possible that dispersals took place between Africa and Asia where Palaeogene faunas remain poorly known.

On the other hand, the European Eocene has yielded terrestrial vertebrates that exhibit clear South American affinities: ceratophryine anurans (Rage 1981) also known in the Cainozoic, Pleistocene, and Holocene of South America, didelphid mammals known from the late Cretaceous to Holocene of this continent (Crochet 1979), and myrmecophagid mammals (Storch 1981) are also present in the South American Neogene, Pleistocene, and Holocene. Iguanid lizards (Estes 1983), aniliid, tropidopheine, and boine snakes (Rage 1978), the distribution of which will be discussed later (see the problem of Madagascar), are also indicative of South American relationships. Phorusrhacid birds, known in the Cainozoic of South America and the Pleistocene of South and North America, have been recovered in the early Oligocene, and perhaps late Eocene, of Western Europe (Mourer-Chauviré 1981*a*). Ziphodont mesosuchian crocodiles are present in the European and Algerian Eocene (Buffetaut 1982) and are also known in the late Cretaceous and Cainozoic of South America; according to Buffetaut (1986), the European and Algerian forms would be reminiscent of the late Cretaceous South American fossils but different from the post-Cretaceous ones of this continent. Two hypotheses that may account for this disjunct distribution have been put forward: (1) a South America–North America–Europe way (Rage 1978, 1981; Buffetaut and Rage 1981); or (2) a South America–Africa–Europe route (Mourer-Chauviré 1981*b*). According to Rage (1981), these groups radiated in South America, and reached North America by using the late Cretaceous land connection; later, from North America, they spread to Europe. The main drawback of this hypothesis is that the Upper Cretaceous and Lower Cainozoic beds of North America lack some of these groups (Ceratophryines, ziphodont crocodiles, phorusrhacids, myrmecophagids). In consequence, Mourer-Chauviré (1981*b*) favours a dispersal by way of Africa. However, in Africa only ziphodont crocodiles represent possible evidence of such a dispersal, but Buffetaut (1985*c*) does not rule out parallel evolution, which casts doubt on the palaeogeographical significance of the group. Therefore, as does North America, Africa lacks ceratophryines, phorusrhacids and myrmecophagids. The African route hypothesis rests on the fact that African localities are less studied than the North American ones and, consequently, that these absences in Africa are less inconsistent with the proposed explanation. On the other hand, it is really surprising that several groups, which dispersed from South America, used two different ways to reach Europe, unless this latter continent was especially attractive to these vertebrates! For the moment, this problem cannot be settled. Only one route, South America–North America–Europe is suggested by fossil distribution; the other possible route, i.e. South America–Africa–Europe, is theoretically possible, and cannot be rejected in spite of the lack of significant African fossils.

In conclusion, trans-Tethyan interchanges are demonstrated between South and North America during the latest Cretaceous. Some interchanges also took place between Africa and Eurasia during the Cretaceous/Eocene period, but tangible evidence remains poor. The scarcity of these exchanges is consistent with the endemic pattern displayed by the African faunas at that time.

Separation and subsequent relations between South America and Africa

Several genera from the early Cretaceous are common to both South America and Africa. The freshwater coelacanthid *Mawsonia*, present in the Aptian of Africa, has been recovered in slightly older beds of Brazil (Campos and Wenz 1982); the giant crocodile *Sarcosuchus* displays similar geological and stratigraphic ranges (Buffetaut and Taquet 1977). Another crocodile, *Araripesuchus*, is known in the Aptian of both continents (Buffetaut and Taquet 1979), whereas the turtle *Araripemys* that is found in the Aptian of Brazil might be also present in the Aptian of Niger (Broin 1980). *Neoceratodus africanus*, an early and late Cretaceous freshwater lungfish widespread in Africa, has been reported from the Cenomanian (or slightly older) beds of Brazil (Martin 1984). This is consistent with the chronology of South Atlantic opening, based on other data. The communication between central and south parts of the proto-Atlantic was established by the middle Albian, according to Moullade and Guérin (1982), or late Albian according to Förster and Scholz (1979). An Albian age is also proposed by Maillot (1983). Reyment (1980) suggested a slightly older age—middle Aptian—for the connection between Central and South Atlantic. This event marked the end of the South American/African community, but it did not abruptly terminate exchanges between the two continents; the presence of the incipient Atlantic probably only hindered interchanges that apparently persisted into at least the late Cretaceous. Apart from the proximity between Africa and South America—that might explain some interchanges, the Walvis and Rio Grande Rises could have acted as a filter bridge during the whole late Cretaceous and perhaps during the early Palaeogene (Rage 1981). Melguen *et al.* (1978) suggest that a part of the Rio Grande/Walvis Rises emerged during the Santonian/Campanian; this has been corroborated by mineralogical and geochemical data (Robert 1981; Maillot 1983). Reyment (1980) suggests that crossing of the Atlantic was allowed by these emerged areas. Pipid frogs, madtsoiine snakes, trematochampsid crocodiles, and didelphid mammals, could have made use of this passage. Trematochampsid crocodiles are known in the late Cretaceous of South America and Africa, but their precise age in South America remains unknown (Buffetaut 1985*c*); their distribution may be the result of a vicariance event or a dispersal by way of the Walvis/Rio Grande emerged areas.

Marsupials have been recently discovered in the Eocene of Africa (Mahboubi *et al* 1983; Crochet 1986). According to Crochet (1984), these African forms are closely related to South American ones. The presence of Eocene marsupials in Africa may be the result of either a vicariance event (Jaeger and Martin 1984) or an eastern extension of the primitive area from South America to Africa across the incipient Atlantic. Apparently, the second possibility is favoured by Crochet (1984). In fact, we are here coming up against the problem of the vicariance definition: did the primitive range of these marsupials split into South American and African areas, or did these marsupials extend their range into Africa while that continent parted from South America? As for pipid frogs and madtsoiine snakes—which do not occur before the early Senonian—a vicariant process seems very doubtful; dispersal after the opening of the South Atlantic appears to be more probable. Moreover, the snake genus *Madtsoia* is known in the Senonian of Africa whereas, in South America, the subfamily is represented by other genera in beds of that age (Albino 1986). *Madtsoia* appeared later in South America, where it is known in the Palaeocene of Brazil and Argentina, and early Eocene of Argentina (Rage 1984*a*). If the absence of the genus in the late Cretaceous localities of South America is not an accident of fossilization, then a trans-Atlantic dispersal from Africa to South America should have occurred during the latest Cretaceous or earliest Palaeocene. Although direct palaeontological evidence is obviously wanting, I suspect that the Walvis and Rio Grande Rises played a part in these crossings.

After the last terrestrial form, that is *Madtsoia*, which probably crossed the South Atlantic during the Palaeocene at the latest, crossing of this ocean remained theoretically possible. Emerged surfaces of the Rio Grande/Walvis Rises vanished at an unknown date, but Tarling (1981) supposes that a more northern way might have acted as a dispersal route; parts of the Céara Rise, Mid-Atlantic Ridge, and Sierra Leone Rise were subaerial during the Eocene and Oligocene, and thus provided a possible filter land bridge—or more probably a route for sweepstakes dispersal—between South America and Africa. Bonaparte (1984*c*) considers that this latter communication may account for the presence of vertebrates with South American affinities in the Eocene of Europe. In particular, he believes that the presence of ziphodont crocodiles in Europe at that time is the result of a South America/Africa/Europe dispersal by way of the Céara/Mid-Atlantic/Sierra Leone emergent areas. However, Buffetaut (1986) shows that the European ziphodont crocodiles are more closely related to the Cretaceous forms of South America than to the post-Cretaceous ones of this continent—which were already more advanced than those from the Eocene of Europe. He logically suggests that dispersal from South America to Europe, by any way, could not take place after the Cretaceous. However, if ziphodont crocodiles, as well as perhaps the other 'South American' forms from the European Eocene, did not make use of this route, the Céara–Sierra Leone route might strengthen the hypothesis of a trans-Atlantic dispersal of hystricognath rodents and simiiform monkeys by the Eocene/Oligocene transition (Hoffstetter 1972).

The problem of India

A pelobatid frog has been discovered in India, in intertrappean continental beds close to the Cretaceous/Palaeocene transition (Sahni *et al.* 1982). Today, pelobatid frogs inhabit North America, Europe, East Asia, and most of north-western Africa. Fossils are known from the late Cretaceous and Cainozoic of North America, Europe, and Asia. In Africa, they are known only from the Pleistocene of Morocco. They originated in Laurasia (Laurent 1975), and subsequently dispersed throughout this supercontinent. Excepting India, they only recently reached a Gondwanan area, that is north-western Africa, perhaps during the latest Miocene interchanges that occurred between Spain and Morocco (see below), or later.

The problem that faces us is this: India was already clearly separated from Africa, and even welded to Asia, when pelobatids entered Africa. Thus, these frogs were unable to reach India by way of Africa. Therefore, these Indian fossils came from one of the northern continents that housed this family. But, by the Cretaceous/Palaeocene transition, India occupied a southern position about 10°/30°S latitude (Patriat and Achache 1984; Besse *et al.* 1984). Moreover, salt-water forms an almost fully effective barrier to amphibian dispersal; the distance that separated Eurasia from India at that time may be considered absolutely impassable by an amphibian (even by rafting) if entirely oceanic. It has been supposed by Sahni *et al.* (1982) that the Iran and Afghan Plates might have taken a part in this dispersal, but, since that time, other data are available. Lin *et al.* (1985) propose a c.4000 km eastward shifting of the South China block; this displacement would be the result of the India/Eurasia collision during the Eocene. On the other hand, Tapponnier *et al.* (1982) suggest a somewhat similar history for the south-eastern Asia block and Sunda shelf: the India-Eurasia collision would have caused the eastward shifting of the south-eastern Asia block/Sunda shelf. Were the South China block and the south-eastern Asia block/Sunda shelf the terrestrial links between Eurasia and India? This is far from demonstrated. Courtillot and Besse (1986) believe that the eastward displacement of the South China block was only 1300 km with respect to northern Eurasia, and Tapponnier *et al.* (1982) suggest only a 800–1000 km south-eastward shifting of the south-eastern Asia block/Sunda shelf. Anyhow, the Eurasia/India collision was not simple (Allègre *et al.* 1984), and the history of the relations between these two continents is still somewhat obscure. As a palaeontologist, I can only indicate that, probably, a terrestrial route was present between India and Eurasia by the Cretaceous/Palaeocene transition.

The problem of Madagascar

The origin of the fauna of Madagascar has been debated for a long time. The proximity of Africa logically leads to the belief that this continent is the main source of the Malagasy fauna.

Some fossils are known from the late Senonian of Madagascar. The crocodile *Trematochampsa* and the snake *Madtsoia*, from the Malagasy Campanian, occurred in the Senonian of Africa (see p. 265). The formation of the Mozambique Channel is apparently older than the Senonian; therefore, this distribution is probably not the result of a vicariance process. Dispersal (quite likely from Africa to Madagascar) across the Mozambique Channel seems possible, and the Davie Ridge might have provided a more or less continuous land bridge (Taquet 1982). Dispersal across the Mozambique Channel, even by way of a land bridge, is supported by the fact that freshwater lungfishes, that easily dispersed during the Cretaceous, were not involved in the late Cretaceous interchanges between Africa and Madagascar (Martin 1981*b*).

Cainozoic faunas from Madagascar are unknown, which hinders speculations about the history of that period, and we are forced to guess from the living fauna. One fact is especially striking: the recent fauna of Madagascar displays, to some extent, a Laurasian Palaeogene pattern. The adapids, probable closest relatives of the lemuriforms, were widespread in the Laurasian Eocene. Some fossils from Africa have sometimes been doubtfully regarded as possibly related to adapids (the Oligocene *Oligopithecus*, Gingerich 1977; the Eocene *Azibius*; Sudre 1979, p. 112), but their referral remains uncertain. Iguanid lizards, today present in South America and Madagascar (as well as in the Fiji and Tonga Islands!), were probably present in the European Eocene (Estes 1983), and perhaps in the late Cretaceous and

Cainozoic of Asia (Ckhikvadze *et al.* 1983); however, these reports from Asia need a corroboration. Living boine snakes are encountered only in South America and Madagascar; they occurred in the Palaeogene of Laurasia. These three groups, adapids/lemuriforms, iguanids and boines are lacking in Africa. Assuming that these absences in Africa are not accidents of fossilization (which is not absolutely certain, especially with regard to adapids), a dispersal route coming from Laurasia, reaching Madagascar but excluding Africa, has to be found. It may be wondered whether the above-mentioned Laurasia/India route extended to Madagascar, which could have acted as a cul-de-sac. No evidence for such a way of dispersal is known, but this route is consistent with the India/Madagascar faunal relationships sometimes brought forward. The opinion I venture here is only a working hypothesis, but I believe that it deserves consideration. Such a route might have been responsible for a part of the colonization, and it does not exclude additional immigrations from Africa across the Mozambique Channel—that apparently acted as a very selective filter.

Post-Eocene events

Apart from intermittent passages between Africa to Madagascar, that probably occurred from time to time during the Cainozoic and Pleistocene, crossing of the Atlantic by hystricognath rodents and simiiform monkeys by the Eocene/Oligocene boundary—if it actually did occur—marks the last exchange between Gondwanan continents (however, South America/ Antarctica/Australia relations remain unknown). After the Eocene, Gondwanan continents have been concerned only with interchanges with Laurasia.

During the Oligocene, marsupials might have crossed the Tethys between Eurasia and Africa (Bown and Simons 1984), but this is not definitely demonstrated (Crochet 1984; Jaeger and Martin 1984). Interchanges between Eurasia and Africa began rather abruptly by early Miocene times; they corresponded with the Africa/Eurasia collision, that is, to the formation of the Mediterranean Sea. The dispersal route was in the Eastern Mediterranean area, and apparently, no trans-Mediterranean exchanges occurred before the end of the Miocene (Bonis *et al.* 1985). By latest Miocene (Messinian), a land communication, of short duration, took place on the western extremity of the Mediterranean Basin, between Morocco and Spain (Thomas *et al.* 1982).

Since the Miocene, some reptiles and mammals of Asian origin have reached Australia, perhaps by island-hopping (Rage 1987). The last important event is the 'Great Interchange', that is the emergence of the Panama area, which took place during the Pliocene and has allowed cross-exchanges between the Americas. This prominent Pliocene phenomenon was heralded by some Miocene forerunners (snakes and mammals; Hoffstetter and Rage 1977; Marshall *et al.* 1982).

Conclusions

The vertebrate evidence suggests that no continent was isolated during the Triassic. Every landmass was connected to Pangaea and was attainable by terrestrial forms. However, faunal provinciality was apparent; the Australian fauna differed from that of the other Gondwanan continents in the early Triassic and, during the late Triassic, the Peri-Tethyan province (which partly overlapped Laurasia) was different from the southern Gondwanan one. The colonization of China by *Lystrosaurus* in the early Triassic may be accounted for even if Chinese blocks were not parts of Triassic Gondwana; south-eastern Asia was connected to northern continents by early late Triassic at the latest. The distribution of Triassic land vertebrates may be explained even if the Earth's diameter did not expand. Moreover, it should be noted, incidentally, that the palaeobiogeographical history of continental vertebrates does not require the existence of a hypothetical Pacifica.

During the Jurassic, exchanges between Gondwana and Laurasia apparently remained easy. After the Jurassic, the Tethys acted as an efficient barrier, although it was crossed from time to time, owing to the establishment of land connections of short duration, or land routes that still exist (Africa/Eurasia, South America/ North America).

During the Cretaceous, one of the major events was the junction of the South and Central Atlantic, which hindered interchanges between Africa and South America but did not end them abruptly. The Rio Grande/Walvis Rises probably played a part in these exchanges up to the latest Cretaceous and perhaps Palaeocene. The Céara/Mid-Atlantic/Sierra Leone Rises might have provided a late, very selective filter route (or more probably sweepstake) between Africa and South America, until the Eocene/Oligocene transition. Exchanges also occurred between Africa and Madagascar by latest Cretaceous times, perhaps on the Davie Ridge. By the latest Cretaceous, a land route allowed important cross-exchanges between South and North America.

By the Cretaceous/Palaeocene boundary, a terrestrial communication was established between Laurasia and India. It is worth considering whether this route extended to Madagascar.

After the Eocene/Oligocene transition, exchanges between Gondwanan continents practically ended, but dispersals between Gondwanan continents and Laurasia still occurred. The most important Cainozoic Gondwana/Laurasia connections are: the establishment of the Laurasia/Africa communication in the Middle East from the early Miocene onwards, the beginning of the colonization of Australia by Asian vertebrates during the Miocene, exchanges of short duration between Spain and Morocco by the latest Miocene, and the establishment of the Panama bridge during the Pliocene.

Finally, it should be noted that a global distinction between a Gondwanan fauna on one hand, and a Laurasian fauna on the other, has never existed.

ACKNOWLEDGEMENTS: I would like to thank E. Buffetaut, C. B. Cox, M. Martin, and J. J. Jaeger for the discussions concerning various palaeogeographical problems. Thanks are due to anonymous reviewers who improved the text.

References

Albino, A. M. (1986). Nuevos Boidae Madtsoiinae en el Cretacico tardio de Patagonia (Formacion Los Alamitos, Rio Negro, Argentina). IV Congreso Argentino de Paleontologia y Bioestratigrafia, Acta 2, Mendoza, pp. 15–21.

Allègre, C. J. *et al.* (1984). Structure and evolution of the Himalaya–Tibet orogenic belt. *Nature* **307**, 17–22.

Attridge, J., Crompton, A. W., and Jenkins, F. A. Jr. (1985). The Southern African liassic prosauropod *Massospondylus* discovered in North America. *J. Vertebr. Paleont.* **5** (2), 128–32.

Battail, B. (1977). Les reptiles thérapsides dans la stratigraphie du Trias continental: les faunes d'âge Stormberg en Afrique et leurs équivalents dans le monde. *Ann. Soc. Géol. Nord* **97**, 343–50.

Battail, B. (1981). Faunes de vertébrés tétrapodes et individualité paléontologique du Gondwana au Trias inférieur. *Bull. Soc. Géol. Fr.* **23**, 571–8.

Battail, B. (1984). Les Cynodontes (Reptilia, Therapsida) et la paléobiogéographie du Trias. *3rd Symposium on Mesozoic Terrestrial Ecosystems. Short papers* (ed. W. E. Reif and F. Westphal), pp. 1–6.

Beltan, L. and Dutuit, J. M. (1977). Apports récents à la géologie du Gondwana. Récapitulation des affinités gondwaniennes anté-jurassiques de Madagascar (Poissons, Amphibiens, Reptiles). *Ann. Soc. Géol. Nord* **97**, 357–62.

Besse, J., Courtillot, V., Pozzi, J. P., Westphal, M., and Zhou, Y. X. (1984). Palaeomagnetic estimates of crustal shortening in the Himalayan thrusts and Zangbo suture. *Nature* **311**, 621–6.

Bonaparte, J. F. (1978). *El Mesozoico de America del Sur y sus Tetrapodos*. Opera Lilloana **26**, Tucumán, Argentina.

Bonaparte, J. F. (1979). Dinosaurs: a Jurassic assemblage in Patagonia. *Science* **205**, (4413), 1377–79.

Bonaparte, J. F. (1982). Faunal replacement in the Triassic of South America. *J. Vertebr. Paleont.* **2**, 362–71.

Bonaparte, J. F. (1984*a*). Late Cretaceous faunal interchanges of terrestrial vertebrates between the Americas. *3rd Symposium on Mesozoic Terrestrial Ecosystems. Short papers* (ed. W. E. Reif and F. Westphal), Attempto, Tübingen, pp. 19–24.

Bonaparte, J. F. (1984*b*). La Formacion Los Alamitos (Campaniano-Maastrichtiano) del Sudeste de Rio Negro, con descripcion de *Kritosaurus australis* n. sp. (Hadrosauridae). Significado paleogeografico de los Vertebrados. *Asoc. Geol. Argentina. Rev.* **39**, 284–99.

Bonaparte, J. F. (1984*c*). El intercambio faunistico de Vertebrados continentales entre America del Sur y del Norte a fines des Cretacico. *Mem. III Congr. Latinoamericano Paleont.*, 438–50.

Bonis, L. de *et al.* (1985). Contribution des vertébrés à l'histoire de la Téthys et des continents périthétysiens. *Bull. Soc. Géol. Fr.* **1**, 781–6.

Bown, T. M. and Simons, E. L. (1984). First record of marsupials (Metatheria: Polyprotodonta) from the Oligocene of Africa. *Nature* **308**, 447–9.

Brett-Surman, M. K. and Paul, G. S. (1985). A new family of bird-like dinosaurs linking Laurasia and Gondwanaland. *J. Vertebr. Paleont.* **5**, 133–8.

Broin, F. de (1980). Les tortues de Gadoufaoua (Aptien du Niger); aperçu sur la paléobiogéographie des Pelomedusidae (Pleurodira). *Mém. Soc. Géol. Fr.* **139**, 39–46.

Broin, F. de, Ingavat, R., Janvier, P., and Sattayarak, N. (1982). Triassic turtle remains from northeastern Thailand. *J. Vertebr. Paleont.* **2**, 41–6.

BUFFETAUT, E. (1982). A ziphodont mesosuchian crocodile from the Eocene of Algeria and its implications for vertebrate dispersal. *Nature* **300**, 176–8.

BUFFETAUT, E. (1985*a*). The palaeobiogeographical significance of the Mesozoic continental vertebrates from South-East Asia. In *Paléogéographie de l'Inde, du Tibet et du Sud-Est asiatique*. (ed. E. Buffetaut, J. J. Jaeger, and J. C. Rage), pp. 37–42, Mém. Soc. Géol. Fr. **147**.

BUFFETAUT, E. (1985*b*). The Mesozoic vertebrates of Thailand. *Scient. Am.* **253**, 80–7.

BUFFETAUT, E. (1985*c*). Présence de Trematochampsidae (Crocodylia, Mesosuchia) dans le Crétacé supérieur du Brésil. Implications paléobiogéographiques. *C. R. Acad. Sci., Paris, Sér. II* **301**, 1221–24.

BUFFETAUT, E. (1986). Un Mésosuchien ziphodonte dans l'Éocène supérieur de La Livinière (Hérault, France). *Geobios* **19**, 101–8.

BUFFETAUT, E. and INGAVAT, R. (1982). Phytosaur remains (Reptilia, Thecodontia) from the upper Triassic of North-Eastern Thailand. *Geobios* **15**, 7–17.

BUFFETAUT, E. and MARTIN, M. (1984). Continental vertebrate distribution, faunal zonation and climate in the late Triassic. *3rd Symposium on Mesozoic Terrestrial Ecosystems. Short papers* (ed. W. E. Reif and F. Westphal), 25–9.

BUFFETAUT, E. and MARTIN, M. (1985). Accretion of the China blocks. *Nature* **317**, p. 750.

BUFFETAUT, E. and RAGE, J. C. (1981). Observations à la note de C. Mourer-Chauviré 'Première indication de la présence de Phorusracidés, famille d'oiseaux géants d'Amérique du Sud, dans le Tertiaire européen: *Ameghinornis* nov. gen. (Aves, Ralliformes) des phosphorites du Quercy (France).' *Geobios* **15**, 667–8.

BUFFETAUT, E. and TAQUET, P. (1977). The giant crocodilian *Sarcosuchus* in the early Cretaceous of Brazil and Niger. *Palaeont.* **20**, 203–8.

BUFFETAUT, E. and TAQUET, P. (1979). An early Cretaceous terrestrial crocodilian and the opening of the South Atlantic. *Nature* **280**, 486–7.

CAMPOS, D. A. and WENZ, S. (1982). Première découverte de Coelacanthes dans le Crétacé inférieur de la Chapada do Araripe (Brésil). *C. R. Acad. Sci., Paris, Sér.* II **294**, 1151–54.

CAPPETTA, H., JAEGER, J. J., SABATIER, M., SIGÉ, B., SUDRE, J., and VIANEY-LIAUD, M. (1978). Découverte dans le Paléocène du Maroc des plus anciens Mammifères euthériens d'Afrique. *Geobios* **11**, 257–63.

CARROLL, R. L. and WINER, L. (1977). Classification of amphibians and list of genera and species known as fossils. *Separate appendix* to Carroll, R. L., Patterns of amphibians evolution: an extended example of the incompleteness of the fossil record. In *Patterns of evolution*. (ed. A. Hallam), pp. 405–37, Elsevier, Amsterdam.

CHARIG, A. (1973). Jurassic and Cretaceous dinosaurs. In *Atlas of Palaeobiogeography*. (ed. A. Hallam), pp. 339–52, Elsevier, Amsterdam.

CHATTERJEE, S. (1982). A new cynodont reptile from the Triassic of India. *J. Paleont.* **56**, 203–14.

CHATTERJEE, S. (1985). The drift of India: a conflict in plate tectonics. In *Paléogéographie de l'Inde, du Tibet et du Sud-Est Asiatique*. (ed. E. Buffetaut, J. J. Jaeger, and J. C. Rage), Mém. Soc. Géol. Fr. **147**, 43–8.

CKHIKVADZE, V. M., SHAMMAKOV, S. SH., and ZEROVA, G. A. (1983). Materials to the formation history of Squamata fauna of the middle Asia and Kazakhstan (in Russian). *Izvestiya Akademii Nauk Turkmenian SSR* **2**, 3–7.

COLBERT, E. (1974). *Lystrosaurus* from Antarctica. *Am. Museum Novitates* **2535**, 1–44.

COLBERT, E. (1982). The distribution of *Lystrosaurus* in Pangaea and its implications. In *Phylogénie et paléobiogéographie. Livre jubilaire en l'honneur de R. Hoffstetter*. (ed. E. Buffetaut, P. Janvier, J. C. Rage, and P. Tassy), Geobios, Mém. Spéc. 6, 375–83.

COLBERT, E. (1984*a*). Triassic tetrapod faunas. *3rd Symposium on Mesozoic Terrestrial Ecosystems, short papers* (ed. W. E. Reif and F. Westphal), Attempto, Tübingen, pp. 53–9.

COLBERT, E. (1984*b*). Mesozoic reptiles: India and Gondwanaland. *Ind. J. Sci.* **11**, 25–37.

COSGRIFF, J. W. (1984). The temnospondyl labyrinthodonts of the earliest Triassic. *J. Vertebr. Paleont.* **4**, 30–46.

COSGRIFF, J. W. and HAMMER, W. R. (1984). New material of labyrinthodont amphibians from the lower Triassic Fremouw Formation of Antarctica. *J. Vertebr. Paleont.* **4**, 47–56.

COSGRIFF, J. W. and ZAWISKIE, J. M. (1979). A new species of the Rhytidosteidae from the *Lystrosaurus* Zone and a review of the Rhytidosteoidea. *Palaeont. Afric.* **22**, 1–27.

COURTILLOT, V. and BESSE, J. (1986). Mesozoic and Cenozoic evolution of the North and South China blocks. *Nature* **320**, 86–7.

COX, B. (1973). Triassic tetrapods. In *Atlas of palaeobiogeography*. (ed. A. Hallam), pp. 213–23, Elsevier, Amsterdam.

COX, B. (1974). Vertebrates palaeodistributional patterns and continental drift. *J. Biogeogr.* **1**, 75–94.

COX, B. (1980). An outline of the biogeography of the Mesozoic world. *Mém. Soc. Géol. Fr.* **139**, 75–9.

CROCHET, J. Y. (1979). Diversité systématique des Didelphidae (Marsupialia) européens tertiaires. *Geobios* **12**, 365–78.

CROCHET, J. Y. (1984). *Garatherium mahboubii*, nov. gen. nov. sp., marsupial de l'Eocène inférieur d'El Kohol (Sud-Oranais, Algérie). *Ann. Paléontol.* **70**, 275–94.

CROCHET, J. Y. (1986). *Kasserinotherium tunisiense* nov. gen., nov. sp., troisième marsupial découvert en Afrique (Eocène inférieur de Tunisie). *C. R. Acad. Sci., Paris, Sér. II* **302**, 923–6.

DERCOURT, J. *et al.* (1985). Présentation de neuf cartes paléogéographiques au 1/20,000,000° s'étendant de l'Atlantique au Pamir pour la période du Lias à l'actuel. *Bull. Soc. Géol. Fr.* **1**, 637–52.

DONG, Z. M. (1985*a*). *The study of dinosaur fossils in China.* Hong Kong Museum of History, Urban Council, Hong Kong.

DONG, Z. M. (1985*b*). A middle Jurassic Labyrinthodont (*Sinobrachyops placenticephalus* gen. et sp. nov.) from Dashanpu, Sichuan Province. *Vertebrata Palasiatica* **23**, 301–6.

DONG, Z. M., ZHOU, S. W., and ZHANG, Y. H. (1983). The dinosaurian remains from Sichuan basin, China. *Palaeont. Sinica* **162**, 139–45.

DUTUIT, J. M. (1976). Où en est la stratigraphie à base paléontologique de la formation rouge d'Argana (Atlas occidental)? *C. R. somm. séances Soc. Géol. Fr.* **6**, 272–3.

DUTUIT, J. M. (1980). Principaux caractères d'un genre de Dicynodonte du Trias marocain. *C. R. Acad. Sci., Paris, Sér. II* **290**, 655–8.

DUTUIT, J. M. (1983). Homogénéité et répartition des Métoposauridés et phytosauridés sont-elles en relation avec les espaces marins du Trias supérieur? *C. R. Acad. Sci., Paris, Sér. II* **296**, 1465–68.

ENAY, R. (1972). Paléobiogéographie des ammonites du Jurassique terminal (Tithonique/Volgien/Portlandien s.l.) et mobilité continentale. *Geobios* **5**, 355–407.

ESTES, R. (1982). The fossil record and early distribution of lizards. In *Advances in herpetology and evolutionary biology. Essays in honour of E. E. Williams*, (ed. A. Rhodin and K. Miyata), pp. 365–98.

ESTES, R. (1983). Sauria terrestria, Amphisbaenia. *Handbuch der paläoherpetologie*, part 10A, G. Fischer Verlag, Stuttgart, xxii + 249 pp.

ESTES, R. and BAEZ, A. M. (1985). Herpetofaunas of North and South America during the late Cretaceous and Cenozoic: evidence for interchange. In *Great American biotic interchange*, (ed. G. Stehli and D. S. Webb), pp. 139–97, Plenum, New York.

FÖRSTER, R. and SCHOLZ, G. (1979). *Salaziceras nigerianum* n. sp. from south east Nigeria: faunal evidence for an open seaway between the northern and southern Atlantic in late Albian times. *Neues Jb. Geol. Paläontol. Mh.* **2**, 109–19.

GALTON, P. M. (1977*a*). The ornithopod dinosaur *Dryosaurus* and a Laurasia–Gondwanaland connection in the Upper Jurassic. *Nature* **268**, 230–2.

GALTON, P. M. (1977*b*). Upper Jurassic ornithopod dinosaur *Dryosaurus* and a Laurasian–Gondwanaland connection. In *Paleontology and plate tectonics*. (ed. R. M. West), pp. 41–54, *Milwaukee Publ. Museum*, Spec. Publ. Biol. Geol. **2**.

GALTON, P. M. (1980). *Dryosaurus* and *Camptosaurus*, intercontinental genera of upper Jurassic ornithopod dinosaurs. *Mém. Soc. Géol. Fr.* **139**, 103–8.

GALTON, P. M. and TAQUET, P. (1982). *Valdosaurus*, a Hypsilophodontid dinosaur from the lower Cretaceous of Europe and Africa. *Geobios* **15**, 145–59.

GINGERICH, P. D. (1977). Radiation of Eocene Adapidae in Europe. *Geobios, Mém. Spéc.* **1**, 165–82.

GINGERICH, P. D. (1985). South American mammals in the Paleocene of North America. In *Great American biotic interchange*. (ed. G. Stehli and D. S. Webb), pp. 123–37, Plenum, New York.

GRANDE, L. and EASTMAN, J. T. (1986). A review of Antarctic ichthyofaunas in the light of new fossil discoveries. *Palaeont.* **29**, 113–37.

HALLAM, A. (1981). Biogeographic relations between the northern and southern continents during the Mesozoic and Cenozoic. *Geol. Rdsch.* **70**, 583–95.

HALLAM, A. (1986). Evidence of displaced terranes from Permian to Jurassic faunas around the Pacific margins. *J. Geol. Soc. London* **143**, 209–16.

HARTENBERGER, J. L. (1986). Hypothèse paléontologique sur l'origine des Macroscelidea (Mammalia). *C. R. Acad. Sci., Paris, Sér. II* **302**, 247–9.

HARTENBERGER, J. L., MARTINEZ, C., and BEN SAÏD, A. (1985). Découverte de Mammifères d'âge Eocène inférieur en Tunisie centrale. *C. R. Acad. Sci., Paris, Sér. II* **301**, 649–52.

HELMCKE, D. (1985). The Permo-Triassic 'Paleotethys' in mainland Southeast-Asia and adjacent parts of China. *Geol. Rdsch.* **74**, 215–28.

HOFFSTETTER, R. (1972). Relationships, origins and history of the ceboid monkeys and caviomorph rodents: a modern reinterpretation. In *Evol. Biol.* (ed. T. Dobzhanskhy, M. K. Hecht, and W. C. Steere), pp. 323–46, **6**.

HOFFSTETTER, R., and RAGE, J. C. (1977). Le gisement de vertébrés miocènes de La Venta (Colombie) et sa faune de serpents. *Ann. Paléont.* **63** (2), 161–90.

HOPSON, J. A. (1984). Late Triassic traversodont cynodonts from Nova Scotia and southern Africa. *Palaeont. Afric.* **25**, 181–201.

HOWIE, A. A. (1970). A new capitosaurid labyrinthodont from East Africa. *Palaeontol.* **13**, 210–53.

INGAVAT, R. and JANVIER, P. (1981). *Cyclotosaurus* cf. *posthumus* Fraas (Capitosauridae, Stereospondyli) from the Huai Hin Lat Formation (Upper Triassic), northeastern Thailand. *Geobios* **14**, 711–25.

JAEGER, J. J. and MARTIN, M. (1984). African marsupials—vicariance or dispersion? *Nature* **312**, p. 319.

JENKINS, F. A. and CROMPTON, A. W. (1979). Triconodonta. In *Mesozoic mammals*. (ed. J. A. Lillegraven, Z. Kielan-Jaworowska, and W. A. Clemens), pp. 74–90, University of California Press.

KALANDADZE, N. N. (1974). O mezkontinental'nykh

svjazjakh faun tetrapod v triasovom periode. *Paleont. Zh.* **3**, 75–86.

KRAUS, M. J. (1979). Eupantotheria. In *Mesozoic mammals*. (ed. J. A. Lillegraven, Z. Kielan-Jaworowska, and W. A. Clemens), pp. 162–81. University of California Press.

LAURENT, R. F. (1975). La distribution des Amphibiens et les translations continentales. *Mém. Muséum Nat. Hist. natur.* **A88**, 176–91.

LEHMAN, J. P. (1979). Nouveaux Trématosaures de Madagascar: les Stégocéphales malgaches et leur paléoécologie. *Ann. Paléontol., Sér. Vertébr.* **65**, 35–53.

LIN, J. L., FULLER, M., and ZHANG, W. Y. (1985). Preliminary Phanerozoic polar wander paths for the North and South China blocks. *Nature* **320**, 444–9.

LOZOVSKIY, V. R. and SHISHKIN, M. A. (1974). First labyrinthodont find in the lower Triassic on Mangyshlak (in Russian). *Doklady Akademii Nauk SSSR* **214**, 169–72.

MAHBOUBI, M., AMEUR, R., CROCHET, J. Y., and JAEGER, J. J. (1983). Première découverte d'un Marsupial en Afrique. *C. R. Acad. Sci., Sér. II* **297**, 691–4.

MAHBOUBI, M., AMEUR, R., CROCHET, J. Y., and JAEGER, J. J. (1984). Implications paléobiogéographiques de la découverte d'une nouvelle localité éocène à Vertébrés continentaux en Afrique nord-occidentale: El Kohol (Sud-Oranais, Algérie). *Geobios* **17**, 626–9.

MAILLOT, H. (1983). Les paléoenvironnements de l'Atlantique sud: apport de la géochimie sédimentaire. *Soc. Géol. Nord, Publ.* **9**, 1–316.

MARSHALL, L. G., WEBB, S. D., SEPKOSKI, J. J., and RAUP, D. M. (1982). Mammalian evolution and the Great American Interchange. *Science* **215**, 1351–7.

MARTIN, M. (1980). *Mauritanichthys rugosus* n. gen. et n. sp., Redfieldiidae (Actinopterygi, Chondrostei) du Trias supérieur continental marocain. *Geobios* **13**, 437–40.

MARTIN, M. (1981*a*). Les Dipneustes et Actinistiens du Trias supérieur continental marocain. *Stuttgarter Beitr. Naturkunde* **B69**, 1–29.

MARTIN, M. (1981*b*). Les Dipneustes mésozoïques malgaches, leurs affinités et leur intérêt paléobiogéographique. *Bull. Soc. Géol. Fr.* **23**, 579–85.

MARTIN, M. (1982). Nouvelles données sur la phylogénie et la systématique des Dipneustes postpaléozoïques, conséquences stratigraphiques et paléogéographiques. In *Phylogénie et paléobiogéographie. Livre jubilaire en l'honneur de R. Hoffstetter*. (ed. E. Buffetaut, P. Janvier, J. C. Rage, and P. Tassy), pp. 53–64, Geobios. Mém. Spéc. 6.

MARTIN, M. (1983). Phylogénie et systématique des Cératodontiformes (Dipnoi). Paléoécologie et paléobiogéographie. *Mém. Sci. Terre Univ. Paris VI* 1–165.

MARTIN, M. (1984). Révision des Arganodontidés et des Néocératodontidés (Dipnoi, Ceratodontiformes) du Crétacé africain. *Neues Jb. Geol. Paläont. Abh.* **169**, 225–60.

MARTIN, M. and INGAVAT, R. (1982). First record of an Upper Triassic Ceratodontid (Dipnoi, Ceratodontiformes) in Thailand and its paleoecological significance. *Geobios* **15**, 111–14.

MAZIN, J. M. and MARTIN, M. (1984). Marine reptiles and fishes as a test for the paleobiogeography of the lower and middle Triassic. *3rd Symposium on Mesozoic Terrestrial Ecosystems, Short Papers*, (ed. W. E. Reif and F. Westphal), Attempto, Tübingen, pp. 141–4.

MCELHINNY, M. W., EMBLETON, B. J. J., MA, X. H., and ZHANG, Z.K. (1981). Fragmentation of Asia in the Permian. *Nature* **293**, 212–16.

MELGUEN, M., LE PICHON, X., and SIBUET, J. C. (1978). Paléoenvironnement de l'Atlantique sud. *Bull. Soc. Géol. Fr.* **20**, 471–89.

MILNER, A. R. and NORMAN, D. B. (1984). The biogeography of advanced ornithopod dinosaurs (Archosauria: Ornithischia)—a cladistic-vicariance model. *3rd Symposium on Mesozoic Terrestrial Ecosystems, Short papers* 145–50.

MOLNAR, R. (1980). Australian late Mesozoic terrestrial tetrapods: some implications. *Mém. Soc. Géol. Fr.* **139**, 131–43.

MOLNAR, R. (1982). Australian Mesozoic Reptiles. In *The fossil vertebrate record of Australasia*. (ed. P. V. Rich and E. M. Thompson), pp. 169–225. Melbourne Monash University Offset Printing Unit.

MOLNAR, R. (1984). Palaeozoic and Mesozoic reptiles and amphibians from Australia. In *Vertebrate zoogeography and evolution in Australia*. (ed. M. Archer and G. Clayton), pp. 331–6, Hesperian Press, Carlisle, Australia.

MONBARON, M. and TAQUET, P. (1981). Découverte du squelette complet d'un grand Cétiosaure (Dinosaure Sauropode) dans le bassin jurassique moyen de Tilougguit (Haut-Atlas central, Maroc). *C. R. Acad. Sci., Paris, Sér. II* **292**, 243–6.

MOULLADE, M. and GUÉRIN, S. (1982). Le problème des relations de l'Atlantique sud et d'Atlantique central au Crétacé moyen: nouvelles données microfauniques d'après les forages D.S.D.P. *Bull. Soc. Géol. Fr.* **24**, 511–17.

MOURER-CHAUVIRÉ, C. (1981*a*). Première indication de la présence de Phorusracidés, famille d'oiseaux géants d'Amérique du Sud, dans le Tertiaire européen: *Ameghinornis* nov. gen. (Aves, Ralliformes) des phosphorites du Quercy, France. *Geobios* **14**, 637–47.

MOURER-CHAUVIRÉ, C. (1981*b*). Réponse à E. Buffetaut et J. C. Rage. *Geobios* **15**, 268–9.

OLSEN, P. E. and GALTON, P. M. (1984). A review of the reptile and amphibian assemblages from the Stormberg of Southern Africa with special emphasis on the footprints and age of the Stormberg. *Palaeont. Afric.* **25**, 87–110.

OWEN, H. G. (1983). *Atlas of continental displacement, 200 million years to the present*. Cambridge University Press, Cambridge. (x + 159 pp.)

PATRIAT, P. and ACHACHE, J. (1984). India–Eurasia collision chronology has implications for crustal shortening and driving mechanisms of plates. *Nature* **311**, 615–21.

RAGE, J. C. (1978). Une connexion continentale entre Amérique du Sud et Amérique du Nord au Crétacé supérieur? L'exemple des Vertébrés continentaux. *C. R. somm. séances Soc. Géol. Fr.* **6**, 281–5.

RAGE, J. C. (1981). Les continents péri-atlantiques au Crétacé supérieur: migrations des faunes continentales et problèmes paléogéographiques. *Cretaceous Res.* **2**, 65–84.

RAGE, J. C. (1984*a*). Serpentes. *Handbuch der paläoherpetologie*, part 11, G. Fischer Verlag. (xi + 80 pp.)

RAGE, J. C. (1984*b*). Are the Ranidae (Anura, Amphibia) known prior to the Oligocene? *Amphibia-Reptilia* **5**, 281–8.

RAGE, J. C. (1986). South America/North America terrestrial interchanges in the latest Cretaceous: short comments about Brett-Surman and Paul's article, with additional data. *J. Vertebr. Paleontol.* **6**(4), 382–3.

RAGE, J. C. (1987). Fossil history. In *Snakes: ecology and evolutionary biology*. (ed. R. A. Seigel, J. T. Collins, and S. S. Novak). MacMillan, New York and London (pp. 51–76).

REYMENT, R. A. (1980). Paleo-oceanology and the paleobiogeography of the Cretaceous South Atlantic Ocean. *Oceanolog. Acta* **3**, 127–33.

RIDD, M. F. (1971). South-East Asia as a part of Gondwanaland. *Nature* **234**, 531–4.

RIDD, M. F. (1980). Possible Palaeozoic drift of SE Asia and Triassic collision with China. *J. Geol. Soc. London* **137**, 635–40.

ROBERT, C. (1981). Santonian to Eocene paleogeographic evolution of the Rio Grande Rise (South Atlantic) deduced from clay mineralogical data (DSDP legs 3 and 39). *Palaeogeogr., Palaeoclimatol., Palaeoecol.* **33**, 311–26.

SAHNI, A., *et al.* (1982). Microvertébrés nouveaux des Trapps du Deccan (Inde): mise en évidence d'une voie de communication terrestre probable entre la Laurasie et l'Inde à la limite Crétacé–Tertiaire. *Bull. Soc. Géol. Fr.* **24**, 1093–99.

SCHAEFFER, B. (1972). A Jurassic fish from Antarctica. *Am. Museum Novitates* **2495**, 1–17.

SCHMIDT-EFFING, R. (1980). Geodynamic history of oceanic crust in southern central America. *Geol. Rdsch.* **68**, 457–94.

SEN, S. and HEINTZ, E. (1979). *Palaeoamasia kansui* Ozansoy 1966, Embrithopode (Mammalia) de l'Eocène d'Anatolie. *Ann. Paléont.* **65**, 73–87.

ŞENGÖR, A. M. C. (1985). The story of Tethys: how many wives did Okeanos have? *Episodes* **8**, 3–12.

SEYFRIED, H. and SPRECHMANN, P. (1986). Über die Frühgeschichte (Campan bis Eozän) der Südlichen mittelamerikanischen Landbrücke. *Neues Jb. Geol. Paläont. Mh.* **1**, 38–55.

SIGÉ, B. (1972). La faunule de Mammifères du Crétacé supérieur de Laguna Umayo (Andes péruviennes). *Bull. Muséum Nat. Hist. Natur.* **99**, 375–405.

SOSSON, M. and MARTIN, M. (1985). Découverte d'une faune de Vertébrés marins du Trias supérieur dans le Hot Spring Range (nord-ouest du Nevada, U.S.A.): implications paléogéographiques. *C. R. Acad. Sci., Sér. II* **300**, 177–80.

STORCH, G. (1981). *Eurotamandua joresi*, ein Myrmecophagide aus dem Eozän der 'Grube Messel' bei Darmstadt (Mammalia, Xenarthra). *Senckenbergiana Lethaea* **61**, 247–89.

SUDRE, J. (1979). Nouveaux Mammifères éocènes du Sahara occidental. *Palaeovertebr.* **9**, 83–115.

SUES, H. D. and TAQUET, P. (1979). A pachycephalosaurid dinosaur from Madagascar and a Laurasia–Gondwanaland connection in the Cretaceous. *Nature* **279**, 633–5.

SWINTON, W. E. (1956). A neorhachitome amphibian from Madagascar. *Ann. & Mag. Natur. Hist.* **9**, 60–4.

TAPPONNIER, P., PELTZER, G., LE DAIN, A. Y., ARMIJO, R., and COBBOLD, P. (1982). Propagating extrusion tectonics in Asia: new insights from simple experiments with Plasticine. *Geology* **10**, 611–16.

TAQUET, P. (1982). Une connexion continentale entre Afrique et Madagascar au Crétacé supérieur: données géologiques et paléontologiques. In *Phylogénie et paléobiogéographie. Livre jubilaire en l'honneur de R. Hoffstetter*. (ed. E. Buffetaut, P. Janvier, J. C. Rage, and P. Tassy), Geobios. Mém. Spéc. **6**, 385–91.

TARLING, D. H. (1972). Another Gondwanaland. *Nature* **238**, 92–3.

TARLING, D. H. (1981). The geologic evolution of South America with special reference to the last 200 million years. In *Evolutionary biology of the New World monkeys and continental drift*. (ed. R. L. Ciochon and A. B. Chiarelli), pp. 1–41, Plenum Press, New York.

THOMAS, H., BERNOR, R., and JAEGER, J. J. (1982). Origines du peuplement mammalien en Afrique du Nord durant le Miocène terminal. *Geobios* **15**, 283–97.

THULBORN, R. A. (1983*a*). A mammal-like reptile from Australia. *Nature* **303**, 330–1.

THULBORN, R. A. (1983*b*). Reply to King. *Nature* **306**, p. 209.

THULBORN, R. A. (1984). Early Triassic reptiles of Australia. *3rd Symposium on Mesozoic Terrestrial Ecosystems, Short papers* (ed. W. E. Reif and F. Westphal), 243–8.

TUCKER, M. E. and BENTON, M. J. (1982). Triassic environments, climates and reptile evolution. *Palaeogeogr., Palaeoclimatol., Palaeoecol.* **40**, 361–79.

WARREN, A. (1984). Triassic amphibians and reptiles of Australia in relation to Gondwanaland. In

Vertebrate zoogeography and evolution in Australasia. (ed. M. Archer and G. Clayton), pp. 299–301, Hesperian Press, Carlisle, Australia.

WARREN, A. and HUTCHINSON, M. N. (1983). The last labyrinthodont, a new brachyopoid (Amphibia, Temnospondyli) from the early Jurassic Evergreen Formation of Queensland, Australia. *Philosophical Trans. Roy. Soc. London* **303**, 1–62.

WOODBURNE, M. O. and ZINSMEISTER, W. J. (1984). The first land mammal from Antarctica and its biogeographic implications. *J. Paleontol.* **58**, 913–48.

JEAN-CLAUDE RAGE, Laboratoire de paléontologie des Vertébrés, Université Paris VI, 4 place Jussieu, 75252 Paris cedex 05, France.

Tectonics from fossils? Analysis of reef-coral and sea-urchin distributions from late Cretaceous to Recent, using a new method

Brian R. Rosen and Andrew B. Smith

ABSTRACT: Biogeographical distribution is seen as a product of small- and large-scale factors, of which only the latter are important in palaeogeographical reconstruction. We review here previous methods for assessing the historical sequence of major events that have caused biogeographical distributions, and we argue that the most rigorous of these is cladistic biogeography. We also advocate our own new method, which uses parsimony analysis of taxa shared between sample localities. In not being dependent on prior phylogenetic analysis, this method is especially suitable for taxonomic groups which lack an adequate phylogeny.

This new method takes contemporaneous distributional data from one or more narrow stratigraphic time planes (including the present), and generates a cladogram of sample localities for each of them. We interpret these cladograms as hypotheses of historical sequences of divergences between biotas, for which the sample localities provide geographical control. Congruence between cladograms from different organic groups and time planes provides an indication of those events which are most likely to have been the result of major (e.g. tectonic) historical factors. This method avoids prejudgement of areas of endemicity. We also argue that because younger major historical events can destroy the biogeographical patterns produced by older events, we can reach a more complete understanding of historical biogeography by using a combination of Recent and fossil data. The method however requires refinement and we summarize some of its present problems.

We present area cladograms for Recent and Lower Miocene corals and echinoids, and for Eocene and late Cretaceous (Campanian–Maastrichtian) echinoids. Although these should be tested against area cladograms based on other taxa, we conclude, amongst our preliminary results, that: (1) the present Indo-Pacific fauna is a young biogeographical entity, not older than the Lower Miocene; (2) prior to this, Pacific faunas had been isolated since a time older than the Upper Eocene; and (3) the faunas of the Atlantic and western Tethys have been distinct from those of the Indian Ocean and eastern Tethys since a time long before the 'terminal Tethyan event' (Miocene), and since as far back, at least, as the late Mesozoic.

Apart from biostratigraphy, the use of fossils to provide information about geological events has usually been based on direct or indirect assumptions about their autecology. These assumptions and the information derived from them are usually closely worked into interpretations which also use non-organic evidence from facies. In this approach, therefore, fossils are used to reconstruct palaeogeographies as part of facies analysis. Although facies analysis has undoubtedly contributed a great deal to our understanding of Earth history (e.g. by drawing analogies from the present to the past between environments or tectonic settings), the *critical* value of fossil evidence within facies analysis is not always clear. There is a fundamental susceptibility to circularity, because evidence from different sources (such as sedimentology, tectonic information, sea-floor magnetic anomalies, and the fossils themselves) is not always kept independent. How, for example, do we infer, and independently test, whether an extinct organism had a tropical shallow-water distribution?

By contrast, most biogeographers working on Recent biotas use presence/absence distributional data in order to discover regional patterns. Although palaeontologists have also used such methods, they have done so relatively infrequently (e.g. Whittington and Hughes 1972). Methods which use only presence/absence data are best, however, because such data lend themselves readily to rigorous methods of analysis, independently of any other evidence.

In presenting this paper, we have borne in mind the context of the original Symposium and the disciplinary range of its likely readership, and have tried to explain our own and other methods so that they might also be of interest to non-biogeographers.

From AUDLEY-CHARLES, M. G. & HALLAM, A. (eds) *Gondwana and Tethys* Geological Society Special Publication No. 37, pp. 275–306.

Historical biogeography

Historical biogeography is the study of the history of geographical distributions of organisms and of the historical factors which affect geographical distributions. Amongst its particular aims (see Patterson 1983 for a discussion), the one most relevant here is its use of patterns of distribution of living and fossil organisms to reconstruct geological history. Unfortunately, there are many historical factors that might affect the distribution of biotas and we should assume, neutrally, that these factors function on different scales of time and space to produce a spectrum of effects. Some may affect the geographical range of only a single species, whereas others, such as continental rifting or collision, isolation of oceanic basins, or global changes in climate or ocean circulation patterns, are historical factors which we might expect to affect large segments of the biota simultaneously. It follows that any method which uses distributions to infer history must be able to distinguish broad-scale events affecting most biotas (extrinsic events), from those which affected only small numbers of taxa.

Over the last decade, there has been considerable controversy about methodology in biogeography and about the factors which govern organic distributions (e.g. Nelson and Rosen 1981). There has been a resultant shift towards the use of analytical methods (Patterson 1983), and towards the recognition of factors other than dispersal and ecology (especially *vicariance*: the appearance of new species as a result of a new biogeographical barrier). There is however no method which establishes the causes of all distributions, and in any case, as we explain later, there are different questions which need to be answered.

Until this controversy arose, biogeographical patterns were interpreted largely in terms of centres of origin and dispersal patterns (but for a history of the subject, see Nelson and Platnick 1981, and Humphries and Parenti 1986). Historical biogeographers relied heavily on palaeontological data on fossil distributions to identify these centres of origin and the associated dispersal pathways through time (for examples, see Wiley 1981; and for a defence and exposition of this approach see Briggs 1984). With the advent of plate tectonics and theories of continental drift there has been a change in emphasis towards scenarios where extrinsic factors, especially plate tectonics, were seen as controlling biogeographical patterns, and the fossil record interpreted to fit in with the currently favoured tectonic model (e.g. in Hughes 1973). More recently still, the role of fossil data as the key to interpreting historical biogeography has come under serious attack, with claims that fossil data cannot add anything to our understanding of the Recent biogeographical distribution of the biota (for a review see Patterson 1981, and for a counter-argument see Ghiold and Hoffmann 1986). Centres of origin hypotheses and scenarios have been criticized as untestable (Nelson and Platnick 1981; Patterson 1981, 1983; Wiley 1981). Add to this the current debate as to whether dispersal or vicariance dominate in biogeographical processes (McCoy and Heck 1983), and it is easy to understand why historical biogeography is in a state of flux.

Fossils in historical biogeography

In view of the current criticisms of the use of fossils in historical biogeography, and the fact that we make use of them here, it is appropriate to include a brief discussion, concentrating on the use of fossils in presence/absence methods (rather than their palaeoenvironmental significance).

Primary biogeographical data consist of the presence or absence of taxa, but these are not necessarily comparable attributes. The presence of a taxon relies upon the competence of the taxonomist but is otherwise unambiguous. By contrast, the absence of a taxon is ambiguous for one of three possible reasons: the taxon may: (1) have never existed in the area (*primitive absence*); (2) exist in the area but may not yet have been recorded (*pseudo-absence*); (3) have existed in the area in the past but may have since become extinct (*reversal absence*). These concepts can also be applied to an entirely fossil biota, by taking the fossil horizon as the vantage point. The likelihood of pseudo-absences obviously diminishes as the biota of the region becomes better known, but is much more of a problem for palaeontological data than it is for Recent data, especially in the less well-studied areas of the world.

It is the problem of distinguishing pseudo-absences from primitive absences that has led some biogeographers to claim that fossil evidence cannot contradict the distributional pattern of modern biotas (namely Croizat 1964). Yet it is the recognition of reversal absences that has led non-vicariance biogeographers to believe that fossil data are important for our understanding of historical biogeography. Clearly though, fossil data can only extend the total geographical range of a taxon, since the ambiguity of absence precludes using absence data to identify patterns of range shifts. Adding to the geographical range of a taxon does little to clarify area relationships and often simply reduces the level of resolution of the study.

The usefulness of fossil data, and the way in which fossils are treated, depends upon the sort of information being sought. For workers interested in the historical biogeography of a group with an excellent fossil record and its distributional history, it is probably valuable to use fossil distributions to identify possible changes through time, always bearing in mind the ambiguity associated with absence of taxa. This approach must always remain narrative and speculative and cannot be used as a test of past palaeogeographies. For workers interested in the historical background that produced modern patterns of distribution, the distribution of living biotas is a much less ambiguous guide which cannot be contradicted by fossil discoveries (D. E. Rosen 1978; Humphries and Parenti 1986).

However, in this paper we are neither interested in reconstructing the historical development of a specific clade, nor in confining ourselves to the particular origins of the distribution pattern of the present-day biota. Our aim is to recognize past patterns of organic distribution and their relevance to testing palaeogeographical reconstructions. This cannot be done without data on the distribution of fossils, especially in the earlier Phanerozoic (see also Young's (1981) discussion, pp. 232–3). Data from living distributions alone are insufficient because, although the geographical distribution of the Recent biota is in part the product of a long and continuous sequence of extrinsic factors that stretches back well into the Precambrian, it must reflect events in the recent past much more strongly than older events. The influence of older events may also be apparent in a Recent distributional pattern, particularly at higher taxonomic levels, but they must have a weaker imprint because more recent events have masked them. We base this view on two factors: (1) *convergence*; and (2) *evolutionary development*, best explained in terms of a descriptive model.

Convergence of biotas results from events like continental collisions (e.g. for terrestrial and shore organisms) or continental rifting (which, for example opens connecting seaways for marine organisms) and therefore overprints previous distribution patterns. Continued evolutionary development of a biota, with the appearance of new sister taxa and the extinction of established taxa, continually creates new biotas, and these must be restricted in their distribution by whatever biogeographical barriers existed at the time. We therefore assume that for heuristic purposes, distributional patterns are at least in part a response, direct or indirect, to geological events. (Why else try to use distributions to infer geological history?) Moreover, with respect to any given time plane, as older taxa which were penecontemporaneous with older events become extinct, and new ones arise reflecting more recent geographical patterns, the older distributional patterns will become less and less distinct. For this reason, we believe that the resolution of historical events using only extant biotas must deteriorate with the geological age of these events, and a Recent distribution (for example) will be much less reliable in assessing pre-Cainozoic, or possibly even Palaeogene, than Neogene palaeogeography. One obvious way to recover this lost information is to choose one or more additional observation datum planes in the geological past, and look for patterns in, say, the Oligocene, or Silurian. We develop this approach further here.

Superimposed on this pattern of diminishing information content, there will also be differences between major taxonomic groups because they have apparently different average species durations (approximately 6 Ma for echinoids, according to Stanley (1979), and 10–15 Ma for reef corals according to Dr A. B. Foster (pers. comm.) and hence, presumably, different average generic durations. The rate at which past extrinsic events will be removed from the biogeographical distribution record will therefore vary from group to group. If corals last on average twice the duration of echinoid species, then the biogeographical pattern of corals will be expected to retain evidence of previous extrinsic events for comparably longer than that of echinoids, while the echinoid pattern will reflect recently past events much more strongly than the coral pattern.

We agree with the common criticism that the incompleteness of the fossil record raises serious difficulties for biogeographical methodology, but it is a reasonable goal to try to find ways of making the best biogeographical use of this record rather than rejecting it outright. How should we do this? Simply to combine Recent and fossil data, effectively superimposes and amalgamates potentially interesting signals from different geological horizons and loses potential stratigraphic resolution (Grande's (1985) 'point 4', p. 236). We therefore assume that, at any given horizon, the fossil record is a vestigial biogeographical sample from which it is still possible to infer historical events, however incompletely, according to explicit, rational methods. Fossil distributional data clearly have an important role in historical biogeography, *if* they are analysed using a method that (1) minimizes the problems associated with absence states; and (2) is based on separate analyses of biogeographical distributions from different, preferably successive, geological horizons. The distribution for each such horizon should have a pattern that most strongly reflects the major biogeographical events that have taken place over the most recent time span prior to that horizon. This approach could be applied to cladistic biogeography, for example, by taking phylogenies of different groups of taxa within a past time-plane, or, synoptically, over a stratigraphic sequence, rather than using Recent biotas alone (B. R. Rosen 1984, and for an example, see Grande 1985). It is a synoptic (though not phylogenetic) approach that we have also adopted here.

Some current methods of analysing distributional data

Importance of endemicity

Biogeographers are universally agreed that endemicity (the exclusive occurrence of a taxon in a particular locality or region) is the most significant feature of distribution patterns. In the simplest instance, every region would have its own exclusive biota without any overlap with other regions, but in practice, even for small taxonomic groups or small areas, this is rarely if ever the case. Endemicity, moreover, is relative, and this is for two reasons. Firstly, in biogeographical practice, distributions (including areas of endemicity) are invariably generalized to a greater or lesser extent; they are syntheses of the raw distributional data gathered from an original set of sample points (Nelson and Platnick 1981). Secondly, a taxon which is shared between two given regions, but which occurs nowhere else, is 'endemic' to both regions when considered together, but 'endemic' to neither when each area is considered separately. Thus for sampling purposes, the biosphere must be assumed to consist of an overlapping mosaic of endemicities or ranges of different species (B. R. Rosen 1984). As we explain below, this overlap has always made delineation of biotas and regions difficult, but it can actually be used to make hierarchical groupings of sample localities.

The way in which distributional data are analysed is crucial to how they can be interpreted, and the following is a critical summary of some of the main methods that are in current use.

Anecdotal treatments

Unfortunately, in many cases, extravagant claims about displaced (allochthonous) terranes and palaeogeography are based on the discovery of one or a few taxa common to two or more regions, without considering the biotas of the regions as a whole. Such shared taxa may reflect similar prevailing ecological conditions in the two areas, rather than any extrinsic (e.g. tectonic) factor. They could even represent disjunct remnant isolates of a previously more extensive range. Such claims are limited in merit and are best treated as preliminary hypotheses.

Biotic provinces

Provinces are not in themselves a method, but rather are a conceptualization about the nature of biotic distributions. There is a long-established history of the use of this concept, going back to Sclater, de Candolle and Wallace (Nelson and Platnick 1981). Provinces have played an important part in the history of the subject because they represent the discovery of the importance of endemicity, and the origin of the idea of relationships between regions.

Provinces are primarily a means of summarizing or generalizing endemic patterns, often by reference to 'key' taxa. Belief in provinces as a real phenomenon has generated a whole gamut of methods which are designed to find them (see below), but the idea meets difficulties with the overlapping nature of endemicity. Notwithstanding this overlap, provinces are invariably conceived as one- or two-tier divisions of the Earth's surface (e.g. Schopf 1979), rather than hierarchically. A common consequence is that studies based on different groups of organisms lead to rival, or apparently unreconcilable, schemes of provinces. For a fossil example, see Fortey and Barnes (1977), Cocks and Fortey (1982), and Schallreuter and Siveter (1985) on trilobites and brachiopods versus ostracods and conodonts in the Ordovician–Silurian.

The boundaries of many provinces, moreover, are defined, like those of Wallace, on mixed criteria, usually a combination of the distribution of certain organisms, and physical features like the present shoreline limits of a continent. This hybrid approach is even more unsatisfactory when we come to fossil distributions. If, as is common, the geographical frame of reference for fossil distributional data is a palaeogeographical reconstruction, physical criteria for province boundaries must, inevitably, be more hypothetical than real (e.g. Tethyan Realm). How might we discover palaeogeography from fossil provinces which are defined in terms of what we are trying to discover? Even those provinces that are defined on the presence/absence pattern of organisms alone (see below) are subjective and arbitrary because of the problem of overlapping endemicities. We reject the idea that objective biogeographical provinces can be identified, and we prefer methods which use the relationships between sample localities alone to develop historical hypotheses.

Range analysis

Range analysis consists of superimposition of a set of organisms' distribution ranges, usually in two dimensions along a geographical sample line. Jablonski *et al.* (1985) stress that species' ranges are a fundamental biological unit, because ends of ranges have an ecological and evolutionary significance. Integration of range end-points can be used to delineate provinces, either in a simple visual way (e.g. Rosen 1984), or by statistical processing of range data (e.g. Pielou 1977, 1978; Jablonski and Valentine 1981).

As with phenetic methods (see below), geographical resolution of range methods is good, being potentially as fine as the set of sample localities. Range methods also answer some of our foregoing criticisms of provinces, in that range-defined provinces are based only on presence/absence data. Being essentially statistical, however, range methods are in fact phenetic and suffer from the same disadvantages. In particular, there is an arbitrary element (e.g. in deciding how many coincident end-points define a provincial limit). There is also the strong likelihood that in the pairs of provinces (regions) delineated by each peak of end-points, one of a pair might be defined on *absence* of taxa (cf. the problem of paraphyly in phylogenetics). These difficulties are more severe for fossil distributions, because of higher probability of pseudo-absences.

There is however a more obvious problem with range methods, because sample localities of range data are (by definition) fixed with respect to each other and a known geography, i.e. they are fixed to a geographical template. There are two consequences. Firstly, for fossil distributions, the only non-hypothetical geography which can be used objectively is the present Earth, but to plot the data on palaeogeographical reconstructions invites circularity. Secondly, for living marine and most fossil distributions, problems also arise if ranges are interpolated through points of absence between neighbouring records. (Some interpolation is usual because the same taxon is rarely present at more than a scatter of localities.) In fact, for distributions on a global scale, there is no way of deciding whether a range should be drawn along the shortest segment of the Earth's surface, or along the complementary longer segment. B. R. Rosen and Darrell (1984) could only devise arbitrary rules for interpolation.

These problems can only be solved by lifting sample localities off their geographical template altogether, as in other phenetic treatments like multivariate analysis, cladistic biogeography, and our own method of parsimony analysis of endemicity. Localities can then be grouped entirely on presence/absence, independently of any existing arrangement of these localities in space. (Parsimony analysis may itself interpolate records, but not according to neighbourhood considerations in a prior geographical scheme.)

Jablonski *et al.* (1985) also noted that ranges occur in overlapping sets, which therefore lend themselves to 'hierarchical analysis', but cited no examples or details. B. R. Rosen's (1984) method represents a simple hierarchical analysis, and our new method here is also a further development of this idea.

Similarity coefficients and multivariate cluster analyses (phenetics)

Rigorous treatment of biogeographical data has generally been based on some form of computerized multivariate analysis program (e.g. Stehli and Wells 1971; Whittington and Hughes 1972) which clusters pairwise calculations of similarity (or dissimilarity) calculated for biotas from a set of localities. Some authors have used similarity measures alone without clustering (e.g. Oliver and Pedder 1979). All these kinds of methods appear to give an objective assessment of biogeographical pattern. They are useful for recognizing areas of endemicity, and in defining provinces in the form of discrete clusters of sample localities. Some valuable features are that these methods can be used for a set of sample localities, without further prejudgement of how these localities correspond with regions or provinces, and no interpolation of ranges is necessary. The methods are explicit, and locality control is as precise as wished.

However, there are many different similarity indices (Cheetham and Hazel 1969; Pielou 1979) and they can yield different results. The concept of similarity, moreover, is questionable. Most calculations use overall similarity, and this has been heavily criticized. (For a summary of criticisms of these techniques as applied to systematics, see Wiley 1981.) Another problem is that small but possibly significant endemic distributions can be lost as 'noise' (Nelson and Platnick 1981; Patterson 1983; and for a coral example, see B. R. Rosen 1984, p. 250). Although the methods themselves are explicit and strict, the choice of discriminant levels for clusters is subjective or arbitrary. Nonetheless, cluster analyses are a useful first step in defining cohesive regions, and clusters can be further evaluated by discriminant analysis (e.g. in systematics, Foster 1984).

Although cluster analysis produces dendrograms which could be taken to reflect a hierarchical pattern of provinciality, most authors usually choose to reduce them to the customary single-tier concept of provinces (e.g. Stehli and Wells 1971, for Recent reef corals). Provinces defined by similarity indices can be interpreted statically to indicate the contemporaneous presence and relative position of (unspecifiable) biogeographical barriers which can then be compared with palaeogeographical reconstructions based on other independent evidence. There is, however, no generally established method by which a cluster analysis, on its own, can be interpreted as a history of the evolution of taxa, biotas, or regions. In systematics, for example, there is no evolutionary theory by which overall similarity of characters in different taxa can be equated with similarity due to phylogeny, unless it is assumed that (1) rates of evolution for all the taxa concerned are the same, and (2) there has been no convergence of characters. (Dr C. J. Humphries, pers. comm.). Events are usually inferred from changing patterns of clusters (or cluster-based provinces) obtained from a sequence of stratigraphic horizons (i.e. a synoptic or stratophenetic approach, as used by Whittington and Hughes, 1972).

Cladistic biogeography

This is a relatively new development which has generated much of the recent controversy in biogeography (e.g. Nelson and Platnick 1981; Nelson and Rosen 1981). There are several methods that are related to this approach, but it is cladistic biogeography that offers a philosophically sound method for the historical analysis of distributions. The method deals only with clades of proven monophyly, whose relationships have been rigorously determined using cladistic methodology (i.e. there must be a well-corroborated cladogram of the group). On this cladogram, the geographical range of each terminal taxon is substituted for the taxon's name, and a hierarchical pattern of areas is worked out using simple principles (see Platnick and Nelson 1978; D. E. Rosen 1978; Nelson and Platnick 1981; Patterson 1981; Wiley 1981; Humphries and Parenti 1986).

This method produces, for each clade, a definitive hierarchical nesting of areas directly interpretable in terms of the group's historical biogeography. Furthermore, although it assumes that evolution occurs, it is otherwise independent of any theory as to how the pattern originated. The method is based primarily on presences alone, and identifies a branching phylogenetic pattern in a chosen taxonomic group. This, in turn, is taken to imply a history of vicariance events (the splitting up of previously unified areas, with contemporaneous divergence of taxa). Vicariance is assumed not on doctrinaire grounds (as is widely misunderstood), but as a falsifiable hypothesis. Where the vicariance pattern does not stand up to testing, other unspecifiable non-vicariant events (which might include dispersal or convergence) are inferred. Testing consists of comparing a number of such cladograms of different monophyletic groups, in a search for distribution patterns that are congruent. Congruence is taken to be evidence for large-scale extrinsic events, reconstructing palaeogeographies, or identifying divergences of biotas. Convergence of areas is inferred from synthesis of studies of numerous groups (Craw 1982; Humphries and Parenti 1986).

The main drawback of this excellent method is that it can only be used on proven monophyletic clades, and the state of taxonomy in many (especially extinct) groups therefore precludes its use. The method has mostly been used by biogeographers concerned with living biotas and, as far as we are aware, the only authors to have applied it to entirely fossil distributions are Young (1981, 1984) for Devonian fishes, Milner and Norman (1984) for Jurassic–Cretaceous ornithischian dinosaurs, and Grande (1985) for Eocene teleost fishes. There is no reason why it should not be attempted more widely by palaeontologists, although pseudo-absences are likely to be a problem.

A new method: parsimony analysis of endemicity (PAE)

Background and summary

From the foregoing discussion, it is possible to summarize some key criteria for achieving more rigour in historical biogeography, with particular respect to palaeogeography (Table 1). These criteria were a guide in developing the method used in this paper, though we have not necessarily met the criteria in full. This new method was first suggested by one of us (B. R. Rosen 1984, p. 251) as a development of range analysis, and as an improved means of discovering and analysing endemic patterns (of present-day reef coral distribution). This involved an intermediate stage of inferring ranges, which was subjective, arbitrary, and tied to existing geography (see p. 279). In place of range methods, it is possible to use parsimony analysis of taxa shared between sample localities, or more briefly, parsimony analysis of endemicity (PAE) without reference to any particular geographical pattern of the localities themselves. One of us (B. R. Rosen 1985) published a preliminary attempt in the following year, but apart from this, the method is largely untried.

Like multivariate analysis, PAE takes as its starting point the presence/absence data for a set of sample localities and a particular group of organisms. This minimizes geographical prejudgement. Each taxon present at a sample locality is then treated analogously to a character in cladistic analysis in systematics. PAE therefore uses the same kind of locality/taxon matrix for its raw data, as is used by cluster analysis, but instead of handling presence/absence in terms of overall similarity, it uses parsimony analysis of shared presence of taxa to identify a nested hierarchical pattern, as in cladistics. In this method, therefore, shared presences (endemicities) are treated analogously to synapomorphies in systematics. The resulting cladograms of sample localities express relationships between biotas, as represented at these localities. For

TABLE 1. *Some criteria for more effective application of historical biogeography to palaeogeography, using fossil distributions*

1. Keep biogeographical data and inferences distinct and independent from other data, and also from palaeogeographical inferences from other data (i.e. distinguish inferred history of the organisms from Earth history).
2. Use a data base of PRESENCE/ABSENCE for taxa from sample localities.
3. Make explicit and minimize ecological assumptions about the organisms used in distributional analysis.
4. Make explicit and minimize geographical extrapolations from the sample localities.
5. Use non-polyphyletic taxa.
6. Minimize the effects of pseudo-absence by using methods which emphasize PRESENCE of taxa and which do not rely on range determinations.
7. Maximize fossil data by analysing distributions over as many stratigraphic horizons as possible (but see also 8.).
8. Use time spans between stratigraphic distributions which are less than the mean duration of the taxa concerned.
9. Use stratigraphically well-controlled distributions.
10. Use a method which provides a consistent and explicit means of inferring historical events (i.e. other than narrative).
11. *Use a testable method.
12. *Use a method which distinguishes small-scale random events from major events.

*The only satisfactory means of achieving these criteria is to use a method which can be repeated for numerous organisms, seeking congruence between them.

reasons explained more fully below, these relationships can, in turn, be read as histories of these biotas, or as histories of the sample localities (which may be generalized, subjectively, as areas). Corroboration of these historical hypotheses is found by searching for congruence in analyses of as many different groups of organisms as possible.

PAE cladograms therefore provide an alternative to the dendrograms of multivariate analysis, avoiding the difficulties of both overall similarity, and dependence on arbitrary decisions about levels of discrimination. They also provide a direct (hypothetical) indication of historical events (i.e. without necessarily using a synoptic approach). Like phenetic approaches, PAE has the advantage that it can be applied to groups whose phylogenetic relationships are unspecified and which need not be clades of proven monophyly. However, the individual taxa used as attributive characters of the sample localities should be, as far as possible, non-polyphyletic (which for the sake of the present paper, we have had to assume). Apart from this, PAE side-steps the lengthy task required by cladistic biogeography, of establishing fully corroborated phylogenies, before even commencing biogeographical analysis proper.

Interpretation: (1) biotic history

For this paper we have used Swofford's (1984) PAUP computer program (version 2.2) to find the most parsimonious hierarchical arrangement of sample localities. When applied to phylogenetics, this method requires that an outgroup taxon be chosen for comparison with the main set of taxa. This procedure is critical to the historical interpretation of cladograms. An outgroup is used to root the cladogram and polarize the character states. For a biogeographical analogy, we would need a locality characterized by a supposedly primitive biota, i.e. a 'centre' or area of origin. This might be possible for a small, well-defined clade with a good fossil record, but it is clearly ridiculous to attempt to identify an area of origin for major groups such as echinoids or reef corals. Moreover, choosing one sample locality as an outgroup prejudges the area relationships which we are trying to discover, and is not feasible for non-monophyletic groups.

To avoid these problems, we preferred to use the program's Lundberg rooting. In systematics, this would be a taxon without any of the characters which occur in the other taxa being analysed. In biogeographical terms, it represents a sample locality without any of the taxa which are present in the rest of the sample locality set. Such an outgroup might be thought of as a locality occurring so far back in geological time that none of the taxa in the sample set of localities had yet evolved. In this interpretation, as taxa appear, their distribution is restricted by prevailing biogeographical conditions, and the presence of shared taxa between two or more sample localities provides a historical dimension. The resulting cladogram of sample localities therefore represents not simply an analysis of shared presences, but also a hierarchical pattern that can be interpreted in terms of the evolution of biotas.

At this point, we should admit that other biogeographical interpretations of distributional patterns derived from the Lundberg rooting are also possible, such as relative effect of adverse ecological factors (e.g. extreme temperatures, high latitude), although there will still be a historical component in such circumstances. It is likely, therefore, that PAE using Lundberg rooting will pick out contemporaneous ecological gradients on a global or regional scale, as well as historical patterns. However, we rejected the alternative of using an outgroup area with all taxa present (which would yield an inverted version of the Lundberg-rooted cladogram), because the resulting pattern implies a history of congruent geographical depletion through time, starting from a cosmopolitan biota, and would group areas according to shared absences. Despite drawbacks, Lundberg rooting seems to be the most appropriate method, so far, for PAE purposes, and we have assumed that this rooting gives cladograms which can be interpreted as historical biogeographical schemes of sample locality relationships, intermingled with some ecological influences.

In a purely historical interpretation, each dichotomy in a PAE cladogram represents a biogeographical divergence of at least one pair of taxa, or more (hence biotas), between the sample localities under analysis. A complete cladogram provides a sequence of such events, in which dichotomies nearer to the origin of the cladogram represent older divergences than those further from the origin. This does not rule out the possibility of convergences also having occurred, but these cannot be read directly from the cladogram itself (see below). A hierarchical pattern of distribution exists because of the history of divergences which have fragmented the biota: convergence acts to destroy this pattern by merging the biotas represented at sample localities.

Since an ecological interpretation cannot also be ruled out, a PAE cladogram might also represent a set of localities more or less affected by adverse ecological conditions. (In the case of marine organisms, as in this paper, these conditions will be related mainly to either latitude or depth.) The likelihood of an ecological rather than a historical interpretation can only be resolved by treating the cladograms as hypotheses of either possibility or a combination of both. They can then be compared with independently derived geological histories and with geographical patterns of ecological parameters (e.g. sea water temperatures). However, for the sake of the remainder of this discussion, we concentrate on purely historical interpretations of PAE cladograms, and defer further ecological discussion of Lundberg rooting until p. 288.

For the purposes of the palaeogeographical and tectonic aspects of historical biogeography, we are only interested in historical events that simultaneously affected large numbers of taxa within an overall biota. Although some divergences in sample locality cladograms will be based on large numbers of taxa (cf. characters in phylogenetics), this will not be so for all divergences. Corroboration of inferred events should therefore be sought in congruences amongst the cladograms of as many different sample groups as possible. Ecological factors arise again here, because there is a higher probability of congruence amongst groups which share similar broad ecological limits and similar dispersal ecology. On the other hand, in order to avoid ecological prejudgement, it is probably better to disregard ecology and to sample different groups of organisms heuristically. As in cladistic biogeography, the recognition of congruent patterns of distribution amongst different taxonomic groups is in itself very strong evidence that they have resulted from the same major historical events, like tectonics or climatic change, regardless of ecological assumptions. The possibility that the same patterns have arisen by chance becomes negligibly small as congruence increases (Nelson and Platnick 1981, p. 503).

So far, we have only considered divergence of biotas, but major historical events can also result in convergence (sometimes also called intermixing or integration of biotas). It might seem that because our method only gives a dichotomous pattern indicative of divergences, it cannot detect convergences. In fact, information about convergences is present in PAE cladograms but is 'cryptic', though Dr Peter Forey (pers. comm.) has pointed out that several PAE cladograms showing congruent polychotomies might indicate convergence. Our method nests

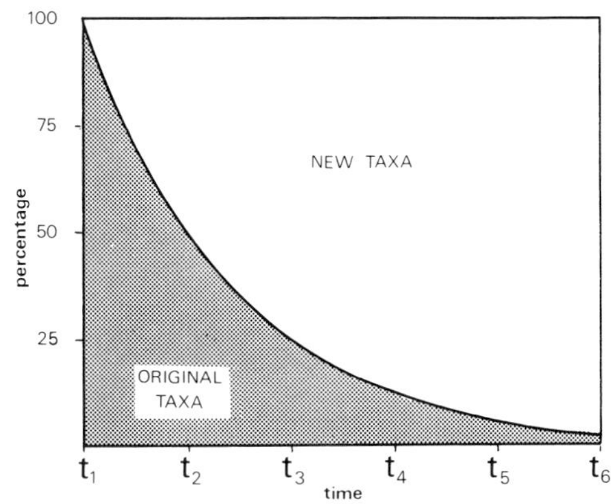

FIG. 1. Model for a constant rate of biotic turnover, shown in terms of a survivorship curve for an original assemblage of taxa (see text below).

sample localities according to how recently they have shared a common biota, irrespective of the real history of convergent and divergent events. To help understand how PAE treats a mixed history of convergences and divergences, it is best to refer to a hypothetical model of changing biotas affected by major historical events (Figs 1 and 2). Note that we are using this model, not as a representation of envisaged processes, but simply to provide a fictitious biogeographical history for demonstrating how our method would perceive it.

As with all models, some initial assumptions have to be made.

1. We have assumed that chance dispersal between areas is insignificant.

2. We have incorporated a major event which affected biotas between each time plane sampled, and have made these alternately convergent (i.e. barrier breakdown) and divergent (i.e. vicariance).

3. We have assumed a constant biotic turnover which is intended to represent the net effect of originations and extinctions of taxa, whether in response to major events, phyletic gradualism or any other factors. Our assumed turnover rate is such that half of the taxa present at one time plane have become extinct and been replaced by new taxa by the next time plane (Fig. 1). For genera of echinoids, this would mean that sampled time planes are approximately 20 to 30 million years apart.

4. We have assumed that each area supports similar-sized biotas.

5. We have assumed that our areas are adequately sampled.

For simplicity, the symbols representing biotas also correspond with areas, which implies that our sample localities are always of the same geographical extent as the biotas. (Such sampling would not be possible in practice.) These 'areas' are therefore not necessarily geographically constant entities between time horizons.

In our model (Fig. 2), each divergence event results in two daughter biotas that have a proportion of taxa in common, inherited from their

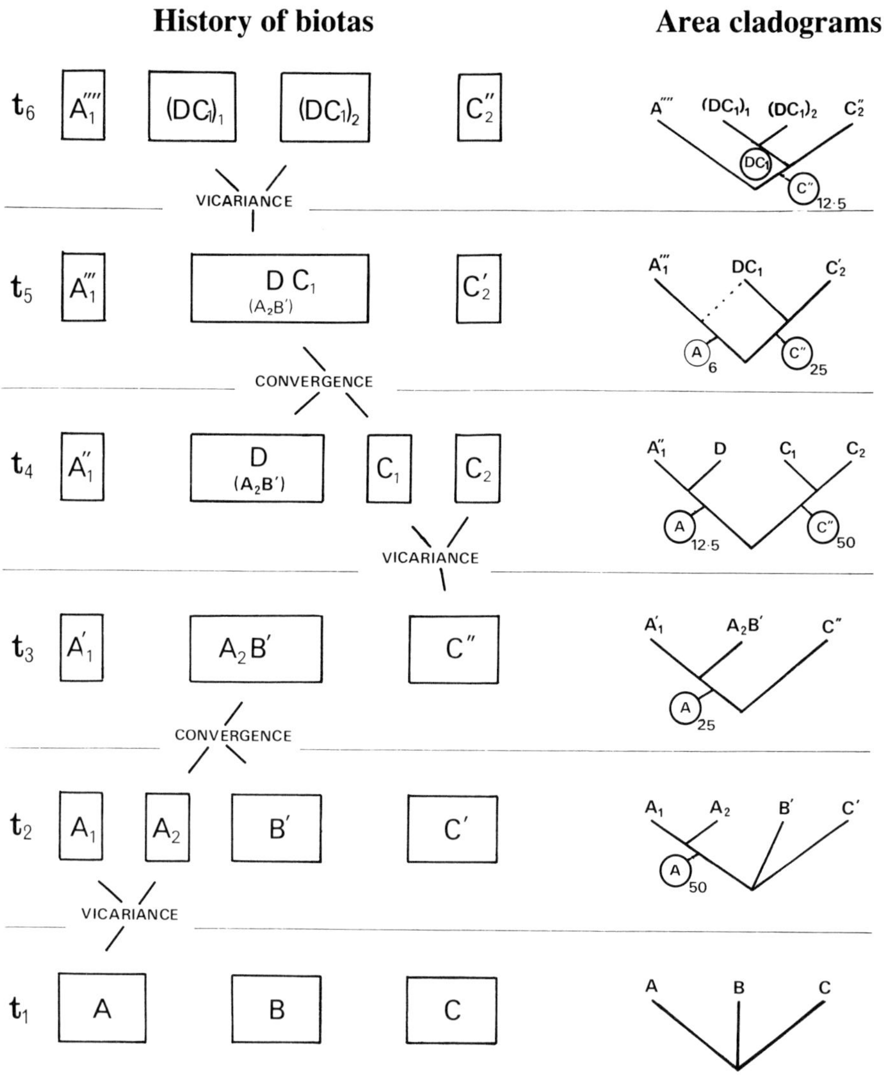

FIG. 2. A hypothetical history of biotas and the corresponding sample locality cladograms, constructed using the rate of biotic turnover shown in Fig. 1. t_1–t_6 = successive time planes. Geographical sample localities are depicted as rectangles, biotas as letters (biotas are assumed to be fully sampled). The initial faunas A,B,C have no taxa in common. Other symbols: C, C′, C″, etc. = turnover of biota C through time; D = endemic element of biota developing after convergence event producing A_2 and B′ faunal mixing. In the area cladograms the common elements of biota shared between sister localities are circled, and the percentage of the total biota shared at the vicariance event that has survived to this time, is also given.

ancestral biota before it became divided. This shared biota acts as a biotic fingerprint to historical kinship of biotas. Through time, however, this fingerprint becomes weaker and weaker, as old taxa become extinct and new ones arise. So for example, the biota common to A_1 and A_2 (i.e. biota A, just prior to divergence t_1/t_2) diminishes through biotic turnover until a later major event (in this case the convergence at t_4/t_5) completely masks the A-fingerprint on the biota DC_1.

Convergence events bring about biotic intermixing, which in geological terms is relatively rapid. Evolution continues within this mixed biota, resulting in the appearance of new taxa within the area which contains this composite biota. New taxa are confined to this area by these most recent biogeographical barriers. Some or all of these taxa may even have originated vicariously in direct response to these barriers. This development of endemicity probably takes place over a longer time span than intermixing, but it remains until disrupted by further events.

Reference to Fig. 2 shows how this process would be seen by PAE cladograms. A newly intermixed biota is linked as a sister-biota to the descendant of whichever pre-convergence biota underwent *divergence* more recently, because this biota bears the stronger fingerprint (i.e. shares more taxa with the newly intermixed biota). For example, in the t_4/t_5 convergence, biota DC_1 shares 25 per cent of its taxa with C'' but only 6 per cent with A, divergence of which (at t_1/t_2) was geologically older than divergence of C'' (at t_3/t_4). Convergences themselves are not shown graphically by the cladograms, but in practice, they can be inferred by comparing cladograms over a time sequence, if between any pair of successive time intervals there is a transposition of sister relationships between three biotas. In Fig. 2, the biota D is sister to biota A_1'' at time t_4, but has switched to being sister of the C-lineage biotas at time t_5, following the convergence at t_4/t_5. The two biotas which have intermixed can be identified as: (1) the transposed biota, e.g. D (in DC_1); and (2) its sister biota, e.g. C (in C_2'), though the cladogram only shows a divergence between them.

It is also possible to relate information derived from data on one stratigraphic time plane, to information from a slightly lower one. This makes it possible for histories inferred for different time planes to be used to corroborate each other, and also recovers information 'lost' through geological time (see p. 277; also Grande (1985) for a cladistic biogeographical equivalent; but for a rejection of Grande's unpublished earlier version of this, see Humphries and Parenti, 1986, pp. 60–4). The minimum time interval between time planes used in this way should be appropriate to the taxonomic level of the data being analysed. Stratigraphic subzones, for instance, would probably be inappropriate for generic data. There will also be a maximum time interval. The main consideration in choice of interval is that of the evolutionary turnover rates for the taxa concerned. This multiple time-plane approach introduces a synoptic approach, in conjunction with the outright historical interpretation of biotas of a single horizon already outlined.

Fig. 2 illustrates how this works. The record of the divergent event which affected biota C'' at time t_3/t_4 is retained through t_5 to time t_5/t_6, where one of its daughter biotas, C_1 (in DC_1) becomes divided again. (Note that in the more recent cladogram (t_6), the older event is to be found nearer the origin of the cladogram than the more recent event.) Thus, the same divergence may appear in cladograms from more than one time plane, with successive divergences displacing each other further down the cladograms through geological time. The same is true for the indirect record of convergences.

Thus, PAE cladograms at all times produce a nested series of biotas, as sampled at particular localities, based only on how recently they have had taxa in common, irrespective of the nature of the historical events. Indeed, it is quite fallacious to use the cladograms as a direct record of vicariance events. A direct reading of the cladogram at time t_6 in Fig. 2, for example, would suggest that all four biotas were united at some time in the past, whereas this is obviously not the case. PAE cladograms only give a graphically direct representation of historical events if there have been no convergences. Nonetheless, each cladogram, interpreted only as a representation of the history (i.e. relative recency of biotic intermixing) of *biotas*, as represented at sample localities, is accurate. It can also be read as an *area* history, by reference to the localities at which the biotas have been sampled, and the pattern in which they are grouped, as discussed in the next section.

Interpretation: (2) areas, time planes, and Earth history

The final step, having inferred a history of biotas as soundly as the data permit, is to translate it into implications for Earth history and palaeogeographical reconstructions. Taken on their own, the cladograms only provide historical hypotheses about biotas—as represented at particular sample localities—in terms of recency of intermixing between these localities. Nevertheless, divergences in cladograms can be read as hypothetical vicariance events, and these imply that barriers to biotic exchange might have appeared somewhere in regions which lie between pairs of sister-sample localities on each side of a divergence.

It is not possible, however, to go further and hypothesize the nature of these barriers or the nature of the events concerned, without introducing major assumptions. Neither can the closeness of relationship between sample localities be used to infer either the 'effectiveness' of the barriers (cf. 'filter barriers') between them, or the palaeogeographical proximity of sample localities. There is, for example, no way of knowing whether a divergence event, affecting shallow-marine groups of organisms, was due to the appearance of a land barrier, to some kind of oceanographic event such as a change in cur rent configurations, or a general widening of oceanic separation between the localities concerned. There is obviously a whole range of possible Earth history events that could account for vicariance and intermixing, these being essentially tectonic, eustatic, climatic, or oceanographic ('TECO' events of B. R. Rosen 1984), singly or in combination (e.g. tectono-eustasy).

Since the nature of barriers and events cannot be specified from parsimony analysis of distributions alone, or in fact from any other analysis of presence/absence distributional data (such as cladistic biogeography or cluster analysis), and since histories inferred from these methods cannot be used as rigorous tests of geological events, or vice versa (Patterson 1983), the main role of biogeographical analysis lies in making test-like comparisons with independently derived histories. This limitation of biogeographical analysis is illustrated by Williams' work (1969, 1973) on the distribution of Lower Palaeozoic brachiopods, in which he was able to fit the same set of cluster analyses into two totally different, independently derived palaeogeographical reconstructions (though part of the problem is attributable to the nature of phenetics; p. 279)). Of course, if no independently derived model of geological events or palaeogeography is available, a biotic event inferred from biogeographical analyses can be used as a palaeogeographical pointer to further research. Well-corroborated biotic distribution histories have stimulus value for those working in other related subjects, and invite explanations (e.g. Humphries and Parenti (1986) on Pacific history and bipolar distributions).

If it seems somewhat discouraging that analytical historical biogeography cannot be used to test geological models, a saving consideration may lie in introducing some basic ecological considerations. For instance, if—as here—we know that we are dealing only with shallow-water marine organisms, then this slightly constrains the range of possible barriers and 'TECO events' implied by the locality cladograms. A different range of possibilities would be implied by results based on terrestrial organisms, though we should expect patterns derived from marine and terrestrial organisms to be complementary (Hallam 1983). Set against this, is that the ecology (especially the habitat and dispersion ecology) of the groups concerned has to be assumed for fossil organisms, although some ecological assumptions will obviously be safer than others.

An important part of interpreting biotic histories is the handling of areas and sample localities, and it can also introduce a subjective element. Although sample localities can be as geographically confined or broad as we wish, each one should be large enough, ideally, to encompass the complete range of local habitats. This would minimize parsimony differentiation between sample localities, which arises from local habitat preferences of organisms rather than historical events. Lambshead and Paterson (1986) have used a method similar to our own for ecological analysis. In practice, complete elimination of habitat factors is probably impossible at the sampling stage, and is best achieved by relying on the congruence approach, in comparing as many different cladograms as possible. Even this may not eliminate regional ecological factors (see p. 288).

Sample localities can therefore be arbitrary in size, and it may be most convenient to use a grid or, as here, outcrop areas for fossil biotas, and island groups for living biotas. Alternatively, a locality set could be devised with a particular geological history in mind, in order to provide a direct, independent test-like comparison with the hypotheses derived from biotas. Tectonic features, like plate boundaries, crustal sutures, suspect terranes, and ocean/continent boundaries (at the present day or conjectural in the past) can be used to define a sample locality set for investigating a tectonic history, and similar schemes could be devised for other TECO events.

If localities have not been grouped beforehand against the background of a particular independently derived history, the problem of grouping localities can arise again after initial analysis, since the resulting sample locality cladograms may be too locality-specific to be useful or comprehensible. Grouping may also be helpful, at the risk of loss of real differences of history—if distributional information from different organisms is based on similar but not identical localities—by designating areas which accommodate differences in the locality sets. The task of grouping may be made easier at this stage by considerations of congruence between the distributional analyses of different organisms, since there will probably only be congruence for parts of the locality sets.

Thus, although choice and grouping of areas and localities introduces an arbitrary and subjective element into PAE, it is offset by the fact that it can be made explicit, and that it should be acceptable so long as it is used experimentally for test-like comparisons with independently derived models.

A similar problem arises with choice of stratigraphic horizons. As with localities, there are arbitrary factors in the choice. The *interval* between horizons is discussed on p. 285. There is also *time span* of the sample horizon to consider (subzone, zone, stage, etc., or an absolute time span), because it is not possible to take a true instant of time as a sample base in geology, other than the present. As with time intervals, choice of span will be affected by available stratigraphic resolution, and the evolutionary rates in the organisms concerned. Note that each cladogram carries historical information, which in theory should cover a span of time as far back as the stratigraphically oldest taxon in the data set (Figs 1 and 2), though in practice probably only as far back as the average survivorship duration within the taxonomic group concerned. On the other hand, of course, inferred events cannot be younger than the uppermost level of the most recent stratigraphic horizon used in an analysis.

Use of time spans implies that the distributional records which make up a particular data set must be an integration of occurrences which are scattered throughout the time span, and this may be as long as several million years, depending on choice of horizon. Use of congruence should help to eliminate some of the effects of stratigraphic scatter on individual cladograms. As with areas, choice of actual horizon, and of time span, can be made for comparison with an independently derived history, or be neutral and heuristic. There is scope for establishing stratigraphic constraints by analysing the ages of individual taxa on which a particular grouping of two or more localities is based, but we have not explored this here.

We have assumed here that all biotic divergences identified in a cladogram followed geologically soon after the major events that caused them, but this may not always have been true. Moreover, vicariant differentiation of taxa acts at the species level and presumably has a progressively longer time-lag in its effect at successively higher taxonomic levels. If, for example, genera are analysed (as here) we may have to bear in mind the possibility of a significant time-lag between a geological event and its resulting biotic event. Indeed, we have no data even for the species-level time-lag, though it is probably related to evolutionary rates—but varies with the actual taxa and events concerned. In other words, although PAE produces a history of possible events, we cannot judge what their actual date and their time resolution are. This needs to be borne in mind when making comparisons with geologically derived histories whose stratigraphic control is often better known and more precise. For convenience we have summarized our assumptions in using PAE in Table 2.

Some currently unresolved problems of PAE

Although PAE has parallels with cladistic analysis in phylogenetic reconstruction, there are a number of significant differences which create problems for interpreting sample locality cladograms in biogeographical terms.

1. by direct analogy with cladistic phylogenetic reconstruction, all absences of taxa from a biogeographical data set should be treated as unknown character states (i.e. the taxon may genuinely be absent or may be present but as yet undiscovered). We have had to assume, however, that all absences of taxa from faunal lists are genuine, and thus our method suffers from the same problems of sampling bias and ambiguity in its data set, as in all methods other than cladistic biogeography. Although it seems to us that our assumption will provide a reasonable approximation to the true state of distributions, when dealing with well-studied groups with a good fossil record, certainly, this cannot always be the case.

2. whereas non-congruence in a phylogenetic cladogram is explained by postulating convergence, parallelism and non-homology in some characters, non-congruence in our biogeographical analyses implies that one or more

TABLE 2. *Summary of parsimony analysis of shared presences method and the assumptions involved*

A. Data

1. Taxon distributions form an overlapping mosaic which changes through geological time due to historical factors.
2. Fossil record, though incomplete, is a biogeographical sample which carries vestigial biogeographical 'signals' and which can be analysed by rational methods to produce testable historical hypotheses.
3. Biogeographical mosaics (present or past) can be analysed by sampling them at particular points (sample localities) within a particular stratigraphic horizon, using taxonomic lists of 'presences' (i.e. without prior need to identify ranges or areas of endemicity).
4. Choice of sample localities can be arbitrary or made for direct comparison with an existing independent biogeographical hypothesis.
5. Taxa used for analysis are non-polyphyletic.
6. Time factors mask older events, making it necessary to use sequences of data based on different stratigraphic horizons (synoptic approach) to maximize use of information in fossil record.
7. Stratigraphic control is adequate.

B. Method

1. Shared presences of taxa between sample localities can be analysed like homologies (synapomorphies) in phylogenetics, and sample localities can be treated like taxa in phylogenetics, to give sample locality cladograms.
2. Lundberg rooting (hypothetical sample locality with no taxa present) gives historical basis for interpreting sample locality cladograms, i.e. presence of a taxon is more biogeographically 'advanced' than absence.
3. Parsimony is the only criterion for choosing between competing biotic histories.

C. Results

1. Sample locality cladograms represent historical hypotheses about relative recency of biotic exchange between sample localities.
2. Suites of shared presences between sample localities represent biogeographical assemblages (or biotas).
3. Dichotomies (nodes) in sample locality cladograms represent divergence events between biotas. (Convergences are not shown but are not precluded.)
4. Age of all events, within a particular cladogram for a particular horizon, must lie within the age of the oldest taxon used in the analysis and the youngest horizon from which data were taken.

D. Interpretation

1. There are two biogeographical models for interpreting divergences: random events and major events disrupting whole biotas, e.g. due to appearance of a new barrier (vicariance).
2. Non-congruence within a single data set is produced either by convergence of biotas, or by localized extinctions and clade-specific dispersal, both of which act to mask or destroy any pre-existing divergence pattern.
3. Congruence of sample locality cladograms between different groups of organisms at the same geological horizon suggests that extrinsic events have affected the biota. Lack of congruence points to other kinds of events (e.g. dispersal) or other factors (e.g. poor stratigraphic control).
4. Probability of congruence is maximized by choosing groups which share similar broad ecological characteristics, though this introduces extra (ecological) assumptions into the analysis of fossil distributions.
5. Hypothesized vicariance events represent the appearance of biogeographical barriers which lie (or lay) in areas between the sample localities concerned.
6. The nature of biogeographical barriers is not deducible from sample locality cladograms, without also using ecological features of the group concerned (which introduces extra assumptions into the analysis).
7. Palaeogeographical distances (e.g. between biotas, or barriers, or areas), whether relative or absolute, cannot be deduced from sample locality cladograms alone.

of our taxa (cf. characters in phylogenetics) may have originated independently at several times or places. Although a few taxa may genuinely be polyphyletic, we doubt if this would be a significant feature of the data base, and have assumed that our taxa are largely non-polyphyletic. We believe that the best solution to this problem in future work is to calculate the most parsimonious cladogram, using Dollo's Law (Farris 1977), an option that is available in some numerical cladistic programs, but one which we have not yet tried. According to this rule, taxonomic characters are transformed from 0 to 1 only once (i.e. in PAE, taxa evolve only once), but may be transformed from 1 to 0, as often as is required (i.e. a taxon can be lost from as many sample localities as necessary).

3. as already discussed, there is the problem of rooting and the interpretation of the sample locality pattern. By using Lundberg rooting, which we feel is the current best option, we cannot eliminate the possibility that PAE cladograms may reflect, either wholly or in part, latitudinal or other major ecological gradients. A nested hierarchy of sample localities showing a pectinated arrangement (e.g. Fig. 3: Indo-West Pacific to New Zealand in the echinoid cladogram) can be produced by an ecological diversity gradient on a large or global scale, such as latitude, to the same degree as by a historical sequence of vicariance events. Thus, these kinds of ecological effects and historical factors are impossible to separate when PAE produces pectinated patterns, without recourse to additional assumptions or reasoning. Furthermore, areas with small biotas, whether due to ecological gradients or collection failure, will inevitably appear as primitive in the sample locality cladograms. A successively dichotomous arrangement of localities, however, indicates that pairs of sister-areas do have their own unique endemicity, which cannot easily be explained in terms of diversity gradients. An interim solution is to group all localities that occur within a pectinated set together—into a larger area which embraces them, or otherwise discount, provisionally, all events suggested by divergences within the set.

4. although PAE seems, in the end, to settle for quantity rather than quality of data, the resultant cladograms do provide the best-corroborated hierarchy of sample localities using available data on shared endemic taxa, and they therefore provide working hypotheses to be tested further.

Data for the present study

In our study, we have chosen to work at the level of genus, as this is the lowest feasible taxon to use on the global scale we are interested in, and higher taxonomic levels are even less reliably monophyletic. Data on echinoid distributions were assembled for the present day, early Miocene, mid- to late Eocene and late Cretaceous (Campanian/Maastrichtian) times. The data on Recent distribution pertain only to Clypeasteroida and are taken from Ghiold and Hoffman (1986). Other data sets are for total echinoid faunas and were compiled from published monographs, supplemented by data from the collections of the British Museum (Natural History) (see Appendix). Where monographs predate the *Treatise on Invertebrate Palaeontology*, generic determinations were amended where necessary.

Reef-coral data were assembled for the Recent, Lower Miocene, and Upper Jurassic. For the Recent analysis, these were almost entirely from an unpublished table kindly made available by Professor John Wells (Cornell University), compiled and continually updated by him from publications and from unpublished lists sent to him by numerous coral workers. The Recent localities used here include the same lists for the eighteen circum-tropical localities that B. R. Rosen (1984, 1985) used in previous papers. For the Lower Miocene, the coral data used here were originally assembled in order to examine the biotic affinities of an important new Iranian collection (B. R. Rosen and Darrell, in press) with respect to Miocene closure of Tethys in the Middle East. It is for this reason that the localities are relatively close together in the southwest Asia area. For other Miocene areas, see Appendix. Upper Jurassic data were taken directly from Beauvais (1973).

Results

Presentation

The results are presented in two forms, firstly as cladograms of the actual sample areas, and secondly as simplified cladograms based on the shared congruent elements of all the cladograms for a particular horizon (Figs 3–6). These congruent cladograms are labelled with broad regional names, representing groupings of the original sample areas, our exact choice of naming being essentially arbitrary (with some awareness of current palaeogeographical hypotheses).

We will also use these names throughout the following sections. It is, however, important to realize that they imply extrapolation from our sample points to broader areas which include them, and this by itself may tend to blur the very relationships that we are trying to establish rigorously. Our use of 'Pacific' and related geographical terms, for instance, are generalizations based on relatively few sample localities, and we hope that these can be made more precise in future work.

The results are no more than a preliminary attempt to use generic distribution data to identify possible palaeogeographical events from mid-Cretaceous onwards, in a rigorous way that can be tested by further work. All results must be treated as provisional. For the Recent and Lower Miocene sediments the two data sets, one for echinoids and the other for corals, together provide a certain amount of congruence. Similarly, there is some corroboration between cladograms from successive time planes. Although in the following section we emphasize only those events for which we have a limited amount of corroboration, we also mention other possible events and show how this kind of area cladogram can be interpreted.

These results can be regarded in two ways. Firstly, statically, as indicators of endemic patterns at each time plane, and secondly, dynamically, as historical hypotheses. The endemic patterns that are revealed by this method are hierarchical and use all of the available data, in marked contrast to many palaeontological faunal and floral provinces (see p. 278).

Recent data (Fig. 3)

Both analyses of echinoids and reef corals gave reasonably high consistency indices (0.61 and 0.65). Comparison of coral and echinoid data gives a high degree of congruence, with the Indo-West Pacific, West Indian Ocean, and Mid-Pacific grouped together and unresolvable on current evidence. An interesting exception, however, is that the affinities of East Pacific echinoids lie with the Caribbean (a pattern that is well established for numerous other organisms; see examples in Humphries and Parenti 1986), whereas the East Pacific corals are apparently more closely allied to the rest of the Indo-Pacific. This may be an example of one group (reef corals) which has had a distinctive history for ecological reasons.

These analyses of the Recent faunas are especially important, because they can be compared with the observable geography of today to give a qualitative notion of how accurately our results reflect modern geography. It is reassuring, for example, that they clearly reveal the importance of the two main land barriers that lie across low latitudes: the Americas and Eurasia–Africa. But, notably, they also show that in each of the two respective major oceanic regions separated by those barriers (the Atlantic and Indo-Pacific) there are also subsidiary endemic areas. These areas presumably reflect biogeographical barriers, but it is important to note that the barriers do not correspond with any likely landmasses, either now or in the recent past, except possibly along the line defined by south-east Asia–Indonesia–Australia (B. R. Rosen 1984). For interpreting past events from fossil distributions, these intra-oceanic barriers suggest that it is possible to have a complex system of partial ('filter'?) barriers, which is not readily specifiable in tangible geographical terms, within a single (albeit large) marine basin. They also demonstrate the possibility of mid-oceanic vicariance, which is of interest with respect to Hallam's (1978) question: 'How is speciation promoted in an equable shallow marine environment?'

Historical interpretation of these Recent area cladograms suggests a sequence in which the Atlantic and Indo-Pacific first became separated from each other. For the echinoids at least, the faunas of the East Pacific and north-eastern Pacific were at this stage still part of the Atlantic group. Following this major divergence there was, within each main oceanic region, a series of as yet uncorroborated vicariance events, in which the Eastern Pacific and north-eastern Pacific also became separated from the Atlantic. For the reef corals and for Indo-Pacific echinoids, these later divergent events were like the splitting away of successively less remote areas, similar to onion layers, from the 'core' regions of the Caribbean and Indo-West Pacific. Some of this pattern, however, may reflect the small sample sizes in the more outlying areas (e.g. for corals, Cocos-Keeling, East Pacific, Bermuda, Brazil, East Atlantic, and Persian Gulf), since PAE interprets small biotas as 'primitive'. The pattern may, therefore, represent instead the effect of ecological gradients (p. 282).

Lower Miocene data (Fig. 4)

The consistency indices are somewhat lower than for Recent data (0.43 and 0.51), but the two area cladograms are largely congruent. Allowing for the smaller number of sample areas, the provincial patterns of the Lower Miocene correspond with the present-day analyses in showing an important break in the Eurasia–Africa region. This separates an Indian Ocean and eastern Tethys fauna (Indonesia, Iran, East Africa, Madagascar, and West Pakistan) from an Atlantic and western Tethys fauna (Mediterranean and Caribbean). However, compared with the present day, the Pacific appears as a very distinct fauna, rather than as part of an Indo-Pacific whole. The eastern Tethys and Indian Ocean, when together, have shared a common fauna at more recent times, than either area has with the Pacific.

Historical interpretation of this pattern suggests that the oldest detectable vicariance event separated the Pacific fauna from a Tethyan fauna (Atlantic, Mediterranean and Indian Ocean). The eastern Tethys plus the Indian Ocean then separated from the western Tethys along a line that apparently lay between present-day Saudi Arabia (with its Mediterranean

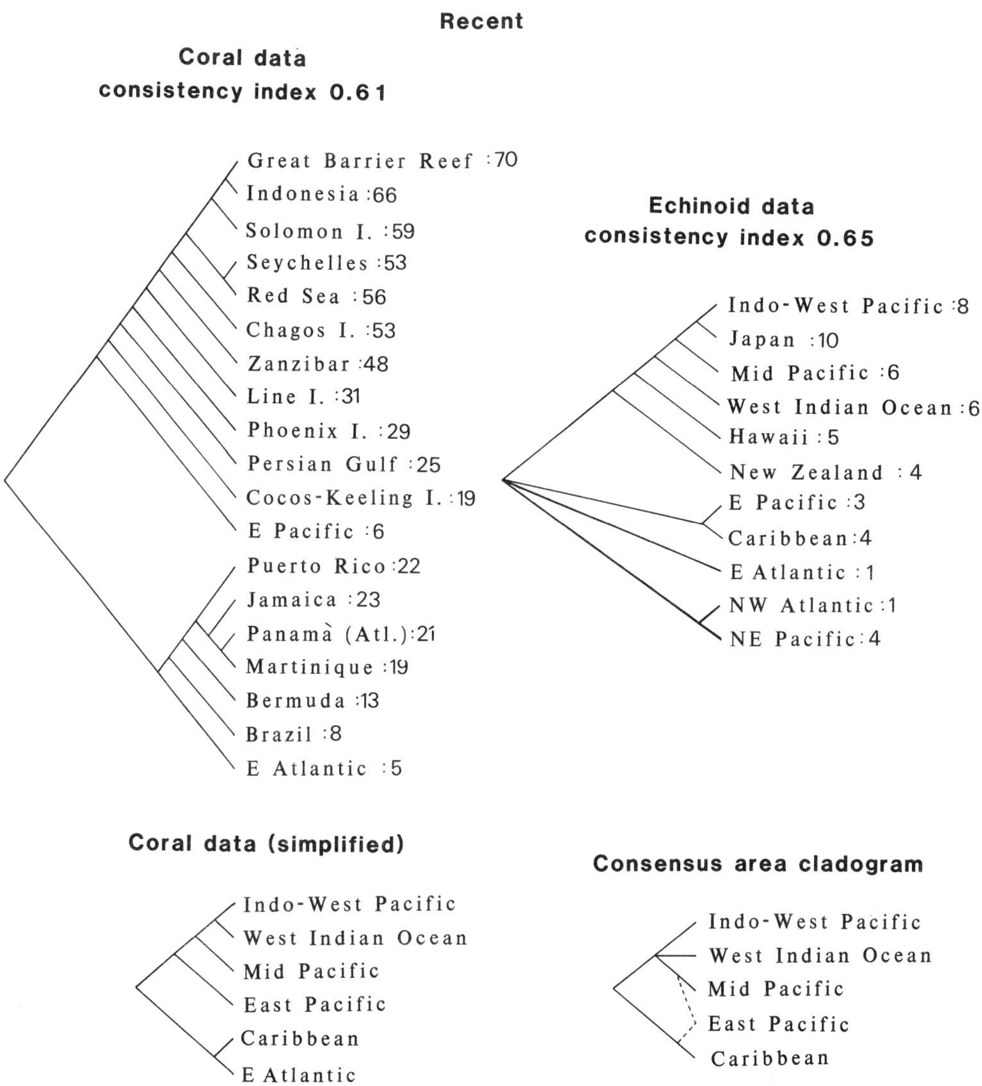

FIG. 3. Sample locality cladograms for Recent distributions of reef corals and clypeasteroid echinoids, and a consensus area cladogram showing only those elements that are common to both cladograms. More than fifty equally parsimonious trees were found for the echinoid data, hence the basal polychotomy in the cladogram. Numbers after sample localities refer to the number of taxa present that are also found in at least one other sample area.

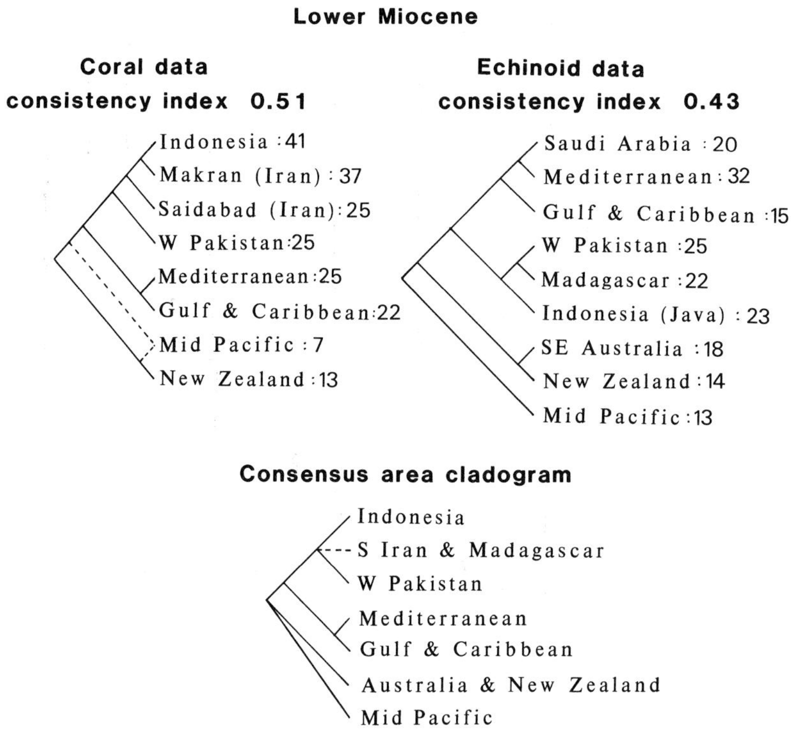

FIG. 4. The most parsimonious sample locality cladograms for Lower Miocene distributions of reef corals and shallow-water echinoids, and a consensus area cladogram showing only those elements that are common to both. Significance of numbers after localities is as for Fig. 3.

echinoid affinities) and Central Iran (with its West Pakistan–Indonesian coral affinities), although this needs further corroboration. The most recent events, with respect to the Lower Miocene, are the divergences within eastern and western Tethys respectively. The most notable of these is between the Caribbean and Mediterranean, a vicariance event that is also seen in the Recent area cladograms.

Although the coral and echinoid data give support to each other through partial congruence, further independent corroboration comes through comparison with the Recent results, since the essentially Atlantic–Indian Ocean, or eastern–western Tethys divergence is common to both sets of results. From the ages of the faunas used, this vicariance event must be older than the end of the Lower Miocene, and probably older than Burdigalian. (Evidence presented later suggests that the history of this divergence may be considerably older.) This in turn implies that, at this taxonomic level, the Recent cladograms have a historical 'depth' of over 20 million years, supporting our earlier surmises based on taxonomic longevity. (Our two geologically older cladograms, however, are for echinoids only, which, on survivorship grounds (p. 277), should have less 'depth' than the coral ones.)

For the Tethys–Pacific vicariance event, there is no congruence support from the Recent analysis. We might have expected to see this below the oldest divergence of the Recent analysis, so it may simply be too old an event to have left its trace on Recent distributions. Indeed, the same divergence does appear as the youngest vicariance event in the Eocene analysis (below), suggesting that it is of late Cretaceous to Palaeocene age. Hence, its absence from the Recent analyses is much more likely to be due to overprinting by later convergence events, which brought the Indian and Pacific Ocean faunas together to make the modern Indo-Pacific grouping. This is supported by the apparent switch in sister-area relationships of the Pacific fauna on the area cladograms between the two time periods. We conclude that the Indo-Pacific fauna (as an assemblage) is itself relatively young, and resulted from a post-Lower Miocene convergence event or events, though of course this does not imply that all the taxa within that region are as young.

Middle to Upper Eocene data (Fig. 5)

Only one data set is available, producing two trees with a consistency index of 0.5. This is the first analysis in our order of treatment to show a vicariance pattern which is not directly recognizable with respect to the modern Earth. It poses a particular problem, in that the Indian Ocean localities are divided into two pairs, one of which (West Pakistan and Madagascar) has had faunal exchange with the Mediterranean region of Europe in the more recent past than it has had with its closer (modern) neighbour pair in the Indian Ocean (Somalia and eastern Arabia). Yet this second pair today lies between Europe and the first pair, thus appearing like some kind of biogeographical 'island'. Madagascar, moreover, is grouped with West Pakistan, whereas today it lies closer to Somalia. These apparent problems, and any relationships which they might reflect concerning plate movements, can only be resolved with further analyses. They may reflect the inadequacy of data taken from older systematic works, or could be an artefact produced by stratigraphic differences between the faunas being compared.

The Indian Ocean and Mediterranean group is separated as a unit from a transatlantic pair (southern USA with the Caribbean, and Senegal), suggesting a latitudinal difference within the Atlantic and its adjacent seas. The two Pacific localities (south-east Australia and Mid-Pacific) are separated from all other regions—as in the Lower Miocene area cladograms—but possibly because they contain only a few taxa. Similarly, a separation of Tethyan localities on either side of the present Middle East, as for the Lower Miocene data, is also apparent.

From a historical point of view, therefore, this Pacific distinctiveness implies an event which we can be slightly more confident about. It represents a divergence, probably pre-Eocene (see above), between the combined eastern Tethys/Indian Ocean and part of the Pacific. Two other divergences correspond less precisely with cladograms of other ages. The intra-Atlantic latitudinal divergence also appears in the late Cretaceous results, and the broadly Middle East, intra-Tethyan divergence occurs in both the late Cretaceous and early Miocene data. The relative timing of these events (if, that is, they are not simply latitudinal gradients) in each area cladogram is, however, not the same, so one of them (the intra-Tethyan event) may have occurred twice, once in the late Cretaceous or before and once between the latest Cretaceous and earliest middle Eocene.

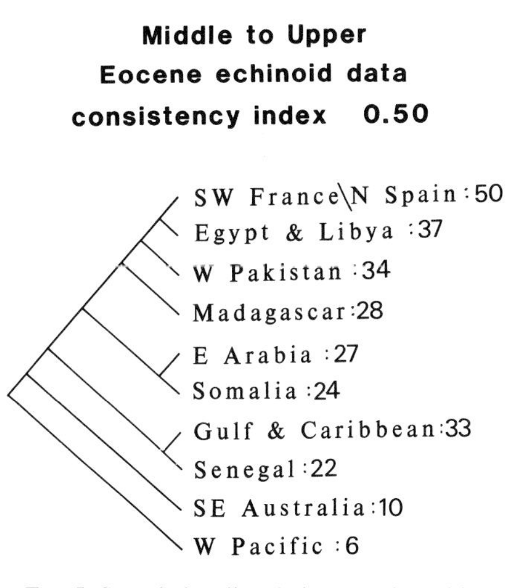

FIG. 5. Sample locality cladograms for Middle to Upper Eocene echinoids. The two alternatives shown are equally parsimonious. Significance of numbers after localities is as for Fig. 3.

Upper Cretaceous (Campanian/Maastrichtian data) (Fig. 6)

Our data for Campanian/Maastrichtian echinoids gave four equally parsimonious cladograms, with the lowest consistency (0.42) so far. This low consistency is largely a result of the ambivalent position of the paired regions, north-east Arabia and southern Iran, the program offering two alternative relationships; one linking it to Asia (Baluchistan plus South Tibet), the other with the Atlantic/Europe/Indian Ocean group. Neither Baluchistan (the Mari Hills) nor South Tibet (south of the Zangbo suture line) is well monographed, and they may therefore appear 'primitive' through pseudo-absences. (Indeed, examination of the Tibetan fauna by A. B. Smith since we analysed the present data has revealed a number of taxonomic errors.) The major ambiguity resulting from this switching obscures almost any major pattern, though two geographical regions are consistent in themselves, namely an Atlantic/Mediterranean/Europe group and a north-east Arabia and Iran group.

The polychotomy prevents any inferences being drawn about relative timing, but as noted already, a similar intra-Atlantic event occurs

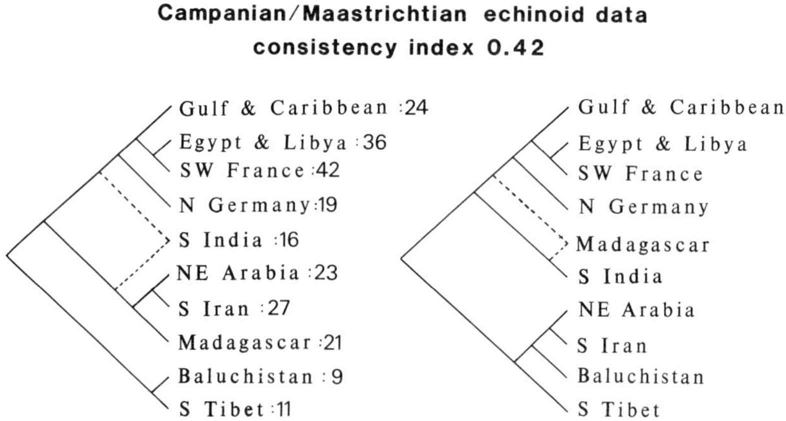

FIG. 6. Sample locality cladograms for Upper Cretaceous (Campanian–Maastrichtian) echinoids. The two cladograms shown are equally parsimonious, and each has two alternative positions for one locality (South India and Madagascar, respectively). Significance of numbers after localities is as for Fig. 3.

both here and in the Eocene data, and points to this being the most certain of the events suggested by the evidence so far. The Indo-Pakistan affinity with Madagascar is eye-catching, implying faunal exchange between these two now widely separated localities at some time earlier in the Cretaceous.

Upper Jurassic data

Our results, for corals alone, were disappointing and are not even reproducible, consisting of almost three hundred equally parsimonious alternative area cladograms! Consistency (0.295) was, not surprisingly, low, suggesting that either the data matrix was bad, due to poor taxonomy or inaccurate distributional information, or that the distributional pattern of corals in the late Jurassic was dominated by chance dispersal and/or convergence events, or a combination of these factors.

Discussion

Geological history?

So far, we have concentrated on inferring divergences and convergences, without further interpretation of these events in terms of geological history. Left like this, they represent a putative faunal history of what is usually regarded as the 'Tethyan' region (with Pacific). In this section we go on to compare some of the least uncertain of these inferred events with selected, independently derived, geological events. It is beyond the scope of this chapter to review all the existing literature on Tethys history that might bear on our own inferred events, so we leave readers, with knowledge or interests not discussed below, to make their own comparisons. We discussed the basis on which we make the comparisons on pp. 285–7.

Intra-Tethyan divergences in the Middle East region

The literature is widely in agreement that, for the time covered by our results, Tethys was largely a continuous marine region connecting the Atlantic to the Indian Ocean until final closure in the Miocene, though opinions differ on final closure date within the Miocene. There is also frequent mention of the cosmopolitanism of Tethyan biotas in both the Mesozoic and Palaeogene. Yet our results show faunal divergence across the modern Middle East region in all four sets of cladograms, from the Campanian/Maastrichtian to the Recent. Does this represent one event or recurrence of divergences in more or less the same region?

We found preliminary evidence (above) for at least two events, both prior to the Oligocene. The similar pattern of divergence found in both of the later area cladograms could also represent further events, or simply be the continuing imprint of the earlier divergences through the rest of the Cainozoic. If so, it would amount to saying that the modern differentiation of faunas (at generic level) on each side of the Middle East region goes back, not to the Miocene, as is widely stated (e.g. for corals, Wells 1956), but to some time in the Palaeogene or even late Cretaceous. This was actually implicit in the analysis of the stratigraphic ages of extant reef coral faunas by Stehli and Wells (1971) and B. R. Rosen (1984), who showed that living Atlantic reef-coral genera are predominantly Cretaceous to Palaeogene in age. In other words, although the modern Indo-Pacific and Atlantic came into being as *physical* entities during the Miocene (by almost universal agreement), the marine *biotas* on either side of the Middle East had become differentiated at generic level a long time before the Miocene. It seems that final closure of the Middle East (itself a modern observable fact) affected shallow-marine organisms principally at species level (e.g. Adams *et al.* 1983) and was not detectable by our analysis.

An alternative possibility is that the Middle East emergence is older than the Miocene age generally attributed to it. Certainly, the precise age of final closure is not widely agreed as yet, so it may be of interest to briefly consider independently derived histories of this region. Taking the pre-Miocene first, Adams (1981) mentions the possibility of a marine regression which left a dry land-bridge area across the Middle-Eastern Tethys in the Palaeocene (Danian), and this might correspond with the later of our two inferred events, as seen in the Eocene results. Adams also mentions a similar event that may have taken place in the Oligocene, and so the divergence seen in our early Miocene results may be this Oligocene event, not the continued imprint of the Danian one. For the Miocene closure of Tethys ('the terminal Tethyan event'), Adams *et al.* (1983) believe that this occurred in the early Miocene (mid-Burdigalian), and other authors (e.g. Schafersman and Frost 1979; Ali 1983) have also suggested a variety of early Miocene dates. Rögl and Steininger (1983, 1984), however, believe that final closure had occurred by 11 to 12 Ma (early Tortonian–Pannonian–late Bessarabian) in the late Miocene, with periods of evaporite deposition and shoaling in the Middle East seaways, at various times between the late Oligocene and late Miocene.

Unfortunately, our data lack the resolution necessary to decide between these competing hypotheses, and any number of events can be hypothesized between our sample horizons. The resolution of biogeographical data could be increased by using lower taxonomic groupings (species rather than genera) and more numerous sample horizons of shorter duration (stages rather than groups of stages). In this way, consecutive identical area cladograms could be taken as evidence of continuing divergence and isolation of faunas, whereas alternate clustering and divergence through time would provide proof of multiple vicariance events. (We chose our taxonomic level and sample horizons to investigate global changes in biotic distributions, showing that it is important to choose a level of taxon and geological time that will be relevant to the question in hand.)

In summary, our present results do not test how many vicariance and convergence events there were in the late Mesozoic to Cainozoic history of the Middle East, but we can state that there has been a longer history of Atlantic/Indian Ocean faunal differentiation than has usually been recognized, and this extended back to the Cretaceous. From the limited resolution gained in our study, there is evidence for at least two and probably three such vicariance events. Faunal divergence is not all due to final closure in the Miocene. Too much emphasis has been placed on the presumed effect of final closure on biotas, though we recognize the converse value of detailed taxonomic evidence at species level for mapping and dating this event.

Isolation of the Pacific faunas until the Neogene

According to Schlanger and Premoli-Silva (1981), there was large-scale volcanicity and uplift in the mid-Pacific during the Cretaceous, and this coincided with, and probably caused, a faunal convergence of the Mid-Pacific with the Caribbean. Their suggestion implies that the Mid-Pacific must previously have been cut off, not only eastwards towards the Caribbean, but also westwards, otherwise, the Earth being round, Caribbean faunas could have been in exchange with the mid-Pacific via other directions than the East Pacific. Unfortunately, we have no relevant late Cretaceous data to compare with Schlanger and Premoli-Silva's suggestion, but our Eocene results do suggest that at some time before or during the Upper Eocene, the Mid-Pacific was in faunal interchange with other regions, although we cannot say in which direction.

Schlanger and Premoli-Silva observe that later, 'by the Tertiary', the Mid-Pacific faunal interchange with the Caribbean was disrupted and convergence occurred instead with faunas to the west (i.e. eastern Tethys and the Indian Ocean). Our results do not contradict this, so long as this western convergence was Eocene or earlier. But, in addition, they also suggest that by a later date (no later than the end of the

Eocene), there must also have been a divergence event or events which isolated those parts of the Pacific (i.e. those actually sampled by us) in all directions. In fact, it is simpler to infer a single event of simultaneous convergence to both east and west of the mid-Pacific in the Cretaceous from our results (as discussed by Kay 1980, 1984), rather than a sequence of switched convergences, as suggested by Schlanger and Premoli-Silva, and for which there is, as yet, no evidence. This event was then followed by divergence, and isolation by no later than the end of the Eocene.

The nature of the barriers that were involved in this Pacific history must remain, on present evidence, speculative. The present form of the Pacific basin makes it difficult, however, to imagine how there could have been any land barriers across its tropical part, in the time span covered by our study, so it seems that sheer isolation of island localities must have been important. In other words, we envisage that any barriers that might have existed in the Pacific consisted of great oceanic expanses, without intermediate islands, and that it was distance, possibly coupled with oceanographic factors, that limited the geographical ranges of shallow-marine benthic organisms. The East Pacific today is usually regarded in this way.

Conversely, the breakdown of these oceanic barriers could arise from changes in oceanography or the emergence of islands and landmasses in previously empty regions, or both. These would have had a 'staging post' role (B. R. Rosen 1983, 1984) for previously confined biotas, providing them with habitats in places where no suitable habitats existed before, and so allowing them to expand their ranges into previously unoccupied regions. ('Staging posts', a dispersion concept, may also have acted as stepping-stones, a dispersal concept, but, by definition, the reverse cannot be true.) The speculative explanation for mid-Pacific biotic convergence, advanced by Schlanger and Premoli-Silva (1981) and widely cited by others, is that a significant land area or dense archipelago emerged in the Mid-Pacific, as a result of large-scale Cretaceous volcanicity, and this effected *eastward convergence* of biotas, while its later subsidence brought about their *eastward isolation and divergence*. Whatever TECO model is adopted for explaining the biotic events, our limited data suggest that a Cretaceous or earlier phase of faunal intermixing was followed by isolation and divergence in the Pacific, in effect by the middle Eocene, and this is compatible with Schlanger and Premoli-Silva's hypothesis.

This isolation of the mid-Pacific continued until the *westward convergence* in the post-Lower Miocene resulted in the appearance of an Indo-Pacific fauna (as distinct from the previously different Indian and Pacific Ocean faunas). The most likely factors which brought isolation to an end are the northward movement of Australia towards south-east Asia, and the associated uplift of additional island masses in the West Pacific, like Papua New Guinea. These regions would have had a 'staging post' role between the Indian Ocean (and Indonesia) and the Mid-Pacific. Audley-Charles (1981) gives a later-early Miocene date for the start of collision between Australia and south-east Asia, and this fits well with the post-early Miocene date of West Pacific convergence, inferred from our own results. In addition, the westward movement of islands on the Pacific Plate could also have caused a convergence of biotas at Mid-Pacific localities with those west of the western plate boundary, but this is beyond the resolution of our study.

Finally, it is also of note that in Springer's (1981, 1982) cladistic biogeography of Recent shore fishes of the Pacific, he concluded that the Western Pacific plate boundary was a consistent vicariance line. Nelson (1985) in his remarks on the cladistic biogeography of anchovies notes a divergence between the 'West Central Pacific' and the 'Indo-West Pacific' which may also represent the same event. These observations could be explained by the same pre-Oligocene divergence inferred here, which isolated the biotas on the Mid-Pacific islands from those around the landmasses to the west. These landmasses would have been concentrated west of the plate boundary (on today's evidence), due, in part, to subduction zone activity. Continuing plate collision, and associated loss of crust between the West and Mid-Pacific since that time, would mask the degree of physical isolation of the Mid-Pacific as it might have been in the pre-Oligocene (though it should be possible to estimate it).

In general, our results corroborate Kay's (1980, 1984) synthesis of the history of the Pacific marine biota, except that our evidence for a western convergence indicates a post-Lower Miocene rather than an Upper Eocene event.

Eastern Pacific

Complementary to the Mid-Pacific isolation in the Palaeogene, there is, today, still a major oceanic barrier between the Mid-Pacific and the Americas. Most authors believe that this barrier is at least as old as the Lower Miocene and, as discussed in the previous section, it may go back through most of the Palaeogene. Since the Miocene, echinoid faunas on the Pacific coastline of the Americas, as well as in the Galapagos and other Eastern Pacific islands, have had the Atlantic fauna as a sister group (prior to this we have no relevant data). The same kind of pattern is well established for numerous organisms, the general explanation being that the seaways between the Caribbean and Pacific remained open until the formation of the Isthmus of Panama in the Pliocene, maintaining faunal exchange between the two areas, while the Eastern Pacific oceanic barrier prevented exchange westward with the rest of the Pacific.

Our data on reef corals, however, prove an exception to this pattern. Dr. Nancy Foster (pers. comm.) states that there are Pliocene corals in the western coast region of the Americas with Caribbean affinities (not included in our samples), suggesting that corals at that time had the same pattern of distribution as other groups. It seems highly unlikely that selective extinction of those corals with Caribbean affinities has taken place, leaving only those with old Pacific affinities. More likely, in our opinion, is that Eastern Pacific corals have suffered a major extinction event after the closure of the Panama seaway, and that what we now see is the colonization of the Eastern Pacific by elements of the Indo-Pacific corals, producing a late Cainozoic trans-Pacific convergence. Heck and McCoy (1978), however, prefer the idea that the corals concerned must have continued to exist in populations not found in the fossil record of the Eastern Pacific. Our method does not resolve these alternative arguments, but their hypothesis is as open to the criticisms of special these alternative arguments, but their hypothesis is as open to the criticisms of special pleading and untestability as they say long-distance dispersal is.

Present distribution

Our results support Kay's (1980, 1984) conclusions that the present patterns of endemicity and cosmopolitanism in shallow-marine biotas reflect a long history of fragmentation and divergence since the Cretaceous, especially for the Pacific, but we would add that Tethyan faunas themselves were not cosmopolitan during much, if any, of this time, even during the Cretaceous. This can be deduced from the intra-Tethyan divergences identified throughout the span of our study. It is clearly an oversimplification to regard modern biotas as having derived from a 'previously widespread pan-Tethyean [*sic*] biota' of Cretaceous age (e.g. McCoy and Heck 1976; Heck and McCoy 1978).

Older ideas of an Indo-West Pacific centre of origin for Indo-Pacific biotas, or even world-wide biotas (e.g. Stehli and Wells 1971) remain difficult to assess, and many authors have pointed out that such an hypothesis is indeed untestable (see p. 276). The centre of origin model is not, and cannot ever be, contradicted by any results using distributional data such as our own. The Indo-West Pacific might as easily have been a 'centre of *refuge*', since our results imply that it is also the only region whose biotas have intermixed with all other Indo-Pacific localities, since some time before the end of the Lower Miocene. It is certainly hard to understand why, if assuming allopatric speciation, a dense pattern of islands should have been a more important source of new species than the more scattered pattern of islands further out in the Indian and Pacific Oceans. A centre of origin model seemingly implies a high level of sympatric or non-geographical allopatric speciation within this Indo-Pacific 'centre'. (For a more detailed 'vicariance and refuge model' for the Indo-Pacific, see B. R. Rosen 1984, 1985, and for a discussion of this and two other historical hypotheses, see Potts (1985)). It is too early to specify events, but PAE results offer a way of testing these various models. First results do show post-early Miocene intra-Indo-Pacific divergences, some of which only make sense if they were preceded by an intra-Indo-Pacific convergence, consistent with Rosen's model.

Conclusions

Parsimony analysis of shared presences provides data on the relative recency of faunal contact between regions, from which it is possible to deduce a history of extrinsic divergent and convergent events. These cannot provide definitive palaeogeographies by themselves, but yield empirical constraints on reconstructions based on independent evidence. This is because barriers to faunal exchange do not need to be physical structures, and there is no way of deducing the nature of a barrier from the faunal distribution pattern alone. The addition of ecological inferences would assist in the process of reconstructing palaeogeographies, but introduces assumptions (i.e. about the modes of life, dispersion capability, and preferred habitat of fossils) at the possible expense of rigour. Biogeographical data can be compared with independent lines of evidence, such as that derived from facies interpretation, to assess palaeogeographical reconstructions.

Such comparisons between parsimony analyses of biotic distributions and palaeogeographies would, in the long term, be more useful in the earliest Mesozoic and in the Palaeozoic, than for the time since then—where we have a much greater wealth of independent geological data (e.g. magnetic lineations). Older palaeogeographical reconstructions are more conjectural, and the constraints provided by biogeographical analysis should therefore have proportionally greater value, as we go back in time.

Essentially, though the palaeogeographical role of biogeographical analysis is inevitably 'narrative' (Ball 1976), a clear distinction ought to be maintained between analysis, test, comparison, and self-consistent explanation. As Patterson (1983) points out, even narrative approaches need a method, and we would claim that parsimony analysis of shared presences, like cladistic biogeography, does at least provide 'a methodology for capturing the pattern' (Eldredge 1981). Even cladistic biogeography can only be compared with independent palaeogeography on a narrative basis. Neither method, however, is in itself narrative, in that they both provide hypotheses about the history of areas in terms of their characteristic biotas, that are rigorous and testable on the criterion of congruence. Ideally, this congruence would be between different taxonomic groups, between the results obtained from both methods, and between data sets from sufficiently closely spaced stratigraphic horizons. We therefore offer our inferences, not so much for their present value as definitive accounts of Tethyan history, but as the first steps towards a testable history of Tethyan marine biotas which is independent of other data. If few of our historical conclusions are especially new or revolutionary, we hope, at least, that we have demonstrated the potential of this kind of biogeographical analysis. We anticipate a refinement of the method and new findings based on it, in the future.

ACKNOWLEDGEMENTS: This paper has benefited very much from extensive advice and discussion offered by our colleagues Dr C. J. Humphries (Department of Botany) and Dr P. L. Forey (Department of Palaeontology), at the British Museum (Natural History), both generally, and through critical reading of the manuscript. J. Ghiold (Louisiana State University) kindly allowed us to use his data on Recent clypeasteroid distribution prior to publication. Professor John Wells (Cornell University) similarly allowed us to use his data on Recent reef-coral distributions. Kay Shaw (British Museum, Natural History) assisted us on computing matters. We should also like to thank the following for advice and discussion on points of detail: Dr Richard Fortey (BMNH), Dr A. B. Foster (Iowa State University), and Dr Alison Kay (University of Hawaii).

References

ADAMS, C. G. (1981). An outline of Tertiary palaeogeography. In *Chance, change and challenge: the evolving earth.* (ed. P. H. Greenwood and L. R. M. Cocks), pp. 221–35. British Museum (Natural History) and Cambridge University Press.

ADAMS, C. G., GENTRY, A. W. and WHYBROW, P. J. (1983). Dating the terminal Tethyan event. *Utrecht micropaleont. Bull.* **30**, 273–98.

ALI, M. S. M. (1983). Tertiary echinoids and the time of the collision between Africa and Eurasia. *Neues Jb. Geol. Paläont. Mh.* (1983), 213–27.

AUDLEY-CHARLES, M. G. (1981). Geological history of the region of Wallace's Line. In *Wallace's Line and plate tectonics.* (ed. T. C. Whitmore), pp. 24–35, Clarendon Press, Oxford.

BALL, I. R. (1976). Nature and formulation of biogeographical hypotheses. *Syst. Zool.* **24**, 407–30.

BEAUVAIS, L. (1973). Upper Jurassic corals. In *Atlas of palaeobiogeography.* (ed. A. Hallam), pp. 317–28, Elsevier, Amsterdam.

BRIGGS, J. C. (1984). *Centres of origin in biogeography.* Biogeographical Monographs, 1. Biogeographical Study Group, University of Leeds, University of Leeds Printing Service. (95 pp.)

CHEETHAM, A. H. and HAZEL, J.E. (1969). Binary (presence–absence) similarity coefficients. *J. Paleont.* **43**, 1130–6.

COCKS, L. R. M. and FORTEY, R. A. (1982). Faunal evidence for oceanic separations in the Palaeozoic of Britain. *J. Geol. Soc. London* **139**, 465–78.

CRAW, R. C. (1982). Phylogenetics, areas, geology and the biogeography of Croizat: a radical view. *Syst. Zool.* **31**, 304–16.

CROIZAT, L. (1964). *Space, time, form; the biological synthesis.* Published by the author, Caracas. (xix + 882 pp.)

ELDREDGE, N. (1981). Discussion [of M. D. F. Udvardy's paper]. In *Vicariance biogeography: a critique.* (ed. G. Nelson and D. E. Rosen), pp. 34–8, Columbia University Press, New York.

FARRIS, J. S. (1977). Phylogenetic analysis under Dollo's Law. *Syst. Zool.* **26**, 77–88.

FORTEY, R. A. and BARNES, C. R. (1977). Early

Ordovician conodont and trilobite communities of Spitsbergen: influence on biogeography. *Alcheringa* **1**, 279–309.

Foster, A. B. (1984). The species concept in fossil hermatypic corals: a statistical approach. *Palaeontogr. Am.* **54**, 58–69.

Ghiold, J. and Hoffman, A. (1986). Biogeography and biogeographical history of clypeasteroid echinoids. *J. Biogeogr.* **13**, 183–206.

Grande, L. (1985). The use of paleontology in systematics and biogeography, and a time control refinement for historical biogeography. *Paleobiol.* **11**, 234–43.

Hallam, A. (1978). How rare is phyletic gradualism and what is its evolutionary significance? Evidence from Jurassic bivalves. *Paleobiol.* **4**, 16–25.

Hallam, A. (1983). Plate tectonics and evolution. In *Evolution from molecules to men*. (ed. D. S. Bendall), pp. 367–86, Cambridge University Press.

Heck, K. L., Jr. and McCoy, E. D. (1978). Long-distance dispersal and the reef-building corals of the eastern Pacific. *Mar. Biol., Berlin* **48**, 349–56.

Hughes, N. F. (ed.) (1973). Organisms and continents through time. *Spec. Pap. Palaeont.* **12**, 334 pp.

Humphries, C. J. and Parenti, L. R. (1986). *Cladistic biogeography*. Oxford Monographs on Biogeography, No. 2. Clarendon Press, Oxford. (98 pp.)

Jablonski, D. and Valentine, J. W. (1981). Onshore–offshore gradients in Recent eastern Pacific shelf faunas and their paleobiogeographical significance. In *Evolution today*. Proceedings of the second international congress of systematic and evolutionary biology. (ed. C. G. E. Scudder and J. L. Reveal), pp. 441–53, Hunt Institute, Pittsburgh.

Jablonski, D., Flessa, K. W., and Valentine, J. W. (1985). Biogeography and paleobiology. *Paleobiol.* **11**, 75–90.

Kay, E. A. (1980). *The Harold L. Lyon Arboretum Lecture. Little worlds of the Pacific. An essay on Pacific Basin biogeography*. University of Hawaii. (40 pp.)

Kay, E. A. (1984). Patterns of speciation in the Indo-West Pacific. In *Biogeography of the tropical Pacific. Proceedings of a symposium*. (ed. F. J. Radovsky, P. H. Ravn, and S. H. Sohmer), pp. 15–31, Spec. Publs. Bernice Pauahi Bishop Mus. **72**.

Lambshead, P. J. D. and Paterson, G. L. J. (1986). Ecological cladistics—an investigation of numerical cladistics as a method for analysing ecological data. *J. Natur. Hist.* **20**, 895–909.

McCoy, E. D. and Heck, K. L. (1976). Biogeography of corals, sea grasses and mangroves: an alternative to the center of origin concept. *Syst. Zool.* **25**, 201–10.

McCoy, E. D. and Heck, K. L. (1983). Centers of origin revisited. *Paleobiol.* **9**, 17–19.

Milner, A. R. and Norman, D. B. (1984). The biogeography of advanced ornithopod dinosaurs (Archosauria: Ornithischia) — a cladistic-vicariance model. In *Third symposium on Mesozoic Terrestrial Ecosystems, short papers*. (ed. W.-E. Reif and F. Westphal), pp. 145–50, Attempto Verlag, Tübingen.

Nelson, G. (1985). A decade of challenge the future of biogeography. *Earth Sci. Hist.* **4**, 187–96.

Nelson, G. and Platnick, N. I. (1981). *Systematics and biogeography: cladistics and vicariance*. Columbia University Press, New York. (567 pp.)

Nelson, G. and Rosen, D. E. (1981). (ed.) *Vicariance biogeography: a critique*. Columbia University Press, New York. (593 pp.)

Oliver, W. A. and Pedder, A. E. H. (1979). Biogeography of late Silurian and Devonian rugose corals in North America. In *Historical biogeography, plate tectonics, and the changing environment*. (ed. J. Gray and A. J. Boucot), pp. 131–45, Oregon State University Press, Corvallis.

Patterson, C. (1981). Methods of paleobiogeography. In *Vicariance biogeography: a critique*. (ed. G. Nelson and D. E. Rosen), pp. 446–500, Columbia University Press, New York.

Patterson, C. (1983). Aims and methods in biogeography. In *Evolution, time and space: the emergence of the biosphere*. (ed. R. W. Sims, J. H. Price, and P. E. S. Whalley), pp. 1–28, Academic Press, London.

Pielou, E. C. (1977). The latitudinal spans of seaweed species and their pattern of overlap. *J. Biogeogr.* **4**, 299–311.

Pielou, E. C. (1978). Latitudinal overlap of seaweed species: evidence for quasi-sympatric speciation. *J. Biogeogr.* **5**, 227–38.

Pielou, E. C. (1979). Interpretation of paleoecological similarity matrices. *Paleobiol.* **4**, 435–43.

Platnick, N. I. and Nelson, G. (1978). A method of analysis for historical biogeography. *Syst. Zool.* **27**, 1–16.

Potts, D. C. (1985). Sea-level fluctuations and speciation in Scleractinia. *Proc. 5th Int. Coral Reef Congr., Tahiti*. Volume 4, Symposia and Seminars (B), pp. 127–32. Antenne Museum—EPHE, Moorea, French Polynesia.

Rögl, F. and Steininger, F. F. (1983). Vom Zerfall der Tethys zu Mediterran und Paratethys. Die neogene Paläogeographie und Palinspastik des zirkum-mediterranen Raumes. *Ann. Naturhist. Mus. Wien.* **85/A**, 135–63.

Rögl, F. and Steininger, F. F. (1984). Neogene Paratethys, Mediterranean and Indo-pacific seaways: implications for the paleobiogeography of marine and terrestrial biotas. In *Fossils and climate*. Geol. J. Spec. Issue **11**, (ed. P. J. Brenchley), pp. 171–200, Wiley, Chichester.

Rosen, B. R. (1983). Reef island staging posts and Noah's Arks. *Reef Encounter* **1**, 5–6.

Rosen, B. R. (1984). Reef coral biogeography and climate through the late Cainozoic: just islands in the sun or a critical pattern of islands? In *Fossils and climate. Geol. J.* Special Issue **11,** (ed. P. J. Brenchley), pp. 201–62, Wiley, Chichester.

Rosen, B. R. (1985). Long-term geographical controls on regional diversity [lecture transcript summary]. *J. Open Univ. Geol. Soc.* **6**, 25–30.

Rosen, B. R. and Darrell, J. G. (1984). Methodology [Appendix to B. R. Rosen's paper]. In *Fossils and Climate. Geol. J.* Special Issue **11,** (ed. P. J. Brenchley), pp. 260–62, Wiley, Chichester.

Rosen, B. R. and Darrell, J. G. (in press). Appendix IIB. Palaeontology. Corals and hydrozoans. Content and significance of the faunas. In *Final report of Makran maps.* (compiler G. J. H. McCall), Geological and Mineralogical Survey of Iran.

Rosen, D. E. (1978). Vicariance patterns and historical explanation in biogeography. *Syst. Zool.* **27**, 159–88.

Schafersman, S. D. and Frost, S. H. (1979). Tropical Cenozoic paleo-oceanography and correlated events in phylogeny and biogeography of scleractinian corals. [Abstract]. *Bull. Am. Ass. Petrol. Geol.* **63**, p. 521.

Schallreuter, R. E. L. and Siveter, D. J. (1985). Ostracodes across the Iapetus Ocean. *Palaeontol.* **28**, 577–98.

Schlanger, S. O. and Premoli-Silva, I. (1981). Tectonic, volcanic and paleogeographic implications of redeposited reef faunas of Late Cretaceous and Tertiary age from the Nauru Basin and the Line Islands. *Initial Rep. DSDP* **61**, 817–27.

Schopf, T. J. M. (1979). The role of biogeographic provinces in regulating marine faunal diversity through geologic time. In *Historical biogeography, plate tectonics, and the changing environment.* (ed. J. Gray and A. J. Boucot), pp. 449–57, Oregon State University Press, Corvallis.

Springer, V. G. (1981). Discussion [of A. Solem's paper]. Comments on Solem's land-snail biogeography, with an hypothesized explanation of the distribution of the Endodontidae. In *Vicariance biogeography: a critique.* (ed. G. Nelson and D. E. Rosen), pp. 225–30, Columbia University Press, New York.

Springer, V. G. (1982). Pacific plate biogeography with special reference to shore fishes. *Smithsonian Contrib. Zool.* **367**, 1–182.

Stanley, S. M. (1979). *Macroevolution: pattern and process.* Freeman, San Francisco. (332 pp.)

Stehli, F. G and Wells, J. W. (1971). Diversity and age patterns in hermatypic corals. *Syst. Zool.* **20**, 115–26.

Swofford, D. (1984). *Phylogenetic analysis using parsimony [PAUP], version 2.2.* Illinois Natural History Survey, Champaign, Illinois.

Wells, J. W. (1956). Scleractinia. In *Treatise on invertebrate paleontology. Part F. Coelenterata.* (ed. R. C. Moore), pp. F328–F444, Geological Society of America and University of Kansas Press, Lawrence, Kansas.

Whittington, H. B. and Hughes, C. P. (1972). Ordovician geography and faunal provinces deduced from trilobite distribution. *Phil. Trans. Roy. Soc. Lond.* **B263**, 235–78.

Wiley, E. O. (1981). *Phylogenetics. The theory and practice of phylogenetic systematics.* Wiley, New York. (439 pp.)

Williams, A. (1969). Ordovician faunal provinces with reference to brachiopod distribution. In *The Pre-Cambrian and Lower Palaeozoic rocks of Wales.* (ed. A. Wood), pp. 117–54, University of Wales Press, Cardiff.

Williams, A. (1973). Distribution of brachiopod assemblages in relation to Ordovician palaeogeography. In *Organisms and continents through time.* (ed. N. F. Hughes), pp. 241–69, Spec. Pap. Palaeont. **12**.

Young, G. C. (1981). Biogeography of Devonian fishes. *Alcheringa* **5**, 225–43.

Young, G. C. (1984). Comments on phylogeny and biogeography of antiarchs (Devonian placoderm fishes) and the use of fossils in biogeography. *Proc. Linn. Soc. N.S.W.* **107**, 443–73.

Brian R. Rosen and Andrew B. Smith, Department of Palaeontology, British Museum (Natural History), London SW7 5BD, UK.

Appendix: Sources of distributional data

Echinoids

Recent

See main text (p. 289 and Acknowledgements).

Lower Miocene

Gulf and Caribbean	Kier 1984 (Cuba); Jackson and Vaughan 1922 (West Indies); Cooke 1959 (south-east United States of America), Poddubiuk and Rose 1985 (Antigua and Anguilla)
Mediterranean	Caria 1975 (Sardinia); Rose 1974 (Cyrenaica, Libya); Rose 1975 (Malta); Negretti 1985 (south-east France, Alpes Maritimes)
Saudi Arabia	(Dam Formation) Clegg 1933 (Bahrain Gulf), Kier 1972 (Riyadh, Saudi Arabia); Roman 1976 (Qatar)
Madagascar	Cottreau and Collignon 1927 (south-west part)
West Pakistan	(Gaj Series) Duncan and Sladen 1885 (Sind)
Indonesia	Jeannet and Martin 1937 (Java)
Japan	Nisiyama 1966, 1968
SE Australia	Clark 1946; Philip 1963, 1964, 1965, 1967, 1969, 1971; Philip and Foster 1971; McNamara 1985*a*,*b*; McNamara and Philip 1980*a*,*b*, 1984
New Zealand	Fell 1954; Henderson 1975
Mid-Pacific	Kier 1964 (Marshall Islands); Cooke 1957 (Saipan, Mariana Islands); Nisiyama 1966, 1968

Middle–Upper Eocene

Gulf and Caribbean	Cooke 1959 (south-eastern USA); Kier 1980 (North and South Carolina); Kier 1966 (Barbados); Kier 1984 (Cuba); Jackson and Vaughan 1922 (West Indies)
Spain and south-west France	Lambert 1927, 1928, 1933 (Barcelona and Catalonia); Castex and Lambert 1920; Castex 1930, 1947 (south-west France, Biarritz)
North Africa	Lambert 1931
Senegal	Roman and Gordinski 1959
NE Arabia	(Dammam Formation) Roman 1976 (Qatar); A. B. Smith's own data (north-east Oman)
Somalia	Kier 1957
Madagascar	Lambert 1929, 1933, 1936*a*,*b*
NW Pakistan	(Khirthar Series) Duncan and Sladen 1884
SE Australia	Clark 1946; Philip 1963, 1964, 1965, 1967, 1969, 1971; Philip and Foster

	1971, McNamara 1985*a*,*b*; McNamara and Philip 1980*a*,*b*, 1984

Campanian/Maastrichtian

Gulf and Caribbean	Cooke 1953, 1955 (south-eastern USA); Kier 1984 (Cuba), Jackson and Vaughan 1922 (West Indies)
NW Europe	Schlüter 1883, 1892; Ernst 1972
N Africa	Lambert 1931; Petitot 1961 (Morocco)
Saudi Arabia	Aruma Formation, Kier 1972
Oman	A. B. Smith's own data
Madagascar	Lambert 1933*a*,*b*, 1936*a*,*b*; Cottreau 1935
Iran	Cotteau and Gauthier 1895 (Louristan)
Baluchistan	Noetling 1897 (Mari Hills)
S Tibet	Wu and Mu 1976 (south of the Indus Zangbo suture)
S India	Stoliczka 1873 (Trichinopoly and South Arcot)

Reef-coral data

Recent

See text (p. 289 and Acknowledgements)

Lower Miocene

The stratigraphic correlation of particular strata at some of these localities has been reinterpreted in the light of more recent literature and also on the advice of Dr C. G. Adams (Department of Palaeontology, British Museum, Natural History), for which we thank him. We also thank Dr S. H. Frost (now of Gulf Research and Development Co.) for making available to us his unpublished compilations (as of 1981). B. R. Rosen is responsible for the taxonomic reinterpretation of various records.

Gulf and Caribbean	Frost 1977; Frost 1981 (unpublished compilations)
Mediterranean	Frost 1977, incorporating Chevalier 1962; Frost 1981 (unpublished compilations)
Iran	Kühn 1933; Sjerp *et al.* 1969 (Saidabad area); Rosen and Darrell in press (Makran)
West Pakistan	(Gaj Series) Duncan 1880 (Sind)
Indonesia	Felix 1921; Gerth 1921, 1933; Gerth 1923 (Borneo); Umbgrove 1943 (Sumba); Osberger 1954, 1955*a*, 1955*b* (Java). See also Indonesia compilations by Gerth 1925, 1931, and Osberger 1956
New Zealand	Hayward 1977
Marshall Islands	Wells 1954, 1964

References

Caria, I. C. (1975). *Gli echinidi del Miocene della Sardegna*. Stabilimento Tipografico Editoriale Fossataro—Cagliari. (95 pp.)

Castex, L. (1930). Revision des échinides du Nummulitique du Départment des Landes. *Actes Soc. Linnéenne Bordeaux* **82**, 5–72.

CASTEX, L. (1947). Notes sur quelques échinides fossiles du sud-ouest de la France. *Actes Soc. Linnéenne Bordeaux* **93**, 25–42.

CASTEX, L. and LAMBERT, J. (1920). Revision des échinides des falaises de Biarritz. *Actes Soc. Linnéenne Bordeaux* **71**, 117–206.

CHEVALIER, J.-P. (1962). Recherches sur les madréporaires et les formations récifales miocènes de la Méditerranée occidentale. *Mem. Soc. Géol. Fr. (N.S.)* **60**, (93), 1–562.

CLARK, H. L. (1946). *The echinoderm fauna of Australia, its composition and its origin*. Carnegie Institution, Washington. (567 pp.)

CLEGG, E. C. G. (1933). Echinoidea from the Persian Gulf. *Mem. Geol. Surv. India: Palaeont. Indica (N.S.)* **22**, (1), 1–35.

COOKE, C. W. (1953). American Upper Cretaceous Echinoidea. *US Geol. Surv. Prof. Pap.* **254–A**, 1–44.

COOKE, C. W. (1955). Some Cretaceous echinoids from the Americas. *US Geol. Surv. Prof. Pap.* **264-E**, 89–112.

COOKE, C. W. (1957). Geology of Saipan, Mariana Islands, part 3: palaeontology, chapter J, echinoids. *US Geol. Surv. Prof. Pap.* **280-J**, 361–4.

COOKE, C. W. (1959). Cenozoic echinoids of eastern United States. *US Geol. Surv. Prof. Pap.* **321**, 106 pp.

COTTEAU, G. and GAUTHIER, V. (1895). *Mission Scientifique en Perse par J. de Morgan; études géologiques, échinides fossiles*. 3. Paris, 142 pp.

COTTREAU, J. (1935). Quelques échinides fossiles de Madagascar. *Arch. Muséum Hist. Natur. Paris* (6), **12**, 259–64.

COTTREAU, J. and COLLIGNON, M. (1927). Paléontologie de Madagascar; fossiles du Miocène marin. *Ann. Paléont.* **16**, 135–40.

DUNCAN, P. M. (1880). Sind fossil corals and Alcyonaria. *Mem. Geol. Surv. India Palaeont. Indica* (7 and 14), **1**, (1), 1–110.

DUNCAN, P. M. and SLADEN, W. P. (1884). The fossil echinoidea from the Khirthar Series of Nummulitic strata in western Sind. *Mem. Geol. Surv. India: Palaeont. Indica* (14), **1**, (3), 101–246.

DUNCAN, P. M. and SLADEN, W. P. (1885). The fossil Echinoidea from the Gaj or Miocene Series. *Mem. Geol. Surv. India: Palaeont. Indica* (14), **1**, (3), 273–367.

ERNST, G. (1972). Grundfragen des Stammesgeschichte bei irregularen Echiniden der nordwesteuropäischen Oberkreide. *Geol. Jb.* **A4**, 63–175.

FELIX, J. (1921). Fossile Anthozoën von Borneo. *Paläont. Timor* **9**, 1–61.

FELL, H. B. (1954). Tertiary and Recent Echinoidea of New Zealand: Cidaridae. *N.Z. Geol. Surv., Palaeont. Bull.* **23**. (62 pp.)

FROST, S. H. (1977). Miocene to Holocene evolution of Caribbean province reef-building corals. *Proc. 3rd Int. Coral Reef Symp.* **2**, 353–9.

GERTH, H. (1921). Die Fossilien von Java auf Grund einer Sammlung von Dr R. D. M. Verbeek und von den anderen bearbeitet durch Dr K. Martin. Coelenterata. *Samml. Geol. Reichsmus. Leiden* (N.F.) **1**, (2), 387–445.

GERTH, H. (1923). Die Anthozoenfauna des Jungtertiärs von Borneo. *Samml. Geol. Reichsmus. Leiden* (1) **10**, 37–136.

GERTH, H. (1925). Jungtertiäre Korallen von Nias, Java, und Borneo, nebst einer Übersicht über die aus dem Kainozoikum des indischen Archipels bekannten Arten. *Leid. Geol. Meded.* **1**, 22–82.

GERTH, H. (1931). Coelenterata (Anthozoa caenozoica). *Leid. Geol. Meded.* **5**, 129–51.

GERTH, H. (1933). Neue Beiträge zur Kenntnis der Korallenfauna des Tertiärs von Java. I. Die Korallen des Eocaen und des älteren Neogen. *Wet. Meded. Dienst Mijnb. Ned. Oost-Indië* **25**, 1–46.

HAYWARD, B. W. (1977). Lower Miocene corals from Waitekere Ranges, North Auckland, New Zealand. *J. Roy. Soc. N.Z.* **7**, 99–111.

HENDERSON, R. A. (1975). Cenozoic spatangoid echinoids from New Zealand. *N.Z. Geol. Surv., Palaeont. Bull.* **46**. (129 pp.)

JACKSON, R. T. and VAUGHAN, T. W. (1922). Fossil echini of the West Indies. *Publ. Carnegie Inst., Washington* **306**. (103 pp.)

JEANNET, A. and MARTIN, R. (1937). Über neozoische Echinoidea aus dem niederlandisch–indischen Archipel. *Leid. Geol. Meded.* (6) **8**, 215–308.

KIER, P. M. (1957). Tertiary Echinoidea from British Somaliland. *J. Paleont.* **31**, 839–902.

KIER, P. M. (1964). Fossil echinoids from the Marshall Islands. *US Geol. Surv. Prof. Pap.* **260-GG**, 1121–26.

KIER, P. M. (1966). Four new Eocene echinoids from Barbados. *Smithsonian Miscell. Collections* **151**, (9), 28 pp.

KIER, P. M. (1972). Tertiary and Mesozoic echinoids of Saudi Arabia. *Smithsonian Contrib. Paleobiol.* **10**. (242 pp.)

KIER, P. M. (1980). The echinoids of the Middle Eocene Warley Hill Formation, Santee Limestone and Castle Hayne Limestone of North and South Carolina. *Smithsonian Contrib. Paleobiol.* **39**. (102 pp.)

KIER, P. M. (1984). Fossil spatangoid echinoids of Cuba. *Smithsonian Contrib. Paleobiol.* **55**. (336 pp.)

KÜHN, O. (1933). Das Becken von Isfahan–Saidabad und seine altmiocäne Korallenfauna. *Palaeontogr.* **79**, (A), 143–221.

LAMBERT, J. (1927). Revision des échinides fossiles de la Catalogne. *Mem. Museo Cienc. Natur. Barcelona (Geol.)* **1**, (1). (102 pp.)

LAMBERT, J. (1928). Revision des échinides fossiles de la Catalogne. *Mem. Museo Cienc. Natur. Barcelona (Geol.)* **1**, (2), 1–62.

LAMBERT, J. (1929). Sur les échinides éocènes de Madagascar. *C. R. Acad. Sci.*, Paris, **189**, 192–4.

LAMBERT, J. (1931). Etude sur les échinides fossiles du

nord de l'Afrique. *Mém. Soc. Géol. Fr.* (N.S. 16) **7**, 1–228.

LAMBERT, J. (1933*a*). Echinides de Madagascar communiqués par M. H. Besairie. *Ann. Géol. Serv. Mines, Madagascar* **3**, 1–49.

LAMBERT, J. (1933*b*). Supplément à la revision des échinides fossiles de la Catalogne. *Bull. Inst. Catalana Hist. Natur.* **33**, 183–95.

LAMBERT, J. (1936*a*). Echinides du Senonien supérieur et de l'Eocène. In *La géologie du nord-ouest. 1: Recherches géologiques à Madagascar.* (ed. H. Besairie), pp. 116–20. *Mém. Acad. Malgache* **21**.

LAMBERT, J. (1936*b*). Nouveaux échinides fossiles de Madagascar. *Ann. Géol. Serv. Mines, Madagascar* **6**, 1–32.

MCNAMARA, K. J. (1985*a*). The spatangoid echinoid *Linthia* from the Late Eocene of Southern Australia. *Trans. Roy. Soc. South Austral.* **109**, 161–5.

MCNAMARA, K. J. (1985*b*). Taxonomy and evolution of the Cainozoic spatangoid echinoid *Protenaster. Palaeont.* **28**, 311–30.

MCNAMARA, K. J. and PHILIP, G. M. (1980*a*). Tertiary species of *Echinolampas* (Echinoidea) from southern Australia. *Mem. Nat. Museum Victoria* **41**. (14 pp.)

MCNAMARA, K. J. and PHILIP, G. M. (1980*b*). Australian Tertiary schizasterid echinoids. *Alcheringa* **4**, 47–65.

MCNAMARA, K. J. and PHILIP, G. M. (1984). A revision of the spatangoid echinoid *Pericosmus* from the Tertiary of Australia. *Rec. Western Austral. Museum* **11**, 319–56.

NEGRETTI, B. (1984). Echinides Néogenes du littoral de la Nerthe. *Travaux Lab. Strat. Paléoécol. Univ. Provence, Marseille.* Nouvelle Série **2**. (140 pp. + 10 pls)

NISIYAMA, S. (1966). The echinoid fauna from Japan and adjacent regions, part 1. *Palaeont. Soc. Japan Spec. Pap.* **11**, 1–227.

NISIYAMA, S. (1968). The echinoid fauna from Japan and adjacent regions, part 2. *Palaeont. Soc. Japan Spec. Pap.* **13**, 1–491.

NOETLING, F. (1897). Fauna of Baluchistan and N.W. Frontier of India, 1(3), Fauna of the Upper Cretaceous (Maëstrichtian) Beds of the Mari Hills. *Mem. Geol. Surv. India: Palaeont. Indica* (16), **1**, (3), (79 pp.)

OSBERGER, R. (1954). Jungtertiäre Korallen von Java, Teil 1. *Neues Jb. Geol. Paläont. Abh.* **100**, 119–58.

OSBERGER, R. (1955*a*). Jungtertiäre Korallen von Java, Teil 2. *Neues Jb. Geol. Paläont. Abh.* **101**, 39–74.

OSBERGER, R. (1955*b*). Beschreibung einiger tertiäre Korallen. *Neues Jb. Geol. Paläont. Mh.* 1955 (B6), 252–6.

OSBERGER, R. (1956). Korallen als Hilfsmittel der tertiär und quartär Stratigraphie Indonesiens. *Publikasi keilm. Djaw. Geol. Republ. Indonesia* **32**, 1–82.

PETITOT, M. L. (1961). Contribution à l'étude des échinides fossiles du Maroc (Jurassique et Crétacé). *Notes & Mém. Serv. Géol. Maroc* **146**, 183 pp.

PHILIP, G. M. (1963). The Tertiary echinoids of south-eastern Australia. 1. Introduction and Cidaridea (1). *Proc. Roy. Soc. Victoria* **76**, 181–226.

PHILIP, G. M. (1964). The Tertiary echinoids of south-eastern Australia. II. Cidaridae (2). *Proc. Roy. Soc. Victoria* **77**, 433–77.

PHILIP, G. M. (1965). The Tertiary echinoids of south-eastern Australia. III. Stirodonta, Aulodonta and Camarodonta (1). *Proc. Roy. Soc. Victoria* **78**, 181–96.

PHILIP, G. M. (1967). Notes on three recently proposed Australian Tertiary echinoid genera. *Proc. Linnean Soc. New South Wales* **91**, 114–17.

PHILIP, G. M. (1969). The Tertiary echinoids of south-eastern Australia. IV. Camarodonta (2). *Proc. Roy. Soc. Victoria* **82**, 233–76.

PHILIP, G. M. (1971). A re-assessment of the Tertiary echinoid genus *Goniosigma* Fell 1964. *Proc. Roy. Soc. Victoria* **84**, 227–8.

PHILIP, G. M. and FOSTER, R. J. (1971). Marsupiate Tertiary echinoids from south-eastern Australia and their zoogeographic significance. *Palaeont.* **14**, 666–95.

PODDUBIUK, R. H. and ROSE, E. P. F. (1985). Relationships between mid-Tertiary echinoid faunas from the central Mediterranean and eastern Caribbean and their palaeobiogeographic significance. *Ann. Géol. Pays Helléniques* **32**, 115–28.

ROMAN, J. (1976). Echinides éocènes et miocènes du Qatar (Golfe Persique). *Ann. Paleont., Invertebr.* **62**, 49–85.

ROMAN, J. and GORDINSKI, A. (1959). Echinides éocènes du Sénégal. *Notes Serv. Géol. Prospect & Mines, Dakar* **3**. (91 pp.)

ROMAN, J. and SORNAY, J. (1983). Ammonites, inocérames et échinides du Crétacé supérieur de Paki (Sénégal). *Bull. Muséum Nat. Hist. Natur. Paris* (4) **5**, 3–23.

ROSE, E. P. F. (1974). Stratigraphical and facies distribution of irregular echinoids in Miocene limestones of Gozo, Malta and Cyrenaica, Libya. *Mem. Bur. Recherches Géol. & Minières* **78**, 349–55.

ROSE, E. P. F. (1975). Oligo-Miocene echinoids of the Maltese Islands *VIth Congress of the Regional Committee on Mediterranean Neogene Stratigraphy, Bratislava, 1975*, pp. 75–9. International Union of Geological Sciences Commission on Stratigraphy. Subcommission on Neogene Stratigraphy. VEDA, Bratislava. Vol. 1.

ROSEN, B. R. and DARRELL, J. G. (in press) Appendix IIB. Palaeontology. Corals and hydrozoans. Content and significance of the faunas. In *Final report of Makran maps.* (compiler G. J. H. McCall) Geological and Mineralogical Survey of Iran.

SCHLÜTER, C. (1883). Die regularen Echiniden der norddeutschen Kreide. I. Glyphostomata.

Abhandl. Königlich Preussischen Geol. Landesanstalt **4**, 1–72.

SCHLÜTER, C. (1892). Die regularen Echiniden der norddeutschen Kreide. II. Cidaridae, Salenidae. *Abhandl. Königlich Preussischen Geol. Landesanstalt* **5**, 73–315.

SJERP, N., ISSAKHANIAN, V. and BRANTS, A. (1969). The geological environment of the Chahar Gonbad copper mine: a study in Tertiary copper mineralization. *Rep. Geol. Surv. Iran* **16** (1064), 68 pp.

STOLICZKA, F. (1873). The Cretaceous fauna of southern India. The Echinodermata. *Mem. Geol. Surv. India: Palaeont. Indica* (8) **4** (3).

UMBGROVE, J. H. F. (1943). Tertiary corals from Sumba (East Indies). *Verh. geol-mijnb. Genoot. Ned.* **13**, 393–8.

WELLS, J. W. (1954). Bikini and nearby atolls. Part 4. Palaeontology. Fossil corals from Bikini Atoll. *Prof. Pap. US Geol. Surv.* **260-P**, 609–17.

WELLS, J. W. (1964). Fossil corals from Eniwetok Atoll. *Prof Pap. US Geol. Surv.* **260-DD**, i–iv, 1101–12.

WU, EN-ZHI AND MU, YONG-RONG (1976). Cretaceous and Tertiary echinoids from the Mount Jolmo Lungma Region. In *A report of the scientific expedition in the Mount Jolmo Lungma Region (1966–1968)*, pp. 357–86, Science Press, Beijing.

BRIAN R. ROSEN and ANDREW B. SMITH, Department of Palaeontology, British Museum (Natural History), London SW7 5BD, UK.

Phytogeography of the eastern end of Tethys

T. C. Whitmore

ABSTRACT: Central to biogeography is the classification of geographical distribution patterns and analysis of their origins. Here we analyse angiosperm ranges in the Malay Archipelago (Malesia) mainly against landmass changes since the late Mesozoic. The interpretation is complicated by continual evolution and extinction, by range alterations due to climatic changes, and by the scanty fossil record.

Malesia has the greatest concentration of primitive angiosperms, and was postulated by Takhtajan (1969) to be the place of origin of the group. This hypothesis later became untenable, after the realization that Malesia was only formed in about mid-Miocene times by collision of Gondwana with Laurasia. The sharp discontinuities in the ranges of many animal and plant taxa in central Malesia, at the biogeographical boundary called Wallace's Line are readily interpreted to be results of this collision. But a few plant taxa, including the 'primitives', have puzzling ranges demanding land before the mid-Miocene—not empty ocean. These enigmas spurred further geological enquiry, which led to the discovery of Gondwanan fragments that have drifted northwestwards before the main collision and which could have been the locus of angiosperm evolution. The early history of angiosperms in Australia is not incompatible with this scenario.

Every class of evidence is incomplete, and all kinds must be considered together. Fossils and palaeogeography are less equivocal than distribution patterns. For the future, more fossil evidence should be sought, detailed taxonomic study and phylogenetic reconstruction of major plant groups is also needed: reinvestigation of the conifers could be very rewarding.

Introduction

One central preoccupation of biogeography is to analyse how present patterns of distribution have originated. My task here is to describe plant geographical patterns in the Malay Archipelago (Malesia to botanists) and surrounding regions, and to look at these together with our understanding of changing dispositions of land masses since about the late Jurassic. The focus is on angiosperms. There are three complicating factors which must be allowed for: evolution, the patchy fossil record, and changes in climate.

Evolution

Flowering plants first appear as fossil pollen in the Barremian (125 Ma) of Africa, England, and America. The group underwent major diversification and achieved dominance in world vegetation by late Mesozoic or early Tertiary time, during the period when Gondwanaland was breaking up and the present geography of the eastern end of Tethys was progressively taking shape. Present patterns of angiosperm distribution are likely to reflect both their evolutionary history of diversification and their migration, taking place on an altering disposition of land and sea.

The fossil record

The fossil record of the angiosperms is very incomplete, and so far very few fossils have been discovered from the eastern end of Tethys (Muller 1984; Walker and Walker 1984). The earliest recognized angiosperm fossils represent several different evolutionary lines, and none are of the most primitive angiosperms. Later microfossils and macrofossils chart evolution and diversification of the group in various parts of the world, but without evidence from our region.

Palaeoclimates

There are two floras, of tropical rain forest and of monsoon (seasonal) forest and savannah plants respectively, and these have complementary geographical ranges. Climate is a major variable that controls range. Today the Malay Archipelago has mainly rain-forest climates with seasonal climates mostly on either side, that is, to the north and south. The pattern of climates is reflected in the flora, with an equatorial zone of

From AUDLEY-CHARLES, M. G. & HALLAM, A. (eds) *Gondwana and Tethys* Geological Society Special Publication No. 37, pp. 307–11.

rain-forest species fringed with seasonal climate species.

In the equatorial belt there has been an alternation of hot, perhumid, equable climates, such as are very extensive today, with periods of cooler, seasonally dry climates (Whitmore 1981*a*; Morley and Flenley 1987).

Our knowledge of past climate is best for the Quaternary, during which the well-known alternation of Ice Ages and Interglacials at high latitudes has been mirrored by the contraction and expansion of rain-forest climates and their plants (and animals). The same climatic fluctuations are now known to extend back into the Tertiary, although knowledge becomes progressively more vague with increasing age (Morley and Flenley, 1987, review current evidence). It is also highly probable that many present-day species' ranges reflect these past climatic fluctuations.

Angiosperm origin and diversification

There are two views current on the place of origin and diversification of the angiosperms.

Primitive angiosperms are today very strongly concentrated in the humid tropics and subtropics of the Malay Archipelago and surrounding regions. From this, Takhtajan (1969) adduced that 'the cradle of the angiosperms' lay 'somewhere between Assam and Fiji', and he has recently reiterated this view, incorporating new data (Takhtajan 1987). Takhtajan argues that the poor and patchy representation of primitive angiosperms in Africa and America is because there have never been many present in those areas. Moreover, he argues, the first fossils are already so advanced that earliest angiosperm evolution must have been pre-Barremian and most probably pre-Cretaceous, and far away from where the Barremian fossils occur (Africa, England, eastern North America, and South America), otherwise traces would have been found. This strengthens the case for a Malesian 'cradle'.

The second view is that western Gondwanaland (Africa and South America) is the region of angiosperm origin and diversification. Moore (1973) argued from the distributions and relationships amongst palms, a big, taxonomically complex tropical plant family, that the palm family originated in western Gondwanaland. Later, Raven and Axelrod (1974) reviewed the ranges of many flowering plant families and also came to the same conclusion, that the spread of flowering plants and their diversification took place from an origin in western Gondwanaland: as the supercontinent broke up and its fragments moved apart. This model, for which others have later found additional evidence, explains the present ranges of many angiosperm groups. There are, however, a few awkward problems, amongst which the oak and beech family Fagaceae is notorious. All Fagaceous genera are now of the Northern Hemisphere, and reach into the tropics only along the Andes and in the Malay Archipelago. The concentration of both genera and species is in East and south-east Asia. There is a single exceptional genus *Nothofagus*, (closely related to, and the southern counterpart of, *Fagus* the beech) which occurs in South America and Australasia, extending north to New Guinea. Fagaceae are entirely absent from Africa south of the Atlas Mountains, and have no fossils in Africa. It is not easy to reconcile Fagaceae distribution patterns with an origin in western Gondwanaland.

Geographical history of the Malay Archipelago

The elucidation of the geographical history of the Malay Archipelago showed that its present form arose from collision between north-west moving Gondwanaland and the south-eastern edge of Laurasia in mid-Miocene times, at about 15 Ma, at or east of the island of Sulawesi (Celebes) (Audley-Charles 1981; Audley-Charles *et al.* 1981). The Sunda continental shelf extension of south-east Asia is Laurasian. The Sahul shelf, extending north from Australia, and bearing New Guinea plus attendant small islands, is Gondwanan. This palaeogeographical reconstruction can be used to explain Wallace's Line, one of the world's major zoogeographical boundaries, which runs north to south through the centre of the archipelago (Whitmore 1981*b*). For example, placental mammals are Laurasian and scarcely extend beyond the Sunda shelf, marsupials are Gondwanan and confined to the eastern islands (Cranbrook 1981). Many plant groups show similar patterns though Wallace's Line is not such a sharp boundary for plants but one can recognize Laurasian families which have apparently extended their range south-eastwards since 15 Ma, and, conversely, Gondwanan families which have apparently extended north-westwards (Whitmore 1981*c*). Especially interesting are those flowering plant groups with bicentric distributions in the archipelago. These could have arrived twice, once from Gondwana and once from Laurasia. The first recognized example was the Proteaceae (Johnson and Briggs 1975). In some cases, drier past climates in central Malesia: leading to extinction of rain-forest plants there, may explain present-day bicentric species concentrations.

Although many present-day angiosperm ranges can be interpreted on the model of a mid-Miocene collision, nagging doubts remained about some which do not easily fit. These were surveyed by Whitmore (1981*c*). Firstly, angiosperms could not possibly have evolved 'between Assam and Fiji' if the region did not exist before 15 Ma ago. This strengthens the case for a western Gondwanaland origin. But how then are we to interpret the strong concentration of primitive angiosperms in the region? Can we really explain such a strong pattern as relict distributions due to extinction elsewhere, for if so, why should extinction have been less complete here than elsewhere in the humid tropics?

Several pairs of northern and southern related families come closest together or overlap in the Malay Archipelago. A simplistic view is that they had a common ancestor there, again impossible if the region is so recent in origin. And what about the Fagaceae, which at present have northern genera concentrated in eastern Asia and only overlap with the southern genus *Nothofagus* in the Malay Archipelago?

There is good evidence that some present-day rain-forest angiosperms entered Australia from the north before the mid-Miocene collision, and vertebrates also, as has recently been reviewed by Truswell *et al.* (1987).

Some pre-mid-Miocene land connections between north and south would assist interpretation of these ranges, a point made strongly by van Steenis (1979). Whitmore (1981*c*) made out a *prima facie* case—for plant geographers to ask geologists whether west and east Malesia really were separated by thousands of kilometres of empty ocean during the whole period of angiosperm evolution and radiation up until the mid-Miocene collision.

New geological evidence was not long in arriving, as has been shown in several papers in this volume. It is now known that there are several substantial areas of Gondwanan rocks, terranes, embedded in continental south-east Asia. These terranes are shards, rifted off Northern Australia and which drifted north to collide with Laurasia. There is some evidence that they were dry land, and hence were available as rafts for plants and animals, or as stepping-stones between Laurasia and Gondwana. Their age remains uncertain. Some would date them as Palaeozoic (e.g. Metcalfe this volume), on the other hand, Audley-Charles (1987 and this volume) believes they could have been as recent as mid-Jurassic in age—useful for the earliest angiosperms.

Takhtajan (1987) uses these shards of Northern Australia to reinstate his case for a south-east Asian origin of the angiosperms, which he now believes took place on northwards-drifting offshore fragments. Archipelagos present good opportunities for evolution to occur. The current interpretation of palm taxonomy, much elaborated from Moore's seminal 1973 paper, is that the family originated not in south-west Gondwanaland, but in Pangaea before that broke up. (Uhl and Dransfield 1987.) There has been subsequent differentiation of separate northern and southern groups since the late Cretaceous, with migration and evolution of elements of both groups of palms into present-day Malesia, some of which could have been on the pre-Miocene fragments acting as an archipelago of stepping-stones (Dransfield 1987).

Early history of angiosperms in Australia

Examination of fossils in Australia (Truswell *et al.* 1987) shows that the angiosperms were there long before the 15 Ma collision. They appear first in the Albian (113 Ma) in Southern Australia, about 10 Ma later than elsewhere in the world. Truswell *et al.* argue that this fact militates against islands off the Northern Australian coast being the place of origin of the angiosperms, unless we adduce that absence from, the admittedly patchy, Australian fossil record is solely the absence of evidence.

The nature of biogeography

The sceptic will argue that those who say that the absence of early fossil angiosperms from Australia is due either to a failure to search fully, or to a failure of preservation, are pleading a special case. But it illustrates a central dilemma in reconstructing biogeographical history. Every class of evidence has missing pieces. (In this case there are few fossils from Northern Australia.) All the different sorts of evidence must be interpreted together as a whole, and have to be reappraised with every new discovery.

Palaeogeography, as constructed by geologists, is hard evidence in this 'morass' but unfortunately palaeogeographical evidence is limited. The story for the eastern end of Tethys, over the last few years shows clearly how palaeogeographical understanding can develop—by biologists and geologists listening to each other.

Plant geographical ranges are notoriously susceptible to different interpretations (as this paper has sought to show) and fossils which do provide hard evidence are still so scanty that they are of little help. In the case of the Malay Archipelago, some of the patterns of flowering plant ranges were so strongly at variance with the perceived palaeogeography of 1981, that botanists needed to stand firm and ask their geological colleagues to investigate further.

What next?

What should be the focus for future attention?

Terrane age

The various different lines of biological evidence point strongly to the existence of some kind of land connection between Laurasia and Gondwana at the eastern end of Tethys during the Mesozoic, long before their mid-Tertiary collision. Geologists have identified terranes, but cannot agree yet on their age. Dating the time of northwards drift of these terranes is vitally important for the reconstruction of biogeographical history.

Fossils

Next, we need to search for more fossils. Consensus on the significance of present-day distributions is often unachievable. Concerning angiosperms there are two problems, their place of origin and their place (or places) of radiation. Is the controversy about south-west Gondwana versus Assam/Fiji perhaps to be resolved along such lines? Fossils are needed to provide firm data for the discussion. In the light of present views about early angiosperms at the eastern of Tethys, it would be of great interest to look for fossils in, for example, the Mesozoic sedimentary basins of Borneo and on the Gondwanan terranes embedded in continental south-east Asia. More fossil evidence is also urgently required from Northern Australia.

Monographs

We need detailed taxonomic study of big, tropical higher-plant groups to elucidate evolution within them. Study of the geological ranges of the different elements then becomes profitable. Most monographs are old and based on the sparse material available when they were written; much modern material accumulated in herbaria has never been studied critically. The practical problems are that finance is seldom available for monographic work, and that the sheer quantity of material and publications presents daunting problems which our forebears did not have to face. It is not surprising, given these two factors, that few modern monographs exist. The new study of the palms has already been mentioned. Another case is *Xanthophyllum*, a genus of trees in the Polygalaceae. Although most species come from Malesia, a new phylogenetic reconstruction shows that the group probably originated in Australia, where it is represented by a few, but primitive, species (Meijden 1982). A third group of flowering plants, which it would be revealing to review against the current palaeogeographical understanding, is the huge genus *Ficus*, the figs, which has some five hundred species in our region, and for which a modern taxonomic outline has been published (Corner 1965).

Conifers

We have so far discussed angiosperms. Another major group of plants whose further investigation would be of great interest to the whole of Tethys, not just its eastern end, is the conifers, which are the major living group of the gymnosperms. Florin (1963) clearly showed that present-day conifer families form two groups, with Northern and Southern Hemisphere ranges respectively. This observation was strengthened by statistical analysis, based on the ranges of all fifty-eight living genera, which grouped together areas of the world's surface that had similar conifer flora (Sneath 1967). The fossil record of conifers, which is copious, goes back to the late Carboniferous and early Permian, when the two geographical groups already existed. The evolution and spread of conifers through time was thoroughly documented by Florin (Florin 1951, 1963) who considered Tethys to have been important. But Florin 'left out of consideration . . . theories much in dispute such as those of large-scale displacement of the continents and wandering of the poles' (Florin 1963, p. 153), and there has been no modern assessment of conifer history in the light of modern palaeogeographical understanding.

References

AUDLEY-CHARLES, M. G. (1981). Geological history of the region of Wallace's Line. In *Wallace's Line and plate tectonics*. (ed. T. C. Whitmore), pp. 24–35, Clarendon Press, Oxford.

AUDLEY-CHARLES, M. G. (1987). Dispersal of Gondwanaland: relevance to evolution of the angiosperms. In *The biogeographical evolution of the Malay archipelago*. (ed. T. C. Whitmore), pp. 5–25, Clarendon Press, Oxford.

AUDLEY-CHARLES, M. G., HURLEY, A. M., and SMITH, A. G. (1981). Continental movements in the Mesozoic and Cenozoic. In *The biogeographical evolution of the Malay archipelago*. (ed. T. C. Whitmore), pp. 9–23, Clarendon Press, Oxford.

CORNER, E. J. H. (1965). Check list of *Ficus* in Asia and Australasia with keys to identification. *Gard. Bull. Singapore* **21**, 1–186.

CRANBROOK, LORD. (1981). The vertebrate faunas. In *Wallace's Line and plate tectonics*. (ed. T. C. Whitmore), pp. 57–69, Clarendon Press, Oxford.

DRANSFIELD, J. (1987). Bicentric distributions in Malesia as exemplified by palms. In *Wallace's Line and plate tectonics*. (ed. T. C. Whitmore), pp. 60–72, Clarendon Press, Oxford.

FLORIN, R. (1951). Evolution in Cordaites and conifers. *Acta Horti Bergiani* **15**, 285–383.

FLORIN, R. (1963). The distribution of conifer and taxad genera in time and space. *Acta Horti Bergiani* **20**, 121–312.

JOHNSON, L. A. S. and BRIGGS, B. G. (1975). On the Proteaceae, the evolution and classification of a southern family. *Biol. J. Linn. Soc.* **70**, 88–182.

MEIJDEN, R. van den (1982). Systematics and evolution of *Xanthophyllum* (Polygalaceae). *Leiden Botan. Ser.* **7**, 159.

MOORE, H. E., JR. (1973). Palms in the tropical forest ecosystems of Africa and South America. In *Tropical forest ecosystems in Africa and America: a comparative review*. (ed. B. J. Meggers, E. S. Ayensu, and W. D. Duckworth), pp. 63–88, Smithsonian Institution Press, Washington.

MORLEY, R. J. and FLENLEY, J. (1987). Late Cenozoic vegetational and environmental changes in the Malay archipelago. In *The biogeographical evolution of the Malay archipelago*. (ed. T. C. Whitmore), pp. 50–9, Clarendon Press, Oxford.

MULLER, J. (1984). Significance of fossil pollen for angiosperm history. *Ann. Missouri Bot. Gard.* **71**, 419–33.

RAVEN, R. H. and AXELROD, D. I. (1974). Angiosperm biogeography and past continental movements. *Ann. Missouri Bot. Gard.* **61**, 539–673.

SNEATH, P. H. A. (1967). Conifer distributions and continental drift. *Nature* **215**, 467–70.

STEENIS, C. G. G. J. VAN (1979). Plant geography of east Malesia. *Bot. J. Linn. Soc.* **79**, 97–178.

TAKHTAJAN, A. (1969). *Flowering plants: origin and dispersal* (trans. C. Jeffrey), p. 310, Oliver & Boyd, Edinburgh.

TAKHTAJAN, A. (1987). Flowering plant origin and dispersal: the cradle of the angiosperms revisited. In *The biogeographical evolution of the Malay archipelago*. (ed. T. C. Whitmore), pp. 26–31, Clarendon Press, Oxford.

TRUSWELL, E. M., KERSHAW, A. P., and SLUITER, I. R. (1987). The Australia southeast Asian connection: evidence from the palaeobotanical record. In *The biogeographical evolution of the Malay archipelago*. (ed. T. C. Whitmore), pp. 32–49, Clarendon Press, Oxford.

UHL, N. W. and DRANSFIELD, J. (1987). *Genera Palmarum*. International Palm Society, Lawrence.

WALKER, J. W. and WALKER, A. G. (1984). Ultrastructure of Lower Cretaceous angiosperm pollen and the origin and early evolution of flowering plants. *Ann. Missouri Bot. Gard.* **71**, 464–521.

WHITMORE, T. C. (1981*a*). Palaeoclimate and vegetation history. In *Wallace's Line and plate tectonics*. (ed. T. C. Whitmore), pp. 36–42, Clarendon Press, Oxford.

WHITMORE, T. C. (ed.) (1981*b*). *Wallace's Line and plate tectonics*. Clarendon Press, Oxford. (91pp.)

WHITMORE, T. C. (1981*c*). Wallace's Line and some other plants. In *Wallace's Line and plate tectonics*. (ed. T. C. Whitmore), pp. 70–80, Clarendon Press, Oxford.

T. C. WHITMORE, Department of Plant Sciences, South Parks Road, Oxford University, OX1 3RB.

Index